WORK MEASUREMENT
Principles and Practice

WORK MEASUREMENT

Principles and Practice

EDITED BY RICHARD L. SHELL, P.E.

Published by
Industrial Engineering and Management Press
Institute of Industrial Engineers

Industrial Engineering and Management Press, 25 Technology Park/Atlanta, Norcross, Georgia 30092   404/449-0460

© 1986 by the Institute of Industrial Engineers

Published 1986
Printed in the United States of America

ISBN 0-89806-085-0

No part of the original material appearing in this publication may be reproduced in any form without permission in writing from the publisher. Articles previously published and appearing in the publication may be reproduced only in conformance with the copyright provisions of the original publisher.

Contents

PREFACE ix

I. INTRODUCTION AND CONCEPTS 1

Industrial Work Standards and Productivity
Irvin Otis
1978 Fall Industrial Engineering Conference Proceedings 5

A Procedure for an Economic Comparison of Work Measurement Techniques, Part I: The Model
S. Keith Adams and Timothy J. McGrath
AIIE Transactions, September 1979 14

A Procedure for an Economic Comparison of Work Measurement Techniques, Part II: An Application
Timothy J. McGrath and S. Keith Adams
AIIE Transactions, September 1979 22

How 'Reliability,' 'Precision' and 'Accuracy' Refer to Use of Work Measurement Data
Mitchell Fein
Industrial Engineering, July 1981 30

Determination of Probabilistic Time Standards for Tasks Performed Under Uncertainty
Don T. Phillips and Donald R. Smith
1979 Annual Industrial Engineering Conference Proceedings 38

Work Measurement: The Flap over MIL-STD-1567 (USAF)
Charles H. Boyer
Industrial Engineering, November 1976 47

Update on Military Standard 1567 and Proposed Applications Guide Describes Contractor Concerns
Lee F. Wade
Industrial Engineering, February 1984 59

Improved Productivity Through Worker Involvement
Mitchell Fein
1982 Annual Industrial Engineering Conference Proceedings 62

II. DIRECT TIMING AND PERFORMANCE RATING 75

Industrial Worker Pace Variability—A Study with Real Time and Posterior Analyses
Shimon Y. Nof, Haim Gershoni, Sherman D. Ansell, and Ayala Cohen
AIIE Transactions, September 1978 79

The Worker's Perception of a Change in Work Rate
Rodney K. Schutz and Thomas H. Campbell
1978 Annual Industrial Engineering Conference Proceedings 89

An Analysis of Performance Rating
Robert M. Wygant
1984 Annual International Industrial Engineering Conference Proceedings 98

New Way to Learn Pace Rating
Vincent G. Reuter
Industrial Engineering, July 1977 103

Performance Rating for Service Jobs
Robert A. Brown and William F. Sowder
AIIE Transactions, June 1979 105

III. ALLOWANCES AND THE IMPACT OF FATIGUE 113

Development of Personal, Fatigue & Delay (PF&D) Allowances
William H. Bostion
1980 Fall Industrial Engineering Conference Proceedings 117

Methods for Estimating Physical Fatigue
Arun Garg
1979 Annual Industrial Engineering Conference Proceedings 125

Determination of Rest Allowances for Repetitive Physical Activities That Continue for Extended Hours
Anil Mital and Richard L. Shell
1984 Annual International Industrial Engineering Conference Proceedings 133

Development of Design Data Base for Manual Lifting Activities for Extended Work-Shifts
Anil Mital and Richard L. Shell
1984 Fall Industrial Engineering Conference Proceedings 142

The Effect of Mental Fatigue on Knowledge Worker Performance
Richard L. Shell and O. Geoffrey Okogbaa
1983 Fall Industrial Engineering Conference Proceedings 149

IV. WORK SAMPLING 157

Past-Present-Future of Work Sampling
Chester L. Brisley
1975 Annual Industrial Engineering Conference Proceedings 161

Selection of Work Sampling Observation Times: Part I—Stratified Sampling
Joseph J. Moder
AIIE Transactions, March 1980 167

Selection of Work Sampling Observation Times: Part II—Restricted Random Sampling
Joseph J. Moder and Henry D. Kahn
AIIE Transactions, March 1980 176

Preparatory Computer Assistance for a Work Sampling Study
David M. Rhyne and Douglas K. Freeman (Gary E. Whitehouse, Editor)
Industrial Engineering, May 1986 182

Work/Activity Sampling—Contemporary Design Analysis Methodology and Applications:
Part II—Work Sampling Calculations Revisited
  Elinor S. Pape
  1979 Fall Industrial Engineering Conference Proceedings 186

V. PREDETERMINED TIME SYSTEMS 191

Comparison of Predetermined Time Systems (PTS)
  Chester L. Brisley
  1978 Fall Industrial Engineering Conference Proceedings 193

Today's International MTM Systems—Decision Criteria for Their Use
  Karl Eady
  1977 Annual Industrial Engineering Conference Proceedings 202

A High Level Predetermined Time Standard System and Short-Cycle Tasks
  Abu Masud, Don Malzahn, and Scott Singleton
  1985 Annual International Industrial Engineering Conference Proceedings 211

Robot Time and Motion System Provides Means of Evaluating
Alternate Robot Work Methods
  Shimon Y. Nof and Hannan Lechtman
  Industrial Engineering, April 1982 216

VI. STANDARD DATA AND MATHEMATICAL APPLICATIONS 225

Standard Data and Its Maintenance Today
  Daniel O. Clark
  1981 Annual Industrial Engineering Conference Proceedings 229

Standard Data Coding Techniques
  William H. Bostion
  1983 Annual Industrial Engineering Conference Proceedings 235

Standard Data Development and Application
  William H. Bostion
  1981 Fall Industrial Engineering Conference Proceedings 240

Meeting the Requirements of MIL-STD 1567 and Lockheed-Georgia
Computerized Standard Data Development System
  Lawrence Aft and Thomas Merritt
  1984 Annual International Industrial Engineering Conference Proceedings 246

An Easier Way to Measure Variable Standards
  Michael D. Stevens
  Industrial Engineering, December 1979 249

Low Cost Measurement of Indirect Labor Productivity
  Blair H. Schlender and Harvey H. Smerilson
  1977 Annual Industrial Engineering Conference Proceedings 252

A Comparison of Alternative Time Slotting Methods for Indirect Time Standards
  E. Emory Enscore, Jr., Kenneth Knott, and Benjamin W. Niebel
  1982 Annual Industrial Engineering Conference Proceedings 257

An Analytical Approach to Designing and Testing Time Slotting Systems
Kenneth Knott, E. Emory Enscore, and Jeya Chandra
1986 International Industrial Engineering Conference Proceedings 263

VII. STANDARDS AUDITING 275

Auditing Standards by Sample
Gavriel Salvendy and George P. McCabe, Jr.
Industrial Engineering, September 1976 277

An Algorithm: The Use of Sequential Sampling in Work Measurement Audits
R.A. Bascom, G.S. Kalemkarian, and J.L. Miller III
1979 Annual Industrial Engineering Conference Proceedings 282

Work Measurement Standards Accuracy and Audit
William F. Fielder, Jr.
1986 International Industrial Engineering Conference Proceedings 289

VIII. THE FUTURE 297

The Impact of Automation on Work Measurement
Richard L. Shell
1982 Fall Industrial Engineering Conference Proceedings 299

Better Use of Better Tools Should Make Work Measurement
Increasingly Valuable in Future
Clifford Sellie
Industrial Engineering, July 1984 305

REFERENCES 309

ABOUT THE EDITOR 320

Preface

The purpose of this book is to present the underlying concepts and theory of work measurement; to cover a wide range of specific techniques and practical applications; and, in sum, to strengthen the reader's sense of the relationship between practice and theory. Directed at practicing engineers and experienced analysts, *Work Measurement: Principles and Practice* should also find a place in the university classroom as a volume of supplementary readings. Topics include, but are not limited to, the following:

- the impact of work measurement on productivity
- how to select the most cost-efficient work measurement technique for a given application
- the influence of worker pace variability on production standards
- the precision and accuracy of various work measurement techniques
- MIL-STD-1567A and its Work Measurement Application Guidance Appendix
- new approaches to determining time standards and performance ratings
- how to estimate physical and mental fatigue allowances
- allowances for work shifts exceeding eight hours
- the correct design and selection of work sampling times
- selecting the best predetermined time system by balancing application time and accuracy requirements
- the development of new predetermined time systems for specialized applications
- improved approaches for standard data development and the mathematical establishment of time standards
- how to correctly audit work standards
- future trends for work measurement

The material has been organized into eight chapters, each beginning with an editorial discussion of the topical area that includes comments about the individual articles. The papers were chosen after reviewing *IIE Transactions*, *Industrial Engineering* magazine, and IIE conference proceedings from 1975 to 1986. In all cases, I strove to find the best, most timely article on a particular subject. Many of the selections appeared in the past two or three years, but some are ten years old or more, proving that copyright date is not always a reliable indicator of currentness. Also, preference was given to articles that were more generic in their presentation, i.e., that were not limited in their application to a single industry. Space limitations necessitated the exclusion of many fine papers. However, for the reader seeking more information on specific aspects of work measurement, an extensive list of references appears at the end of this volume.

My deepest appreciation is extended to the authors of the forty articles reprinted here. Their wide range of experience and expertise has contributed to making a volume that will serve the needs of work measurement professionals for years to come.

1. Introduction and Concepts

REQUIREMENTS FOR WORK MEASUREMENT

Work measurement may be defined as the application of techniques designed to estimate the time for a qualified worker to conduct a specified task at a defined level of performance to produce a minimum acceptable quality output. Industrial Engineering Terminology Standard Z94.12 defines work measurement as "a generic term used to refer to the setting of a time standard by a recognized industrial engineering technique, such as time study, standard data, work sampling, or predetermined motion time systems."

Properly practiced, the field of work measurement encompasses correct methods definition that specifies the human interface with all necessary tools and equipment. In addition to determining "how and with what to perform the task," the engineer should make sure that the workplace meets acceptable standards of ergonomic design and of occupational safety and health. A final requirement of professional work measurement is to ensure worker cooperation in the measurement process and involvement in the creation of the total job environment. The importance of applying interpersonal skills by the work measurement engineer or analyst cannot be overvalued; labor standards are not psychologically limiting if the worker is involved and motivated.

WORK MEASUREMENT TECHNIQUES

There are five fundamental work measurement techniques: 1) judgment estimating; 2) historical data (includes records and self-logging); 3) direct observation timing with performance rating; 4) work sampling; and 5) predetermined time systems. Chapters in this book are devoted to the last three of these techniques. Another chapter covers the important subject of standard data and mathematical applications—an area that cannot technically be called fundamental, as it relies on one or more of the five listed techniques. While judgment estimating and historical data are often used to approximate standard time values, they have been omitted here. These techniques have little underlying theory or standardized procedures, and consequently produce poor estimates of standard times; they are thus not considered engineered work measurement practices.

GOVERNMENT INVOLVEMENT

The U.S. Department of Defense for many years has shown great interest in work measurement, primarily through the efforts of the Air Force Systems Command. This is perhaps best evidenced by the establishment of MIL-STD-1567 (USAF), work measurement, released June 30, 1975, and superseded by MIL-STD-1567A, released March 11, 1983. More recently, the application guidance and verification plan was initiated in 1984, drafted during 1985, and finalized May 3, 1986, as MIL-STD-1567A Work Measurement Guidance Appendix. The military standard requires the application of a disciplined work measurement program as a management tool to improve productivity on those contracts to which it is applied. It establishes criteria which must be met by the contractor's work measurement programs and provides guidelines for using the techniques to assure cost-effective development and production of systems and equipment.

The standard defines Type I engineered labor standards as those established using a recognized technique such as time study, standard data, a predetermined time system, or a combination thereof to derive at least 90 percent of the normal time associated with the labor effort covered by the standard and meeting specified requirements. All Type I standards must reflect an accuracy of plus or minus 10 percent with a 90 percent or greater confidence at the operation level. For short operations, the accuracy requirement may be better met by accumulating small operations into super operations whose times are approximately one-half hour. Type I standards must also include 1) documentation of an operations analysis; 2) a record of standard practice or method followed when the standard was developed; 3) a record of rating or leveling; 4) a record of the standard time computation, including allowances; and 5) a record of observed or predetermined time system values used in determining the final standard time. All other work measurement standards are defined as Type II and have no specified accuracy requirements.

BENEFITS OF WORK MEASUREMENT

A comprehensive, professionally developed work measurement program has considerable value to any manufacturing or service organization. MIL-STD-1567A outlines the following benefits of employing such a program:

1) achieving greater output from a given amount of resources;

2) obtaining lower unit cost at all production levels through more efficient operations;

3) reducing the amount of waste time in performing operations;

4) reducing the number of operations and the amount of equipment needed to perform these operations;

5) encouraging continued attention to methods and process analysis because of the necessity of achieving improved performance;

6) improving the budgeting process and providing a basis for price estimating, including the development of government cost estimates and "should cost" analyses;

7) providing a basis for long-term planning of manpower, equipment, and capital requirements;

8) improving production control activities and delivery time estimation;

9) focusing continual attention on cost reduction and cost control;

10) helping solve layout and materials handling problems by providing a relevant data base;

11) providing an objective and measured base from which management and labor can project piece work requirements, earnings, and performance incentives.

SUMMARY OF ARTICLES

Although the lead article in this chapter, Irv Otis's "Industrial Work Standards and Productivity," is now almost a decade old, it is as timely today as when it was orginally published. Its key points are basic and important. Stressing that productivity is our biggest underdeveloped resource, the article expresses concern about the U.S. losing its position as the leading industrial power, discusses the impact of imports, and argues that reliable work measurement standards understood by both employees and management can profoundly influence profitability and competitive market position.

"A Procedure For an Economic Comparison of Work Measurement Techniques, Part I: The Model," by S. Keith Adams and Timothy J. McGrath explains the use of the Delphi method to determine the expected net present value of any given labor standard established by any work measurement technique. "Part II: An Application" illustrates that each technique has both advantages and costs. The advantages are reflected in expected savings; the costs are expressed in terms of the industrial engineering time required to develop and maintain the standard, as well as the inherent inaccuracies that tend to result in wasted direct labor hours. The article reports the results of a study of the work measurement techniques in use in a particular firm, and then suggests a methodology for selecting the work measurement techniques for your organization that best balance costs and benefits.

In "How 'Reliability,' 'Precision' and 'Accuracy' Refer to Use of Work Measurement Data," Mitchell Fein stresses the importance of correct data collection. After defining terms, he discusses problems involving stop watch time study, standard data, and predetermined time systems. The article presents guidelines to assure that the data collected is reliable, explains the amount of data required for time studies, and provides information on correctly completing the standard. (For additional discussion of this subject, see "Balancing Cost and Accuracy in Setting Up Standards for Work Measurement," by Chester L. Brisley and William F. Fielder, Jr., in the *Industrial Engineering* magazine of May 1982.)

"Determination of Probabilistic Time Standards for Tasks Performed Under Uncertainty," by Don T. Phillips and Donald R. Smith, presents analytical procedures for determining a probabilistic time standard for classes of industrial-type tasks. The approach taken is based upon the assumption that individual components of the standard may exhibit uncertain behavior, and that components of the standard time can be treated as a random variable. The article emphasizes classes of industrial tasks that involve the human operator performing a sequence of work elements where stochastic events occur. The methodology is illustrated by an example problem, with comparisons made to the application of traditional time study methods.

In "Work Measurement: The Flap over MIL-STD-1567 (USAF)," Charles H. Boyer summarizes the plusses and minuses of the military standard released in 1975. Points of view are expressed by Department of Defense and military personnel, industrial engineers and managers within contractor firms, and consultants. The extreme positions relative to this initial military standard are summarized by the following statements:

1) "Experience has shown that excess manpower and lost time can be identified, reduced, and continued method improvements made regularly where work measurement programs have been implemented and conscientiously pursued."

2) "The question is not the adequacy of proposed MIL-STD-1567, the question is whether any customer, including the Government, has a right to coerce private industry by a system of checks and balances on their internal management practices."

The article "Update on Military Standard 1567 and Proposed Applications Guide Describes Contractor Concerns" by Lee F. Wade summarizes the major requirements of MIL-STD-1567A. The article also comments on selected problems, acceptance by contractors, and what flexibility of application may exist for a specific work measurement program.

The "Introduction and Concepts" chapter could not be complete without some reference to worker participation and the behavioral viewpoint. "Improved Productivity Through Worker Involvement" by Mitchell Fein is an excellent overview of the subject. This article presents ideas on how most effectively to reshape the world of work so that everyone benefits. Discussed are the issues which deter managers, workers, and unions from adopting changes in traditional worker-employer relations; the conditions at the workplace which effect employees' attitudes toward their work and especially toward increased productivity; and the benefits of linking productivity teams—Labor Management Committees (LMC), Quality of Work Life Programs (QWL), Quality Circles (QC), and Productivity Teams (PT)—with work measurement and incentive programs. The article also contains information on several productivity sharing plans and includes a list of further readings.

INDUSTRIAL WORK STANDARDS AND PRODUCTIVITY

IRVIN OTIS
MANAGER, INDUSTRIAL ENGINEERING
AMERICAN MOTORS CORPORATION

Five years ago, I took a long, hard look ahead at business prospects for the ensuing ten years. I didn't realize at the time how serious they could become before the decade had run half its course.

But I did acknowledge some well-publicized problems confronting the world -

- Inflation
- Overpopulation
- Pollution
- Energy supply
- Urban blight
- Inadequate housing
- Racial stress
- International tensions

I feel solutions will be found eventually, or at least begun for alleviating our national problems.

However, when I speak of productivity output, I have to stop and speculate. The foreign competition challenge facing the United States manufacturers is real - the emergence of strong and increasingly sophisticated foreign competition, and the deterioration in this country's international economic position that has accompanied this phenomenon. Productivity disparities between the U.S. and industrial nations like Germany and Japan are being projected for the balance of the 70's and the 80's.

Productivity is our biggest undeveloped resource.

- Productivity Growth Increases
  - Economic Growth
  - Social Progress
  - Political Freedom

- The United States is still the world's most productive nation, but ...
  - Today's social aspirations and competitive conditions require greater growth.
  - Foreign nations are catching up.

Figure I shows the Vehicle World Production, which indicates the U.S. share in 1950 was 76%; for the 1973 model year, our share of the world production dropped to 31.1%.

As indicated by a special report published by Industry Week Magazine, the following cold facts indicate the inroads of foreign industry. (1)

IS THE U.S. GOING TO BLOW ITS POSITION AS THE NO. 1 INDUSTRIAL POWER?

THESE QUESTIONS ARE RAISED BY EVIDENCE THAT OUR POSITION AS A WORLD ECONOMIC AND PRODUCTIVE POWER IS SLIPPING.

— THE U.S. SHARE OF WORLD STEEL PRODUCTION WAS 47% IN 1950; IN 1971 IT WAS ONLY 20%.

IMPORTS INTO THE U.S. IN 1971:

- 9 OUT OF 10 HOME RADIOS
- 1 OF EVERY 6 NEW CARS SOLD
- 2 OF EVERY 5 PAIRS OF SHOES
- 1 OF EVERY 2 BLACK AND WHITE TV SETS

IN ADDITION, THE FOLLOWING IMPORTS CAN ALSO BE CITED:

- 96% OF OUR MOTORCYCLES
- 30% OF CERAMIC TILE
- 90% OF BASEBALL GLOVES
- 30% OF BICYCLES
- 76% OF TENNIS RACKETS

MANAGEMENT AND LABOR MUST SEARCH FOR NEW WAYS TO INCREASE U.S. PRODUCTIVITY AND BEAT FOREIGN COMPETITION.

In business and industry, the emphasis is, and has been, changing from "Production" to "Productivity," that is, to a much greater concern over the achievement of the desired output coupled with a maximum utilization of resources. This problem, given additional importance because of the current battle against inflation, has prompted President Ford to continue the activity of the National Commission on Productivity; a commission that had been appointed by President Nixon on June 17, 1970. The Commission, when established, included six members each from business, labor, the public at large, and five members from the Federal Government.

In President Ford's recent economic message to Congress, he said,

"I thank the Congress for recently revitalizing the National Productivity Commission. It will initially concentrate on problems of productivity in government; it will develop meaningful blueprints for labor-management cooperation at the plant level. It should look particularly at the construction and health industries."

The following indicates progress by the Senate:

SENATE PRODUCTIVITY HEARINGS (4)

The Senate Committee on Government operations held a two-day hearing (December 16 and 17, 1974) on two bills addressing productivity - S.4130 introduced by Senator Nunn and S.4212 introduced by Senator Percy. The main thrust of both bills is to strengthen national-level leadership with a primary aim of improving productivity in both the public and private sectors.

Both bills provide for: (a) the establishment of a Federal or National Policy on Productivity; (b) removal of existing statutory and regulatory impairment of a National Productivity Center; and (c) provisions for grants to foster research and development of ways to improve productivity.

Key witnesses appearing before the Committee included representatives from business, labor,, academic, federal agencies and the Japanese Productivity Center. The key government witnesses were (a) Frederick Dent, Secretary of Commerce; (b) Elmer Staats, Comptroller General; (c) Stephen Gardener, Deputy Secretary of the Treasury; (d) Bernard Rosen, Executive Director of the Civil Service Commission, and (e) Jerome Mark, Assistant Commissioner in the Bureau of Labor Statistics. Jack Jericho testified for the AIIE.

A few points stressed by several witnesses were: (a) Productivity improvements require long-range planning; (b) service industries and public sectors have not improved productivity at the same rate as other segments of the economy; and (c) many uncoordinated efforts for evaluating and improving productivity are under way. Mr. Jericho's testimony highlighted the contributions of the profession and the AIIE to productivity and stressed inclusion of industrial engineering skills in the staffing of the new agency.

Besides foreign competition, the level of productivity conforming to anyone's concept of a Fair Day's Work depends on many factors. The following seven broad categories influence productivity. (2)

FACTORS INFLUENCING PRODUCTIVITY	
Category	Factors
Technological	- Ingenuity of engineers
Management	- Attitude and behavior of businessmen
Financial	- Availability of capital for financing innovations
Labor	- Characteristics of the labor force
Government	- Policies, taxation
Economic	- General economic climate
Natural	- Uncontrollable items "Act of God"

In today's competitive environment there is an ever-increasing need for the establishment of improved methods of labor cost control through work measurement.

What is Work Measurement

Work measurement is a means of establishing what a fair day's work should be through the development of sound work standards. My definition has always been, "a method of establishing a fair relationship between work performed and manpower used." Or, putting it more simply, how much work is there to do and how many people should it take to do the job.

However, I must point out that in measuring work, the results must be accurate and achieved economically. The measurement must also be auditable for accuracy to warrant support of executives and first-line supervisors - those responsible for carrying out the workload goals.

As shown in Figure II, the kind of measurement that is appropriate for measuring a particular unit of work must be determined, keeping in mind the cost of measuring in relationship to the results of the measurement.

Work measurement, defined in its broadest sense, is a means of establishing an equitable relationship between man-hours used and units produced. This relationship is normally expressed as a unit/time work standard - a specification of the amount of time required by a qualified worker performing under normal conditions to produce an item of quality output.

```
WORK STANDARD = STANDARD TIME
         FOR A
     QUALIFIED OPERATOR
 WORKING AT A NORMAL PACE
         UNDER
                      ⎧ TOOLING
STANDARD CONDITIONS - ⎨ EQUIPMENT
                      ⎪ WORKPLACE
                      ⎩ LAYOUT
           AND
 FOLLOWING A STANDARD METHOD
       TO PRODUCE A
   QUALITY PART OR ASSEMBLY
```

The gradual but persistent increase in wages, fringe benefits and, more recently, income security - resulting from collective bargaining processes over the past ten years - has tended to alter management's problems only in form rather than in substance. Efficient managements have always planned and worked to accomplish these same objectives for their employees through the promotion of a healthy economy, which we know from past experience provides a higher standard of living for everyone.

At the risk of oversimplification, we can say that progressive managements have always striven, and will continue to strive, to reduce labor and manufacturing costs through sound cost reduction programs, whether faced with fixed obligations in the form of increased employee benefits or not.

And to the extent that these increased benefits can be absorbed through effective management, nothing new has been added. The imposition of a contractual obligation to increase labor costs alters the basic problem only to the extent that a greater effort is required to stay even - although merely staying even in most American industries today is a slow death, and not too slow at that.

Just as the need for cost reduction is not new, so also no new dramatic way to effect such reductions has been devised. Cost reductions do not just happen: they must be caused. For many years a rather haphazard application of individual ingenuity resulted in cost reductions of sufficient magnitude to sustain most businesses. Today, however, a planned, systematic approach is essential. A sound work standards program embodies such an approach and, in my opinion, represents the best, though certainly not the newest way to handle the problem. Supported by careful engineering and planning, such a program provides the requisite data for an evaluation of where a company stands and discloses the areas where a concentration of effort is needed.

No matter what type of industry is involved, there is the need for a daily feedback performance. Figure III indicates a flow chart for the preparation of Direct Labor performance; however, similar reports indicating performance for Indirect Labor activities should also be provided by progressive-minded management.

Providing for cost reduction in any program at the planning stage cannot be too strongly recommended. Many problems and much grief can be avoided by keeping high-labor-cost processes from ever getting into the actual production processes. In the automotive industry in general, and in the company I represent in particular, careful planning is mandatory preceding each new model year to insure that the most economical methods of manufacturing will be utilized. Once the wheels are set in motion for production of a model, the opportunities for making further changes are limited because of production requirements. However, after the model is in production, careful follow-up is required to insure that the production processes are in operation according to plan. And at the same time, the company can investigate whether a better way of manufacturing can be developed and put into use at the next appropriate time.

In our particular case, the responsibility for increased productivity is largely management's.

Work standards based upon work measurement will permit management to produce and operate with more satisfactory results. Work measurement may be applied by management to determine how well its employees are performing, not only in production operations but also in nonproduction, engineering, clerical and administrative tasks.

The basic philosophy outlined in Figure IV indicates the process of achieving corporate objectives in realizing efficient utilization of employee effort.

The development of a <u>sound work standard</u> is essential to any <u>cost control program</u>. With the advent of contractual definitions of management's rights to establish and enforce standards of production there came, at least in our case, the accompanying obligation that such standards be fair and equitable to all parties involved, and that some system of measurement be employed in their establishment. All systems of measurement worthy of the name are essentially composed of these ingredients. But the quality of the resulting standard is largely dependent upon the care and the skill with which the prescribed techniques are employed.

How good a job is done up to this point is entirely within management's control. But the overall success of a work standard program is obviously dependent upon the employee's acceptance of the standards as being fair and equitable and his willingness to work to them. No system of work measurement, no matter how carefully constructed or how painstakingly executed, can be successful without intelligent application, administration, and follow-up of corrective action.

> Management's responsibility for follow-up:
>
> The fundamental tool for effective aids to members of the management team is the recording and reporting of current and past performance and progress to plan. Without adequate <u>reports</u> and <u>feedback</u>, an organization would have no way of knowing progress made to date.
>
> Cost control assists in the systematic improvement of management in all phases of activity through the feedback network as shown in Figure V.

Since the introduction of the scientific approach to cost reduction by Frederick W. Taylor in 1881, the setting of production or work standards has had a checkered career. Alternately the procedure has been praised and damned by some management and workers alike. As happens too often, at first many overzealous advocates of the procedure expounded its virtues until it came to be regarded as something of an essence for the reduction of labor costs. And an all-too-prevalent feeling developed that a work standard established through the use of time study was to work measurement what the thermometer is to the measurement of temperature. There was little or no appreciation of the limitations of time study or its subjective aspects.

Early emphasis in methods improvement was centered largely upon elimination of obvious waste. And productivity increases resulted from either increased worker effort or short cuts devised by workers. As might be expected, the looseness with which the methods analysis was made prior to setting of the standard soon led to the phenomenon of what has come to be regarded as creeping changes.

Too often the expedient and simple solution to this was an arbitrary adjustment of the standard, with an attendant increase in production requirements. Because so many applications were associated with piecework wage payments, this practice came to be called rate cutting, and time study fell into wide disrepute. The worker looked upon time study with suspicion because the "carrot" was constantly being moved farther out front. And many managers discredited work measurement because workers defeated its purpose by artificially limiting production.

However, I am happy to report that enlightened representatives from labor, in my experience, have recognized that a work measurement program establishing work standards is the only sound way to determine a fair day's work.

At Jeep, we have learned to live together with our UAW representatives and through careful, conscientious application and administration have established mutual goals to increase productivity. Through experience, we have developed a set of governing principles for settling our grievances in an intelligent manner.

The following guidelines have helped us to be effective in establishing work standards that are reliable. They must be realistic and capable of achievement under normal, reasonable working conditions. They must be understood by both employees and by management.

<u>FIRST</u>, work standards must be realistic. A good standard covers all items of work for which the employee is responsible as well as all delays over which he has no control. Of course, any delay allowed for in the standard must be consistent with the concept of efficient operation after considering the facilities available. This is not to imply that inefficiencies are to be covered by the standards. Rather, it means that once having established the method for which management is responsible, the standard must fairly reflect the method and be attainable by a qualified employee using normal skill and effort.

<u>Qualifications of Individuals Establishing Standards and Rating Performance</u>

1. The establishment of work standards requires a well-qualified and well-trained individual to rate the performance of the employee performing at an acceptable pace.

2. The methods and standards engineers at Jeep have been carefully selected. They have been trained in the Corporate Work Standards Program regarding rating an employee's performance. Industrial engineers must possess certain characteristics that enable them to qualify for their position. Some of these characteristics are: the ability to deal effectively with hourly rated employees and others with whom they have contact; and possession of good judgment and a sense of fairness. Industrial engineers must be observant, and have an analytical ability to uncover high cost areas and then arrive at practical plans to reduce excessive cost.

3. Rating the operator's performance is one of the most important parts of establishment of a standard. At Jeep, weekly training sessions are held to review the concept of normal pace. These sessions insure that the ratings applied on the floor by the industrial engineers are consistent with the Corporation's concept of a normal pace. (Figure VI.)

<u>SECOND</u>, the standard is not a secret document. Management must stand ready to discuss any or all information relative to the standard with the employees or their representative. It is only through willingness to explain the development of the standard and what it covers that the confidence of the employees is gained and held.

Management has also recognized that there is a statutory obligation under the Federal Labor Law to furnish all work measurement information to union representatives. (3)

<u>THIRD</u>, the work standard applies to a given set of conditions. The conditions referred to here are those which are described in the standard as being factors which have a significant effect on the time required to do the work.

Check the Study for Proper Method

FOURTH, changes in operations must be promptly reported and just as promptly acted upon for adjustment in the standard, because any delay in taking advantage of methods improvement represents a loss. The most important loss is employee resistance to the new standard when considerable time is allowed to elapse before the revision. After an employee has grown accustomed to performing an improved operation against an old standard, both production and employee confidence in the system have suffered. And the production lost cannot be regained when the standard is finally revised.

FIFTH, work standards should not be indiscriminately changed. Generally speaking, a standard should be changed only when warranted by a change in operating conditions. This is not to suggest that no attempt at correction should be made in isolated cases where errors have resulted in gross inequities to either party. Such inequities are almost as intolerable to the employee as to management. And if the employee is openly approached, the situation in all probability can be rectified with no real loss of prestige.

SIXTH, work standard disputes must be intelligently handled. All disputes must be promptly investigated. And if a revision is indicated, it should be made promptly. Disputes can be most effectively settled on the floor where the job and the employee involved are located. The more people who are called into a dispute, the farther the discussion gets away from the essential question, and the more difficult it becomes to resolve it.

SEVENTH, the work standard is an aid to, not a substitute for, effective supervision. The intent of the standard is to provide supervisors with a guide to be used to maintain the required flow of production. The use of the standard as a club to cover ineffective supervision can lead in only one direction; to dispute over the standard itself.

It must be remembered that the standard, in effect, defines the equivalent of normal production capacity when related to a human being. And since motivation and mental outlook are factors, we are dealing with unmeasurable variables, if we use the term "measurement" in the strict sense. Therefore, establishing a standard on a job does not magically transform that job into a smooth-running, trouble-free operation. Good supervision remains the only real answer to efficiency. The work standard will help the good supervisor to do a better job. But the poor supervisor nullifies its effectiveness.

As shown in Figure VII, the production foreman participates in the preparation, analysis and establishment of a work standard.

In a rather brief way we have taken a quick look to see what has happened in the past in the field of standards and cost reduction. Now what about the future? What will organized labor's reaction be to continued and, in all probability, accelerated labor-cost-reduction programs?

For the company that has built a sound work-standards system and conducted an intelligent, forceful cost-reduction program in the past, there is no reason to anticipate any additional labor relations difficulties. Organized labor's position in the past few years, judging by statements of responsible leaders, has been one of acceptance of continuing technological improvement as essential to our economic progress. It is reasonable to hope that in the future, where sound programs are conducted, labor will not choose to resist, but rather assist in implementing a work measurement program.

We can be quite sure that labor will continue to press for greater benefits. But realizing these benefits on a sound basis will be possible only if there are continuing reductions in unit manhour requirements. It is also safe to say that, with the ever-increasing emphasis upon cost reduction, management will not be static in its efforts to improve techniques. Improvements in management sciences are just as essential to economic progress as is the development of new manufacturing techniques or new products.

Recent history supports this idea. Witness the development in production control, the introduction of statistical quality control, and improved personnel and training procedures. And these are only a few examples of some of the latest developments.

Substantial progress has also been made in some industries to improve the liaison between designing and manufacturing. In the field of work measurement, various tabular time-study procedures - such as methods time measurement, basic motion, time and work factors - have been developed to assist and to expedite the standard-setting program. Work sampling or ratio delay is gaining wide acceptance in the nonproductive service areas. In other words, there has already been some progress in the development and improvement of management techniques in this field.

Dynamic management, however, will not be content to rest on its laurels. Much more needs to be done and much more will be done. For example, the industrial engineer is constantly searching for ways to reduce, if not eliminate, the need for the subjective judgment that is still required in all current work measurement techniques.

Similar strides may be expected in all areas of management. Management will continue searching for the better way - the better way to build the product, the better way to employee relations (i.e., labor-management relations, "make the worker feel like part of the team"), and the better way to do its own job.

Productivity has been a catchword for years and improvement of productivity a virtual catechism for business management. However, today's economic climate, beclouded by the devils of inflation, recession, and foreign competition, makes it essential that productivity improvement move out of the realm of words and be translated into action.

A Work Standard is a Key to Successful Operations

Work standards, in the economic sense, have a profound influence on profitability and competitive market position. Therefore they represent a strong factor in the growth and security of a manufacturing organization.

Increasing productivity may thus be regarded as the keystone of an improved standard of life and environment for all of society. It is with this broad view in mind that the National Commission on Productivity has set its task of finding ways to continue or accelerate the historical rates of productivity gains in the United States.

REFERENCES

(1) INDUSTRY WEEK MAGAZINE, "Is There Still Time to Save the U.S. Industry?", November, 1975.

(2) GOTTLIEB, B., "A Fair Day's Work: Fact or Fiction," IMS Clinic Proceedings, 1968, page 130.

(3) GOTTLIEB, B., and WERNER, C. A., "Statutory Obligations of an Employer to Furnish Information to a Union," Monograph Series #1, Industrial and Labor Relations Division, American Institute of Industrial Engineers, May, 1971.

OTIS, I., "Why - Unionize Industrial Engineers!", Technical Papers, 1972, 23rd Annual AIIE Conference and Convention, May, 1972, page 569.

(4) POWER, R. J., Editor AIIE News, Government Division, Volume VIII, Number 2, January 2, 1975, "Senate Productivity Hearings," page 1.

WORDLDWIDE MOTOR VEHICLE PRODUCTION

YEAR	U.S. PRODUCTION	WORLD TOTAL PRODUCTION	UNITED STATES SHARE
1950	8,006	10,577	75.7%
1955	9,204	13,743	66.9
1960	7,905	16,375	48.3
1965	11,138	24,543	45.4
1970	8,263	29,700	27.8
1971	10,672	33,204	32.1
1972	11,292	35,500	31.8
1973	12,682	40,730	31.1
1974	9,984	36,000	27.7
1975	8,970	33,193	26.9

SOURCE: WARD'S STATISTICAL DEPT.
ADD 000's TO ALL VOLUMES

Figure I

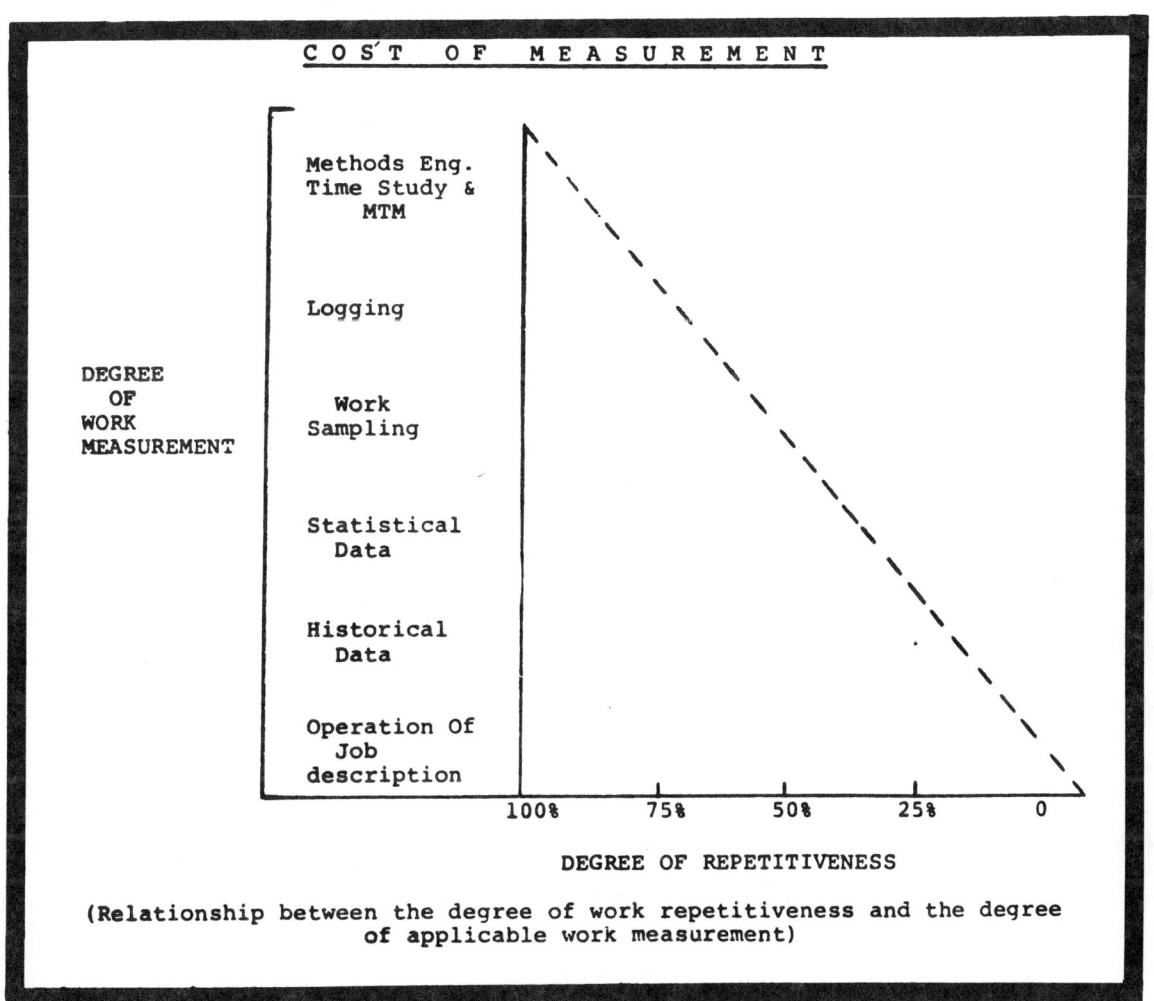

Figure II

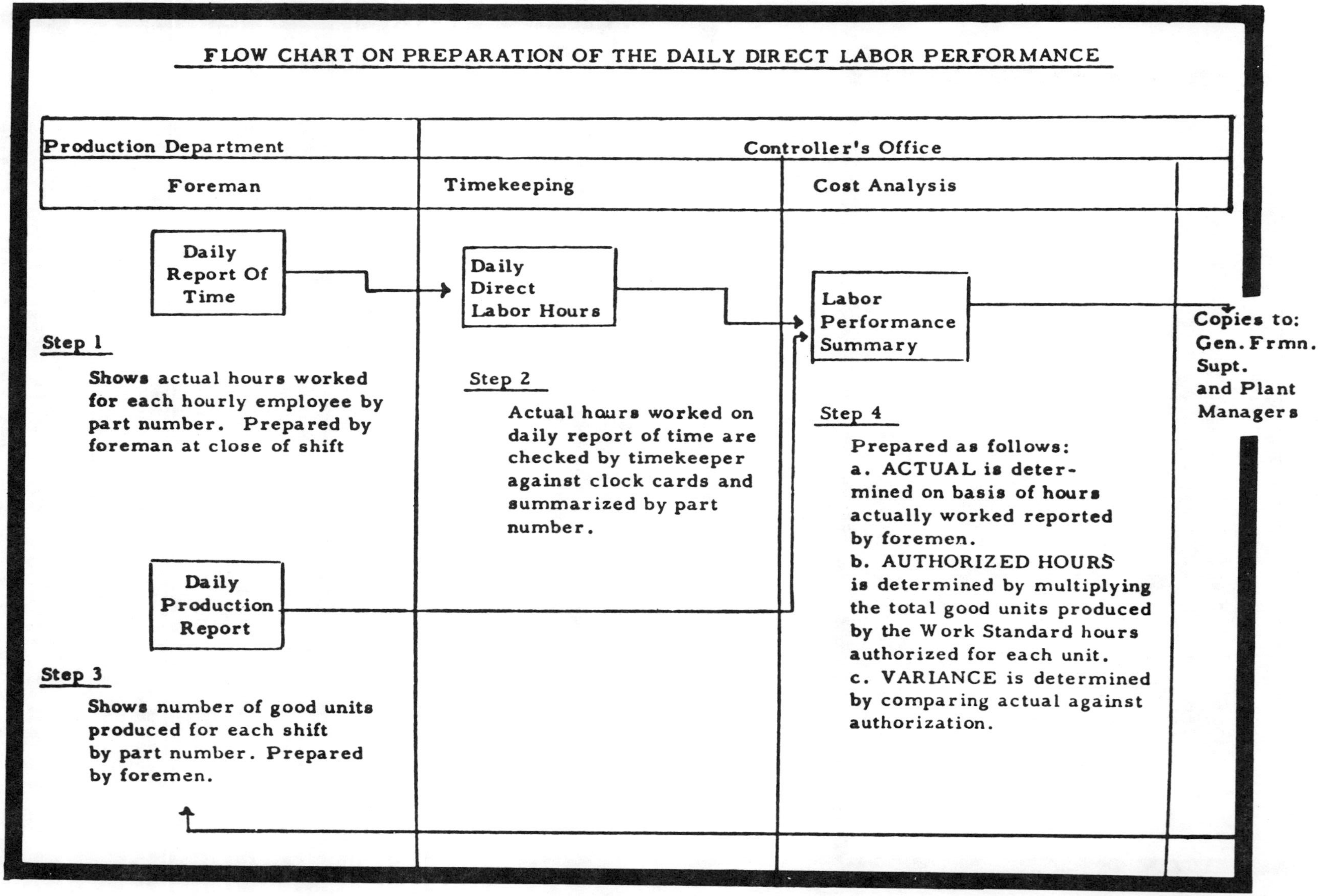

FLOW CHART ON PREPARATION OF THE DAILY DIRECT LABOR PERFORMANCE

Production Department		Controller's Office	
Foreman	Timekeeping		Cost Analysis

Step 1

[Daily Report Of Time]

Shows actual hours worked for each hourly employee by part number. Prepared by foreman at close of shift.

Step 2

[Daily Direct Labor Hours]

Actual hours worked on daily report of time are checked by timekeeper against clock cards and summarized by part number.

Step 3

[Daily Production Report]

Shows number of good units produced for each shift by part number. Prepared by foremen.

Step 4

[Labor Performance Summary] — Copies to: Gen. Frmn., Supt. and Plant Managers

Prepared as follows:
a. ACTUAL is determined on basis of hours actually worked reported by foremen.
b. AUTHORIZED HOURS is determined by multiplying the total good units produced by the Work Standard hours authorized for each unit.
c. VARIANCE is determined by comparing actual against authorization.

Figure III

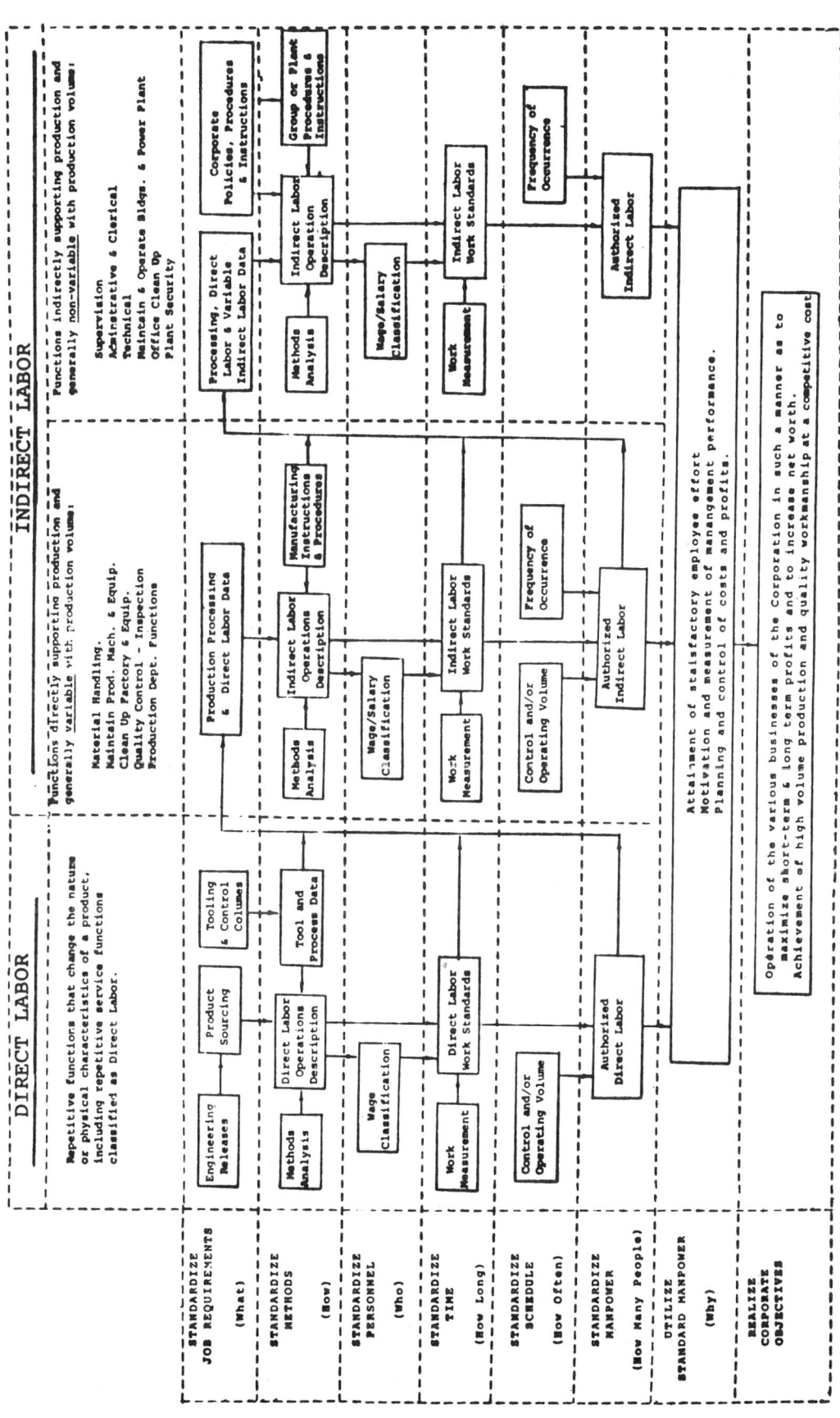

PROCESS OF DEVELOPING AUTHORIZED MANPOWER TO ACHIEVE CORPORATE OBJECTIVES

Figure IV

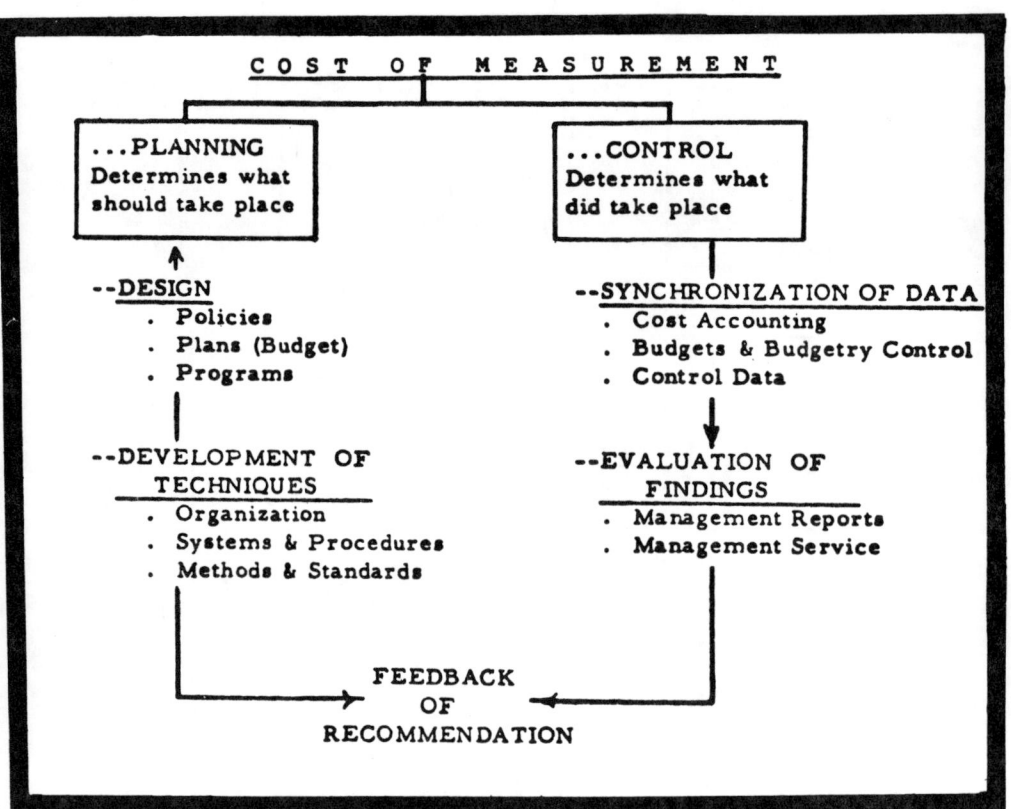

Figure V

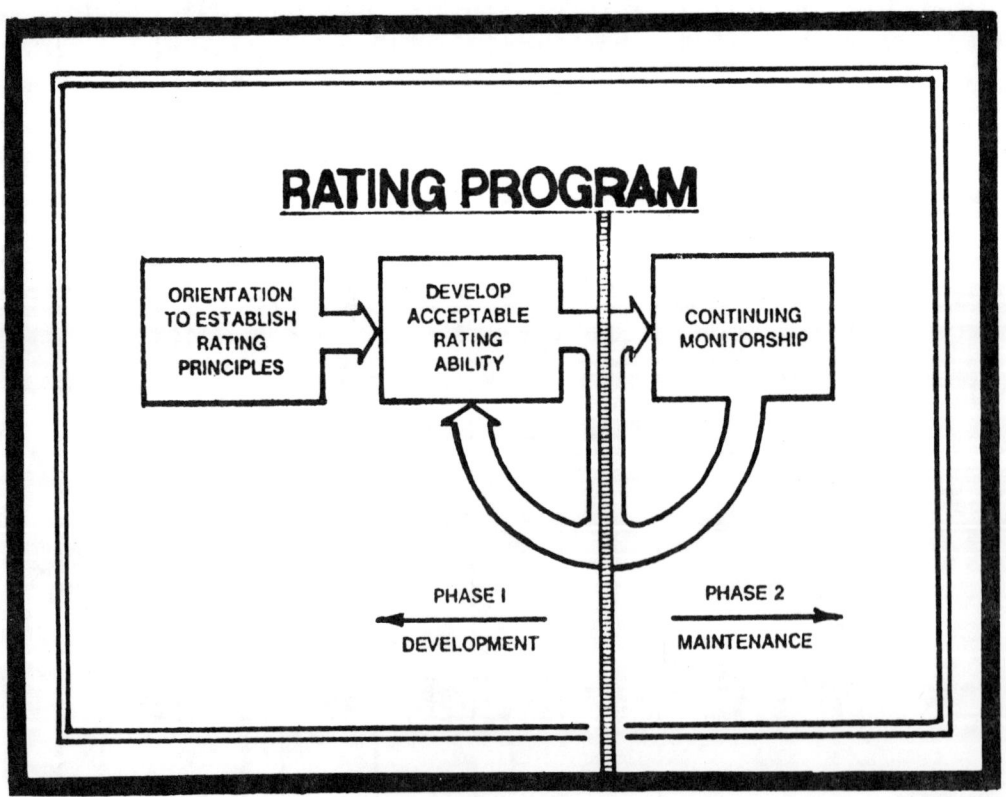

Figure VI

WORK MEASUREMENT FLOW CHART
WORK STANDARDS

Figure VII

A Procedure for an Economic Comparison of Work Measurement Techniques, Part I: The Model

S. KEITH ADAMS
SENIOR MEMBER, AIIE
Department of Industrial Engineering
and
Engineering Research Institute
Iowa State University
Ames, Iowa 50011

TIMOTHY J. McGRATH
SENIOR MEMBER, AIIE
Industrial College of the Armed Forces
Fort McNair, Washington, D.C. 20319

Abstract: A procedure is developed whereby an organization is able to determine, within the constraints of its own practices and environment, the most economical work measurement technique to use in studying a particular work operation. Application of the Delphi Method permits the costs and benefits of each available technique to be estimated. An economic model is then developed to determine the expected net present value of any given labor standard established by any of the available work measurement techniques.

■ Throughout industry many resources are devoted to the establishment and maintenance of labor standards. A predictive economic model to be used in selecting the optimum work measurement technique for any given job study has been a recognized need for some time. Overshadowing this need has also been the appreciation that both the costs and the benefits of a work measurement program are quite complex. However, there is a definite economic return, positive or negative, to the organization from this program. This return can be estimated on a predictive basis according to the procedures developed in this article.

In reviewing recent accomplishments in the area of labor standards costs and benefits, one can note the emergence of several new approaches to this problem. The use of statistical sampling procedures to economize the auditing of labor standards has been recommended by Salvendy and McCabe [21]. The study of analysis of variance interactions, standard deviations, and confidence limits provides the ability to isolate specific job functions or departments in which more accurate (or more valid) work standards would produce the greatest savings. Kaganowicz and Krususki [19] have developed optimization techniques for MTM designed to minimize relative error or to select an acceptable balance between relative error and the hourly cost and volume of activity associated with a given job. This model takes into account the cost of the MTM analyst as well as that of the operator on the job being studied. Hancock and Langolf [15] have performed an extensive analysis of the relative costs of using the several MTM systems now available. They point out that in addition to the industrial engineering costs, one must consider the effect on productivity, product quality, and worker grievances of each of the available techniques.

The objective of the research described in this paper was to develop a procedure for evaluating the relative economy of the work measurement techniques being used by an organization. It is based on the use of survey data which lead to statistical distributions of the costs and benefits of deriving, implementing and maintaining a given labor standard. It is useful for any work measurement technique, including work sampling. It allows an industrial or governmental organization to predetermine, within the constraints of its own unique operating environment, the most relatively economical work measurement technique to use to develop the labor standard for a specific job.

The Value of a Labor Standard

A labor standard itself is simply a measure of the amount of time required by a qualified operator to perform a specified

Received November 1977; revised December 1978. Paper was handled by Work Measurement/Methods/Ergonomics/Health Systems.

task under a given set of conditions using specified equipment and materials. There are, of course, many potential sources of error in the process of establishing this standard. Fankhauser [8] lists a number of mistakes which can occur in establishing normal times. These include such items as errors in measured time, performance rating, defining methods, observing method changes, use (by the operator) of incorrect speeds, feeds, tools, materials, etc. He lists other errors that occur in estimating allowances such as the use of incorrect allowances, unnecessary (redundant) allowances, and improperly based delay allowances. Frederick [11] points out a number of problems which can arise when a predetermined motion time system is being used. These include inexperience of the analyst, incorrect classification of motions, faulty interpretation of motions, and tendencies to make the system fit the production expected of the operator being observed. Such errors in measurement can damage the statistical reliability or overall credibility of the standard. In so doing, they affect its potential value.

While the information provided by a labor standard may have many uses, the actual value of the standard itself depends both upon its own cost and the reduction of operating costs obtained through its use. It is a basic premise of this paper that the only real measure of the value of a particular labor standard is the number of direct labor man-hours that are saved because the standard is being used.[1] This is directly related to the credibility (validity) of the standard and the credibility of the work measurement technique used to develop it. The relative values of two labor standards, based upon the same or different work measurement techniques can be compared. An absolute measure of the value of a particular labor standard is probably not obtainable. However, it is possible to compare the costs and savings associated with a particular standard to the costs which would result if no formal standard was in effect.

As an example, consider a production situation where the cost of direct labor is $6 per hour. Even where no formal work measurement program exists, someone must still estimate the amount of time that it will take to complete a job. Assume that in this case it is the foreman and that he estimates the job under consideration to take one man-hour. The logic that went into this estimate is of little importance, since barring any information to the contrary, it is the figure that must be used. Now assume that the job is studied by a work measurement analyst, and, because of better work methods design among other factors, the standard time for the task is found to be 40 minutes. Using the foreman's estimate, the cost of the job was $6 per repetition. By application of work measurement, this cost is now reduced to $4 per repetition. The value of the labor standard, due to the savings in direct labor time, is the difference in these amounts, or $2 per repetition of the task.

---

[1] It will be assumed throughout this paper that a product of acceptable quality results upon completion of each repetition of the task irrespective of the work measurement system used to establish the labor standard for the task.

It should be pointed out that the engineered standard is probably not perfect; if more time had been spent by the analyst studying the job and seeking improvements in the work methods, the standard time could probably have been reduced even further. This is, therefore, a "cost of ignorance" inherent in this labor standard, a cost that can be taken into account through estimates. However, it is readily seen that even though the standard and the associated work method are not perfect, there is real value accruing to the organization from the knowledge this information supplies.

Use of the Delphi Method

A direct approach in ascertaining the economic parameters of a work methods and standards program may be impractical because of the time and effort required to completely investigate the workers and time study personnel involved. It would be very difficult to directly examine and analyze all of the relevant variables. The Delphi Method presents an alternate approach. The source of the information here is the experience (supported by data collected over time) of the members of the organization itself. The actual persons chosen to take part in the study would be a unique determination within each individual organization. Members of the methods and standards group, first line supervision, cost accounting, and quality and production control are a few of the obvious candidates for consideration. As in any other situation, the choice of the experts to provide the data must be based on the stipulation that they do indeed have appropriate knowledge of the quantities they are estimating and that their judgements will follow a rational pattern. Concerning the precision of the data, it is only necessary that it be precise enough to allow relative evaluations to be made.

Reliance on experts has been accepted by many authors in the field of operations research [2, 13, 18]. Helmer and Rescher [16] provide an excellent justification for the use of experts as well as criteria to be applied in their selection. The industrial engineer, too, frequently relies on opinions and judgements in obtaining data. Applied studies in inventory control, queueing theory, equipment replacement strategy, and depreciation scheduling often require estimates of time, value and costs which cannot be verified through direct measurement. When expert judgement is the source, the problem becomes one of extracting the maximum amount of useful information from a group of people with a minimum of distortion from misinformation and psychological pressures [16, 17]. It is for these reasons that the group response survey technique known as the Delphi Method was developed. Dalkey and Helmer are generally credited with originating this procedure [4].

The Delphi Method is based upon several important assumptions. It is assumed that the individuals in the group are qualified to make rational estimates of the data being elicited, that their individual estimates are independent of one another, and that a statistical combination of these estimates will provide a more valid answer than would be obtained from any one member of the group. This procedure should obviously not be used when a more direct source of the information is available

An important aspect of the Delphi Method is the use of controlled feedback. The respondents are usually given a series of four questionnaries. On the first, each is asked to provide estimates based on his or her own individual knowledge and experience. Communication between respondents regarding the survey is not allowed. Answers obtained from the first round are arranged in numerical order and grouped into four quartiles. On the second round, the same questions are given to each respondent along with the mean or median answers, the central ranges (the central 50% of the estimates), and the respondent's answers (or quartiles into which his or her answers fell). The respondents can revise their previous estimates, if they wish, in providing answers during the second round.

Individuals who provide second round estimates outside the central range of first round estimates are asked to provide their reasons. These reasons are included with the new set of tabulated estimates (grouped into quartiles as before) and sent back to the participants on the third round questionnaire. On the third round they are again asked to reconsider their estimates and to offer comments regarding the reasons submitted in the second round. On the fourth round, these comments are included in the questionnaire along with the statistical outcome of the third round. Since the fourth round is usually the final round, each participant is asked to consider all previous information and response data and to provide a final best estimate.

Obviously, many factors influence the outcome of the final results. The size of the group, the dispersion of individual estimates, the difference between the mean or median estimate and the true answer, and the willingness of individuals to change their estimates when provided with feedback are important influences affecting the iteration process. Despite these problems, the method has been validated experimentally by Campbell, Dalkey and Helmer [1, 4, 5]. It generally provides good results. Meetings of the participants are not necessary, and yet a final collective judgment or estimate is obtained.

The Economic Parameters of a Work Measurement Program

Determining the economic parameters of a work methods and standards program is a problem with aspects that can be studied using the Delphi Method. The costs and benefits of such a program are so complex that it would be quite difficult to determine them by the more direct and precise techniques used to study some other production operations. However, the capability to estimate the return per dollar invested in a particular labor standard is highly desirable. A procedure is needed, therefore, that will allow the organization to select, for each job to be studied, the work measurement technique that will result in the greatest economic benefit.

There are, obviously, many factors that must be taken into account. As stated earlier, the only real measure of the value of a labor standard is the number of direct labor man-hours that are saved because of the standard. A first requirement, then, is to establish estimates of the level of productivity or, conversely, the number of wasted man-hours that would result if no labor standards were in use. This is an extremely important set of data, since it is the expected increase in productivity that justifies the work measurement function itself. The Delphi survey can be very helpful here, but, as is true for all aspects of the investigation, the design of the questions and the selection of participants to provide valid data must be accomplished with great care.

In a similar manner, each of the work measurement techniques being used by the organization has an inherent prediction accuracy that must be estimated. This accuracy is not a constant. It will vary between studies of the same kind of work as well as between studies of different kinds of work. In addition, the accuracy of a particular work measurement technique can be a function of the length of the work unit measured (work cycle). Tedious, time consuming techniques may be necessary to achieve reasonable accuracies for short cycle jobs, whereas more rapidly applied systems may be quite acceptable for long cycle tasks. If, therefore, the organization has a wide range of work cycle lengths and/or many differing kinds of work, several surveys will be required to establish a number of sets of data.

It is important that the individuals estimating the level of productivity or, again conversely, the number of wasted man-hours that will result from the use of a particular technique understand that the purpose of work measurement is to determine how long it should take a properly trained and supervised worker, working at a normal pace, to accomplish a particular task. Time lost due to personal needs, official break periods, cleanup, minor maintenance, or other recognized delays and allowances is not to be considered unproductive. What is being estimated here is the amount of time the worker would get that is over and above what he really should have to perform his duties either because of the poor method definition or inaccuracy of the particular technique used to set the standard. Concerning these estimates, in some cases objective data are available. For example, the accuracy of the various MTM systems has been quantified in different situations [7, 9, 10]. In addition, the organization itself will probably have historical records that may be used. But it is the opinion of the authors that experienced work measurement analysts and foremen, among others, will have formed reliable and valid judgments over the years about the productivity that is obtained with labor standards developed from the various available techniques or the productivity that will result if no standards are in force. These judgments, if properly exploited using the Delphi Method, can provide the basis for a sound economical evaluation of a methods and standards program.

One further but quite important point that must be kept in mind by the survey respondents, in addition to the individual technique characteristics, is the propensity of the workers to file grievances against standards that are "too tight" and the effect of this on productivity. For example, a given technique may be neither loose nor tight on the average, but its inherent inaccuracy leads to individual standards that are too demanding. These standards result in grievances and must be corrected to require a more equitable level of performance. Since it is unreasonable to expect grievances over those standards that are too loose, correcting the tight standards actually lowers the overall productivity obtainable from the technique.

The gain (benefit) to the organization resulting from a particular labor standard is the dollar savings in man-hours that result from the use of the standard as compared to no standard at all. Concerning the costs involved perhaps the most obvious is the initial standard development cost. Each technique has an associated procedure for determining the standard time. The time required to follow this procedure is a cost in terms of the salary of the person who establishes the standard and is a function of the procedure and detail inherent in the technique used as well as the length and complexity of the job studied.

Frequently, after a labor standard has been developed, the need arises to change the standard. This change may come about because of a methods improvement action, a change in the process itself, or, as mentioned earlier, as a result of a grievance. Regardless of the origin of the change, part or all of the original standard must be restudied to update the standard time. These revisions require the time of a paid employee to accomplish and are, again, directly related to the length or complexity of the part (or all) of the task to be restudied, the detail of the methods analysis, and the work measurement technique used.

Labor standards also require periodic auditing to assure compliance and accuracy. How often an audit is performed may depend on the standard's frequency of use, company policy, or other factors, such as excessively high or low paychecks, or when it is apparent that workers can control their earnings artificially. This periodic auditing is an additional cost in the work measurement program and is in direct proportion to the same set of variables as the cost of labor standard revision.

All of the above costs represent a monetary investment in a particular labor standard. Depending on the situation, there are probably many other costs associated with the work measurement program of an organization. If the standards-setting personnel undergo initial or periodic training, this cost must be considered. Those organizations with a large group of people setting labor standards probably have a person designated as a supervisor, who may or may not actively determine standards himself. In addition to the supervisor, there may be a secretarial staff assigned to this function. These and similar overhead costs are part of the investment in the overall work measurement program of the organization, as are such items as keypunch operators and electronic data processing costs if the firm maintains its standards by machine records. Although many of these costs may not be readily apparent in the shop, if an adequate analysis of the investments due to labor standards is to be made, these additional costs must be searched out and considered.

In the establishment of the economic parameters of its work measurement program each company must develop its own data. Care must be taken to insure that the problems inherent in establishing prediction system accuracies are not oversimplified or that some of the more subtle components are not overlooked. The authors do not suggest that careful cost analyses, where practical, be discarded in favor of Delphi Method surveys. However, in those instances where these analyses are not feasible for one reason or another, Delphi does offer an alternative. Such a study was conducted by the authors. The following sections describe the model that was used to analyze the resulting data.

The Statistical Model

Nearly all parameters associated with the development, use, and evaluation of a work methods and standards program are subject to statistical variation. No two jobs are exactly alike. The time taken to develop a standard for a particular production task using a given technique will depend upon the length and complexity of the task. Task complexity also varies. Hence, a statistical model is necessary to study such a problem even though the final results may be used in more deterministic budgeting and cost control equations.

All parameters associated with a given labor standard were assumed to be independent random variables whose distributions are approximated by the beta model of the form

$$f(x : \alpha, \beta) = \frac{(\alpha + \beta + 1)! \, (x-L)^\alpha (H-x)^\beta}{\alpha! \, \beta! \, (H-L)^{\alpha+\beta+1}} \quad (1)$$

where

x = the particular variable of interest,

H = the upper limit of the value of x,

L = the lower limit of the value of x, and

α and β = beta parameters.

Expressions for the mean (μ), variance (σ^2), mode (M), and α and β parameters of this distribution are given by Cole and Mikasa [3], as

$$\mu = \frac{(H-L)(\alpha+1)}{(\alpha+\beta+2)} + L, \quad (2)$$

$$\sigma^2 = \frac{(H-L)^2 (\alpha+1)(\beta+1)}{(\alpha+\beta+2)^2 (\alpha+\beta+3)}, \quad (3)$$

$$M = \frac{(H-L)\alpha}{(\alpha+\beta)} + L \quad (\text{for } \alpha, \beta \geq 0), \quad (4)$$

$$\alpha = \frac{(\mu-L)^2 (1 - \frac{\mu-L}{H-L})}{\sigma^2} - \left(\frac{\mu-L}{H-L}\right) - 1, \quad (5)$$

and

$$\beta = \frac{\left(\frac{\mu-L}{H-L}\right)(H-\mu)^2}{\sigma^2} + \left(\frac{\mu-L}{H-L}\right) - 2. \quad (6)$$

The beta distribution was chosen because its properties describe the basic characteristics that would be expected in the variables themselves. It has a finite range ($L \leq x \leq H$), continuity, and unimodality (given that α and β are both greater than zero). It can also assume various degrees of skewness and peakedness, a desirable property in this application.

Subjects participating in the study were asked to specify values for L, M, and H, and then to select the distribution from the nine represented in Fig. 1 which they felt best depicted the pattern of variability for each particular numerical estimate. For all nine of the distributions shown, the mode is at the first quartile, midpoint or third quartile of the range, depending upon whether the distribution is skewed or symmetric. Any calculated value of a high or low point on one of the curves may differ somewhat from that specified by the respondents in the Delphi survey, but since the inputs to such calculations are also estimates, the discrepancy is not considered critical.

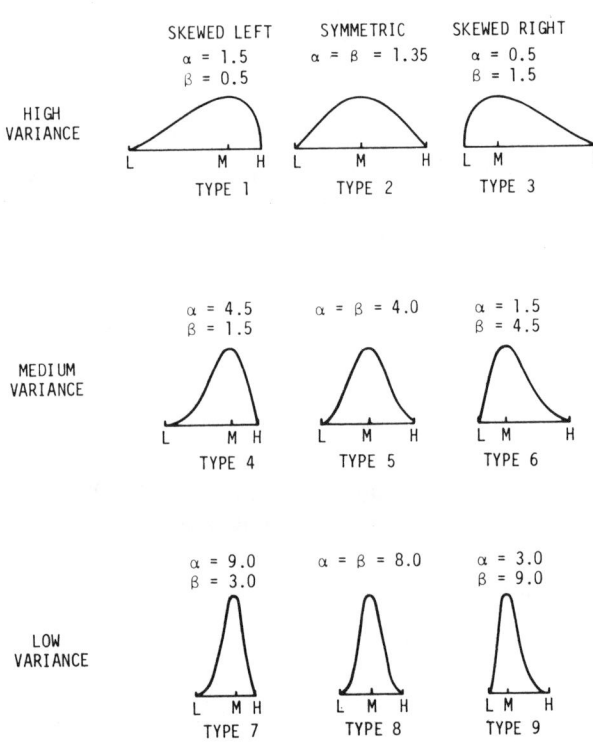

Fig. 1. Beta distributions used with the Delphi Method. Curve types are as defined by Dienemann [6].

It is very unlikely that complete agreement will be obtained among respondents regarding the lowest, highest, and most likely values in a given distribution. It is expected that upon completion of the Delphi process, each estimated numerical value will have a distribution formed by variability among the respondents in their final estimates. A common Delphi procedure is to use the median value as the predicted value. But this does not allow for unequal expertise among the respondents. In order to allow for variations in expertise, the participants were asked to provide confidence ratings for their answers. This rating was actually a tolerance interval and had no relationship to a willingness to bet against given odds that the true value fell within specified limits. The following scale was used to weight the various responses:

5 - I feel that this answer is probably within ± 20% of the true value.

4 - I feel that this answer is probably within ± 40% of the true value.

3 - I feel that this answer is probably within ± 60% of the true value.

2 - I feel that this answer is probably within ± 80% of the true value.

1 - I feel that this answer is probably within ± 100% or more of the true value.

At the conclusion of the Delphi survey procedure, then, the analyst has obtained various estimates of the lowest, most likely, and highest values (L, M, and H) of certain variables along with individual confidence ratings for each estimate. Assume for the moment that all participants have agreed upon the same curve-type from Fig. 1 to represent the distribution of some variable. Since the α and β parameters of a distribution are implicit in the selection of that distribution, it is apparent that the specification of any two of the remaining three parameters, L, M, or H, will completely describe the distribution in question. The analyst should, therefore, use those two parameters that have the least "disagreement" or "relative dispersion" in order to obtain the final function. The procedure proposed here for accomplishing this task is to make use of the confidence ratings to obtain weighted means and variances for each of the three estimated parameters.

Let $w_1, w_2, \ldots, w_n$ be the confidence ratings obtained from n experts concerning their estimates $h_1, h_2, \ldots, h_n$ of the value H. The weighted mean, μ_H, and the weighted variance, σ_H^2, of these estimates for this highest value are:

$$\mu_H = \frac{w_1 h_1 + w_2 h_2 + \cdots + w_n h_n}{w_1 + w_2 + \cdots + w_n} \quad (7)$$

and

$$\sigma_H^2 = \frac{w_1 (h_1 - \mu_H)^2 + w_2 (h_2 - \mu_H)^2 + \ldots + w_n (h_n - \mu_H)^2}{w_1 + w_2 + \cdots + w_n}. \quad (8)$$

In a similar manner, the weighted means and variances of L and M may be obtained.

To determine which two of the three estimated values have the least "disagreement," the ratio of the standard deviation to the mean for each parameter is calculated,

$$r_L = \frac{\sigma_L}{\mu_L} \quad , \tag{9}$$

$$r_M = \frac{\sigma_M}{\mu_M} \tag{10}$$

and

$$r_H = \frac{\sigma_H}{\mu_H} \quad , \tag{11}$$

and the two parameters with the smallest ratios are selected. Equation (4) may then be used to solve for the third value in terms of the other two.

The possibility also exists that all experts surveyed about a particular cost or benefit value will not agree upon a common distribution from the nine presented. When agreement is obtained, the analyst should use the procedure outlined above. When the participants do not agree on the same curve-type, a distribution other than the nine presented must be used to describe the variable in question. To determine the characteristics of this new distribution, the procedure previously described is followed for each different curve-type chosen. Then, using Eqs. (2) and (3), the mean and variance of each different curve-type selection can be calculated.

Let $W_1, W_2, \ldots, W_m$ be the overall sums of the individual confidence ratings of the two parameters L, M, or H with the smallest ratios found from Eqs. (9), (10), and (11) for m different curve types selected to represent a particular variable. Then, the weighted mean, variance, lower limit, and upper limit of the composite distribution, μ_c, σ_c^2, L_c, and H_c are found from

$$\mu_c = \frac{W_1 \mu_1 + W_2 \mu_2 + \cdots + W_m \mu_m}{W_1 + W_2 + \cdots + W_m} \quad , \tag{12}$$

$$\sigma_c^2 = \frac{W_1^2 \sigma_1^2 + W_2^2 \sigma_2^2 + \cdots + W_m^2 \sigma_m^2}{(W_1 + W_2 + \cdots + W_m)^2} \quad , \tag{13}$$

$$L_c = \frac{W_1 L_1 + W_2 L_2 + \cdots + W_m L_m}{W_1 + W_2 + \cdots + W_m} \quad , \tag{14}$$

and

$$H_c = \frac{W_1 H_1 + W_2 H_2 + \cdots + W_m H_m}{W_1 + W_2 + \cdots + W_m} \tag{15}$$

With this information, Eqs. (5) and (6) may be used to solve for the α and β parameters, and Eq. (4) may be used to find the mode of the composite beta distribution that best represents the overall weighted estimates of the variable in question.

One further situation must be considered: when only one individual has selected a particular curve type, or when only one expert can be found to provide estimates. In this case, the analyst should use those two values of L, M, or H with the highest confidence ratings in order to find the third. When ties occurred in these confidence ratings, or when ties occurred in the ratios obtained through Eqs. (9), (10), and (11), the authors assumed that the estimated values for M, L, and H were the most accurate, second most accurate, and least accurate, respectively, in order to determine which parameter should be recalculated.

Time Value of Costs and Benefits

The expected duration of a labor standard is important in determining its net present value. Present and future expenditures needed to establish and maintain the standard must be evaluated in terms of savings that occur in the future. Another factor is the frequency of occurrence of the task being studied. These variables should be included in a cost/return model. Selection of an appropriate interest rate is also necessary.

All initial or present costs must be converted to dollars per standard hour developed. Let μ_x and σ_x^2 be the mean and variance, respectively, of the cost of technician time, and μ_y and σ_y^2 be the mean and variance, respectively, of the time required to develop a labor standard by the yth technique. Then, the mean and variance of the technician cost of developing a one standard hour labor standard using the yth technique are [12, 20]:

$$\mu_{xy} = \mu_x \mu_y \tag{16}$$

and

$$\sigma_{xy}^2 = \mu_x^2 \sigma_y^2 + \mu_y^2 \sigma_x^2 + \sigma_x^2 \sigma_y^2 \quad . \tag{17}$$

Implicit in the use of these expressions is the assumption that the time required by a technician to develop a labor standard is independent of his pay scale.

In a similar manner, all future costs and savings may be expressed in terms of dollars per standard hour per appropriate time period. With the exception of the initial cost of developing a labor standard, all cash flows (costs and savings) occur in the future. It was necessary to specify a procedure for translating these to present costs.

It was assumed that all individual cash flows concerning a particular labor standard occur at the end of their respective time periods. Furthermore, all future cash flow elements were assumed to be both independent and identically distributed from one time period to the next. Let

$$\mu_r = \sum_{s=1}^{t} \mu_s \tag{18}$$

and

$$\sigma_r^2 = \sum_{s=1}^{t} \sigma_s^2 \tag{19}$$

where μ_s and σ_s^2 are mean and variance, respectively, of the sth individual future cash flow distribution, for a particular work measurement technique, in dollars per standard hour set per time period ($s = 1, \ldots, t$). The expected net present value (ENPV) and the variance of the net present value of a particular labor standard developed by a particular work measurement technique may then be written

$$\text{ENPV} = \mu_{xy} + \sum_{r=1}^{n} \frac{\mu_r}{(1+i)^r} \qquad (20)$$

and

$$\text{Var(NPV)} = \sigma_{xy}^2 + \sum_{r=1}^{n} \frac{\sigma_r^2}{(1+i)^{2r}}, \qquad (21)$$

where n is the number of time periods that the standard will be in use, i is the appropriate interest rate, and the means of all cost distributions are taken to be negative quantities while the means of all benefit distributions are taken to be positive. Since the future cash flow distributions were assumed to be independent and identically distributed, Eqs. (20) and (21) may be rewritten

$$\text{ENPV} = \mu_{xy} + \mu_r \left[\frac{(1+i)^n - 1}{i(1+i)^n} \right] \qquad (22)$$

and

$$\text{Var(NPV)} = \sigma_{xy}^2 + \sigma_r^2 \sum_{r=1}^{n} \frac{1}{(1+i)^{2r}}. \qquad (23)$$

Although the above expressions provide the mean and variance of the net present value of a particular labor standard developed by a particular work measurement technique, they say nothing about the type of distribution that results.

The Bounded Liapounov Theorem provides a necessary and sufficient condition for the sum of finite independent random variables to have a limiting normal distribution. In essence, this theorem states that if X_k ($k = 1, \ldots, n$) is a set of independent random variables with each X_k having both an upper and lower limit ($-\infty < X_k < +\infty$ for all k), and a variance that exists ($\sigma_k^2 \neq 0$ for all k), and if

$$Z = \sum_{k=1}^{n} X_k, \qquad (24)$$

then a necessary and sufficient condition for the distribution function of the random variable Z to approach a normal distribution function is that

$$\lim_{n \to \infty} \sum_{k=1}^{n} \sigma_k^2 = \infty. \qquad (25)$$

Since the net present value of a labor standard results from the sum of supposed mutually independent random variables, since these random variables are in fact bounded by upper and lower limits, and since their variances do exist, the constraints of the theorem are satisfied. It was assumed, therefore, that the individual cash flows were present in a large enough number, n, such that the effect of any one cash flow was slight with respect to their sum and that the distribution law followed by this sum did not differ significantly from the normal distribution law. Commonly tabulated values of the standard normal deviate could then be used, if desired, to make probability statements concerning the net present value of each of the work measurement techniques at hand.

Conclusion

The major applications of the Delphi Method to date have been in the area of long-range forecasting of expected technological and societal developments concerned with such subjects as political alliances, market forecasts, war prevention techniques, economic indices, and medical advancements. Its major contribution to the information-gathering process appears to be its tendency to produce a convergence of opinion in the majority of cases to which it has been applied [1, 4, 13, 14, 17, 18]. The results have been described as generally favorable; a reasonable consensus was obtained which provided the basis for subsequent analysis, planning, and action.

The success of the Delphi Method in obtaining a consensus of expert opinion in the field of forecasting makes it a potentially valuable technique for the industrial engineer. Even though this technique has met with success and is accepted in some circles, it should not be viewed as a device that produces the "truth." "The Delphi Method is designed to produce consensus judgments in inexact fields; it would be a mistake to consider such judgments as complete or precise descriptions..." [14]. The procedure described in this article did, however, in an actual application by the authors, provide good data for developing estimating functions of the costs and benefits of a particular work measurement program and its associated techniques.

Acknowledgments

The authors would like to thank the referees of this paper for their many helpful suggestions.

Support for this research was provided by the Engineering Research Institute of Iowa State University.

References

[1] Campbell, R. M., "A Methodological Study of the Utilization of Experts in Business Forecasting," unpublished PhD Dissertation, University of California, Los Angeles (1966).

[2] Churchman, C. W. and Eisenberg, H. B., "Deliberation and Judgment," in M. W. Shelly II and G. L. Bryan (editors), *Human Judgments and Optimality,* John Wiley and Sons, New York (1964).

[3] Cole, P.V.Z. and Mikasa, G. K., *A Probability Distribution for the Reliability of Complex Weapon Systems,* U.S. Naval Ammunition Depot, Oahu, Hawaii (July 1969).

[4] Dalkey N. C. and Helmer, O., "An Experimental Application of the Delphi Method to the Use of Experts," *Management Science,* **9**, 3, 458-467 (April 1963).

[5] Dalkey, N. C., "The Delphi Method: An Experimental Study of Group Opinion," Rand Corporation Report No. RM-5888-PR (June 1969).

[6] Dienemann, P. F., *Estimating Cost Uncertainty Using Monte Carlo Techniques,* The Rand Corporation, Santa Monica, Calif (January 1966).

[7] Fankhauser, G., "How Good Are MTM Standard Times?" *Advanced Management Journal,* **31**, 3, 53-60 (July 1966).

[8] Fankhauser, G., "The Introduction of MTM into the Plant." *Advanced Management Journal,* **32**, 1, 65-75 (October 1966).

[9] Foulke, J. A. and Hancock, W. M., "Summary of U.S./Canada Experiments on the Precision of GPD and MTM-3," U.S./Canada MTM Association (December 1970).

[10] Foulke, J. A. and Hancock, W. M., "Summary of Comparisons Between MTM-1, MTM-2, GPD, and MTM-3," U.S./Canada MTM Association (December 1971).

[11] Frederick, C. W., "On Obtaining Consistency in Application of Predetermined Time Systems," *The Journal of Industrial Engineering,* **11**, 1, 18-19 (January-February 1960).

[12] Goodman, L. A., "On the Exact Variance of Products," *Journal of the American Statistical Association,* **55**, 292, 708-713 (December 1960).

[13] Gordon, T. J., "New Approaches to Delphi," J. R. Bright (editor), *Technological Forecasting for Industry and Government,* Prentice-Hall, Englewood Cliffs, New Jersey, Part 2, 134-143 (1968).

[14] Gordon, T. J. and Ament, R. H., "Forecasts of Some Technological and Scientific Developments and Their Societal Consequences," Institute for the Future, Report No. R-6 (September 1969).

[15] Hancock, W. M. and Langolf, G. D., "Productivity, Quality, and Complaint Considerations in the Selection of MTM Systems," MTM Association, Fairlawn, New Jersey (August 1974).

[16] Helmer, O., and Rescher, N., "On the Epistemology of the Inexact Sciences," *Management Science,* **6**, 1, 25-52 (October 1959).

[17] Helmer, O., *Social Technology,* Basic Books, New York, Part 2, 116-122 (1966).

[18] Helmer, O., "Analysis of the Future: The Delphi Method," in J. R. Bright (Editor), *Technological Forecasting for Industry and Government,* Part 2, 116-122, Prentice-Hall, Englewood Cliffs (1968).

[19] Kaganowicz, J. and Krususki, S., "Mathematical Criteria to Develop and Select Optimal Derived MTM Procedures for Various Specified Conditions of Manual Work," *AIIE Transactions,* **8**, 1, 134-145 (March 1976).

[20] Mosteller, F., Rourke, R. E., and Thomas, G. B., Jr., *Probability and Statistics,* Addison-Wesley, Reading, Mass (1961).

[21] Salvendy, G. and McCabe, G. P., "Auditing Standards by Sample," *Industrial Engineering,* **8**, 9, 25-29 (September 1976).

Dr. S. Keith Adams is an Associate Professor of Industrial Engineering and Member of the Engineering Research Institute at Iowa State University, Ames, Iowa. Current research interests include applications of human factors engineering to problems in industry and agriculture. He has consulted for numerous corporations, government agencies, and attorneys in topics involving human factors engineering, occupational safety, and work design, and has authored papers and articles in these areas. He is also currently involved in research on the engineering economics of municipal solid waste recovery. He holds a BMgtE degree from Rensselaer Polytechnic Institute and MSE and PhD degrees from Arizona State University. He is a registered professional engineer, a senior member of AIIE, and a member of the Editorial Board of *AIIE Transactions*. Other memberships include the Human Factors Society, Sigma Xi, ASEE, Alpha Pi Mu, and the Iowa Academy of Science.

Dr. Timothy J. McGrath is a lieutenant colonel in the United States Air Force, currently serving as a faculty member of the Department of Resource Management Studies, Industrial College of the Armed Forces. He has previously held positions as the chief of software development and maintenance for the Air Force Satellite Control Facility and as a staff officer with the Office of the Secretary of the Air Force. He received a BChE degree from New York University, and MS and PhD degrees in Industrial Engineering and Management from Oklahoma State University.

A Procedure for an Economic Comparison of Work Measurement Techniques, Part II: An Application

TIMOTHY J. McGRATH
SENIOR MEMBER, AIIE
Industrial College of the Armed Forces
Fort McNair, Washington, D.C. 20319

S. KEITH ADAMS
SENIOR MEMBER, AIIE
Department of Industrial Engineering
and
Engineering Research Institute
Iowa State University
Ames, Iowa 50011

Abstract: In a previous article, the authors developed a model for the economic analysis of various options in choosing a work measurement technique. Each technique has its own economic advantages in terms of the savings to be expected in the work place. Likewise, each technique has its own costs in terms of the industrial engineering time required to develop and maintain the standard, as well as the inherent inaccuracies that tend to result in wasted direct labor man-hours. This article reports the results of an examination of the work measurement techniques in use in a particular organization. It is concluded that the expected net present value of any available technique can be determined in advance of the actual work study itself and that this determination can have a significant impact on the economic effectiveness of an organization's work measurement program.

■ Throughout industry many resources are devoted to the establishment and maintenance of labor standards. For years administrative decisions and judgments have been made concerning which of the many techniques available should be used to set the standard for a particular task, but the individual attempting to economically optimize the work measurement program within his own organization has found little help in the literature. Typical past publications concerning labor standards tend to stress either the need for accuracy or the need for economy. For example, Kadota [8] states:

> Standards must be as accurate in measured daywork as in wage incentive plans. One reason is that measurement by an incorrect scale leads to wrong judgments and actions. A second reason is more psychological. In measured daywork. . . unless operators and foremen have confidence in the standards, they **are not** strongly motivated to increase their performance. Thus, the standards must be set accurately, and this accuracy must be "sold" to the operators and foremen.

Received November 1977; revised December 1978. Paper was handled by Work Measurement/Methods/Ergonomics/Health Systems.

The question of whether, in specific individual applications, the costs of wrong judgments, actions and lower motivation may be greater than or less than the cost of obtaining the "accurate" standards is not considered.

Fankhauser [4] reports the results of an experiment comparing the accuracy of MTM-1 and the stopwatch procedure originated by the German Work Study Society (REFA). He states:

> *Therefore, based on the listed comparisons it may be concluded that the MTM procedure and the REFA stopwatch procedure yield practically the same normal times* [author's own emphasis].

In his summary, he concludes:

> We furthermore have seen that the stopwatch procedures known to us are fraught with more than one source of potential error and that they do not yield more accurate results than the MTM procedure. *Substituting MTM for these procedures could not, therefore, involve disadvantages for either the employers or the workers,* contrary to the apprehensions of some stopwatch advocates [emphasis added].

Fankhauser does not present the costs, in terms of application time, for either of the two techniques compared. If, as he indicates, the two procedures result in practically the same normal time, and if the stopwatch procedure can be applied in some instances more quickly and easily than MTM-1, then it is quite reasonable to assume that under these circumstances the added cost of MTM-1 could outweigh that caused by the "potential error" of the stopwatch. It should be pointed out that in a subsequent article he states, without verification, that one of the advantages of MTM is "quicker and more accurate development of standard data and time formulas" [5]. From his experience this may be true. However, the results presented in this paper do not substantiate this relative speed of MTM application when compared to other techniques available.

[At the time this study was conducted, MTM-2 and -3 and the more recently developed MOST (Maynard Operation Sequence Technique) approach were not used by the organization. These newer methods offer considerable advantages over both MTM-1 and stopwatch time study in many situations and would undoubtedly rate favorably in the comparative method to be illustrated.]

The "economy" viewpoint of work measurement application is to be found in the writings of those practitioners who advertise the allegedly high reductions in cost with the use of a particular procedure. For example, both Beck and Gibson report on the advantages of work sampling in specific instances. Beck [2] states, "productivity was found to improve at approximately the same rate as if standards had been set for every operation but at one-tenth the cost." Gibson [6] states that the many

> ... complications and variables in the work of repair shops would make a system of true engineered standards inadequate, costly, and subject to unfair measurement practices. The work measurement personnel would be so involved in revising old standards and developing new ones that little time would be available for improving methods, work area layouts, or analyzing performance problems.

It is not clear why the option of increasing the number of work measurement personnel to handle heavier workload was rejected. Granted that the decision to "engineer" all labor standards would be more "costly" than the described procedure, the savings to be realized, in terms of reduced production costs, could far outweigh the additional cost of the standards themselves.

Other examples of the attention to economy may be found. Klein [10] describes a type of desk calculator designed to calculate element times "at the push of a button" and "take the man-hours out of time study and the drudgery out of industrial engineering." Kopp [11] and Ross [16] describe computer programs to analyze time study data, and Stukey [18] predicts that in 1980 there will be a centralized computer bank of universal time data that all industrial engineers will be able to draw upon to save time in their work measurement applications. The Department of Defense has already developed an eleven volume automated standard data system based on the Dictionary of Occupational Titles published by the Department of Labor and upon numerous established standard data systems including MTM [15]. The system covers a vast array of work activities from bench work, machine trades, and clerical functions to services and professional and technical work. Material handling, warehousing, packing, and even farming, fishing, and forestry are also included. A reduction of 20 man-hours of work measurement personnel time has been claimed for each hour of standards developed under this program. Initial cost benefits of extending the use of standard data were estimated at $7.5 million, based only on the saving of time spent by industrial engineering personnel in establishing and maintaining standards. This system is the first major attempt to develop such a universal data bank covering virtually all types of work activity. An assumption basic to all of these innovations is that the savings in terms of industrial engineering time will more than offset the cost of mechanized data analysis.

More recently, Kaganowicz and Krususki [9] developed mathematical criteria for evaluating MTM methodologies. Explicit in their optimization procedure is the prior decision that some form of MTM is, indeed, an economically justified technique in itself. Salvendy and McCabe [17] point out the economic impact of inaccurate labor standards, but their paper presumes that the standards themselves have been established and, therefore, that the organization has already invested resources in their development. It should be apparent that the person charged with the responsibility for practical application of a work measurement program is caught between the desirability of accurate labor standards, with a probable high cost in engineering time but improved productivity in the work area, and the possibility of savings in engineering time, with the concurrent probability of reduced productivity from direct labor. In a previous article [1] the authors developed a method whereby an industrial organization could predetermine, within the constraints of its own production environment, the most economical work measurement technique to use to study a particular human task. That model was actually applied to a working industrial organization. The purpose of this article is to report the results of that investigation.

It is important to note that the specific results reported here are applicable only to the specific situation described. The results are, however, presented in detail since the authors believe that the trends established and the lessons and conclusions that may be drawn may be generally applicable to other organizations.

The Organization Studied

One branch of a large organization operating under a measured daywork standard and specializing in the maintenance and repair of aircraft parts and accessories was the subject of this study. At the time, 24 industrial engineering technicians out of a total of 134 employed by the parent organization itself were assigned to this branch. These 24 technicians were responsible for the establishment and maintenance of over 4,000 active labor standards, which are classified according to one of the following four types:

Type A: A labor standard developed by a recognized work measurement technique and backed up by sufficient data to statistically support an accuracy of plus or minus 10% of the mean, with 95% confidence.

Type B: A labor standard developed by a recognized work measurement technique and backed up by sufficient data to statistically support an accuracy of plus or minus 25% of the mean, with 95% confidence.

Type C: A labor standard developed by a recognized work measurement technique, but which lacks sufficient data to satisfy the requirements for classification as either a Type A or Type B standard.

Type D: An estimate of the standard time arrived at through coordination between an industrial engineering technician and one or more representatives from production control, quality control, shop supervision, etc.

The industrial engineering technicians employed by the organization have the ability to establish labor standards by any of six basic techniques: MTM, standard data, stopwatch time study, work sampling, engineered estimates, or coordinated estimates. For clarification of the terminology used in this article, it should be pointed out that the standard data is maintained by an electronic data processing procedure, known as the Automated Standard Data (ASD) System. A labor standard developed using this information is termed an "ASD standard."

When the technique options are combined with the various type classifications the result is that any given labor standard would have been developed by one of the following twelve technique types:

MTM (Type A),
ASD (Types A, B and C),
Stopwatch (Types A, B and C),
Work Sampling (Types A, B and C),
Engineered Estimate (Type C), or
Coordinated Estimate (Type D).

There are two basic sources of costs associated with the work measurement program of the subject organization. The first is attributable to the time required by the industrial engineering technicians to develop and maintain the labor standards. The second stems from the practice of processing the labor standards by digital computer. This cost may also be divided into two categories: (1) the cost of initially establishing the labor standard and (2) the periodic cost of maintaining that standard as part of the mechanized records.

The benefits of the work measurement program to the organization are derived from the savings in unproductive direct labor that results from the use of the labor standards themselves.

Eight industrial engineering technicians assigned to the branch studied were administered Delphi Method questionnaires in order to obtain the best estimate of the amount of time required by them to establish and maintain a one-standard-hour labor standard by the various technique type options available. Historical records formed the basis for deriving the costs associated with processing the standards by digital computer. It should be noted that, as defined in the study, a "one-standard-hour labor standard" is a standard set on a job that will be charged for one man-hour of work. In other words, after the task has been studied, and the appropriate fatigue, delay, and personal allowances have been added the resulting labor standard shows how much of a task should be accomplished in a standard time of one man-hour. The actual questionnaires used in this study may be found in [12]. Questionnaires concerned the costs of (a) establishment, (b) maintenance for two years, and (c) unproductive labor, based on 20 men working eight hours, under a given method used to set standards. This was done for each of the standard setting techniques listed in the tables provided in this article.

Initial Standard Development Costs

The amount of pay received by the industrial engineering technicians is determined by each individual's level on his respective wage scale. This variable, therefore, has a distribution. Let

μ_x = mean cost of technician time = 8.842 \$/hr,

σ_x^2 = variance of the cost of technician time = 0.796 (\$/hr)^2

and

μ_y and σ_y^2 = mean and variance, respectively, of the time required to develop a labor standard by the yth technique-type.

Then, the mean and variance of the technician cost of developing a one-standard-hour labor standard using the yth technique-type is [7, 13]

$$\mu_{xy} = \mu_x \mu_y \qquad (1)$$

and

$$\sigma_{xy}^2 = \mu_x^2 \sigma_y^2 + \mu_y^2 \sigma_x^2 + \sigma_x^2 \sigma_y^2 . \qquad (2)$$

Implicit in the use of these expressions is the assumption that the time required by a technician to develop a labor standard is independent of his pay scale.

Using Eqs. (1) and (2) the initial technician cost of establishing a one-standard-hour labor standard may be found for each available technique-type. These costs are shown in Table 1 along with the mean and variance of the technician time required to develop the standards, obtained from the survey. Since the initial electronic data processing (EDP) costs were obtained directly in the units required, no

Table 1: Initial costs of establishing a one-standard-hour labor standard by various technique-types.

Technique-Type	Industrial Engineering Time		Industrial Engineering Cost		EDP Cost	
	μ_y (hrs/std-hr)	σ_y^2 (hrs/std-hr)^2	μ_{xy} ($/std-hr)	σ_{xy}^2 ($/std-hr)^2	μ ($/std-hr)	σ^2 ($/std-hr)^2
MTM (A)	60.4	92.81	534	10,240	154.0	30.32
ASD (A)	8.9	0.61	79	112	27.0	1.32
ASD (B)	6.2	0.39	54	61	16.3	0.87
ASD (C)	4.7	0.18	42	32	6.8	0.08
Stopwatch (A)	12.2	2.48	108	314	27.0	1.32
Stopwatch (B)	9.1	1.46	80	180	16.3	0.87
Stopwatch (C)	2.4	0.02	21	6.13	6.8	0.08
Work Sampling (A)	35.2	73.5	312	6,791	2.0	0.01
Work Sampling (B)	17.8	12.35	158	1,228	2.0	0.01
Work Sampling (C)	10.3	2.48	91	280	2.0	0.01
Engineered Estimate (C)	1.3	0.05	11	5.44	6.8	0.08
Coordinated Estimate (D)	0.9	0.01	8	1.1	2.0	0.01

conversion was necessary. These values are also shown in Table 1.

Quarterly Standard Maintenance Costs

The economic analysis used here requires that all initial costs be presented for final analysis in terms of dollars per standard hour developed. Likewise, all anticipated future costs and benefits must be in terms of dollars per standard hour developed per appropriate time period. Due to the characteristics of the subject organization, three months was considered to be a suitable base unit of time for subsequent analysis of the work measurement techniques. Some of the information collected on the questionnaires could be applied directly in these analyses; other data required conversion to the appropriate units.

The costs associated with the maintenance of the various labor standards were obtained on the questionnaires using a two-year reference period. This basic time unit was selected because organization policy required that each labor standard be audited at least once every two years. Even if no other effort on the part of the industrial engineering technician was necessary, he would at least have had to perform an audit on the labor standard during this time.

If it is assumed that the estimated values for the two-year period represent the sum of eight independently and and identically distributed quarterly random variables, then the following equations may be used to find the mean and variance of these quarterly values [13]. Let

μ_t and σ_t^2 = mean and variance, respectively, of the estimated value for a two year period,

and

μ_q and σ_q^2 = mean and variance, respectively, of the estimated value for a three-month period.

Then,

$$\mu_q = \mu_t/n \quad (3)$$

and

$$\sigma_q^2 = \sigma_t^2/n \quad (4)$$

where

n = 8 quarters/two-year period.

Using Eqs. (3) and (4), the two-year industrial engineering technician time estimates obtained from the survey were converted to quarterly time values. Equations (1) and (2) were then used to convert these times to the quarterly costs which are included in Table 2 along with the values obtained from the survey and the quarterly electronic data processing (EDP) costs.

The Costs of Unproductive Direct Labor

To determine the costs of unproductive direct labor that might result if each of the various work measurement technique-type options were used, an "average" shop consisting of 20 direct-labor employees working eight hours per day was used as a base of reference. This figure was selected under the assumption that the industrial engineering technicians would be better able to estimate the amount of overall "wasted time" inherent in the particular labor standard classifications for a group working a full day than, for example, a single individual working only one hour. This assumption is based strictly on the judgment of the authors; there is no way to determine whether it provided more or less valid data than would another reference figure.

Table 3 shows the estimates of the number of unproductive direct labor man-hours, out of the 160 man-hours available in the 20-man shop each day, that might result if all labor standards in that shop were set in the manner indicated. As stated in the study, these values were to be

	Table 2: Costs of maintaining a one-standard-hour labor standard by various technique-types.					
Technique-Type	Two-Year Industrial Engineering Time		Quarterly Industrial Engineering Cost		Quarterly EDP Cost	
	μ_t (hrs/std-hr)	σ_t^2 (hrs/std-hr)^2	μ_{xq} ($/std-hr)	σ_{xq}^2 ($/std-hr)^2	μ ($/std-hr)	σ^2 ($/std-hr)^2
MTM (A)	22.7	20.22	25.1	206.06	103.8	52.68
ASD (A)	4.8	1.64	5.3	16.40	20.2	5.92
ASD (B)	3.2	0.22	3.5	2.34	12.0	0.95
ASD (C)	3.4	0.23	3.8	2.44	5.1	0.37
Stopwatch (A)	3.8	1.82	4.2	18.18	20.2	5.92
Stopwatch (B)	2.8	0.88	3.1	8.71	12.0	0.95
Stopwatch (C)	1.8	0.41	2.0	4.07	5.1	0.37
Work Sampling (A)	8.3	4.21	9.1	42.39	1.5	0.06
Work Sampling (B)	5.8	3.36	6.4	33.51	1.5	0.06
Work Sampling (C)	3.0	0.58	3.3	5.88	1.5	0.06
Engineered Estimate (C)	2.2	0.06	2.4	0.61	5.1	0.37
Coordinated Estimate (D)	0.8	0.01	0.8	0.05	1.5	0.06

determined by considering the amount of time that the workers would get that was over and above what they should need to perform their duties either because of the poor method definition or the "looseness" of the technique used to set the standards. Time lost due to personal needs, official breakperiods, cleanup, minor maintenance, or other recognized delays and allowances was not to be considered unproductive. In effect, these values represent estimates of the inherent accuracy or credibility of each of the various work measurement technique types, as applied in the subject organization branch.

Assuming that the estimated values represent the sum of 160 independent and identically distributed random variables, let

μ_y and σ_y^2 = mean and variance, respectively, of the number of unproductive direct labor man-hours per repetition of a one-man-hour task established by the yth technique type,

μ_{ty} and σ_{ty}^2 = mean and variance, respectively, of the total number of unproductive direct labor man-hours (out of the 160 man-hours available in the 20-man shop each day) that would result if all standards were established by the yth technique type,

μ_x = mean cost per direct-labor-hour = 9.23 $/hr,

and

σ_x^2 = variance of the cost per direct-labor-hour = 0.055 ($/hr)^2.

Then, using Eqs. (3), (4), (1), and (2), and letting n = 160 man-hours per day, the second set of values in Table 3 is derived. Again, an assumption implicit in these calculations is that the number of unproductive man-hours per repetition of the task is independent of the worker's pay scale. The percent of the mean cost per direct-labor hour represented by the mean cost of unproductive direct labor for an MTM standard is:

$$\frac{\mu_{xy}}{\mu_x} = \frac{0.47}{9.23} = 0.051 = 5.1\% . \quad (5)$$

The interpretation of this value is that, in the long run, this shop can expect to receive approximately 94.9% labor effectiveness when MTM is used to establish labor standards.

The Net Present Value of a Labor Standard

From the economic characteristics of the organization, it was determined that a suitable discount rate was 10% per year or 2.41% per quarter [19].

Assume that it is desired to determine the most economical work measurement technique to use to study a task that has an expected life of 18 months (six quarters). Further assume that the task will be performed twice a week, or 26 times per quarter, and that the technique type being considered is Stopwatch (Type C).

In order to determine the value of a labor standard, the costs of that standard must be compared to the costs of having no standard at all. These values may be obtained from Tables 1, 2, and 3. The mean and variance of the initial development costs for a Stopwatch (C) labor standard are

$$\mu_p = \sum_{j=1}^{k} \mu_j = -21.0 - 6.8 = -27.8 \ \$/std\text{-}hr \quad (6)$$

and

Technique-Type	Total Unproductive Direct Labor		Cost per Repetition of a One-Man-Hour Task		Direct-Labor Effectiveness (%)
	μ_{ty} (hrs/160 hrs)	σ^2_{ty} (hrs/160 hrs)^2	μ_{xy} ($/rep)	σ^2_{xy} ($/rep)^2	
No Standard	39.7	77.65	2.3	41.38	75.2
MTM (A)	8.1	0.63	0.5	0.34	94.9
ASD (A)	8.8	0.53	0.5	0.26	94.5
ASD (B)	12.3	6.74	0.7	3.58	92.3
ASD (C)	20.4	8.25	1.2	4.44	87.3
Stopwatch (A)	9.5	0.74	0.5	0.43	94.1
Stopwatch (B)	12.6	2.21	0.7	1.19	92.1
Stopwatch (C)	21.3	12.36	1.3	6.57	86.7
Work Sampling (A)	9.6	0.65	0.5	0.34	94.0
Work Sampling (B)	14.2	4.84	0.8	2.56	91.1
Work Sampling (C)	22.6	33.35	1.3	17.75	85.9
Engineered Estimate (C)	25.0	10.61	1.4	5.63	84.4
Coordinated Estimate (D)	29.0	13.74	1.7	7.34	81.8

Table 3: Cost of unproductive direct labor by various technique-types.

$$\sigma^2_p = \sum_{j=1}^{k} \sigma^2_j = 6.13 + .08 = 6.21 \, (\$/\text{std-hr})^2. \quad (7)$$

The value of a labor standard is derived from the savings in unproductive direct labor costs as a result of the use of the standard. Assuming that these values are independent and identically distributed from one repetition of the task to another, and remembering that the task is to be repeated 26 times per quarter, the mean and variance of the quarterly distribution of the value of the Stopwatch (C) labor standard becomes:

$$\mu_s = n(\mu_2 - \mu_1) = 26 \, (2.3 - 1.3) = \$26.0 \, \$/\text{qtr} \quad (8)$$

and

$$\sigma^2_s = n(\sigma^2_2 + \sigma^2_1) = 26 \, (41.38 + 6.57) = \$1247 \, (\$/\text{qtr})^2. \quad (9)$$

The mean and variance of the quarterly cash flows is obtained from

$$\mu_r = \sum_{s=1}^{t} \mu_s = -2.0 - 5.1 + 26.0 = \$18.9 \, \$/\text{qtr} \quad (10)$$

and

$$\sigma^2_r = \sum_{s=1}^{t} \sigma^2_s = 4.07 + 0.37 + 1,247 = 1,251.44 \, (\$/\text{qtr})^2. \quad (11)$$

Then, setting $i_q = 0.0241$ and $n =$ six quarters, the expected net present value and the variance of the net present value of the labor standard are:

$$\text{ENPV} = \mu_p + \mu_r \left[\frac{(1+i_q)^n - 1}{i_q(1+i_q)^n} \right] = 85.58 \, \$/\text{std-hr} \quad (12)$$

and

$$\text{VAR (NPV)} = \sigma^2_p + \sigma^2_r \sum_{r=1}^{n} \frac{1}{(1+i_q)^{2r}} = 6,382 \, (\$/\text{std-hr})^2. \quad (13)$$

In like manner each of the other competing technique types may be evaluated. Table 4 is a summary of this information obtained for all options available. It is interesting to note that in this case, if MTM-1 were the only technique available, the organization would stand to lose more than $800 in expected net present value by developing the standard for this task. Although the ENPV's obtained by

Table 4: Comparison of the value characteristics of a labor standard with an expected life of 18 months for a task to be repeated 26 times per quarter by various technique-types.

Technique-Type	ENPV ($/std-hr)	Var(NPV) ($/std-hr)^2
Stopwatch (C)	85.6	6,382
ASD (B)	70.4	6,036
Coordinated Estimate (D)	64.7	6,457
ASD (C)	63.4	6,117
Engineered Estimate (C)	62.7	6,239
Stopwatch (B)	44.3	5,871
Work Sampling (C)	22.6	8,144
ASD (A)	9.0	5,743
Work Sampling (B)	7.8	7,221
Stopwatch (A)	-19.0	5,977
Work Sampling (A)	-123.0	12,535
No Standard	-328.9	5,483
MTM (A)	-1,139.1	17,116

this example are relatively small, this is not always the case. The expected life of the labor standard and the number of times that the task is to be repeated are the two factors that primarily influence the magnitude of these values. Table 5 shows selected work measurement situations for the subject organization, along with the value characteristics of the technique-type with the greatest ENPV.

Table 5: Value characteristics of technique-types with highest expected net present values for selected combinations of labor standard life and frequency of task repetition.

Life (Quarters)	Repetitions per Quarter	Technique-Type	ENPV ($/std-hr)	Var(NPV) ($/std-hr)^2
2	60	ASD (B)	82.2	5,094
4	21	Stopwatch (C)	29.6	3,603
6	81	ASD (A)	550.8	17,413
8	2	Coordinated Est(D)	-18.1	635
10	26	ASD (B)	153.9	9,169
12	15	Stopwatch (C)	63.3	6,465
14	4,131	MTM (A)	86,243	1,732,418
16	1	No Standard	-30.1	452
18	62	ASD (A)	1,124	30,844
20	3,795	MTM (A)	105,879	2,007,434
22	13	Stopwatch (C)	85.7	8,363
24	20	Work Sampling (B)	228.5	13,969
26	50	Work Sampling (A)	1,146	37,780
28	30	ASD (B)	572.6	20,478
30	1	Coordinated Est (D)	-46.6	762
32	39	Work Sampling (A)	949.6	33,564
34	17	Work Sampling (B)	233.7	14,062
36	133	ASD (A)	4,948	93,580
38	3,430	MTM (A)	150,362	2,468,381
40	15	Work Sampling (B)	200.8	13,316

Conclusions

Although the economic importance of a general work measurement program has often been cited in the literature, a method for comparing the costs and benefits of specific application procedures within each individual production environment has been virtually ignored. The model developed in the previous article [1] and applied here adds a measure of objectivity to what has been a rather subjective decision, by providing a procedure whereby the person responsible for this function is aided, quantitatively, in selecting the most economically effective technique to apply to a given job study. A discrete choice must be made in selecting a work measurement technique. As long as the measures provided in the model allow for this choice, the model is useful. The method developed provides such measures. Much of the guesswork involved in technique selection can be removed, and the best choice can be made on a consistent basis.

Although the results of the investigation presented in this article are applicable only to the specific organization studied, the following conclusions may be drawn from the information obtained:

1. The Delphi Method may be used advantageously to evaluate the costs and benefits of a work measurement program. The results obtained from this study strongly support the desirability of this procedure when making such an investigation.

2. There is real value to be obtained from a properly applied work measurement program. It is possible to determine the expected net present value of a labor standard. The value of a single labor standard can reach hundreds of thousands of dollars.

3. Although a properly applied work measurement program can be quite valuable, on the other hand, an improperly applied program can be quite costly. An emphasis on either accuracy or economy of application alone can result in a significant economic loss to the organization with the development of each labor standard.

4. The results of this study tend to support the contention that the MTM-1 procedure produces quite accurate labor standards. However, in many cases the added cost of MTM-1 can outweigh that of the potential errors in the other techniques, and, contrary to Fankhauser's assertions [4, 5], substituting MTM-1 for the stopwatch (or any other technique) might be disadvantageous to the organization. The primary cause of this disadvantage is that, at least in the organization studied, MTM-1 does not produce quicker time studies than the other options available; in fact, it is five times as costly, on the average, as the best stopwatch labor standard (see Table 1).

5. The credibility or accuracy of a given work measurement technique is not necessarily proportional to the amount of time or cost required to develop the standard.

6. The accuracy and economy of the Automated Standard Data (ASD) System included in this study and of other recently developed systems such as UnivEl and the Department of Defense standards tend to justify them for many large organizations. The selection of such a system should be based on the expected returns on investment. These depend upon the number of labor standards to be developed each year and the suitability of the work in question to formal standards.

7. Results of this study suggest that stopwatch methods, although frequently criticized, can provide accurate labor standards. This fact would, in many cases, justify investment in better measuring devices such as the electronic data recorders and processors used in DATAMYTE Systems [3] or a video tape system.

Recommendations

The results of this study suggest additional areas of investigation. These include:

1 — Studies of the relative accuracy of alternative work measurement techniques under actual production conditions.

2 — Studies into the causes of unproductive direct labor as being related to poorly defined methods, inaccurate timing, or both.

3 — Development of improved methods of identifying the characteristics of jobs which make them best suited for analysis by a given work measurement technique; also, the determination of the best combination of techniques to use in cases best suited for several techniques.

4 — Improved methods of identifying jobs or tasks that should be redesigned from the standpoint of methods improvement, and of predetermining how much can be saved in terms of direct labor costs prior to deciding upon the improvement.

Recent developments in Work Unit Analysis by Mundel [14] suggest further extensions of the Delphi Method into the area of overall organizational productivity studies and decentralized management by objectives. Work Unit Analysis clarifies the concept of a unit of work and ties it more closely with identifiable units of product or service. It also assures that work methods and measurement programs will deal with units of work which are affordable in terms of the detail and cost required in using them. Another powerful technique which combines mathematical models of human behavior with computerized network analysis is known as SAINT (Systems Analysis of Integrated Networks of Tasks), and was developed by Pritsker and Associates [20]. The Delphi Method could be used to obtain data needed in developing models of behavioral characteristics and in specifying network relationships.

Acknowledgments

The authors would like to thank the referees of this paper for their many helpful suggestions.

Support for this research was provided by the Engineering Research Institute of Iowa State University.

References

[1] Adams, S. K. and McGrath, T. J., "A Procedure for an Economic Comparison of Work Measurement Techniques, Part I: The Model," *AIIE Transactions*, **11**, 3 (1979).

[2] Beck, C., "Work Sampling Sets Direct Labor Standards," *Industrial Engineering*, **1**, 2, 22-25 (February 1969).

[3] Electro/General Corporation, DATAMYTE Operation Manual, Hopkins, Minnesota (1976).

[4] Fankhauser, G., "How Good Are MTM Standard Times?" *Advanced Management Journal*, **31**, 3, 53-60 (July 1966).

[5] Fankhauser, G., "The Validity of Certain Criticisms of the MTM Procedure," *Advanced Management Journal*, **31**, 4, 45-55 (October 1966).

[6] Gibson, D., "Work Sampling Monitors Job-Shop Productivity," *Industrial Engineering*, **2**, 6, 12-19 (June 1970).

[7] Goodman, L. A., "On the Exact Variance of Products," *Journal of the American Statistical Association*, **55**, 12, 708-713 (December 1960).

[8] Kadota, T., "PAC - Performance Analysis and Control," *The Journal of Industrial Engineering*, **19**, 8, 407-411 (August 1968).

[9] Kaganowicz, J. and Krususki, S., "Mathematical Criteria to Develop and Select Optimal Derived MTM Procedures for Various Specified Conditions of Manual Work," *AIIE Transactions*, **8**, 1, 134-145 (March 1976).

[10] Klein, K., "A New Way to Take Time Studies," *Industrial Engineering*, **2**, 4, 17-19 (April 1970).

[11] Kopp, K. K., "A Computer Program for Time Study Analysis," *The Journal of Industrial Engineering*, **18**, 2, 147-152 (February 1967).

[12] McGrath, T. J., "A Procedure for the Economic Analysis of Work Measurement Techniques," unpublished PhD dissertation, Oklahoma State University (May 1971).

[13] Mosteller, F., Rourke, R. E. K., and Thomas, G. B. Jr., *Probability and Statistics*, Addison-Wesley, Reading, Massachusetts, 329-331 (1961).

[14] Mundel, M. E., "Work-unit Analysis," *Proceedings, Spring Annual Conference, American Institute of Industrial Engineers*, Dallas, Texas, 431-434 (May 1977).

[15] Power, R. J., "Development of a Defense-wide Standard Time Data Program," *Proceedings, Spring Annual Conference, American Institute of Industrial Engineers*, New Orleans, Louisiana, 462-471 (May 1974).

[16] Ross, R., "Computer Analyzes Data – Man Analyzes the Job," *Industrial Engineering*, **2**, 5, 22-25 (May 1970).

[17] Salvandy, G. and McCabe, G. P., "Auditing Standards by Sample," *Industrial Engineering*, **8**, 9, 25-29 (September 1976).

[18] Stukey, A. E., "Work Measurement in Perspective-universal Time Data," *The Journal of Industrial Engineering*, **15**, 1, 37-41 (January-February 1964).

[19] Thuesen, H. G. and Fabrycky, W. J., *Engineering Economy*, 3rd edition, Prentice-Hall, Englewood Cliffs, New Jersey (1964).

[20] Wortman, D. B., Pritsker, A. A. B., Sevin, C. S., Seifert, D. J., and Chubb, G. P., *SAINT: Volume II. User's Manual*, AMRL-TR-73-128, Aerospace Medical Research Laboratory, Wright Patterson Air Force Base, Ohio (1974).

Dr. Timothy J. McGrath is a lieutenant colonel in the United States Air Force, currently serving as a faculty member of the Department of Resource Management Studies, Industrial College of the Armed Forces. He has previously held positions as the chief of software development and maintenance for the Air Force Satellite Control Facility and as a staff officer with the Office of the Secretary of the Air Force. He received a BChE degree from New York University, and MS and PhD degrees in Industrial Engineering and Management from Oklahoma State University.

Dr. S. Keith Adams is an Associate Professor of Industrial Engineering and Member of the Engineering Research Institute at Iowa State University, Ames, Iowa. Current research interests include applications of human factors engineering to problems in industry and agriculture. He has consulted for numerous corporations, government agencies, and attorneys in topics involving human factors engineering, occupational safety, and work design, and has authored papers and articles in these areas. He is also currently involved in research on the engineering economics of municipal solid waste recovery. He holds a BMgtE degree from Rensselaer Polytechnic Institute and MSE and PhD degrees from Arizona State University. He is a registered professional engineer, a senior member of AIIE, and a member of the Editorial Board of *AIIE Transactions*. Other memberships include the Human Factors Society, Sigma Xi, ASEE, Alpha Pi Mu, and the Iowa Academy of Science.

© 1982 by Mitchell Fein

How 'Reliability,' 'Precision' And 'Accuracy' Refer To Use Of Work Measurement Data

By Mitchell Fein
Mitchell Fein Inc.

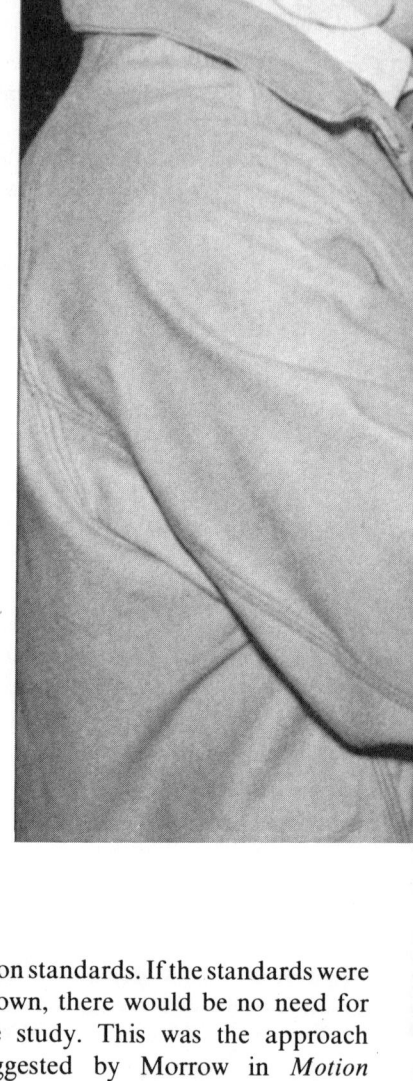

When work measurement data is used for shop loading and manpower determination, the accuracy and reliability of time standards is not critical. When used to establish fair day's work standards or for wage incentives, closeness to the "truth" is important. When an arbitrator is given the authority to determine whether a reprimanded employee should be retained, the subject is critical.

Users of work measurement data, and especially measurement analysts who create the data, should fully understand the meanings of reliability, precision and accuracy terms as they relate to work measurement. Unfortunately, meanings and definitions from physical sciences are carried over into work measurement without the user's realizing that physical science measurement criteria have little or no relevance to work measurement.

Modifying time study data by performance rating introduces subjective bias which forecloses scientific validity. Regardless of the care used to collect the original time data and the tests performed to assure statistical validity and control of the data, performance rated time values cannot be objective.

The standard statistical tests for validity of data have meaning only to the extent the data represents the population from which it was taken. Time study data after performance rating no longer represent, in a statistical sense, that population, but instead represent a "normalized" population. If work measurement consisted only of obtaining the time taken by an operator to perform discrete blocks of work, it would be relatively simple to devise data collection methods paralleling the sound procedures used in quality control; there would then be no problems.

Stop-watch time study procedures are very different from quality control. The data obtained in a time study is mentally compared to a defined work pace concept, and the data is then adjusted to the mental criteria. Benchmark films may be used for comparison but the comparisons are still in the analyst's mind. No other measurement system which tests data for statistical validity alters the observed data. There is no way of determining whether the performance rated data is biased because that would require testing against agreed-upon standards. If the standards were known, there would be no need for the study. This was the approach suggested by Morrow in *Motion Economy and Work Measurement*.

Abruzzi made concerted efforts to develop statistically sound approaches to collect, analyze and treat time study data from which to establish valid time standards. Most of his criticisms of time study practices are still valid today and practitioners have much to learn from his writings, *Measurement, New Principles and Procedures* and *Work, Workers and Work Measurement*. But Abruzzi's main emphasis was on the methods used to collect and treat data. He did not develop a way to obviate the need for performance rating, nor how to

Despite all proposals, performance rating is still indispensable to stopwatch time study. (Photo courtesy of Weirton Steel Division, National Steel Corp.)

test for the validity of "should take" time standards.

Gomberg's sharp critique in 1955 of work measurement practices was very complete and pertinent in *A Trade Union Analysis of Time Study*. His comments and conclusions are still valid. Though Gomberg, at the time, represented trade unions, he aimed more to remove unsound work measurement practices than to criticize management or the profession. His efforts in large measure contributed to an increasing understanding of work measurement practices among unions and arbitrators. He prodded industrial engineers to revise their practices. However, his pinpointing of work measurement shortcomings did not eliminate the weakness stemming from performance rating.

The searching analysis of Davidson in his *Functions and Bases of Time Standards* is germane even today and should be must reading for practitioners. Many of his criticisms and challenges have still not been answered by the profession.

The search for validity has pinpointed shortcomings in data collection techniques and in the statistical methods used to treat the data. No one, however, has yet devised a way to remove the greatest source of error: the need to performance rate time study observed data. The most exacting care in taking time studies, which observe sound statistical procedures, will not later remove the bias introduced by rating. Not having a precise and verifiable method of performance rating, however, does not excuse practitioners from using unsound data collection methods.

The mathematical purists who insist that work measurement must be mathematically valid from beginning to end should heed Shewhart in his *Statistical Method from the View of Quality Control*:

". . . . if we were forced to choose between the formal mathematical description and the physical description, I think we should need to look for a new mathematical description instead of a new physical description because the latter is apparently what we have to live with."

Despite all proposals, performance rating is still indispensable to stopwatch time study.

Meaning of terms

There is often misunderstanding regarding the meaning of reliability, precision and accuracy of time study data and the time standards which are set from the data. Definitions by Churchill Eisenhart of the National Bureau of Standards well apply to work measurement:

"The precision of a measurement process refers to and is determined by the degree of mutual agreement characteristic of independent measurements of a single quantity yielded by repeated applications of the process under specified conditions; and its accuracy refers to and is determined by the degree of agreement of such measurements with the true value of the magnitude of the quantity concerned. In brief, 'accuracy' has to do with closeness to the truth; 'precision,' only with the closeness together.

"Systematic error, precision and accuracy are inherent characteristics of a measurement process and not of a particular measurement yielded by the process. But these terms are not defined for a measurement operation that is not in a state of statistical control.

"Experience shows that in the case of measurement processes the ideal of strict statistical control that Shewhart prescribes is usually very difficult to attain.... Many measurement processes simply do not and, it would seem, cannot be made to conform to this ideal of producing successive measurements of a single quantity that can be considered to be 'observed values' of independent identically distributed random variables."

Reliability and precision of the data refers only to the degree to which the observed and recorded data represent the operation and work performed during the time study. The reliability concept is readily illustrated by quality control procedures which are well standardized and agreed upon in industry. The various AQL sampling plans designate the degree to which a sample represents the lot from which the sample was drawn. As the AQL sampling plan is tightened, the probability is increased that the sample represents the lot from which the sample is taken.

Work measurement practitioners prefer a reliability level of 95% ± 5%, meaning the chances are that, in 95 out of 100 instances, the sample mean will not deviate from the lot mean by more than ± 5%. Employing statistical procedures, it is possible to compute the size of a sample that must be drawn from the lot to meet the reliability level desired. For a given operation, a reliability level of 95% ± 5% will require a larger sample size than 95% ± 10%.

Time study is a sampling process, very much like QC sampling. Recording time study data corresponds in QC to selecting a sample which will be inspected. The ongoing operation is the lot from which a time study sample will be selected. Just as the sample in QC only represents the lot being sampled, so the time study is the sample of the operation being performed in front of the time study analyst.

While time study analysts may take liberties in the assumptions they make in their time study procedures, it must be clearly understood that the reliability of the data refers only to what operation being observed, not to what occurred yesterday or may occur tomorrow. Most important, reliability does not specify any degree of accuracy regarding the data or the final time standards which are developed from the data with respect to criteria for fair and equitable standards established for the measurement system.

Accuracy of time standards can be described in Eisenhart's *(Realistic Evaluation of the Precision and Accuracy of Instrument Calibration Systems)* terms as that it:

"... is determined by the degree of agreement of such measurements with the true value of the magnitude of the quantity concerned ... 'accuracy' has to do with closeness to the truth; "precision" (reliability) only with the closeness together (of the data)."

Words in parentheses were added.

This definition causes no difficulty in physical measurements because clear and reliable measurement standards are readily available and there are no differences of opinions on the accuracy of the measurement standards. In work measurement, however, measurement standards always represent a consensus of opinions in a particular plant. To complicate the measurement process, over a period of time the measurement standards may change.

The definition of accuracy for physical measurements has no meaning in work measurement because the measurement standards against which time standards are judged are subjective yardsticks. Work measurement standards for performance rating are usually expressed in the minds of work measurement analysts, based on measurement criteria for each plant.

Standard data

Practitioners know that data from individual time studies may deviate significantly from the average of a number of studies of the same operation. The most studies that are averaged, especially if taken by different observers, the greater the probability that random errors are eliminated and that the data represents the criteria for standards for that company.

The operational advantages of standard data far outweigh the evidence from research studies that standard data elements are not fully additive. Approximately 85% of companies studied used standard data. Phil Carroll, a staunch advocate of standard data, stated in *Better Wage Incentives* that standards derived from standard data are far more consistent than when set from individual studies. Gomberg expressed workers' preference for standard data:

". . . . workers feel that standard data at least reduces to writing an implied bargain between them and the management . . . which is not likely to be changed daily with every new time study."

Mundel, in *Motion and Time Study*, candidly observes that:

"Many standard times in plants may appear to function well, even when they are incorrect . . . because the workers have learned that it is advantageous to have them function as if they were correct."

Standard data has appeal on all sides.

A main advantage to standard data is that once the data has been agreed to, time standards for operations are set with relative ease. Performance rating is left behind in setting the original data. (A full discussion of the advantages and limitations of standard data is contained in the texts of Mundel, Nadler, Barnes and Niebel.)

Predetermined time standards

Predetermined time standards (PTS), such as MTM, Work Factor (WF) and other systems, were created to provide reproducible and accurate standards. Reproducibility for finite micromotion elements is obtained. However, the accuracy goal is not fully reached because the PTS data were originally established by stop-watch time study; and the additivity of microelement time values is questioned; building an operation standard by identifying the micromotion elements is not as easy nor as accurate as claimed by PTS proponents.

For example, in electronics the operation of picking up a precut wire, forming a prehook on one end, positioning the end to a terminal, wrapping the hook to the terminal and clipping the end of the wire consists of over 30 distinct motions in either MTM or WF; the total time is about 0.10 minutes. Recognizing and clearly identifying all elements is not as

> **The most exacting care in taking time studies, which observe sound statistical procedures, will not later remove the bias introduced by (performance) rating. Not having a precise and verifiable method of performance, however, does not excuse practitioners from using unsound data collection methods.**

easy as it seems on the surface. There are advantages and values to using PTS systems, but increased accuracy is by no means assured.

A predetermined time standards system is a form of standard data, most often of microwork elements. The main objective of the PTS systems is to create measurement standards and procedures to set standards so work measurements can be made which are reproducible. The development of standard motion times for discrete motions permits measurements to be made without a stop watch and performance rating. PTS proponents state that when differences arise with employees, these are usually around readily defined motions.

A small but vocal group opposes the foundations of PTS with claims that the various motion times are not additive and that different arrangements of motions result in different times. (A summary of the discussions on PTS shortcomings is contained in Nadler, *Work Design*, p. 465; Mundel, *Motion and Time Study*, 4th ed., pp. 410, 411.) There are arguments for and against PTS systems. Practitioners and managers have their own reasons for using a PTS system or the stop-watch. This material does not delve into such issues and the arguments for or against PTS do not affect the essence of this presentation.

PTS systems are discussed because the basic time data used to establish the elemental time values of each PTS system were originally established from conventional stop-watch time studies, which were performance rated by groups of experienced analysts who agreed on the basic concepts and criteria underlying their system.

It is in regard to the performance ratings of the original PTS data that PTS systems must be examined together with stop-watch time study methodology. If there are shortcomings in the stop-watch time study process, these may have been built into the PTS systems in the original derivation of the microstandard data. Users of the PTS systems should understand all aspects of stop-watch work measurement.

A useful contribution of the PTS innovators is the shorthand symbols and definitions which are used to describe the motions of an operation. The verbal descriptions used in stop-watch time study are usually inadequate.

Assuring reliable data

The main steps to assure reliable data are:

1.) Study representative workers and conditions.
2.) Study adequate number of cycles.
3.) Properly treat time study data.

4.) Calculate representative times from study.

The degree of care employed in standardizing work conditions for a study depends on the nature of the work. In a job shop with short runs, the operator usually makes his own set-ups, and precise standardization of work methods is difficult and not important. Large volume, highly repetitive tasks should be standardized before the study so that a precise description of work methods and conditions is recorded as part of the study.

The analyst should assure that the work conditions, equipment and parts used in the study, and the service provided to the operator during the study, are representative of normal conditions. For example, do not create ideal conditions by providing the operator with undue attention, perfect parts, newly sharpened tools and so forth, if such conditions do not normally prevail. Similarly, abnormally poor conditions should not be allowed, unless conditions are usually poor.

Selection of operators to be studied sometimes causes disagreements with employees. Worker and management representatives readily understand that neither the poor nor the super operator should be studied. Theoretically, it should not matter which worker is studied, providing the worker is experienced and can properly perform the operation.

Time study analysts usually prefer above-average performance operators because they then have greater assurance that the employees are fully experienced with the operation and perform it properly. Experience and skill with the operation are vital to the study because, in rating the performance of the operator during the study, the analyst must judge the operator's performance against a mental image of how an "average" experienced employee would perform the operation. It is difficult to performance rate an inexperienced or unskilled operator. Wherever possible, study operators who can properly perform the operation and thereby minimize questions relating to proper methods, etc.

Study an adequate number

The number of readings taken for a study will affect the reliability and precision of the data. Generally the greater the number of readings, the more reliable the sample. However, increased readings raise the cost of the study, and beyond a point additional readings are not useful.

The number of readings required for a particular time study may be determined by "feel" based on the analyst's experience, an understanding of the variability of the elements in the study and the consistency of the observed operator's performance. Most experienced time study analysts use this approach.

Should the analyst wish mathematical verification for the number of cycles to take, from probability theory it is possible to determine the number of readings required so that in 95 out of 100 instances the readings are within ± 5% of the average for the element at the observed work pace. A looser requirement would be 95 out of 100 at ± 10%.

A statistical method for determining time study data reliability was developed by Mundel. It works with the range of the sample from high to low and shows the number of time study readings necessary for the study, based on the range from a sample of either five or ten readings. To use Table 1 divide (H − L) by (H + L). H is the highest reading and L is the lowest. Look up the closest decimal in the table and read off the total readings required, based on whether the sample was five or ten readings. For example, if in dividing the H and L factors the answer is 0.16 in a ten-reading sample, the table shows that 17 readings are necessary; therefore take seven more readings to obtain a total of 17.

The element with the highest value in dividing (H − L) by (H + L) determines the number of study readings required. In the example shown it is 0.20; refer to the table and locate 0.20 in the column (H − L)/(H + L). In the column of the ten-piece sample it is seen that the total is 27. Since the other elements require fewer observations, the element with the highest total is the one to check. In this time study the total number of time study observations to take would be 27. Since ten have already been taken, 17 more are needed, taken the same way as the original time study sample.

If this method is followed, the probability is 95 out of 100 that the mean of the 27 readings taken in the above time study will be within ± 5% of the true values of the operation being studied. This does not in any way reflect on the validity of the time standards which are finally established from these observations. For example, if the performance ratings are not valid, the final time standards must be defective, regardless of the validity of the samples.

The actual number of readings taken in practice often depends on the overall cycle time of the operation. If the overall cycle time is one hour, it may be difficult or impractical to take 27 cycles. With cycle times of 0.10 minute, 100 cycles will only take ten minutes, which is quite reasonable. Common sense should be used, keeping in mind that sample reliability is increased with sample size.

Several short studies, preferably of different operators, are more reliable than one long study. Combining the data from several studies to calculate a time standard, or combining data for standard data, accomplishes this result.

Properly treat the data

When time study data is examined by the analyst in preparing to calcu-

Table 1. Number of Readings Required for ±5%; 95/100 Probability*

$\frac{H-L}{H+L}$	Data from sample of 5	Data from sample of 10	$\frac{H-L}{H+L}$	Data from sample of 5	Data from sample of 10	$\frac{H-L}{H+L}$	Data from sample of 5	Data from sample of 10
0.05	3	1	0.21	52	30	0.36	154	88
0.06	4	2	0.22	57	33	0.37	162	93
0.07	6	3	0.23	63	36	0.38	171	98
0.08	8	4	0.24	68	39	0.39	180	103
0.09	10	5	0.25	74	42	0.40	190	108
0.10	12	7	0.26	80	46	0.41	200	114
0.11	14	8	0.27	86	49	0.42	210	120
0.12	17	10	0.28	93	53	0.43	220	126
0.13	20	11	0.29	100	57	0.44	230	132
0.14	23	13	0.30	107	61	0.45	240	138
0.15	27	15	0.31	114	65	0.46	250	144
0.16	30	17	0.32	121	69	0.47	262	150
0.17	34	20	0.33	129	74	0.48	273	156
0.18	38	22	0.34	137	78	0.49	285	163
0.19	43	24	0.35	145	83	0.50	296	170
0.20	47	27						

Example: Suppose you took a time study in which the ten elapsed time values for each of three elements were as follows:

	el 1	el 2	el 3
	0.07	0.12	0.50
	0.09	0.13	0.55
	0.06	0.15	0.53
	0.07	0.16	0.51
	0.08	0.1	0.54
	0.08	0.14	0.49
	0.07	0.12	0.52
	0.08	0.15	0.51
	0.09	0.11	0.55
	0.07	0.12	0.52
H to L	0.09 – 0.06	0.16 – 0.11	0.55 – 0.49
H – L	0.03	0.05	0.06
H + L	0.15	0.27	1.04
$\frac{H-L}{H+L}$	0.20	0.185	0.077
Total readings	27	23	4	Number cycles obtained from above chart

*For ±10%, 95/100 probability, divide above required readings by four.
H = high value   L = low value

late element time values, often high and low element time values are encircled and dropped from the averages; most often high times are eliminated, rarely low values. The rationalization of the analyst is that these high values are not representative of what *should* occur.

One interpretation of the encircled times is that the operator has attempted to fool the analyst and has interspersed several fumbles and excessive times to raise the average time. This game goes on constantly; employees want to increase the time allowed by the standard.

Time study analysts cannot control the actions of employees, and insist that they perform their work to the linking of the analyst. Attempts by employees to mislead the analyst do not justify the handpicking of time study values by the analyst, even though many time study text books contain examples of time study sheets which show encircled time values that were excluded in calculating the average.

Since time study is a sampling procedure, recording the data and calculating the results must follow accepted QC procedures. In quality inspection, after a sample lot has been selected and inspected, an inspector is not permitted to discard some of the rejected or accepted pieces just because these did not look right to him. He could under multiple inspection procedures select an additional sample lot and add those to the original sample lot. He could for valid reasons reject the entire lot and start over. But he could not reject some of the sampled pieces to suit his fancy.

Admittedly time study is not performed as rigorously as quality control inspection. Pieces when inspected are static and do not change. The operator being "inspected" in the time study is quite alive and his performance and methods of work may change from cycle to cycle. Pieces do not attempt to fool the inspector; the operator has every incentive to mislead the analyst.

To protect against instances when the operator may have intentionally fumbled or increased an element time, the analyst should key off such time recordings on the time study form with a letter A, B, C, etc., and then note in the infrequent element box on the sheet what the analyst believed occurred during that element to alter the time. During the calculation of the study, these noted time values should be treated following company or agreed company-union policy for such occurrences. Time values should not be dropped during the calculations just because some values are higher than most of the other values.

The statistical method of determining the number of time study cycles required based on the range of low to high values should take care of random highs and lows. The greater the range, the greater the number of cycle readings necessary for a statistically valid study. It must be remembered that time values are randomly distributed and there will be high and low values around central time values. Some studies may have no central tendency and time values may be scattered with no correlation. Such studies require greater numbers of cycles to assure data reliability.

Where high values result from poor methods, inadequate training or deliberate employee actions, these will partly be reflected in the analyst's performance ratings. Sound practice is not to study such employees or operations.

Calculate the times

The only sound method of calculating a representative time from a column of observed time values is to average the entire column. There is no justification for the arbitrary dropping of some occurrences. The use of the mode or median of a column is not proper because these exclude time values on both sides. It is important to remember that the times in a column are random occurrences and all values must be included in calculating a representative value.

Some analysts will select several element time values from a column of cycle times and performance rate these selected values to determine a standard time. The rationale offered for this method is that since performance rating is a judgment, the analyst should be able to select what he considers a representative time value and then performance rate this value. These analysts may point to their record of few worker grievances, good labor relations and management satisfaction with their standards. I feel that such practice is highly questionable and is not recommended.

The reliability of time study data is increased when a time standard is calculated from the combined data of several studies; generally more data will improve reliability. When combining the data from several studies, especially when the studies were made by different analysts, it is very important that the element breakdowns are the same and conditions are comparable. Practitioners who set standards from a single study are more likely to take short cuts and drop values they believe are not representative.

Data reliability is especially important for long term protection of the data against erosion. This requires that definitions and element descriptions are adequately spelled out. Data files should be orderly and clean; avoid sketchy notes on scraps of paper. All records should be dated and initialed so that several years later someone not familiar with the files will have no difficulty in determining the sequence of the papers, when changes were made and why, and numerous events which caused standards revisions. When deals and compromises are made to settle grievances, the affected standards should clearly note the reasons for the changes to prevent the erosion of other standards not involved in the compromise.

Agreement on standards

The concept of performance rating is always involved in setting time standards, whether or not performance rating is actually used. This occurs when management unilaterally decides on given benchmarks and criteria from which to establish standards, or when management and employees jointly agree on standards and that a given work pace level is acceptable. Using agreed standards or benchmarks as the bases from which to establish other standards also involves the concept of performance rating.

Work measurement users, which include managers, union representatives and workers on the plant floor, usually do not experience difficulty in translating their opinions and concepts of a fair day's work into acceptable time standards. The very fact that millions of standards are in use, involving hundreds of thousands of workers in the United States and elsewhere, attests to the utility of the practices and procedures used to establish standards which are agreed to be fair and equitable. Most differences are settled on the plant floor through well established procedures. When differences are not resolved, most companies and unions agree to impartial and binding arbitration.

Terms denoting degrees of reliability used in physical measurements may not be used in work measurement because performance rating introduces bias which negates the statistical meanings of the terms. While work measurement users and practitioners may feel more comfortable with mathematical processes that have defined reliability, the subjective bases established by work measurement do not permit using such terms. It is improper to attempt to use terms designating degrees of accuracy, such as are used in quality control and probability calculations to cover work measurement with a degree of objectivity which the technique does not have.

Despite the subjective nature of work measurement, it is universally used and agreed to, even under adversary labor-management relations. Benchmarks and guidelines are agreed to in thousands of companies which permit work measurement to operate on an ongoing basis.

For further reading:

Abruzzi, Adam; *Work, Measurement, New Principles and Procedures*; Columbia University Press, 1952.

Abruzzi, Adam; *Work, Workers and Work Measurement*; Columbia University Press, 1956.

Barnes, Ralph M.; *Motion and Time Study*; 7th ed., John Wiley & Sons, 1980.

Carroll, Phil; *Better Wage Incentives*; McGraw-Hill, 1957.

Davidson, Harold O.; *Functions and Bases of Time Standards*; AIIE, 1957.

Eisenhart, Churchill; "Realistic Evaluation of the Precision and Accuracy of Instrument Calibration Systems"; reprinted in *Precision Measurement and Calibration; Statistical Concepts and Procedures*; Vol. 1, National Bureau of Standards, Feb. 1969.

Gomberg, William; *A Trade Union Analysis of Time Study*; Prentice-Hall, 1955.

Morrow, Robert L.; *Motion Economy and Work Measurement*; Ronald Press, 1957.

Mundel, Marvin E.; *Motion and Time Study*; 4th ed., Prentice-Hall, 1970; 5th ed., 1978.

Nadler, Gerald.; *Work Design*; Richard D. Irwin Inc., 1963.

Niebel, Benjamin W.; *Motion and Time Study*; 5th ed., Prentice-Hall, 1978.

Shewhart, W.A.; *Statistical Method from the View of Quality Control*; The Graduate School, Dept. of Agriculture, Washington, DC, 1939.

Mitchell Fein is a consulting IE who primarily deals with solving the people problems that arise between employees and management to retard productivity. Since 1937, he has served over 500 companies in a variety of industries, concentrating on developing ways of increasing productivity. He is a fellow of AIIE and was adj. associate professor of industrial engineering at New York University.

DETERMINATION OF PROBABILISTIC TIME STANDARDS FOR TASKS PERFORMED UNDER UNCERTAINTY

Don T. Phillips
and
Donald R. Smith
Department of Industrial Engineering
Texas A&M University
College Station, Texas

ABSTRACT

This paper presents analytical procedures for the determination of a probabilistic time standard for classes of industrial-type tasks. The approach taken is based upon the assumption that individual components of the standard may exhibit uncertain behavior, and that components of the standard time can be treated as a random variable. The emphasis will be on classes of industrial tasks that involve the human operator performing a sequence of work elements where stochastic events occur. The methodology is illustrated by an example problem, with comparisons made to the application of traditional time study methods.

INTRODUCTION

A basic task of industrial engineering is the determination of the amount of time required under assumed conditions for the completion of work elements involving human activity. In conjunction with this effort, the IE seeks to systematically determine preferable work methods, evaluate the value of work where the human operator is involved and to constantly strive to improve productivity while simultaneously reducing worker effort. This paper is concerned with the problem of analytically determining a "probabilistic time standard" of classes of industrial tasks that involve the human operator performing a sequence of work elements where stochastic events occur. In other words, the operator is involved with a sequence of tasks - some of which involve the worker's individual judgment as to what the next step is to be or situations in which the next step is an "unpredictable" random factor. In the production environment, this situation occurs when the worker is involved with composite tasks, i.e. assembly, self-inspection, rework, sorting, etc. This kind of worker activity is often observed when production tasks are "grouped" as under a job enrichment program or where disabled workers are involved in a vocational rehabilitation setting.

The approach developed in this paper is contrasted with "conventional" time study (stopwatch or predetermined methods). These methods - while still viable and productive approaches - are somewhat limited when the production tasks involve stochastic or unpredictable occurrences. In addition, the conventional approach collapses all variable data to a single time estimate. This approach tends to omit (or ignore) random influences which may occur. The fundamental premise of this paper is that a standard time should be viewed as a random variable, possessing a finite mean and variance, and described by some relevant probability distribution function. In summary, methodologies are illustrated which allow an analyst to consider probabilistic behavior of work elements and uncertainty in task results. These results form an operational framework from which stochastic time standards can be determined directly, suitable confidence limits be established on a task structure, and the mean and variance of the task sequence accurately determined.

To illustrate these techniques, a hypothetical example involving random and deterministic elements is presented and analyzed via four different approaches: 1) conventional analysis, 2) a brute force solution, 3) Markov Chain analysis and 4) GERT methodologies. The GERT approach is found to derive the greatest information about the variability of the time standard.

CLASSICAL WORK MEASUREMENT

The time dimension of a work element is a major input in determining the cycle time for any type of operation. In the manufacturing environment, a worker's motion times, machining times and appropriate allowances are added to derive the standard time for a given task. Because of the areas of application, the amount of information obtained from time data may vary according to their needs. For instance, only the expected cycle time value plus certain allowances are necessary to set a piece payrate for workers who are paid under a piecerate incentive system. However, the variance of time is as important as the mean value in planning and scheduling activities. Most of the machining times such as cutting, drilling, lathe and milling operations are treated as constant as suggested by Barnes [1]. Standard data and nomographs have been developed to obtain the machining times for various operations. But even for a basic machining operation, the time value will be governed by random factors such as machine failure, tool wear, power failure, etc.

The current popular work measurement systems include work sampling, stop watch study and the use of predetermined time systems (e.g. MTM, Work Factor, MOST). Each system has its advantages and disadvantages in terms of attempting to set a standard time. A brief overview of each system will now be presented.

Work Sampling

The main application of work sampling is to measure the content of work or rather the frequency of occurrence of work based on categories of work that have been determined by prior analysis. Under certain circumstances, work sampling can be used to establish a time standard for certain operations. Barnes [1] suggested that for short cycle repetitive operations, time study and predetermined time methods may be preferred.

Classical Time Study

Time study by means of stopwatch or other timing devices is the most widely utilized technique in measuring cycle time. Stopwatch studies are conducted on jobs that are in existence and where the worker being studied is considered to be properly trained and equipped. A major criticism of conventional stop watch study is the need for subjective leveling of the worker being studied by the time study analyst.

Predetermined Time Systems (PTS)

Predetermined time systems are gaining in popularity in that it is possible to establish a reasonably accurate time standard without the need to study an operator by stopwatch. To apply a PTS, one needs to define the task in terms of motions, for example, reach 10", grasp 1 pound object, transport 12" to the work station. Given the basic motions, distance, weights, etc., one can assign times to the motions consistent with the PTS being used. One advantage of using PTS is that the resultant times are pre-leveled and minimizes the subjective judgment as required by classical time study. One disadvantage of this system is that the analyst needs to be well trained in the selection system. For example, to be an "expert" in MTM-1, one requires approximately 300 hours of concentrated training. Other systems such as MODAPTS and Brief Work Factor (BWF) require from 40 to 80 hours of training.

Predetermined time systems are based on the assumption that a given motion is independent of the basic motion preceding or following it and thus the times are additive. The validity of this assumption has been questioned by Nanda [5] and Niebel [6]. All predetermined systems to date are based solely on the expected or average value of the motion performance. The time value for a given operation or work element is treated as constant, where in reality, they are better represented by random variables as suggested by Dudley [2] and Hicks [3].

As previously discussed, certain industrial tasks exist that contain random or unpredictable events that can significantly alter the process time. Classical time study approaches attempt to handle such events by estimating this frequency of occurrence and allocating an amount of time to the normal time. While the process of allocating the times of certain predictable events is commonly done, the classical approaches have difficulty in handling random or unpredictable events that may occur.

The next section presents a hypothetical problem involving stochastic events. In this problem it is assumed that a single operator is required to assemble and test a printed circuit board composed of six electronic components. The assembly process is comprised of deterministic events (inspection, insert and solder components, etc.) and stochastic events (failure to pass a test, removal of previously assembled components, rework, etc.). The problem is formulated and analyzed via 4 different approaches.

ESTABLISHING A STANDARD TIME
INVOLVING STOCHASTIC ELEMENTS

Consider a single operator assembling, inspecting and testing components that are inserted in a printed circuit board. In this process, six electronic components are inserted and soldered into the board. The process is comprised of two stages. In the first stage, three of the components are inserted into the board, the leads are cut to length if necessary, and soldered by hand to the back of the board. The three components are tested by VOM (Volt-Ohm-Milliamp meter) for continuity and to determine if all solder points are good. If the operator detects any failure, the appropriate remedial steps are taken. For example, one, two or three of the components could fail under the test. When detected, the operator might be faced with replacing one or more components.

Figure 1 illustrates the sequence of operations, inspections and operator controlled decision points for the example problem. The process consists of eight operations (designated 0-1,...,0-8), five inspections (I1,...,I5) and five decision points (D-1, ...,D-5). The relevant processing times in minutes per board are shown in parenthesis in Figure 1. Note that decision points D-2, D-3 and D-4 result in probabilistic events. For example, decision D-2 requires the operator to perform a visual inspection of the leads. Past experience has shown that in 3 percent of these inspections, problems exist with the leads which require an adjustment. Thus, if the operator determines that an adjustment is warranted, he will proceed with operation two-"leads adjustment". Records have shown that the probability of one lead requiring adjustment is 0.7; two leads-0.2; and three leads-0.1. Operations 0-4 and 0-7 exhibit the same behavior and are depicted in a similar manner.

Assuming the first stage of assembly inspection and testing is satisfactory, the operator proceeds to the second stage. In this stage, three more components are installed and the same process of inspection and testing is completed. However, due to the complexity in diagnosing the causes of failure, it is assumed that the entire assembly will be discarded if the VOM test yields one or more failures. The remaining discussion will assume that the assembly process is performed totally by a single worker. Since the above process involves stochastic events - inspection and testing, the establishment of a standard time could become a difficult task.

Procedure I: Conventional Time Standards

For illustrative purposes, assume that the instability and stochastic behavior of the total task at points of inspection are ignored. In other words,

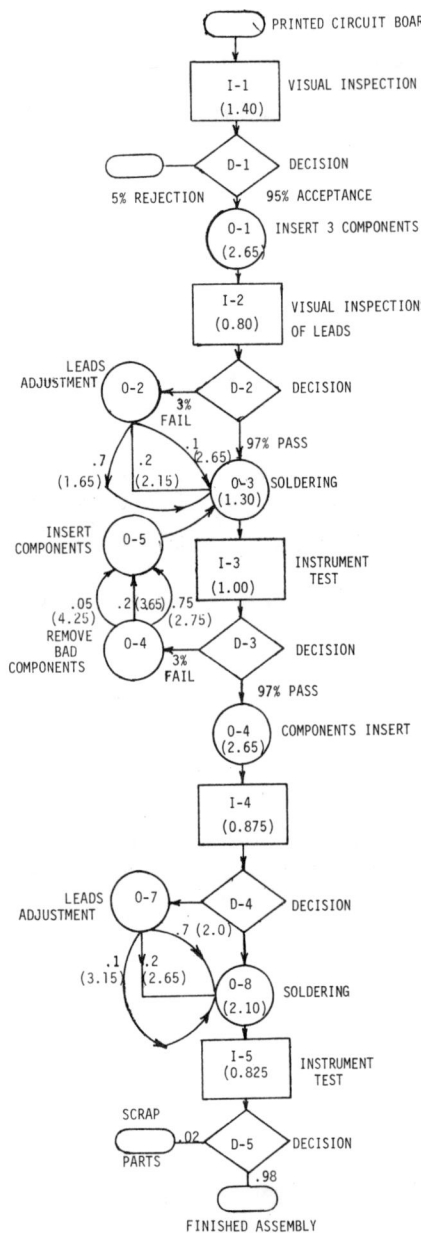

Figure 1. Stochastic Assembly Chart

the only times included in the sequence of tasks given in Figure 1 are those associated with the vertical sequence of activities on the main branch. Utilizing conventional additive time elements, the task standard of 13.60 minutes per circuit board is easily verified. If one assumes a 20% allowance for this type of work, the resultant standard would be 16.32 minutes per board. Clearly, this result is a simplified approximation of the standard time. The next method will incorporate the probabilistic behavior of the inspection processes and their associated delays.

Procedure II: A "Brute Force" Approach

The simplification of Procedure I essentially ignores the effects of three feedback loops: one at the visual inspection of the board, a second at the first instrument test, and a third at the second instrument test. Of secondary importance in terms of impact are the two feedforward loops at the first and second visual inspection of the leads. The feedforward loops are easily handled, but the feedback loops require further consideration. The central question is, "how many times will each feedback loop be taken?". Once known, the times which are involved (repeated) can be adjusted (inflated). Consider the following feedback loop.

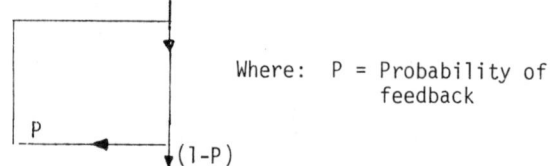

Where: P = Probability of feedback

The expected number of times through this loop is given by:

$$T = \sum_{j=1}^{\infty} j \, P^{j-1} \qquad (1)$$

One can easily verify that this is a convergent series whose closed form solution is:

$$T = \frac{1}{1-P} \qquad (2)$$

For example, the first visual inspection has a probability of failure P = 0.05. Hence, one can expect $T = (1-.05)^{-1} = 1.05263$ inspections per arriving printed circuit board. From the last instrument test, the expected number of boards subject to the final test are $T_F = (1-.02)^{-1} = 1.02041$. Hence, on the average $1.02041 \cdot (1.05263) = 1.074112$ boards pass through the first inspection. Using the above formula, one can verify that on the average, 1.03093 boards pass through soldering and the second instrument test. The standard time associated with the entire sequence of tasks, considering feedback and feedforward loops, is calculated in Table 1. The adjusted (total) standard time is 14.03171 minutes. Using the same 20% allowance factor, the standard becomes 16.8381 minutes per board.

Procedure III: A Markov Chain Approach

Procedure II can be utilized anytime feedforward or feedback loops are present in the sequence of tasks which constitute a standard procedure. However, if there are several loops with many branch points, the procedure can be tedious. An alternate method which will produce the same results under completely general structures is that of Markov chains. Since the total sequence of tasks ultimately terminates, the process is characterized as an <u>absorbing</u> Markov chain. A complete theory can be found in Phillips, Ravindran and Solberg [7]; for purposes of this discussion the following definitions will suffice.

Table 1. Adjusted Times for the Example Problem

Task	Time	Adjustment Factor	Adjusted Time
Visual Inspection	1.40	1.07411	1.50376
Component Insert	2.65	1.02041	2.70409
Inspect Leads	.97(.75)+.03(.85)	1.02041	0.76837
Leads Adjustment	.70(1.65)+.2(2.15)+.1(2.65)	0.03061	0.05663
Soldering	1.30	1.05197	1.36756
Instrument Test	0.90	1.02041	0.91837
Fail	3.005	0.03156	0.09983
Insert	1.10	0.3156	0.03472
Component Insert	2.65	1.02041	2.70409
Inspect Leads	.96(.8)+.04(.95)	1.02041	0.82245
Adjust Leads	2.218	.04082	.09164
Soldering	2.10	1.02041	2.14286
Instrument Test	.80	1.0	0.80
	.85	0.02041	0.01435
		TOTAL:	14.03171

The <u>state</u> of the Markov chain will be the particular task being conducted. A <u>transition</u> from state i to state j will occur with probability P_{ij}. A state which can only be entered, but not left, is called an <u>absorbing</u> state. Define the <u>Transition Matrix</u> as the matrix which exhibits all transition probabilities from each state to every other state. For an absorbing Markov chain, this transition matrix can always be written in the following form.

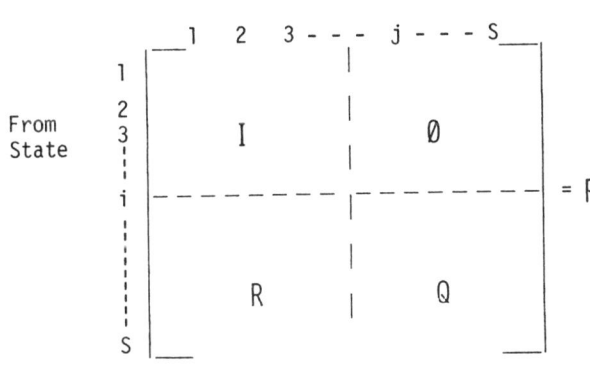

P is a square matrix which contains an <u>Identity</u> sub-matrix (I) in the upper left-hand corner, a <u>null</u> sub-matrix in the upper right-hand corner (∅) and sub-matrices of nonzero entries (R) and (Q). The I sub-matrix is generated by the absorbing state probabilities; the ∅ matrix by states which cannot be reached from absorbing states; the R matrix by transition probabilities, P_{ij}, from nonabsorbing to absorbing states; and the Q matrix by transition probabilities P_{ij} from nonabsorbing to nonabsorbing states.

Define the matrix E as the inverse of the difference between the Identity sub-matrix I and the transition sub-matrix Q, specifically,

$$E = (I-Q)^{-1} \qquad (3)$$

which provides the <u>mean number of times</u> that each state will be visited prior to absorption starting at <u>any state</u>. In other words, the c_{ij}th entry in the E matrix gives the mean number of times that the state j (row entries) will be visited given that the process started in state i (row label). Based upon the Stochastic Process Chart (Figure 1), define the states as shown in Table 2.

Table 2. State Assignments for the Sequence of Operations and Inspections

State	
1	Visual Inspection #1
2	Component Insert #1
3	Visual Inspection #2
4	Leads Adjustment #1
5	Soldering #1
6	Instrument Test #1
7	Remove and Replace Components
8	Component Insert #2
9	Visual Inspection #3
10	Leads Adjustment #2
11	Soldering #2
12	Instrument #2
13	Finished Product

The Transition matrix P is shown in Table 3.

Table 3. Transition Matrix

From State \ To State	1	2	3	4	5	6	7	8	9	10	11	12	13
1	.05	.95											
2			1										
3				.03	.97								
4					1								
5						1							
6							.03	.97					
7				1									
8									1				
9										.04	.96		
10											1		
11												1	
12	.02												.98
13													1

Note that state 13 is considered to be an <u>absorbing state</u>. (Finished Product). Although P could be formed by rearranging this matrix, it is clear that since the matrix Q is the transition probabilities of each nonabsorbing state to the other nonabsorbing states, Q is given by the first 12 rows and columns of the Transition Matrix. For purposes of setting time standards, if the final state is "finished product" then the first (S-1) rows and columns will <u>always</u> form Q. Hence;

Table 4. The E Matrix

$$[I-Q]^{-1} =$$

	1	2	3	4	5	6	7	8	9	10	11	12
1	.95	-.95										
2		1										
3			1	-.03	-.97							
4				1	-1							
5					1	-1						
6						1	-.03	-.97				
7							1					
8								1	-1			
9									1	-.04	-.96	
10										1	-1	
11											1	-1
12	-.02											1

Note that since the sequence of tasks is defined to start in State 1, the only entries of interest in $[I-Q]^{-1}$ are those of the first row. One can verify that the first row of $[I-Q]^{-1}$ is given by:

	1	2	3	4	5	6	7	8	9	10	11	12
1	1.0741	1.0204	1.0204	0.0306	1.002	1.052	0.0316	1.0204	1.0204	0.0408	1.024	1.024

The standard time is recovered as shown in Table 5.

Table 5. Expected Times

Process	Expected Number per Finished Item	Mean Time of Operation	Expected Time per Finished Item
Visual Inspection	1.07411	1.40	1.50375
Component Inserting	1.02041	2.65	2.70409
Successful Inspection of Leads	.98980	0.75	.74235
Failing Inspection of Leads	.03061	0.85	.02602
Leads Adjustment	.03061	1.85	.05663
Soldering	1.05197	1.30	1.36756
Pass on Instrument Test	1.02041	0.90	.91837
Fail on Instrument Test	.03156	1.10	.03472
Remove & Replace Component	.03156	3.005	.09983
Inserting Component	1.02041	2.65	2.70409
Successful in Inspecting Leads	.97959	.80	.78367
Fail in Inspecting Leads	.04082	.95	.03878
Adjust Leads	.04082	3.305	.09164
Soldering	1.02041	2.10	2.14286
Pass on Instrument Test	1	0.80	.80
Fail on Instrument Test	.02041	0.85	0.1735
		TOTAL:	14.03171

This is, of course, the same result obtained previously by Procedure II. Although complex at first exposure, the entire sequence of calculations is easily mastered with a little practice, and the procedure is easily computerized.

Procedure IV: GERT Analysis

It is evident that both Procedure II and Procedure III correct the simplifying assumptions of Procedure I, but all of the preceding results fail to characterize the real problem in setting accurate time standards - the <u>total</u> impact of random variation. Clearly, all three approaches only generate the expected value of the standard which is then adjusted to "account" for variation. A desirable procedure would be one that not only generates the expected value of the standard but the variance as well. Such a procedure exists, based upon <u>GERT</u> (Graphical Evaluation Review Technique [10]).

In order to generate variance estimates, assumptions must be made regarding the stochastic behavior of each element (Task) conducted in the standard. Although this is a more complex description of a task, it is clearly more accurate than a single time. Except for highly automated processes, the time to perform an activity is always a random variable. Table 6 defines the mean and variance for each element of the standard. For this example, the stochastic behavior of each element could be described by a normal density function. However, in the discussion which follows, the time to perform a given task can be <u>any</u> continuous density function with defined moments about the origin.

The GERT methodology was developed to study <u>semi Markov</u> processes. A semi-Markov process is one which adheres to all the assumptions of a Markov process, except the transition time from one state to the next is described by a known probability density function. In essence, the GERT methodology allows one to generate the expected value and the variance of a

Table 6. Branch Parameters for GERT Network of the Assembly Operation

from/to	Description	Prob.	Time Distribution	Mean	Variance
1 / 2	W_1; Success on inspecting circuit board	0.95	normal	1.40	0.35
1 / 1	W_{24}; Fail on inspecting circuit board	0.05	normal	1.40	0.35
2 / 3	W_2; Inserting components	1.0	normal	2.65	0.30
3 / 5	W_3; Success on inspecting leads	0.97	normal	0.75	0.20
3 / 4	W_4; Fail on inspecting leads	0.03	normal	0.85	0.25
4 / 5	W_5; Adjust component; one component's leads unqualified	0.70	normal	1.65	0.35
4 / 5	W_6; Adjust components; two components leads unqualified	0.20	normal	2.15	0.40
4 / 5	W_7; Adjusted components; three components' leads unqualified	0.10	normal	2.65	0.55
5 / 6	W_8; Soldering operation	1.0	normal	1.30	0.10
6 / 8	W_9; Pass on instrumental test	0.97	normal	0.90	0.25
6 / 7	W_{10}; Fail on instrumental test	0.03	normal	1.10	0.30
7 / 6	W_{11}; Remove and replace component; ore component failed	0.75	normal	2.75	0.55
7 / 6	W_{12}; Remove and replace two components	0.20	normal	3.65	0.68
6 / 6	W_{13}; Remove and replace three components	0.05	normal	4.25	0.79
8 / 9	W_{14}; Inserting components	1.0	normal	2.65	0.30
9 / 11	W_{15}; Success on Inspecting leads	0.96	normal	0.80	0.25
9 / 10	W_{16}; Fail on inspecting leads	0.04	normal	0.95	0.30
10 / 11	W_{17}; Adjust component; one component's leads unqualified	0.70	normal	2.0	0.5
10 / 11	W_{18}; Adjust components; two component's leads unqualified	0.20	normal	2.65	0.60
10 / 11	W_{19}; Adjust components, three component's leads unqualified	0.10	normal	3.15	0.75
11 / 12	W_{20}; Soldering operation	1.0	normal	2.10	0.40
12 / 13	W_{21}; Pass on instrumental test	0.98	normal	0.80	0.15
12 / 1	W_{22}; Fail on instrumental test	0.02	normal	0.85	0.15

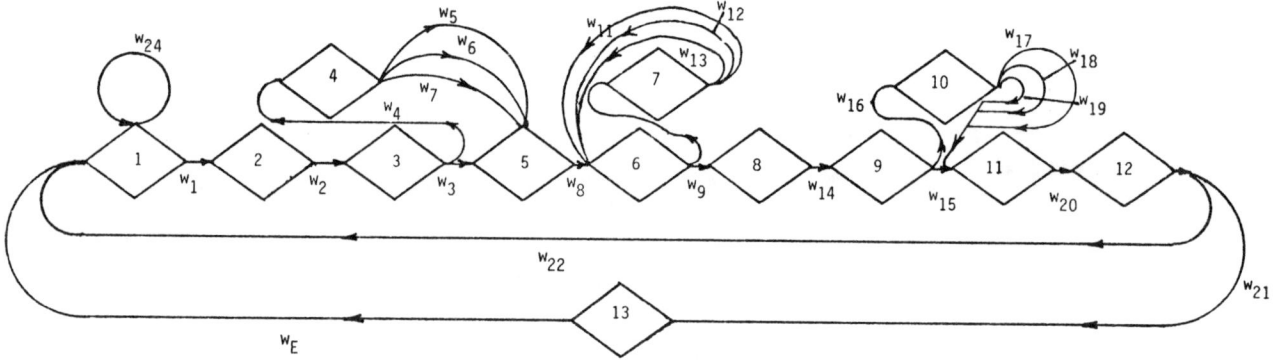

Figure 2. The GERT Network of the Example Problem

Table 7. GERT W-Functions for the Example Network

Start	End	Branch	P_{ij}	W-Function
				$W = P(\text{Exp } \mu t + t^2/2 \; \sigma^2)$
1	2	1	.95	.95 Exp(1.4t + .175t²)
2	3	2	1.0	1.0 Exp(2.65t + .15t²)
3	5	3	.97	.97 Exp(.75t + .10t²)
3	4	4	.03	.03 Exp(.85t + .125t²)
4	5	5	.70	.70 Exp(1.65t + .175t²)
4	5	6	.20	.20 Exp(2.15t + .20t²)
4	5	7	.10	.10 Exp(2.6t + .275t²)
5	6	8	1.0	1.0 Exp(1.3t + .05t²)
6	8	9	.97	.97 Exp(.9t + .125t²)
6	7	10	.03	.03 Exp(1.1t + .15t²)
7	8	11	.75	.75 Exp(2.75t + .275t²)
7	8	12	.20	.20 Exp(3.65t + .34t²)
7	8	13	.05	.05 Exp(4.25t + .395t²)
8	9	14	1.0	1.0 Exp(2.655 + .15t²)
9	11	15	.96	.96 Exp(.8t + .125t²)
9	10	16	.04	.04 Exp(.95t + .15t²)
10	11	17	.70	.70 Exp(2t + .25t²)
10	11	18	.20	.20 Exp(2.65t + .325t²)
10	11	19	.10	.10 Exp(3.15t + .378t²)
11	12	20	1.0	1.0 Exp(2.1t + .20t²)
12	13	21	.98	.98 Exp(.8t + .075t²)
12	1	22	.02	.02 Exp(.85t + .075t²)
13	1	23	1.0	$1.0/W_E$
1	1	24	.05	.05 Exp(1.4t + .175t²)

sequence of interrelated tasks (serial, feedback, or feedforward components). The theory of GERT development is well documented [9], [10], [12] and will not be treated in this paper. Only formulas and results central to this application will be presented.

The GERT methodology requires that the sequence of events which constitute a task structure be represented as a network diagram. The network will be constructed from <u>nodes</u> which represent the beginning and ending points of all individual tasks and <u>arcs</u> which represent the actual time duration of each task. Attached to each arc is an <u>arc parameter</u> called a <u>W-Function</u> which is defined as the product of the probability the arc will be traversed (the transition probabilities of Procedure III) and the <u>Moment Generating Function</u> (MGF) of the Probability Density which describes the task time. Every continuous density function possesses a unique MGF. Referring to Table 1, assume that every time element in our example is <u>Normally Distributed</u>. Although this assumption might seem unrealistic, the selection of a normal distribution for component tasks is a reasonable choice as defended by Hicks [3], Dudley [2] and Thomas [13]. The MGF for a normal density is given uniquely in terms of μ and σ^2 as follows.

$$\text{MGF} = \text{EXP}[\mu t + (1/2)t^2\sigma^2] \quad (4)$$

Accordingly, the W-function for each activity is defined as follows:

$$W_k = P_{ij} \cdot \text{MGF}_{ij} \quad (5)$$

Where: i = Start Node
j = End Node
k = arc identifier

Figure 2 is the GERT network for the example problem. The associated W-Functions are given in Table 7. Note that an additional arc has been added from Node 13 to node 1. This arc is necessary to form a <u>closed loop</u>, which is required for analytical solution. The W-Function will be <u>defined</u> for this closed loop as $(1/W_E)$. W_E is the <u>equivalent</u> W-Function for the <u>entire sequence of tasks</u>. Since $P_{13,1} \equiv 1$, then $W_E = (P_{13,1})(\text{MGF}_{13,1}) = \text{MGF}$. Hence, if W_E can be calculated, then <u>all</u> central moments relating to the standard time can be recovered, which is turn will generate the mean and variance of the standard time - our end objective.

Once the GERT network is constructed and the W-Functions defined, the mean and variance of network completion time (the standard time) can be recovered in two steps.

Step 1: Combine all W-Functions through use of a single equation known as the Network Analysis Equation, given by:

$$H = 1 - \sum_{j=1}^{J} L_{1j} + \sum_{k=1}^{K} L_{2k} - \sum_{\ell=1}^{L} L_{3k} + - + - + \sum_{m=1}^{M} (-1)^{i} L_{im} \quad (6)$$

Define L_{im} as the $m\underline{th}$ loop of order i. A first order loop (L_{1m}) is one in which a consecutive path of arcs leave one node and return to that node without retracing a single arc. L_{1m} is calculated by multiplying together all the W-Functions associated with each arc in the loop. A Second Order Loop is the Product of two disjoint (nontouching) first order loops. In general, an ith order loop is defined to be i-nontouching first order loops, and L_{im} is calculated by the product of all first order L_m functions composing L_{im}. For example, using Figure 2, a first order loop is given by $L_{11} = W_{10}W_{11}$, another by $L_{12} = W_{24}$, and a third by $L_{13} = W_1W_2W_3W_8W_9W_{14}W_{15}W_{20}W_{22}$. One can verify that there are 18 first order loops in Figure 2. Note that since L_{11} and L_{12} are nontouching (no nodes in common) a second order loop is defined by $L_{21} = L_{11} \cdot L_{12} = W_{10}W_{11}W_{24}$. One can also verify that there are only three second order loops in Figure 2. There are no loops of third order or higher. Hence, in summation form Equation 6 is applied to the example in the following form.

$$H = 1 - \sum_{i=1}^{18} L_{1i} + \sum_{i=1}^{3} L_{2i} = 0 \quad (7)$$

Step 2: Solve the Network Analysis Equation for W_E, the equivalent W-Function for the entire standard.

In this example, all 18 first order and the three second order loops are represented in terms of activity W-Functions in Equation 7. Solving for W_E in Equation 7, one obtains after some simplification,

$$W_E(t) = \frac{(W_1W_2W_8W_9W_{14}W_{20}W_{21})[W_3+W_4(W_5+W_6+W_7)][W_{15}+W_{16}(W_{17}+W_{18}+W_{19})]}{1-(W_1W_2W_8W_9W_{14}W_{20}W_{22})[W_3+W_4(W_5+W_6+W_7)][W_{15}+W_{16}(W_{17}+W_{18}+W_{19})]-W_{24}-W_{10}(W_{11}+W_{12}+W_{13})(1-W_{24})} \quad (8)$$

Note that $W_E(t)$ is a moment generating function since $W_E(t) = MGF_{13,1}$. Since it is a moment generating function, expressed in terms of only the transformation variable "t", the first two central moments μ_1 and μ_2 about the origin can be obtained by differentiating the expanded form of the right-hand side of Equation 8 with respect to t and evaluating the first and second derivatives at $t \equiv 0$ to obtain μ_1 and μ_2 respectively. Since μ_1 is actually the expected value of the standard, and $\mu_2-[\mu_1]^2$ is by definition the variance of the standard, the result we seek is achieved. After considerable numerical calculations, one can verify that μ_1 = 14.032 minutes and σ^2 = 7.15 minutes.

Note that this result provides a great deal of useful information about the standard. Using Chebechevs inequality, we can assert that the actual processing time per part will vary between 6.104 minutes and 21.96 minutes over 99% of the time. In this particular example, much stronger statements can be made. Since all component times are normally distributed, then the standard time is also normally distributed. With this knowledge, exact probability statements concerning the time to completion can be made. In addition, if a standard is set against this result, accurate hypothesis tests governing changes in standards can be run to maintain standards in the future. It is also instructive to note that the GERT approach presents a unique alternative to the traditional processes of setting time standards. Using conventional methods, the individual elements of time are treated as constants, and after summation, an adjustment is made to simulate random fluctuations or absorb imbalances in the actual processing times. On the other hand, the GERT approach incorporates random variation and uncertainties directly into each time element. Hence, the resultant standard already includes all random imbalances and needs no further adjustment except for the appropriate allowances. A bonus is the by-product of the amount of variance in the standard, enabling construction of confidence intervals on the standard time.

Finally, although the calculations presented in alternative IV were complicated and time consuming, hand calculations are entirely unnecessary. The GERT numerical procedure has been coded in FORTRAN-IV and is maintained/distributed by Pritsker and Associates [8]. Hence, alternative IV is executed routinely by simply preparing standard FORTRAN data cards for any continuous density functions.

SUMMARY AND CONCLUSIONS

Three procedures have been presented which can be used to obtain standard times for processes which exhibit random behavior and probabilistic uncertainty. All approaches correctly generate the expected value of the standard time. It was also shown that the GERT approach can be used to generate time standards and produce both the mean and variance of the standard. When each time element is normally distributed, it is also possible to describe the exact distribution of this standard time; facilitating exact probability statements, confidence intervals, and hypothesis tests. It is believed that the GERT approach is unique to the process of setting accurate time standards, and its benefits suggest that it should be closely examined.

REFERENCES

[1] Barnes, Ralph M. *Motion and Time Study*. New York: John Wiley & Sons, Inc., 1968.

[2] Dudley, N.A. "Work Time Distribution," *International Journal of Production Research*. Vol. 1, No. 2, 1963.

[3] Hicks, C.R. and Young, H.H. "A Comparison of Several Methods for Determining the Number of Reading in a Time Study," *The Journal of Industrial Engineering*, March-April, 1962.

[4] Karger, Belman W. and Bayta, Franklin H. *Engineered Work Measurement*. New York: The Industrial Press, 1966.

[5] Nanda, R. "Developing Variability Measures for Pre-determined Time Systems," *The Journal of Industrial Engineering*. January, 1967.

[6] Niebel, Benjamin W. *Motion and Time Study*. Homewoods, Illinois: Richard D. Irwin, Inc. 1976.

[7] Phillips, D.T., A. Ravindran, and J.J. Solberg, *Operations Research: Principles and Practice*, John Wiley and Sons, Inc., New York, N.Y.

[8] Pritsker and Associates, West Lafayette, Indiana - private consulting firm.

[9] Pritsker, A.A.B. and W.W. Happ, "GERT: Graphical Evaluation and Review Technique, Part I, Fundamentals," *The Journal of Industrial Engineering*, May, 1966.

[10] Pritsker, A.A.B. and G.E. Whitehouse, "GERT: Graphical Evaluation and Review Technique, Part II," *The Journal of Industrial Engineering*. June, 1966.

[11] Raouf, A. and A. Ela Maged, "Human Performance Prediction for Inspection Type Task," *Management Service*. Vol. 21, No. 10, 1977.

[12] Raouf, A. and M.L. Mehra, "Experimental Investigations Related to Combined Manual and Decision Tasks," *International Journal of Production Research*. Vol. 12, No. 3, 1974.

[13] Thomas, M.U., W.M. Hancock, and D.B. Chaffin, "Performance of a Combined Manual and Decision Task with Discrete Uncertainty," *International Journal of Production Research*. Vol. 12, No. 3, 1974.

[14] Whitehouse, Gary E., *System Analysis and Design Using Network Techniques*. Englewood Cliffs, N.J.: Prentice-Hall, Inc., 1973.

BIOGRAPHICAL SKETCH

Dr. Don T. Phillips is a Professor of Industrial Engineering at Texas A&M University. He is a member of ORSA, TIMS, and AIIE, and serves on the Editorial Board of the *AIIE Transactions*. His publications number over fifty technical articles and four books, including the Book-of-the-Year Award for his textbook; *Operations Research: Principles and Practice*. Dr. Phillips has served for a number of years as a consultant to both government and industry, and is a registered Professional Engineer in the State of Texas. He is actively directing research in the fields of materials handling, digital simulation techniques and network flow analysis.

Dr. Donald R. Smith is an Associate Professor of Industrial Engineering at Texas A&M University, College Station, Texas. Dr. Smith has previously served on the faculty of the University of Arkansas and Louisiana Tech University at Ruston and worked as a staff engineer for Phillips Petroleum and the Sun Oil Company. He is a senior member of AIIE, Alpha Pi Mu and a registered Professional Engineer.

SPECIAL REPORT

Work measurement: The flap over MIL-STD-1567 (USAF)

"Experience has shown that excess manpower and lost time can be identified, reduced, and continued method improvements made regularly where work measurement programs have been implemented and conscientiously pursued."

"The question is not the adequacy of proposed MIL-STD-1567, the question is whether any customer, including the Government, has a right to coerce private industry by a system of checks and balances on their internal management practices."

CHARLES H. BOYER
Assistant Editor, Industrial Engineering

A year ago last June, the United States Air Force published a Military Standard (1567) establishing minimum requirements for mandatory work measurement programs in the private aerospace industries. This standard would affect, generally, all companies working under major Air Force weapons production contracts — $20 million annually, or $100 million total, and development contracts totalling $100 million.

To date, the standard has not gone into effect because it has not been applied to any contracts. It is anticipated that it will be applied to the new B 1 bomber, should those contracts be forthcoming.

Even before the document was officially released on June 30, 1975, questions concerning how it should be implemented, and, perhaps more importantly, should it be implemented, were being raised by the people ultimately responsible for work measurement — industrial engineers. Industry representatives, divided amid a spectrum of cries ranging from "deterrent to free enterprise," "galloping socialism," and "cost prohibitive-administrative monster," on one side, and "long overdue," "badly needed-necessary and acceptable," on the other, Reference [1], began a series of panel discussions, presentations, and informal talks with the Government, namely personnel from the Air Force Systems Command, their Contract Management Division, and Air Force Procurement Officers.

This continuing dialogue has been centered on the standard's three primary requirements:

1. A plan to achieve 80% coverage of direct manufacturing (touch) labor with engineered standards, with sufficient data to show an accuracy of at least 25% at a 90% confidence level,

2. Written variance analyses of labor performance reports, and

3. A system of audits to verify coverage and accuracy. (The audits to be available for Government inspection.)

All(?) industrial engineers. Of real interest to industrial engineers outside the aerospace industry is a complex problem raised in arguments both for and against the MIL-STD. One industrial engineer with many years of experience in the aerospace industry explains it this way: "The very important issue here, particularly as it relates to industrial engineers both inside and outside of the aerospace industry, and particularly in view of the increasing number of people affected by government-contract business, is the question of **how do you control costs in an environment that is substantially segregated from the normal pressures of the free market place.**"

The dialogue concerning the benefits and detriments of MIL-STD-1567 provides a cogent and fertile background for exploring answers to this question. Experts representing the major parties involved with the MIL-STD were asked to submit their views. Their report:

Behind the MIL-STD

Donald J. Heacox
Major, USAF
Chief, Manufacturing Support Division
Manufacturing Operations Directorate
HQ, Air Force Contract Management Division

During the last ten years all aspects of the procurement and production of new weapon systems have been studied. The motivation for these studies has been concern over the increasing costs of new weapon systems. The MIL-STD-1567 (USAF), Work Measurement, is but one manifestation of these studies.

Let's review some milestones on the evolutionary path that eventually led to the military standard. In 1970, The Manufacturing Committee of the AIA (Aerospace Industries Association) conducted a survey and recommended increased use of work measurement to achieve cost reductions, see box, "AIA Labor Productivity Report." Also in 1970, the Lyons Study found that "Contractual instruments have not contained definitized requirements for effective control of the production process.

As a result of the Lyons Study, MIL-STD-1528 (USAF), Production Management, was created. One unelaborated requirement of this standard was for the "Maintenance of a work measurement program." The contract for the production of the A-10 aircraft contained a somewhat more elaborate requirement for work measurement.

In 1972, the Sagamore Study found 45% of the time charged against aircraft 1,000 was unproductive. This influenced Project ACE (for Acquisition Cost Effectiveness) which tasked Air Force Contract Management Division (AFCMD) with drafting a military standard on work measurement. After several revisions and dialogue with industry and professional association people, the standard was published with a date of June 30, 1975, without the complete concurrence of industry.

One of the problems

The aerospace business has become highly structured. There appear to be barriers to entering and to leaving the business. The market does not appear to be sufficiently attractive to induce major new participants. The degree to which the Government absorbs risks and supports the existing large companies helps keep them in business. Efforts in the past to "disengage" the Government from monitoring contractors after award have often been disappointing. **Normally, the relaxation of government scrutiny has created an atmosphere in which more and greater problems were produced than when the Government was "engaged."**

In essence, the environment has not been conducive to cost control. The ability of aerospace contractors to control costs may have eroded. For this, neither Government nor contractors are to blame. This environment is a logical result of the past emphasis on technological performance rather than effective cost controls.

The benefits

There are many benefits from a disciplined work measurement system, particularly in the area of planning, scheduling, and loading.

The magnitude of the savings potential from a disciplined, as opposed to undisciplined, system is largely judgmental. There are several specific examples, however, that support a savings potential of from 10 to 20% of the direct labor measured. One contractor revitalized his work measurement system and improved his shop productivity over 20% in ten months. The burdened value of this savings was an estimated $2.8 million. Another example is the North Island Naval Air Rework Facility. According to a GAO study, they improved their direct labor performance by 10 to 15% by converting estimates to engineered standards. The unburdened value of this savings was estimated to be $3.7 million.

The costs

Work measurement programs cost money. There is no doubt about that. There are costs associated with applying standards, maintaining standards, developing standard data, if needed, and administering and operating the program and the audits. How much this costs depends on a number of factors. Usually, a range of 1 to 4% of the direct labor hours measured should bracket the cost of such a program. For an absolutely new program, the range might be as high as 3 to 7%.

The standard is designed to be applied to contractors who already have work measurement systems. Therefore, the extent to which additional costs would be incurred indicates the extent of potential savings. **The military standard did not create the need for work measurement. The need for work measurement is**

Courtesy McDonnell Douglas

AIA Labor Productivity Report

In 1970, the Manufacturing Committee of the Aerospace Industries Association of America, Inc., sent out a questionnaire to 53 companies. Their objective was to develop a source of information for reviewing "aerospace manufacturing labor planning and control from a time standard base." Forty-seven of the questionnaires were returned. This is what they found:

1. Almost universal use of standards for labor planning and control of production labor — nearly two-thirds reported using standards for measuring some portion of indirect labor.
2. Ratios of savings to costs vary from 2:1 to 5:1.
3. Companies reporting the greatest savings tended to measure by group and to use standard time data.
4. Most common frequency for submitting labor performance reports is weekly.
5. Allowances for personal fatigue and delay time average around 13 percent — none reported lower than 5 percent, and none reported higher than 22.5 percent.
6. Most respondents envisioned continued and increased use of standards for such purposes as computerized shop loading, product and equipment design evaluation, and indirect labor measurement.
7. Standards are normally communicated to supervision and production workers, and one-third of the respondents use operator performance against standards for disciplinary action.
8. Seventy-five percent of the respondents use electronic data processing terminals for data collection. Seventy-five percent also provide means for reporting nonproductive labor and delays.
9. Among the respondents, 72 percent of production labor hours are covered by engineered standards. Many have plans to increase coverage, none plan to decrease coverage.
10. Data from respondents indicates that personnel responsible for establishing and maintaining standards systems tend to be somewhat satisfied with current techniques and collectively envision a need for more of the same.

An additional sentence, added to the end of conclusion number 10, appears to be an opinion expressed by the authors of the survey: *"This [more of the same] may be an unrealistic approach particularly with the low production quantities associated with most current aerospace programs."*

---

inherent in production management. In fact, it is difficult to envision efficient manufacturing without a disciplined work measurement program at the heart of work planning and control.

Why the basic requirements?

Why 80% coverage? There are two rational reasons. The first is that according to the Pareto Distribution, or ABC Curves, we could expect 20% of the time standards to cover 80% of the direct labor hours. The second is that approximately 80% coverage of the direct labor hours appears to be a reasonable minimum to give credibility to and promote confidence in a work measurement program. There is a third, perhaps irrational, reason. That is, that 20% of noncoverage should provide sufficient flexibility by a reasonable margin to accommodate the true anomalies that may not justify engineered time standards. When one looks at the coverage achieved in the commercial market, 20% of noncoverage seems indefensibly high. Perhaps, then, in reality, the 20% represents a maximum compromise position.

Why an accuracy requirement for engineered standards of ±25% with a 90% confidence level? This is, in fact, a maximum compromise position. The requirement is a minimum common denominator that should be exceeded when warranted. Industry resistance to the accuracy requirement resulted in revision downward of requirements proposed earlier. There is still some industry suggestion that 90 ±25 as a common denominator is excessively stringent. This appears to be somewhat irrational. Most standards routinely exceed the 90 ±25 requirement. Most standards in commercial industries without incentives probably exceed 95 ±10. Using time studies, 35½ times fewer observations are needed to demonstrate 90 ±25 as are needed to demonstrate 95 ±5. Using MTM systems, cycle times more than 35 times as great are required to demonstrate accuracy of 95 ±5 compared to 90 ±25. The key to an effective work measurement program is confidence in the standards. In order to maintain that confidence, economics and other factors would probably dictate that the time standards exceed 90 ±25.

Why an audit requirement? In short, to help assure system discipline. We, the Government, are not asking to audit the contractor. We are asking him, the contractor, to review himself. We view audit as the key to system discipline. I have heard others outside the Air Force express the same opinion. Audit is the key to confidence in the system and the time standards. Demonstrating accuracy and coverage is important not so much for the Air Force as for the contractor. If the contractor's workforce doesn't believe in the standards, the value of those standards will be largely lost.

The Air Force view

It should be obvious from the Air Force perspective why MIL-STD-1567 (USAF) is advocated. The answer is threefold. First, industry initiatives have thus far been largely inadequate. Second, the savings potential of 10 to 20% of the direct labor measured compares favorably with the costs of 1 to 4% of the direct labor measured. Third, and perhaps less obvious, public support of defense expenditures could be significantly enhanced by the capability to demonstrate effective cost control.

The Air Force believes (as does industry) that understanding is the key to effective implementation of work measurement programs. **Hopefully, Air Force-industry growth of understanding will converge in effective cost control. If so, the vehicle for achieving that cost control would be irrelevant.**

Where the buck stops — an IE...

William F. Fielder, Jr.
Head, Standards Administration
El Segundo Manufacturing Division
Hughes Aircraft Company

We at Hughes were actively involved in industry's responses to the Air Force, starting with the first draft, many months prior to publication of MIL-STD 1567-(USAF) in 1975. The spirit and the letter of that first draft set the stage for possibly some of the strongest industry reaction I have seen.

But first, let's review some of the natural, built-in problems in the basic subject. Work measurement has always been faced with negative reactions. Additionally, advances in the management sciences in recent decades have been moving away from work measurement as a basic tool of management, with a great deal of criticism of some of our techniques.

In the aerospace industry, where the complexity and dynamics of our manufacturing environment are so different from typical commercial work-measurement applications, managements have frequently been especially wary of the appropriateness and economics of work measurement. There are many valid reasons for these concerns, but we have made real progress in overcoming this natural resistance. It has taken very creative approaches to our problems to accomplish the progress we have made. It has not been easy, though the results have been rewarding.

Starting with this background, the initial draft — referred to as "MIL-STD XXXX" (symbolic of McGregor's Theory X to many of us) — was indeed a radical thrust. The proposed requirement (since removed) that industry be required to use basic standard data to be furnished by the Department of Defense Data Bank was only one of the "red flags" indicating a serious threat of unwarranted Government intervention in the management of our industry.

What's past is prologue

Those who have experienced some degree of real success in our industry know that the most critical requirement for success is management understanding and commitment. Such understanding and commitment are dependent upon the responsiveness of each work measurement system to the specific conditions which exist, and most importantly, to the specific philosophy of management existing in each application. Without such responsiveness, success is impossible. This is our most consistent experience.

Conversely, **one approach which has always failed is the arbitrary, external imposition of a system which is contrary to management's philosophy and which is not suited to the specific conditions existing.**

While most of the current requirements of MIL-STD-1567 do allow (actually require) management to establish its own system, many of the words and requirements relative to variance analysis and system audit, for example, still imply deep penetration into specifying the way managers manage.

A year ago, the issue of standards accuracy was one of the least understood subjects covered by the military standard. The progress made both within the Air Force and within our industry has been significant in my view. Don Heacox has personally worked hard at bridging the huge gap in this area. We are beginning to come together.

I feel, however, there is more that must be done, on a professional level, to bring the two sides together. We need the professional support and direction of AIIE in helping to resolve these differences. Certainly, in the controversy involving work measurement, the Government, and private industry, AIIE must be active at the leading edge of the matter.

A key issue

Implicit in the military standard requirement seems to be the assumption that the labor performance reports present simplistic reflections of simplistic perfor-

Courtesy Boeing Aerospace

ON THE LEGALITY OF MIL-STD-1567...
Simply put, it's not a factor. What transpires between a contractor and a contracting officer is up to them. If a contracting officer insists on imposing requirements that the contractor refuses to accept, the contractor has the basic right, of-course, to refuse the contract, [3].

mance determinants — for example, the obvious relationship between pressure applied to the throttle, and miles per hour reflected on the speedometer.

Seldom are the casual conditions this simple. When they are, management's identification of the problem and correction of it are simple. There is less need for formal management procedures.

Where the problems are complex — which is the most typical situation — the approaches to problem identification and solution are complex and, frequently, outside the area of direct influence by immediate supervision. **To expect operating management to reduce complex problems and their solutions to writing, as required by MIL-STD-1567, can easily be distorted into a highly bureaucratic, superficial, and costly procedure with the sole purpose of satisfying periodic Government audits. This is window dressing at its worst.** The greatest cost of such bureaucracy is the distraction of vital management attention from effective use of their most precious resources — time and energy.

Every manager has his own techniques of analyzing variances and assuring appropriate actions. Some of these techniques include written documentation — action-item lists, minutes of meetings, get-well schedules, and the like. However, imposition of a system of documentation similar to typical quality-control cause-and-corrective-action reports, which is implicit in the language of the military standard, would be very costly and highly ineffective.

The key to success in this area of the MIL-STD is the same as that in all other aspects of work measurement: management's own initiatives to employ its own systems of managing, consistent with its own fundamental philosophies and practices.

A bottom line problem

The cost of measuring can easily exceed the benefits obtained. I think the answer lies in considering the so-called "deltas." As an example, for the additional cost of periodically measuring (sampling) standards accuracy, what would be the additional savings to be accrued through better performance? In other words: Will we save money, or just add costs, if we institute a formal sampling plan to periodically measure the statistical accuracy of all of our standards? (My own conviction with respect to our shop is that we would just add costs and detract valuable industrial engineers from more productive work. The MIL-STD disagrees.)

The good and the bad

At worst, my greatest concerns are over the clear possibility — almost probability — of a creeping invasion of our profession with cost-ineffective, deadening bureaucracy. The threat is real. The natural result of unwanted, imposed requirements is degeneration into bureaucracy. One of the most alarming statements I have heard in professional meetings on this subject was the statement, "... They howled at first about the imposition of MIL-Q-9858 A (the basic quality assurance standard). They don't howl anymore, now that it's in effect." This scares me. How many volumes of manuals and files of documentation and so-called "spec trees" does it now require the thousands of quality engineers and other administrative personnel to administer MIL-Q-9858 A? Do we really know the cost effectiveness of this Government requirement? Is our work measurement profession headed toward this form of "system disciplines"? Words like, "cause and corrective action" (paragraph 5.12.1 or MIL-STD-1567), frighten me into believing that this could be the ultimate fate of our present embryonic start!

At best, on the other hand, I know that work measurement is in fact not only a valuable tool of management; it is an absolutely essential, integral part of cost-effective management.

Because our profession is as complex as it is, few people — in industry as well as in Government — really understand many of its intricacies. And those of us who think we do have some significant measure of understanding are thoroughly challenged to understand the overall management systems, philosophies, and practices of which our own specialization is simply an important component.

The greatest mystery to all of us is the understanding of human motivation. This is a basic point. I have the distinct feeling that the Government's concentrated focus on work measurement stems from failures to achieve the desired responses from other Government-industry relationships.

It feels a bit to me like whipping the dog's tail to try to get the dog himself to "wag." What I do mean is that the key to real and substantial improvement in aerospace productivity must originate in sources outside of the work measurement profession. And I fully agree: a well disciplined work measurement system — that is an appropriately disciplined work measurement system — is a potentially powerful tool for achieving our desired objectives.

The consultant: in between

Stan Wolfberg
Administrator, Engineering & Applied Science Div.
California Institute of Technology

Certainly no industrial engineer can argue with the purpose and the benefits which can accrue from the successful implementation of a work measurement program as described by MIL-STD-1567 (USAF). The objectives are sound, they reflect the "motherhood of industrial engineering concepts," and they very much follow an outline of the academic approach to one of the solutions to improving productivity and more effective cost control.

Some significant points

There are important key items to be noted in the standard that will have a significant bearing on the successful outcome of any installation.

"Active support of the program by all affected levels of management, based on the appreciation of work measurement and its objectives,..." In order to achieve this support at all levels — management, supervision, and the "touch labor" being measured — an effective "selling job" must be done by the Government to industry management, who in turn must sell it to supervision, who in turn must sell it to the "touch labor." Suffice to say, the document does not "spell out" how this is to be accomplished.

If the program is successful, one of the benefits is *"encouraging continued attention to methods and process analysis because of the necessity for achieving improved performance."* I have no argument with the benefits obtainable by methods and process improvements, which far outweigh the productivity improvements of "getting people to work faster" — the most common response by management, supervisors, and labor to the introduction of a work measurement program. Somhow during this almost two year period of presenting and discussing the MIL-STD-1567 (USAF), I have neither heard nor read about the significance of these benefits, which in the majority of individual instances will provide productivity improvements far in excess of the 33% targeted (from 60% estimated performance level at present to 80%) for the aerospace industry.

Who's qualified? Who defines?

With respect to corrective action — mandatory requirement, if the program is to be successful — I am concerned with the adequacy of the qualifications of the persons in Government to interpret the system, even though they may have designed it. For example, as I understand it, much of the Government's data bank is predicated on Methods-Time-Measurement (MTM), which requires a considerable amount of training and application — and the emphasis is on application in the specific area where it is to be used — to understand how it is to be applied and how to "bend it or straighten it out" to fit the situation. Whenever you consider this phase of application and interpretation, you get away from the written word and into the area of qualified personal judgment. To the extent that I have reviewed MIL-STD-1567 (USAF) I find little or no mention of the judgment required to effectively operate the program.

While the definitions included in the write-up of MIL-STD-1567 (USAF) cover the basics, there is, I believe, a need for greater detail and the inclusion of guidelines to facilitate more consistent application and interpretation, particularly where qualified judgment is so important to the success of a work measurement program.

For example, what level of effort is inferred? Are we talking about low task or high task? What is the equivalency in walking: 3.0 or 3.6 miles per hour? What are the benchmarks for personal and fatigue allowances? What is a reasonable range of delay allowances for specific types of operations? Each of us practicing industrial engineers

Courtesy McDonnell Douglas

> "Some industry executives argue that they are in the best position to evaluate productivity and cost control and should not be... restricted... by MIL-STD-1567. This is not a satisfactory answer if there are valid reasons to insist upon and require adherence to the use of a sound management tool from the standpoint of the public, who are paying the bills."
>
> Stan Wolfberg
> California Institute of Technology

may have our own answers to the above. But before the installation of a successful work measurement program, somebody is going to have to "spell out" the answers.

Some specifics

While demonstrating accuracy of ±25% with a 90% confidence level is far from onerous, I question whether or not this is sufficient accuracy to effectively monitor the program, pinpoint the problem areas, and respond with timely corrective action. Starting with ±25% and adding allowances for personal, fatigue, unavoidable delay, learning, realization, etc., it will be difficult, to say the least, to recognize the problem, not to mention the corrective action to be taken.

Inclusion of the provision that the Work Measurement Coverage Plan shall be based on cost-trade off analysis which relates savings to be accrued through improved productivity and simplification of work methods to the cost of attaining Type I standards coverage should certainly make industry prime contractors a little happier. The original draft provided this opportunity to subcontracting operations only. Incidentally, it makes me happier, too.

Who picks up the tab?

Borrowing a slogan from Western Airlines, "Standard Data is the only way to fly." As for the industry and/or a company taking full advantage of available standard time data of known accuracy and traceability, the answer has to be a strong affirmative. There is no reason to "reinvent the wheel." However, the only way the industry and/or company can adapt the available standard data to their specific operations, even though it is documented, is to be thoroughly educated and trained in the particular system. As I read the MIL-STD-1567 (USAF) I do not see any indications that the Government will provide for the education and training of the persons that are to understand, apply, and monitor the data. It's going to cost some agency, Government or industry, "a lot of bucks." But without this training and education, the chances for the success of the MIL-STD Work Measurement program are nil.

As for the stipulation that the contractor shall not charge costs directly to the contract for the development of basic or general purpose data, I assume this to mean if it is already available, packaged, and presented to the contractor. Standards, including basic or general purpose data, are a dynamic and flexible means of control and must be subject to continual change. In my opinion charges for these functions should be recognized by the Government as an appropriate operating expense.

One palatable suggestion

I will offer one suggestion that would make the MIL-STD-1567 (USAF) more palatable to industry: Reduce the required scope of the initial installation to a significant cost and/or production center rather than total operations. I have no objection to the long term goals of 80% coverage of all categories of touch labor; anything less than 80% provides too many opportunities for "working around" a work measurement program. However, projecting ahead in terms of full coverage to this extent, without having some specific preliminary milestone to evaluate the effectiveness of the program before going "all out" to satisfy the proposed coverage, appears to me to be a major stumbling block to the introduction of MIL-STD-1567 (USAF) to the aerospace industry.

I am very much in favor of the concept of MIL-STD-1567 (USAF) and firmly believe that if the program is properly and adequately "sold" to contractors — not "rammed down their throats" — it can be successfully installed and the benefits accrued as advertised.

Boeing makes their move

W. E. Selby
Director of Manufacturing

Ernie Shaw
Methods Engineering Manager
Boeing Aerospace Company

In the early '70's new equipment and new processes driven by technologically new products contributed to the need for additional standards development in Boeing Aerospace Company (BAC). During the same time period, the requirement for modern computer support became apparent and a Computer Business System Development Program was initiated. Early in 1975, a complete review of work measurement activities was made in support of the system development program. Basic requirements for an integrated Work Measurement System were defined emphasizing planning and control. An estimate of resources required for this system to integrate labor standards with estimating, budgeting, and forecasting was being prepared at mid-year 1975, when MIL-STD-1567 (USAF) was issued. Since MIL-STD-1567 embodied commonly accepted concepts of work measurement, BAC incorporated its criteria in the system planning. Representative companies were surveyed to evaluate development and use of work measurement data in large, integrated management systems.

Work measurement review

From those studies, estimates and plans for an expanded work measurement system were submitted to management. This culminated in a phased plan starting in 1975, which provided for refinements in the use of labor standards as a work measurement tool; integration of the work measurement system with new business systems being developed and provisions for standards to be used for estimating in conjunction with other estimating techniques.

Air Force-Boeing seminar

During January 1976 the work measurement plan was presented to the Air Force Plant Representative Office (AFPRO) and Boeing Corporate Management. As a result of the AFPRO presentation, it was proposed that BAC and the AFPRO jointly sponsor a working session to develop authoritative answers to questions on MIL-STD-1567 implementation. It was recognized that work measurement was a highly specialized area of management, subject to differing interpretations, definitions and understandings; also, that there were many groups within Boeing and the Government affected by MIL-STD 1567 and that growth in differing interpretations and understanding of the requirements must be avoided. During February and early March 1976 a joint Boeing/AFPRO panel resolved interpretations of most of the MIL-STD-1567 definitions and requirements. Panel results were reviewed by Boeing and Government spokesman at a one-day seminar on March 18, 1976. Results of this seminar were published by the Seattle Air Force Plant Representative. These results have been of substantial assistance to us in implementing our Integrated Work Measurement Program.

Contract implementation

A document containing the BAC Work Measurement System Description and Phasing Plan has been prepared and approved by the local AFPRO. It provides a base line plan which meets the intent of MIL-STD-1567 and may be used to satisfy contractual requirements of subsequent Government procurements from BAC.

Courtesy Boeing Aerospace

The BAC Work Measurement System includes the functions of standards development and application, performance reporting, methods analysis, system maintenance, and audit. It also provides for the use of standards in prediction of performance for budgeting, estimating, forecasting, load and capacity planning.

Impact of MIL-STD-1567

When MIL-STD-1567 was released in June 1975 there was concern that the requirement for sufficient data to statistically support specified accuracy levels would cause a very expensive redevelopment of established Boeing Labor Standard Time Data. Boeing was confident that the work measurement disciplines followed in original data development were sufficient to achieve the element accuracy required for a viable Work Measurement System. During the joint Air Force/Boeing Seminar existing Boeing Labor Standard Time Data was accepted as compliant with the intent of the MIL-STD. However, Boeing agreed that as new standards were developed, data to support statistical accuracy checks at the elemental level would be maintained. Additional audit disciplines were implemented and file retention schedules established to comply with MIL-STD-1567 requirements.

An initial cadre of Boeing Analysts has been trained in MTM disciplines and a training program to develop qualified MTM instructors is underway. Use of 4M, MTM3 and MTM V is being considered. Disciplined Time Study techniques are used for Work Sampling, Long Cycle, and Multi-Man applications and for data development where work load exceeds trained MTM analyst time available. Use is made of DOD 5010 data where the elements described fit BAC environment. Other predetermined time data of known accuracy will be acquired from Government agencies and other companies as a prudent method of reducing development cost.

Type I standards coverage, currently limited to fabrication and sub-assembly in BAC mechanical areas, will be extended to include major assembly when contracted production quantities provide sufficient development cost payback opportunity.

The requirement to provide a plan for use of standards as an input to planning, budgeting, and estimating was satisfied by AFPRO sign-off of the aforementioned BAC Work Measurement System Description and Phasing Plan and additional detail review with the AFPRO of plans for standards input to estimating. While the cost of this documentation was high, its preparation was used as the occasion to strengthen Work Measurement System discipline through a command media audit and to provide a base line for business system design planning.

The MIL-STD requirement that each element contributing to the modification of labor standards by realization factors be identified is still an open discussion item with the AFPRO. It is expected that this issue will be resolved in time to support business system development schedules contained in the BAC Work Measurement Plan.

Comments: AF Systems Command

H. H. Driessnack
Brigadier General, USAF
Deputy Chief of Staff/
   Procurement & Manufacturing
HQ, AF Systems Command

We have been experiencing decreased buying power of the defense dollar. At the same time, we are adding several new major weapon systems to our forces and must obtain maximum efficiencies to insure that we get the optimum quantities for a given level of expenditure. There are two basic ways to do this: upgrade our facilities through use of new technology, and increase the efficiency of our manpower resources. Work measurement is aimed at the latter and has become a major implementation goal for Air Force Systems Command. My own personal experience with the technique is that it works *when applied in the proper manner*. It certainly assists in developing a disciplined methodology for effective cost control.

Current language in MIL-STD-1567 appears to focus on work standards when in fact the intent is that equal attention needs to be given to methods improvement and work sequencing. Clearly, our intent is to reduce in-plant costs. The ultimate test of the effectiveness of MIL-STD-1567 is not the methodology used or the procedures that are generated but how well it drives down costs. If our implementation gets bogged down in statistical excursions on work standard tolerances, we will have missed the mark. I don't intend to let that happen.

Implementation of the standard will be evolutionary and will span an extended time frame. It is our intent that we maintain a continuing dialogue with Industry.

Michael A. Nassr
Colonel, USAF
Director of Manufacturing
HQ, AF Systems Command

MIL-STD-1567 (USAF) was established and published as a result of previous studies and after extensive review by Air Force and industry personnel. The resulting standard captured the minimum principles and criteria considered essential for our defense contractors to meet. It very decidedly steers away from the "how to" of implementation. Virtually all of our contractors have some type of work standard programs, but we have noted considerable variance in terms of effectiveness. Our intent in implementing the standard is to assure that contractors have a disciplined system of measuring labor productivity and that it is used effectively to drive down costs.

We don't view the work measurement standard as a panacea which solves or even addresses all the problems of productivity. Production processes, work sequencing, and other improvements in manufacturing technology also play key roles, but these are much more difficult to measure or define in terms of criteria.

To date there has been some industry apprehension toward the MIL-STD. I believe it is basically a communications problem and one we are gradually surmounting as our dialogue improves. We are meeting with industry groups and individual companies to dispel these concerns and to better understand and appreciate the common goals and mutual benefits that are attainable through positive and rational implementation.

In my discussion with industry representatives I've discovered that most concerns center around semantics; the only real differences expressed appear to be matters of relatively minor detail. Since we encourage the tailoring of all specifications and standards, I am convinced that there should be no major roadblocks to application on specific contracts and in individual plants.

DoD: They already do it

Richard J. Power
Director, Defense Industrial
Resources Support Office

USAF MIL-STD-1567 was developed by the Air Force to meet a perceived need to improve the productivity of contractors supporting Air Force production needs. The need for the improvement documented by surveys of contract or production operations has been compounded by a decline in procurement dollars and inflationary pressures increasing the costs of weapon systems and components. Both experience in the results of application of work measurement within the Department of Defense (DoD) and industry and surveys such as that documented in *Industrial Engineering*, September 1973, identify the potential for productivity gains that could be expected through more effective use of work measurement by defense industry contractors.

DoD policy and sound management theory dictate a minimum of paper work and interference with the contractor's operations. While the MIL-STD does require that

---

AIR FORCE SYSTEMS COMMAND:
"Their responsibility is to advance aerospace technology, adapt it into operational aerospace systems, and acquire qualitatively superior aerospace systems and material needed to accomplish the United States Air Force mission, [4]."

contractor documents be available for review, it does not require that additional specific reports be developed. It must be recognized, however, that if a contractor does not already produce a variance analysis and perform a system audit, improvement in the contractor's internal management system will be required. The benefits of these improvements are well documented in industrial engineering literature and have been achieved internally in DoD where work measurement has been applied.

The USAF MIL-STD 1567 contains a qualification criteria for engineered standards. The minimum level of accuracy required to classify a labor standard as an engineered standard is stated as ±25% at a 90% confidence level. This same criteria appears in the DoDI 5010.34 as the minimum criteria that must be met in the classification of a standard as engineered. The DoD standards program described in DoD 5010.15.1-M prescribes five categories of engineered standards from the ±25% at 90% confidence level to a level of ±5% at a 95% confidence level. The standards are classified using standard statistical computational techniques found in texts on statistics or work measurement. For internal program use, computations have been simplified through the use of monographs and other approximations.

The requirements to assure the integrity of the work measurement program through periodic audits is sound industrial engineering practice. It should both benefit the contractor and provide assurance to the government contracting personnel that the system is relevant and current. It will provide identification of the strong and weak points of the program, thus enabling industrial engineering management to assure continued soundness of the system. Comments on the cost effectiveness of such an audit have been raised. Yet, we continually perform inventories of the production materials and audits of financial records to assure their validity and assess the accuracy of basic records. The increasing importance of labor resource data to management requires that it should be audited. Hopefully, the audit plan developed in each individual case will consider both costs and benefits

AMETA: Rising to the challenge

Albert C. Adlfinger
Edward P. Kindinger
Industrial Engineers
AMETA

Since the contractor will audit his own operations, it was immediately apparent that successful implementation of the MIL-STD depended on the Air Force's ability to recognize and understand the concepts and operations of good work measurement systems as it applied to contractor evaluation. The Air Force Plant Representative Offices would perform this evaluation by examining and evaluating the contractor's required audit. Thus, the Air Force became concerned over a need to provide the AFPRO Manufacturing Operations personnel with the capability for evaluating contractor work measurement systems.

In 1975, the United States Army Management Engineering Training Agency (USAMETA) was selected as the appropriate activity to assimilate information and develop a course to train Air Force personnel in this area. USAMETA is the Army Proponent Activity for providing training in managerial analytical techniques including work methods and measurement, Methods-Time-Measurement (MTM 1, 2, and 3), standard time data, economic analysis, work planning and control, and cost/schedule control system criteria (C/SCSC). Its basic mission is to provide professional level management training, research and consulting for a wide range of Government agencies as well as the Department of the Army.

Training Package

In addition to in-depth research at USAMETA, information was collected through visits to the Air Force Contract Management Division (AFCMD) and several aerospace industries, including Hughes Aircraft Company, Rockwell International Corporation, the Northrop Corporation, and the Boeing Company. The final product was the development of an 80-hour intensive training course entitled "Evaluation of Contractor Work Measurement Systems."

The course is designed for on-site AFPRO Manufacturing Operations personnel including AFCMD engineers, production officers, and industrial specialists. The students must possess a working knowledge

Courtesy McDonnell Douglas

> "In the past, contractors rarely imposed requirements upon themselves to annually audit their own work measurement systems. Some of the contractors have never performed such an audit. With the advent of MIL-STD-1567, plans and methodologies for performing annual in-house work measurement systems audits must now be developed."
> A. Adlfinger, E. Kindinger, AMETA

of methods and work measurement, although the course is *not* designed for supervisory personnel nor staff who require only an appreciation of the subject.

The specific course objectives are to provide the trainee with the skills necessary to:
• Conduct comprehensive evaluations of the methods and work measurement systems utilized by the United States Air Force contractors, and
• To improve the contractor work measurement programs in order to reduce the weapon system acquisition costs.

Attaining these objectives led to the development of a topical outline covering subjects considered critical for proper evaluation of work measurement systems.

The methods of instruction includes lecture conferences, practical exercises, case studies, and examinations. The first week of the two-week course is a review designed to update industrial engineers in current methodologies and technological aspects of work measurement. The second week utilizes the case study technique for interfacing the governmental MIL-STD auditing procedures with the contractors' audits. The case studies are based on actual industrial projects in order to provide realism and assure proficiency in dealing with real-world problems.

Innovations

Construction of the training package required the development of several technical innovations in the field of work measurement, including the following:
• New checklists were incorporated in the course to aid the AFPROs in their review of an audit regardless of the contractor's approach.
• Sequential Sampling, a familiar technique used in quality assurance work, was adapted for sampling the work measurement time standard population. It tests the existing engineered standards coverage and provides an economically feasible approach for the contractor's use in establishing the present standards coverage for their annual audits.
• Learning curves and their applications, considering both the unit value straight line technique and the cumulative average straight line technique, were developed relating actual times, standard times, and variances.
• Accuracy and confidence values for time studies, work sampling, and MTM were developed for tables and charts. This was accomplished by converting existing accuracy tables and charts to show the minimal acceptable engineered data

required for 25% accuracy and the 90% confidence limit in accordance with the MIL-STD.

Training results

The initial course was taught at the Space and Missile Systems Organization (SAMSO), Los Angeles, 17-28 May 1976, followed by a second class at SAMSO, 21 June - 12 July. The third class was conducted at Wright-Patterson AFB, OH, 26 July - 6 August. The fourth class was 13-24 September at L.G. Hanscom AFB, MA. The average class size of the first three courses was 20 AFPRO industrial engineers. Initial feedback from these students indicate that the course has been successful in achieving learning objectives. However, the real test will occur during actual operations involving specific Contractor Work Measurement Evaluations.

Courtesy Boeing Aerospace

References

[1] Draft, *Industrial Engineers Debate the Air Force's Proposed Military Standard for Work Measurement*, Editor, William F. Fielder, Jr., AIIE Chapter 23, Los Angeles, June 10, 1975.

[2] Summary, *Industrial Engineer's Survey Regarding The Proposed Military Standard for Work Measurement*, AIIE Chapter 23, Los Angeles, Guy C. Close, Jr., Survey Chairman, June 10, 1975.

[3] From the author's notes. Phone conversation with David A. Webster, outgoing head of the Aerospace Industries Association, Inc., Manufacturing Committee, Sept. 24, 1976.

[4] *United States Government Manual 1976/77*, Office of the Federal Register, National Archives and Records Service, General Services Administration, Superintendent of Documents, U.S. Government Printing Office, Washington, D.C. 20402. **IE**

Should AIIE get involved?

Shortly before MIL-STD-1567 (USAF) was published in June of 1975, the Los Angeles Chapter (Number 23) of AIIE surveyed 27 industrial engineers — chosen for their experience in commercial aerospace work, specifically in the work measurement area — to determine their feelings regarding the standard and its specific requirements, [2].

The 21 IE's who responded to the survey represented 220 years of experience in aerospace industries, 215 years of experience in work measurement, and job experience in 57 different companies.

Here is the response to the 11th and last question on that survey.

Question: **What position, if any, do you think AIIE should take with respect to MIL-STD-1567?**

Position	No. of Respondents	Responses
Strongly positive	2	AIIE is a neutral and concerned professional organization. AIIE should aid Government in writing MIL-STD. Supportive 100 percent.
Positive	12	AIIE should be very active. Support it. Should encourage industry to actively utilize work measurement disciplines removing need of Government surveillance. Should work with AIA. Should provide expertise. Should establish or certify criteria. Support it. AIIE should object to...onerous sections. Too stringent. Should be backed...with modification. See that it is properly written. Assist with guidelines; reduce government intervention.
Neutral	3	AIIE should not take any position. Launch aggressive professional campaign to advance the industry — strongly oppose MIL-STD-1567 AIIE must support use of I.E. where proper — do not believe federal control is proper.
Negative	1	(No comments)
Strongly negative	3	Tell Government to "shelve it!" Oppose any Government control of inner company functions. MIL-STD should be opposed legally. Oppose strenuously.

Industrial Engineering Challenges In The Defense Industries

Update On Military Standard 1567 And Proposed Applications Guide Describes Contractor Concerns

By Lee F. Wade
Northrop Corp.

Approximately eight years after the Defense Department program management specialists installed their criteria-based Cost Schedule Control System (C/SCS), the U.S. Air Force industrial engineering specialists released their criteria-based performance measurement system—military standard (MIL STD) 1567—in 1976. Although MIL STD 1567 is applicable only to direct fabrication and assembly hours, it too has procedural needs, reporting needs, variance analysis needs, earned values relative to Type I standards, and audit requirements relative to the criteria concept.

This similarity of MIL STD 1567 to C/SCS is noted because C/SCS started with only four pages of criteria, but has now grown into a specialist industry of its own, complete with reporting formats, system descriptions, check lists, implementation guides, subsequent application review teams, analysis guides, validation teams, surveillance guides, and numerous consulting firms specializing in contractor system and computer software/hardware installations.

The proposed first draft of the MIL STD 1567 applications guide indicates that another specialist activity is about to emerge within the defense industry. It may be the criteria nature of such systems that opens them to such expansions, especially because constant changes among trained monitoring personnel within the individual military services make uniform application and monitoring of the criteria difficult.

What may be acceptable and reasonable today in systems application may not be tomorrow; so what starts out as a well intentioned performance measurement system can slowly begin to take on an existence of its own without regard to corresponding benefits.

MIL STD 1567

The following specific requirements are clearly stated in the standard:

5.1 Type I standards accurate to ±10% with 90% confidence (was ±25% in June 1975 release).

5.4 80% coverage of all categories of Touch Labor Type I standards. Type I standards are to be established using a *recognized technique* to derive 90% of the normal time associated with the labor type. Type II standards are labor standards other than Type I.

5.4.3 Schedule for upgrading all Type II standards to Type I.

5.6 Personal fatigue and delay (PF&D) to be included in labor standards.

5.7 Labor standards' relationship to price proposals.

5.13 Performance reports prepared at least weekly for each work center.

5.13.1 Written variance analysis for significant departures from standards.

5.14 Audit program.

However, human subjectivity can come into play among the various contractors during MIL STD 1567 initiation with regard to tailoring any part of the specification to speed its acceptance, subcontract flowdown, government system surveillance, documentation scope, minimum requirements, variance thresholds, cost center definition and scope of audit requirements.

It may be the lack of clarity in these application areas that leads to differences between contractors and hinders acceptance. Acceptance to date by contractors has not been without controversy. This is obvious from the time it has taken (from 1975 to 1983) to get all the services to accept the U.S. Air Force-initiated specification, and only after the U.S. Navy excluded all its ship building programs from MIL STD 1567, at that.

Negative contractor comments were summarized in the June 1980 Government Accounting Office (GAO) Report on Work Measurement to Congress and have been stated by representatives of both the National Security Industrial Association (NSIA) and the Aerospace Industries Association (AIA) committees during normal contacts with their counterpart DoD systems specialists.

June 1980 GAO report

Four major defense contractors were identified in the Government Accounting Office report as having achieved significant benefit-to-cost ratios (5 to 1 and 13 to 1, for example) after application of MIL STD 1567. Although these are the kinds of data that are needed to accelerate MIL STD 1567's wholehearted acceptance by contractors, this acceptance has not yet developed, except in relation to new contracts for which it is a requirement and any associated implementation costs are automatically rolled into total contract cost performance. Of course, few defense contractors will fail to respond in such a manner relative to a major new program acquisition.

The problem with these stated savings lies in the GAO use of U.S. Air Force assessments of savings values reported as due to the application of a USAF-initiated regulation. It would have been much more effective if the using contractors had quantified the savings based on better performance relative to historical learning curve data to show that historically normal learning was not tangled up in such assessments and that no other dynamic program occurrences had affected the application of the standard.

Representatives of two of the four companies mentioned in the GAO report stated in private discussions that although they might not agree with the report, they were in no position to mount opposition on such a sensitive issue with management groups within their plants who would remain there long after such claims were forgotten.

However, the cost-effectiveness of this major new regulation will remain of prime interest to everyone in the defense industries, even if the government may be installing it as much for price analysis as for cost managing, because hardware cost improvements are the only means available to offset the increased non-hardware costs associated with implementation and continued maintenance of MIL STD 1567.

C/SCS interface

There is no specification in either C/SCS or MIL STD 1567 which would keep it from being compatible with the other. Although some specialists have tried to separate the two systems, MIL STD 1567 technically cannot be detached from C/SCS involvement, because C/SCS Criterion 2-4b(7) contains provisions that require the two systems to be considered together.

Installation and acceptance

The U.S. Air Force has especially pushed for the acceptance of MIL STD 1567 by all its contractors over the past few years. During an April 1983 AFCMD/Industry Joint Executive Conference in Denver, General Stukel again emphasized USAF support of MIL STD 1567 as an asset to the Air Force's "War on Cost." His enthusiasm will continue to be transferred through his plant representatives into each contractor's plant.

Where the regulation is not on contract, various methods can be employed to encourage its acceptance. CMSEP (Contract Management Systems Evaluation Program) government administrators can equate non-conformance with MIL STD 1567 with non-conformance with their PD-5 series CMSEP indicators, and code the contractor an undesirable color. Air Force industrial engineers can pressure their contractor counterparts, and plant representative officers can chat with contractor executives until MIL STD 1567 gains acceptance.

Where the regulation is required by contract, progress payments can be withheld to get the attention of contractors who wish to argue over the specific merits of any part of MIL STD 1567.

Technically, MIL STD 1567 is applicable to full scale acquisition programs which exceed $100 million and production programs that exceed $20 million annually or $100 million cumulatively when put on contract. As a guide, it is also applicable to subcontractors who exceed $5 million annually or $25 million cumulatively when the standard is applied to a prime contract.

Some contractors may have poor work measurement programs, and MIL STD 1567 will help them correct their deficiencies. The USAF continues to apply MIL STD 1567 to new contracts, so its application appears to be gaining momentum. It is up to industrial engineering professionals to get out in front of its application and help steer it in the right direction, even if this exposes them to some criticism.

Tailoring

Tailoring is a method whereby government representatives may alter or soften certain MIL STD 1567 requirements where cost-effectiveness justifies such actions and the basic intentions of the standard are not lost. Although the MIL STD 1567 foreword encourages tailoring, it can create problems later on, because tailoring agreements reached with one government representative may not be acceptable to other government representatives who participate in later reviews of your criteria-based system.

At present, it will be advantageous for each contractor to stay abreast of the tailoring that has been permitted other contractors in the initial application of MIL STD 1567 to keep them from overreacting to all the cost impacts of its installation and operation.

MIL STD 1567 applications guide

A proposed first draft copy of the

MIL STD 1567 applications guide released during 1983 for review appears to be attracting much comment. Some of the more interesting sections of this guide deal with the following:

2. Tailoring processing becoming more formalized.
5. Coverage for full scale engineering development (FSED) type hardware.
6. Contractor reviews and annual reviews thereafter.
7,8. More detailed accuracy measurement and verification data.
9. Variance analysis content and frequency.
14. Use of standards in responses to requests for proposals for spares and hardware acquisitions to the degree that all proposals are to be returned to the contractor for non-compliance when standard data are not submitted with the proposal.
16. Costs of MIL STD 1567 should be charged to indirect pools.
17. Notices of non-compliance including an agreement that 2% is a reasonable cost for maintaining a work measurement system.

This draft is still in review, but few changes are expected if no more is done with contractors' comments (regarding accuracy or application, for instance) than was done relative to the initial MIL STD 1567 release.

When the application guide is added, "criteria-based-system" will probably become an inaccurate description of MIL STD 1567. In any case, the proper Council of Defense and Space Industry Associations (CODSIA) group is currently addressing guide issues in a businesslike manner.

Of great importance to industry is whether this application guide, when released, will become a contractual appendix or remain a non-contractual guide to assist government reviewers. If it becomes a contractual appendix, it will function more as a doctrine than as a guide.

No one in the defense industries argues against cost improvements, but there is much discussion regarding their attainment—especially in connection with better program definitions, multiyear obligations for smoother production planning, reduced concurrency, manufacturing technology emphasis and technology modernization.

The jury remains out on the true cost-effectiveness of MIL STD 1567, which concentrates more on standards than on methods, although the methods area is where most cost savings are currently being generated, because most large defense contractors have had reasonably good standards for decades.

Contractors with no standard programs, or poorly maintained ones, can benefit from a standards program, but implementation of MIL STD 1567 should be monitored closely to assure that the mechanics of the system do not overshadow its cost improvement purpose.

This standard, in its current state of development, could become an administrative balloon, complete with an elusive accuracy syndrome. The plant representatives who will audit the contractors will surely be audited themselves by other government representatives, and the tendency to build surveillance files will add further to its administrative complexity.

Of major concern to defense contractor cost analysts today is the growing number of support personnel needed to support the people who actually perform touch labor and the reasons for this growth. If MIL STD 1567 application costs are charged to indirect pools (other peculiar program problems being successfully nulled out), and given how difficult it is to measure true performance improvements due to better standards, the jury is expected to remain out for some time regarding the standard's industry-wide cost-effectiveness.

In the mean time, implementation of MIL STD 1567 is expected to be expanded, and more requirements placed on industrial engineers. While our heads are being turned toward work measurement, industrial engineering resources should not be withdrawn from methods improvement and programs which stimulate worker involvement in improving individual performance. A reasonable balance between these efforts must be maintained to enable continuing cost improvements in the defense area.

It has been my intent in this update to sound a note of caution regarding the potential of excessive detail that is inherent in the application of any new administrative system. On the other hand, an improved work measurement system can be an effective tool when properly administered. The cost-effectiveness of MIL STD 1567 rests in the hands of administrators, who will need to exercise good judgment in carrying out the intent of its criteria. IE

Lee F. Wade has spent 30 years in the defense industries in research and development administration, manufacturing management, program office and division controller roles working with such management systems as PERT, PERT COST, C/SCS and MIL STD 1567. He is currently manager of product cost management for the aircraft division of the Northrop Corp. and also a member of the NSIA management systems subcommittee.

IMPROVED PRODUCTIVITY THROUGH WORKER INVOLVEMENT

Mitchell Fein, PE, President
Mitchell Fein, Inc.
Hillsdale, NJ 07642

Abstract

Worker involvement programs which offer only job satisfaction as the prime reward for involvement will be supported by only a small proportion of the work force and tap a fraction of the potential for improvement in the organization. Involvement programs which offer financial rewards by sharing productivity improvement with employees through formal productivity sharing plans create high levels of involvement, produce results quickly, and raise productivity to much higher levels than are attained by non-financial reward programs only.

This paper examines the ideas, judgments, and biases on how most effectively to reshape the world of work so that everyone benefits in win-win relationships. The issues which deter managers, workers and unions from adopting changes in traditional worker-employer relations are examined against many successes and failures. Decision makers on all sides need to know what works and what does not, and why.

The factors and conditions at the work place which affect employees' attitudes to their work and especially to increased productivity are described and examined. Much research and writing has been devoted to the subject of worker motivation, yet there is a wide gulf between how workers and managers view the world of work.

The organization psychology school of major theorists such as McGregor, Maslow and Herzberg postulates that involvement by workers in their work in itself is satisfying and rewarding; that though people work for money, the role of money as a motivator to improved work performance is minimal; the major motivation to work is provided by the work itself and other nonfinancial factors.

These concepts were the foundations for the much publicized job enrichment experiments starting with Texas Instruments, AT&T and others, the job restructuring and design efforts through organization development, on through the Quality of Work Life and the more recent Quality Circle programs. All have a common base: The main reward offered to workers for involvement in these programs is the satisfaction of having contributed to improvements in productivity and how work is performed. None offered financial rewards either as an inducement to participate or as a reward for improvements in work and in productivity.

This paper disputes the claim that the work itself is a primary motivator for most workers and supports the proposition that financial incentives, at least for manufacturing plants, is a more powerful means to encourage workers to increase their productivity and to accept managements' productivity goals as congruent with theirs. The work itself is attractive to a minority of the work force. Increased earnings through productivity sharing plans is attractive to all.

This paper proposes that:

- worker involvement programs which offer financial rewards by sharing productivity improvement with the potential for improvement in the organization.

- worker involvement programs which offer financial rewards by sharing productivity improvement with employees through formal productivity sharing plans create high levels of involvement, produce results very quickly, and raise productivity to much higher levels than are attained by non-financial reward programs only.

There are no data on entire plants to show the effect on worker satisfaction and on productivity through the operation of psychological needs rewards programs as against financial rewards. Data are not available simply because no company will sponsor or permit studies to be made which would split a company to operate both programs side by side. The two reward

---

Copyright 1982 Mitchell Fein

programs can be compared by examining two aspects of results: (1) how long it takes to establish an involvement program, and (2) the extent to which productivity is improved. The reasoning to using these two measures is that a program which is established quickly probably has more worker support than a program which takes longer to get into operation.

Obtaining higher productivity gains compared to lower gains probably indicates that workers in the high gain plants are more involved in helping to raise productivity since productivity gain is measured as the increase obtained over management's best efforts in the base period before the program starts. These two indicators may be challenged as oversimplified and not reliable, but as will be shown, the differences in results obtained between the psychological needs rewards and the financial rewards are so great that there can be no question that the differences are significant.

Encouraging worker involvement

Much of the literature favoring employee involvement in raising productivity and improving workplace conditions frightens many managers who fantasize that employees will take over the shop, get on the Board of Directors and order managers around. The leadership of the AFL-CIO and the major unions in the United States have on numerous occasions stated that they are unquivocally against a partnership with management in operating companies; they prefer the adversary relations of collective bargaining.

Despite strong denials from union leaders that they want to share the board rooms with management, apprehension continues at top management levels that any dilution of management's traditional rights is a step closer to worker control of the shop. Feeding management's fears are the vaguely defined terms used in management and behavioral science literature in this country which create misconceptions concerning worker involvement efforts. The term "participation" is applied to a wide range of activities in which workers voluntarily become involved with management to solve problems, and "participative management" is used to describe the attitudes of managers who favor the involvement process.

Participation does not adequately describe the true relationship between nonmanagement employees and management in these voluntary efforts. Participation implies that those involved have a share of responsibility for results, which they do not have. Involvement without accountability is not participation, but consultation. That is what really occurs when employees are encouraged to suggest improvements; management may accept or reject employees' ideas, because only management has authority to make decisions.

"Involved in decisions" is another term used vaguely in the literature and this, too, is misleading. Workers can make suggestions and are consulted; they may influence decisions, but they are not involved in decision making in the sense the term is used in managerial practice. Using these terms loosely can harm cooperative efforts in two critical areas:

- Managers may assume the literal meaning of "participation" and "involved in decision making" and reject cooperative efforts, apprehensive that employees will take over the shop.

- Workers may be disappointed when they find they have less influence than they were led to believe they would have.

The term "participative managing" should be replaced by the more appropriate and descriptive "consultative managing."

Consultative managing encourages worker involvement in day to day operations, in questioning how work is performed and suggesting changes. This significant change from traditional managing affects many other business policies. For consultative managing to be successful, it must be accepted by top management, believed in, and followed down the line by supervisors.

Consultative managing does not require management to give up any of its rights to manage; management still has the last word in all decisions. Managers give up only the right not to listen, and take on an obligation to consider employees' proposals. Most important, managers' actions and attitudes must show they mean what they say.

Encouraging workers to become concerned with improvements and aid in raising productivity alters their perceptions of their roles in the company. Management cannot tell workers to become half involved, to think only about work and to ignore everything else. When workers can question how an operation is performed, they may question other things. Tough traditional managers may not take kindly to questioning from below, but it is necessary for consultative managing. Democratizing the work place will frustrate many managers who are accustomed to making decisions and giving few explanations for their actions. Having to give reasons to worker groups for not making changes may be time consuming, but that may be part of the price for gaining worker support.

Psychological needs reward programs

Psychological needs reward programs are based on several major premises:

- The work itself is a powerful motivator to involve workers in their work.

- Work tasks in factories have been so simplified that workers' desires to work are turned off. The remedy proposed is that skills and judgments should be added to denuded jobs; enriched jobs are more desired by workers.

- The mechanical nature of work and the low challenge of the work place has created worker dissatisfaction which reduces worker productivity. Restructuring jobs with greater responsibility and judgments will remedy the disenchantment process and result in more effective work performance.

The role of financial incentives as a reward for improved performance is minimized by supporters of this school, who propose that psychological and social needs are very important to workers and vital to their well-being. Behavioral science writings highlight studies which demonstrate that workers value money and job security lower than opportunities for work involvement activities.

Much of the support for the interesting work concept is based on the 1970 SURVEY OF WORKING CONDITIONS prepared by the Survey Research Center (SRC) of the University of Michigan, widely cited as the authority that interesting work is preferred by workers over pay.1/ The study found that of the five work features rated most important by workers, only one had to do with tangible or economic benefits and that one, good pay, was ranked number five.

An analysis by Fein of the SRC data found that the SRC researchers had erroneously averaged all the data for executives, professionals, technicals, clericals, sales service, factory workers, truck drivers and farm workers, and called the average a "composite worker." When the data were analyzed by occupation, good pay and job security moved up and interesting work dropped for factory workers; professionals and managers may place a higher priority on interesting work.2/

A study by Edwin A. Locke, et al, found that money is a more powerful motivator than is generally believed:

> Our findings may surprise or even shock many social scientists. For the last several decades ideological bias has led many of them to deny the efficacy of money as a motivator and to emphasize the potency of participation. The results of research to date indicate that the opposite viewpoint would have been more accurate.3/

The notion that a lack of interesting work causes dissatisfaction among workers and so reduces their will to work is not solidly supported by research and data. A study of 300 research studies on the relation between job satisfaction and motivation by a New York University team, supported by the National Science Foundation (NSF/NYU), headed by Katzell and Yankelovich, of which this author was a member, found that:

> .....if there is any one fact that stands out clearly from the massive accumulation of data - the hundreds of studies encompassed in this report - it is that worker job satisfaction and productivity do not necessarily follow parallel paths. This does not mean that the two objectives are incompatible, for there is evidence that it may be possible to achieve them together. Nor does it mean that the two goals are totally independent of one another. Under certain conditions, improving productivity will enhance worker satisfaction and improvements in job satisfaction will contribute to productivity. What it does mean is that there is no automatic and invariant relationship between the two.

> Indeed, the two objectives are so loosely coupled, there are so many intervening links between them, and the relationship is so indirect, that efforts which aim primarily at improving worker satisfaction on the assumption that productivity will thereby automatically increase are more likely than not to leave productivity unchanged, or at best to improve it marginally, and may even cause it to decline.4/

Four combinations of satisfaction/motivation are possible:

- A satisfied worker may be highly or weakly motivated.

- A dissatisfied worker may be highly or weakly motivated.

The data do not show a negative or a positive relationship, rather no relationship. The NSF/NYU study found that though common sense and logic suggest that improved job satisfaction leads to raised motivation, workers do not necessarily react that way. A satisfied worker may just be a satisfied worker, not any more motivated to produce than before.

Studies and data I have examined over the years point to two main groups at work. About 15% of the work force are active involved people who identify with their work; interesting and judgment-laden work is important to them. The other 85% of the work force come to work to eat; the nature of their work is not critical to them; they find their involvement and gratification outside the workplace. The 15% achievers get into the skilled jobs. The 85% nonachievers take on jobs they prefer, but interesting work and fulfillment from their job is not necessary for their well-being. Obviously there are gray areas between the groups. 5/

Real life experience in plants bears out this 15/85 concept of workers' desires. A study by Simons and Orife reported that:

> In 51 of the 71 pairs of jobs, employees transferred to jobs with increased pay; only 4 moved to lower pay. The majority of the higher-paid jobs to which workers moved were more enriched, but when the job shift did not provide a pay increase, no preference was shown for the more enriched job. No statistically significant preference for less routine (more enlarged) jobs was shown either with or without increased pay. Required physical effort proved not to be a significant factor.

> The assumption that job enrichment and/or enlargement is sought and desired by employees has become widespread in recent years. This is despite the fact that empirical studies of what workers want in their jobs have been limited, have often been based on questionable methodology, and have at times given conflicting results. 6/

Job bidding records in companies provide ample data of which employees desire to change their jobs. Invariably the most senior employees, who normally have first claim to job openings, are not those most frequently bidding. Significantly, few employees shift to higher skilled jobs without a pay increase. Data on worker job preferences show that transfers to higher rated jobs reflect individual worker desires. Everyone wants more money, but as jobs become more demanding, fewer employees aspire to these jobs.

The various reports of experiments with nonfinancial psychological needs programs provide sparse data, if any, on the productivity gains achieved through the programs. Results attained in creating better labor relations, improved job satisfaction, reduced spoilage, fewer grievances, and so on, are also sketchy and difficult to reduce to hard data. A key indicator of the accomplishments can be inferred from the fact that these programs have been terminated by management with no record of workers in these plants have objected to their discontinuance.

The arguments in favor of psychological needs reward programs do not stand up when all the experimental data and cases are examined. Involvement in the work is a need of the minority of the work force who benefit from it. A large majority of workers have other needs and wants which come before work involvement. They, too, may find satisfaction from involvement in their jobs, but it is of a lower order than for the achievers.

The inequity of nonfinancial reward programs

Practically no QWL, Quality Circle or behavioral science-supported worker involvement programs provide financial gains to groups of workers or entire plants when operations are improved. Their only reward is a feeling of satisfaction from having done a good job and helped the company.

Not rewarding workers for improvements they create is questioned on two grounds: equity and fairness; both are linked in workers' minds. Many companies that try hard for years to develop credibility with their employees may find that workers resent performing work for which they are not compensated.

Management violates universally accepted principles for establishing equitable pay rates by encouraging employees to work at higher skill levels without commensurate extra compensation. Pay scales in the private and public sector are set to reflect the skill and effort needs of the job. Many companies determine job rates through formal job evaluation plans. The procedure always followed is to prepare a job description delineating the principal requirements of a job and then evaluate the job against a point evaluation plan. Production jobs never require employees to use judgment and initiative to improve the way work is performed and to reduce costs; these tasks are the responsibility of technicians and engineers, who are properly paid to do such work. When higher level tasks are added to a job, it is reevaluated and a new job rate is set.

When workers are encouraged to use higher skills to innovate work improvements they should receive extra compensation for their efforts. If workers feel good because they did a good job, they will feel better with extra pay, which is an excellent way for management to thank these employees who did more than was expected of them.

As their efforts reduce operating costs and workers do not share, management's credibility and good faith are questioned.

It is one thing for workers to cooperate and work hard to prevent a plant from closing; everyone gains as the plant and jobs are saved. It is quite another for a company not in dire straits to pocket the gains from cooperative efforts.

Rewarding improved performance

Managers wishing to improve productivity with money as a motivator will find support in the findings of a behavioral science study which agrees that money can be a prime way to increase employee motivation. The NSF/NYU study team developed six critical ingredients of effective systems to raise job satisfaction and worker motivation, headed by:

> Financial compensation of workers must be linked to their performance and to productivity gains. 7/

The NSF/NYU study found that when workers' pay is linked to their performance, the motivation to work is raised, productivity is higher, and the workers are more likely to be satisfied with their work.

A study by Fein of over 400 plants in the United States found that when these plants instituted work measurement, productivity rose an average of 14.6%. When plants instituted traditional wage incentives where previously there was only work measurement, productivity rose an added 42.9%. The average increase from no measurement to incentives was 63.8%. 8/

Most managers believe that pay tied to productivity will motivate to higher performance. From two-thirds to three-quarters of all the sales forces in the United States are on incentives. 9/ Approximately 89% of manufacturing companies have executive bonus plans; the median bonus for the three top executives averages 48% of their base pay. 10/ A study of executive compensation of 1100 companies listed on the New York Stock Exchange found that companies which had formal incentive plans for their executives earned on the average 43.6% more pretax profit than did the nonincentive companies. 11/

By any measure, pay tied to productivity is the most powerful motivator to improve work performance. Strangely, it seems as if proponents of nonfinancial reward systems are saying that money is unwholesome and debasing. Compensation practices in the United States are based on the principle of equal pay for equal work, in which workers expect and receive higher pay as the skill and demand characteristics of their jobs increase. Japanese workers, who are held up as models for American workers, work under an entirely different system. Primary pay differentials are based on seniority and education with education assumed to reflect job skills, while seniority is related to job performance.

Within the American culture workers strive for increased earnings and benefits. Executives of large American corporations with salaries around $350,000 aspire to bonuses which bring their total pay to over $1 million a year. When top managers seek more, why are workers, who may barely eke out a living wage, expected to place their work above pay and job security?

The work itself serves as a motivator to some workers and its importance should not be overlooked. Money has been shown to have universal appeal; pay by performance definitely encourages higher performance.

The rationale of productivity sharing

A productivity sharing plan divides productivity gains between employees and the company when productivity exceeds the level of the base period. A prime objective of productivity sharing is to create conditions under which workers and management benefit by moving on parallel paths to a common goal: more units of product made in fewer man hours of work; only finished good units are counted. The entire plant or company is grouped.

The rationale for traditional wage incentives stems directly from traditional managing practices; it is rooted in the adversary relations environment of the work place. Time standards under traditional managing, whether for measured day work or for incentives, are probably the greatest single cause of worker grievances. The we-they relations on the plant floor are sharpened by traditional incentives.

Productivity sharing is not an incentive plan; it is a philosophy of managing that encourages employees to become involved in productivity improvement. Productivity sharing creates a work environment in which employees see improved productivity as beneficial to them. Under the philosophy of productivity sharing, worker productivity goals and management goals become congruent.

Under productivity sharing, measurement standards are established as the average of a past base period for an entire product. Using the past base eliminates disputes on the plant floor over loosening operation time standards. A major difference between traditional incentives and productivity sharing is that under incentives, individual workers are motivated only to make more money; they have no interest in raising productivity or in management's productivity goals. Under productivity sharing, workers may still be primarily motivated to make more money, but since productivity sharing is based on achieving overall improvement in productivity, more money is earned only by raising productivity. Employee interests expand to the plant when all employees are rewarded as a group for their gains. Counting output only as finished units ready for shipment, and rewarding all workers for productivity gains over a base period, focuses their attention on the need to reduce labor input and increase product output.

As workers' interests shift toward management's, the rationale for traditional managing fades. When workers become concerned with final outcomes, they are also more interested in how operations proceed throughout the plant; they become attentive to production impediments that occur around them which in the past they ignored or even encouraged. Since productivity gains are shared, whether innovated by workers or by management, it is conceivable that industrial engineers, who are now disparaged by workers, will instead be welcomed because workers will gain from engineers' improvement efforts. That would indeed signal a significant change in the attitude of workers and managers toward each other's interests and needs.

Productivity sharing plans in use

Three productivity sharing plans are used in this country: the Scanlon Plan since about 1936, the Rucker® Plan since World War II, and the Improshare® Plan since 1974. All are company-wide sharing plans with significantly different productivity measurement systems.

The Scanlon Plan

The Scanlon Plan, developed by Joe Scanlon, then research director of the Steelworkers Union, is the most widely known plan in this country. The term "Scanlon Plan" is often used generically, referring in general to productivity sharing. After Scanlon's death in 1956, several associates carried on his work; Frederick Lesieur and Carl Frost are the most prominent.

The Scanlon Plan measures productivity improvement by a change in the computed ratio of total payroll dollars divided by the total dollar sales value of production. Most Scanlon Plans distribute 75% of the gains to employees and 25% to the company, though the percent can be changed. Twenty-five percent of the monthly gains are placed into a pool to absorb loss months; at the end of the year the entire pool is distributed. Other plan details relating to how and when sharing gains are paid, the highly structured suggestion plans and labor management committees do not significantly differentiate the plan from others.

Moore and Ross present considerable operating details and different ways to measure productivity changes to overcome the single ratio measure.[12] Frost, Wakely and Ruh describe Scanlon Plans in operation.[13] Both books are excellent as source materials and contain broad bibliographies of published materials on Scanlon Plans.

The Rucker Plan

The Rucker Plan was developed by Allan W. Rucker of the EddyRucker-Nickels Company in Cambridge, Mass., in the late 1940's. Measuring productivity change as a change in the ratio between dollar payroll and dollar value added provides a more reliable measure than the Scanlon measure of payroll dollars per dollar sales value because, under value added, all purchased materials are excluded. The calculations are similar to Scanlon, except that instead of sales dollars, the figures used are sales less all purchased materials. Productivity measurement under Rucker with a single ratio presents the same problems as under Scanlon. Several papers describe details of the Rucker measurement system.[14,15]

Improshare®

Improshare is derived from "improved productivity through sharing." The plan was developed by Mitchell Fein and was first used in 1974; parts of the plan were used for over 20 years. Improshare productivity measurements use traditional work measurement standards and practices modified to a selected base period. Productivity measurement with Improshare is considerably different from Scanlon or Rucker. Productivity gains are shared 50/50 between employees and company. Improshare details are contained in "IMPROSHARE®: An Alternative to Traditional Managing."[16]

The GAO study of productivity sharing programs

The General Accounting Office conducted a study of productivity sharing programs [17] to determine (1) how productivity sharing plans operate and what benefits are obtained, and (2) what long term increases in productivity can be realized through productivity sharing. The report estimates that about 1000 productivity sharing plans are in use. The GAO researchers solicited 78 firms and interviewed the managers of 54 to determine their experiences with productivity sharing and other incentive plans. Thirty-six of the firms interviewed had productivity sharing plans of which 17 were Scanlon, 8 were Rucker, 11 were Improshare, and 2 were other. Of the 9 firms interviewed which did not have productivity sharing plans, 5 had individual or group incentives based on engineered standards, 5 had profit sharing, 4 had Quality of Work Life programs, and 2 were "other"; the total adds to 16 because several of the firms had more than one type of plan. The size of the firms contacted ranged from a small company with less than 100 employees to one with over 100,000.

The GAO report briefly describes traditional individual and group incentives, suggestions systems and profit sharing. The prominent features of Scanlon, Rucker and Improshare plans are described, including how productivity is calculated.

The GAO reports that:

> Among the 24 firms providing financial data, those with a productivity sharing plan in effect the longest showed the best performance. Firms that had plans in operation over 5 years averaged almost 29 percent savings in work force cost for the most recent 5-year period, with individual firms' average savings ranging from 13.5 to 77.4 percent. Those firms with plans in operation less than 5 years averaged savings of 8.5 percent. To cite some specific examples: A company with 360 employees reported an average savings in cost of 15.5% over a 5 year period. Another company employing 2000 saved an average of 24% over 5 years, ranging from 20-35% a year. A company of 215 employees averaged 14% savings over 5 years, ranging from 11-18% per year. 18/

The GAO cites the nonmonetary benefits of productivity sharing programs: Improved labor relations were reported by 80.6% of the companies, 47.2% reported fewer grievances, 36.1% had less absenteeism, 36.1% had reduced turnover, and 47.2% reported other benefits.

> The vast majority of firms expressed satisfaction with their productivity sharing plans and believed that the current benefits to the firm from their plans warranted their continuation. Officials at 22 firms said that the benefits originally anticipated were realized......For the most part, firms.....believed that their productivity sharing plan gave them a competitive advantage in marketing their products or services. 19/

The rapid start of an IMPROSHARE© Plan

Reports from nonfinancial worker involvement programs show that it takes from a year and half to three years to establish a program which will encourage widespread involvement of workers to improve productivity and labor relations.

The rapid start of an Improshare Plan in the first 12 weeks creates substantial productivity improvement which usually puzzles people who have not been involved in productivity sharing plan activities. The reason for the short starting time is illustrated by a discussion that took place in a meeting called by management of a highly mechanized machine shop to discuss introduction of an Improshare Plan, at which were present operating managers and union committee representatives.

In response to a question posed by the plant manager as to how long it would take for an Iproshare program to start operating, I turned to a union steward from one of the large machine departments and asked him how long it would take for the workers in his department to "rev it up." The response was "15 minutes." This was not a flippant remark, but a considered statement by a knowledgeable worker of how long it would take him to turn up his production - if he wanted to.

When workers want to make suggestions to increase productivity and to increase their own output, all they have to do is do it. It does not require a one and a half to three year program to convince workers that increased productivity under a productivity sharing plan benefits them financial and nonfinancially. When workers realize the financial benefits the plan can provide, and the opportunities open if they wish to become involved, they get into the spirit of the plan. In times of keen competition, helping the company to reduce costs will save jobs, which workers readily recognize. Turning on the 85 percenters who come to work to eat permits the 15% achievers to get involved in work improvements to their hearts' content. The magic ingredient that makes productivity sharing work is financial reward, reinforced as increasing numbers of workers join in the efforts.

Results obtained by IMPROSHARE® companies

The data in Table 1 show the change in productivity in 72 companies after Improshare was established. No hard data is available on employee attitudes before, during and after the installation of an Improshare Plan simply because attitude studies were not made.

Improved results

Discussions with employees, union representatives and managers at the companies showed decided improvements in employee-company relations. Practically all companies reported reduced scrap and spoilage; some quite substantial. Waiting for work and equipment downtime, which cause considerable productivity loss, was significantly reduced. Surprisingly, equipment downtime, which is a responsibility of maintenance mechanics, was often reduced by machine operator ingenuity and diligence.

A prime measure of the effectiveness of an employee involvement program and the degree of employee participation is reflected in the increase in productivity achieved after the program starts. Manufacturing plant productivity is usually on a plateau and fairly flat before the start. The rise that occurs after the start must be attributed to the efforts and cooperation of the employees.

The productivity improvement obtained by 72 companies operating with Improshare since 1974 is shown in Table 1. As of February 10, 1982, 57 plans were operating; five plans were discontinued because the plans did not work; ten plans were discontinued for other reasons: three plants were closed, one plant is being rebuilt and the plan will restart, the business volume in three plants was drastically reduced, two plans were discontinued because of labor difficulties, one plan was suspended and will restart after major production changes have been made.

Thirty-eight of the companies are nonunion and thirty-four are represented by 18 different international unions. The companies range from heavy machinery manufacturing, steel fabrication, foundries, automotive and truck parts, electrical goods, appliances and consumer products, chemicals, wood and upholstered furniture, plastics, lumber and wood products, textiles and fabric, rubber, oil equipment, construction, transportation, and warehousing. The companies represent a broad cross section of American industry with the exception of primary steel and nonferrous production. Three plants are in Canada, one in the UK and the rest in the United States, fairly spread around the country. A total of 19,860 employees work in the 72 companies.

Early Improshare Plans up to about 1980 usually covered only hourly employees; occasionally plant supervisors and some plant salaried were included. After 1980 almost all companies, union and nonunion, included hourly and salaried up to managers, excluding only those covered by a management bonus plan. The reasoning is that productivity improvement should be a companywide activity, with no we-they distinctions.

Table 1 shows the percent productivity gains achieved by each plant in the first 12 weeks of the plan's operation and then after a year. The weighted average increase after 12 weeks for all plants, including plans that were discontinued, was 9.4%. The average increase for all companies after one year was 22.2%. Eliminating companies that discontinued plans increases the first 12 weeks improvement to 10.3% and the yearly figure to 24.4%.

A distribution of productivity improvement percent is shown in Table 2, grouping by five percentage points. Most of the first 12 weeks' gains ranged between 5% to 15%, with a mean of 10.3%. The distribution of one year gains concentrated between 15% to 20%, with a larger number of plants skewed to the higher gain groupings, also reflecting the mean of 24.4%.

Statistical data hides some outstanding gains made by some of the top plants:

- A highly mechanized brick plant raised productivity by 30% in three weeks; savings in fuel for the kilns exceeded labor cost savings. The Steelworkers in the plant cooperated.

- A plywood plant raised productivity by 24% in four weeks, by 34% in 12 weeks, and averaged 34% at the end of a year. Output of this plant now exceeds the productivity of other similar plants in this large company. The Woodworkers Union wholly supported the program.

The financial benefits to employees and company can be seen from the following calculations. Since an Improshare Plan shares productivity gains 50/50 between employees and the company, the gain to both is the same. Using an hourly wage rate of $7.00 as an example, a 1% increase in productivity is worth $70 per year, per employee. The mathematics are: $7.00 x .005 (½ of 1%) x 2000 hrs per year). An increase of 10.3%, when shared, equals 5.15%. At this level, an employee's hourly earnings are increased by $.36, equal to $721 for the year. With a 24.4% increase, productivity sharing to the employee is worth 12.2%, equal to $.85 per hour and $1708 for the year. The savings for the company equals the total savings for all employees. Assuming a plant of 500 employees, at the 10.3% level the increased productivity will save $360,500; at 24.4% the company saves $854,000. Since employees are guaranteed not to make less than their regular earnings no one can lose under the plan.

<u>Poor results and failures with productivity sharing</u>

There are no data showing clearly why some productivity sharing plans produced meager results or were discontinued. Analysing plans that work well does not clearly show which factors contributed to the good results. In reviewing the plans that do poorly, it becomes clear that management's attitude toward worker involvement has a greater effect on results than any other factor, including the initial attitudes of the employees.

It would seem that managers anxious to improve their competitive position and their profits, who know the results that have been obtained by productivity sharing plans, would embrace this approach and proceed to bank the savings. It does not work that way; managers operating their companies for many years in a particular fashion often find it difficult to adjust to changes. Productivity sharing is not like piece work, under which a money hungry worker can work hard at his own job, caring little for those around him.

Productivity sharing is a way of life, a way of people working with other people; it recognizes that people are individuals with different needs and desires in life and at their work. A tradition-steeped manager may have trouble in adjusting to working in this people oriented system.

Table 1 shows that all five plants that discontinued Improshare had no productivity teams; three were union, two were nonunion. The two that discontinued their plans because of labor problems also had no teams; one was union and one was nonunion. Lacking clear data, it seems logical that managers in these five plants probably also had poor labor relations. From my contacts with the plants, I know there was low credibility and trust between management and employees in these plants. Managers who lack confidence in the capabilities of their work force and pay lip service to the involvement process, are perceived as not sincere by the workers, who then work as usual and remain uninvolved.

Productivity teams

Labor management committees (LMC), Quality of Work Life Programs (QWL), Quality Circles (QC) and Productivity Teams (PT), all have common goals, yet there are fine differences which depend more on what is meant in a given plant than on definitions of the terms.

Productivity teams afford management and the workers an excellent means for communication, especially important during depressed business conditions. In a number of plants, when workers were shown the economics of costing and how bid proposals are developed and the effect even small cost reductions can have on a company's ability to obtain orders, workers became more conscious of the need for cost reduction. Showing workers where customer orders stand, the backlog of open orders, and the problems and decisions facing management in determining how to operate a plant gives workers an understanding of management priorities. Open two way discussions remind managers that workers also have serious personal problems that need to be considered and production problems at their workplace which affect productivity. Open communication is one of the most important contributions made by productivity teams.

The concepts of LMC, QWL, QC, and PT are different in some respects and are similar in others. The major difference is whether the teams operate under nonfinancial programs or share in the productivity gains they produce. There is nothing about any of the team structures or methods of operation that prescribes whether or not employees should share in the gains; that is a management decision. The prime point of this paper is that when teams share in productivity gains, their actions, efforts and outlook are substantially different from when they do not share; superior results are attained.

Productivity Teams without rewards

In examining the workings of nonfinancial reward labor-management committees, starting with World War II and on, I found that these were not as effective in raising productivity as some writers reported, especially for periods beyond several years. Committees to solve crises have been successful as, for example, the LMC set up in Jamestown, New York, under Stanley Lundine, to create stable labor relations in the area. Similarly successful were the efforts by the union and management to save the plant and jobs at the Tarrytown General Motors plant. Such joint efforts can have long lasting effects. A classic joint effort was the mechanization and modernizing agreement between the Longshoremen's and Warehousemen's Union under Harry R. Bridges with the Pacific Maritime Association in 1961, which has stood the test of time.

In the absence of clear data on why productivity teams are more effective in raising productivity with financial reward plans than without them, I can only offer my opinion, which is simply that when workers are given a stake in the outcome, and they stand to gain materially through productivity improvement, they will operate at higher levels, cooperate more with management, and do things they would not consider doing in the absence of the rewards.

Productivity Teams under IMPROSHARE®

Of the 72 plants in Table 1, 51 had some form of worker involvement ranging from highly structured labor-management committees with steering and departmental committees, on through weak arrangements with informal meetings at irregular intervals.

Hard data are not available by plant to show the extent to which formal productivity teams increased productivity over the levels attained in plants where worker involvement was weak or nonexistent. Since I was personally involved in almost all the plants and am familiar with what transpired in all plants, I can report that productivity team activities definitely help to improve productivity, if only by creating and spreading receptive and supportive attitudes among the workers. Suggestions and improvements made by team members, together with technicians and engineers, are, of course, what team involvement is all about. Active teams goad managers to operate more effectively. The reinforcement process becomes contagious, success encourages greater activities. The most effective teams become important adjuncts to the efforts of the company's technicians, specialists and engineers.

The forerunner of Quality Circles

An entirely different approach was developed in the early 1930's by Allan H. Mogensen,

Lillian Gilbreth, David B. Porter and Erwin H. Schell, who suggested that workers should be involved in raising productivity by encouraging their participation in work improvement efforts. Their concepts, formalized in 1937 as Work Simplification, proposed training workers in how to make improvements using industrial engineering techniques and to work as teams in diagnosing and solving problems. The story of Mogensen and his associates is contained in "An Amazing Oversight, Total Participation for Productivity."[20] These far-sighted, involvement minded engineers were the pioneers of what later became known as QC Circles, as practiced in this country.

A Work Simplification program sought to train workers and supervisors in the principles of motion economy and how to make improvements in work methods, flow, planning, and other essential aspects of work. The concept was not opposed by workers in shops or by unions in the 1930's; it just did not catch on. During World War II I produced a training manual on Work Simplification principles for labor-management committees. The manual, PRODUCING FOR VICTORY, was widely circulated by unions and companies.[21] My recollection is that it did not do much to raise productivity.

We now know that the major reason Work Simplification was not widely accepted and used was that workers did not benefit financially from the gains that were made.

Productivity sharing works

Whether something works or not is often more important to a pragmatic manager or union official than why it works. Behavioral scientists are not so easily satisfied; they want to dissect the social forces at work and assign numeric values to their findings. Since all such evaluations are subjective and there is no way to assert with assurance that the numeric values assigned to factors and degrees in a matrix are valid, the values only show the author's opinion.

Disentangling each of the many factors which influence the attitudes of workers and managers, even in a small plant, is extremely complex if at all possible. Frost, Wakely and Ruh discuss the difficulty of dissecting a plan: "....the Scanlon Plan is a complex system of interdependent elements, rather than a single, unidimensional variable. The effectiveness of the Plan is, therefore, very likely beyond simple proof or disproof."[22]

We have ample evidence from company figures, discussions at conferences, and personal reports of people who have visited and talked with workers and managers at productivity sharing plants, that attitudes on both sides change, often dramatically. Concern for each other's welfare increases. Workers definitely want to know more about company problems and help to solve them. While psychologists may have difficulty in drawing proofs from their data, managers and workers know productivity sharing promotes improvements they can readily see and which benefit everyone.

In the face of these glowing reports, one may wonder why there are not many more companies working with productivity sharing plans. My answer lies in a simple story: I visit about 100 different companies a year. Each time I return from a trip my wife asks me how the trip went. I invariably give the same response: "I met some wonderful people." When she recently said, "You always say that," I suddenly realized that only people-oriented people will even consider sharing productivity gains with workers. A tough traditional boss prefers to rely on traditional managing.

Improshare is not an incentive plan in the conventional sense, though it shares productivity gains with employees. It is a philosophy of managing which is significantly different from the traditional way in which management views productivity improvement as a unilateral responsibility. In sharing productivity gains, employees accept management's productivity goals. The we-they adversary relationship changes so both gain together as they cooperate to produce more product or services in fewer man hours. Productivity impediments and losses affect both.

REFERENCES AND BIBLIOGRAPHY

1. SURVEY OF WORKING CONDITIONS, Survey Research Center, University of Michigan, US Government Printing Office, November 1970.

2. Mitchell Fein, "The Real Needs and Goals of Blue Collar Workers," The Conference Board RECORD, February 1973.

3. E. A. Locke, D. B. Feran, V. M. McCaleb, K. N. Shaw, and A. T. Denny, "The Relative Effectiveness of Four Methods of Motivating Employee Performance." PROCEEDINGS OF THE NATO INDUSTRIAL CONFERENCE, August 1979, edited by K. D. Duncan, M. M. Bruneberg, D. Wallis, John Wiley & Sons, London, 1980.

4. R. A. Katzell, D. Yankelovich, M. Fein, O. A. Ornati, A. Nash, WORK, PRODUCTIVITY, AND JOB SATISFACTION, The Psychological Corp., 1975, p. 12.

5. Mitchell Fein, "Motivation for Work," Section 11 in HANDBOOK OF WORK, ORGANIZATION AND SOCIETY, edited by Robert Dubin, Rand-McNally College Publishing Co., 1976.

6. Rollin H. Simonds and John N. Orife, ADMINISTRATIVE SCIENCE QUARTERLY, December 1975, Vol. 20.

7. Katzell, WORK, PRODUCTIVITY, AND JOB SATISFACTION, p. 38.

8. Mitchell Fein, "Work Measurement and Wage Incentives," INDUSTRIAL ENGINEERING, American Institute of Industrial Engineers, Norcross, GA 30092, September 1973.

9. David A. Weeks, COMPENSATING SALESMEN AND SALES EXECUTIVES, The Conference Board, Report No. 579, 1972, p. 6.

10. H. Fox, TOP EXECUTIVE COMPENSATION, The Conference Board, Report No. 793, 1980, p. 11.

11. L. J. Brindisi, "Survey of Executive Compensation," WORLD, Peat, Marwick, Mitchell & Co., Spring 1971, p. 52.

12. Brian E. Moore and Timothy L. Ross, THE SCANLON WAY TO IMPROVED PRODUCTIVITY, John Wiley & Sons, 1978.

13. C. F. Frost, J. H. Wakely and R. A. Ruh, THE SCANLON PLAN FOR ORGANIZATION DEVELOPMENT, Michigan State University Press, 1974.

14. A. W. Rucker, PROGRESS IN PRODUCTIVITY AND PAY, The EddyRucker-Nickels Co., 1952.

15. A. W. Rucker, GEARING WAGES TO PRODUCTIVITY, The EddyRucker-Nickels Co., 1962.

16. Mitchell Fein, IMPROSHARE®: AN ALTERNATIVE TO TRADITIONAL MANAGING, American Institute of Industrial Engineers, 25 Technology Park/Atlanta, Norcross, GA 30092, 1981.

17. General Accounting Office, PRODUCTIVITY SHARING PROGRAMS: CAN THEY CONTRIBUTE TO PRODUCTIVITY IMPROVEMENT?, US GAO, Document Handling & Information Services Facility, PO Box 6015, Gaithersburg, MD 20760, AFMD-81-22, March 3, 1981.

18. Ibid, p. 15.

19. Ibid, p. 18.

20. B. S. Graham, Jr. and Parvin S. Titus, THE AMAZING OVERSIGHT, TOTAL PARTICIPATION FOR PRODUCTIVITY, Amacom, 1979.

21. Mitchell Fein, PRODUCING FOR VICTORY, A LABOR MANUAL FOR INCREASING WAR PRODUCTION, Congress of Industrial Organization, 1942.

22. Frost, Wakely and Ruh, 1974, p. 151.

Table 1: **PRODUCTIVITY IMPROVEMENT OBTAINED IN IMPROSHARE® COMPANIES**

Company	Manufacturing	Total Empl	Union	Year estab	% Gain (1) Start	% Gain After 1 yr	Still oper(2)	Prod team
1	metal stampings	75	IUE	1974	12	32	yes	no
2	automotive parts	200	AIW	1976	22	70	yes	yes
3	steel fabrication	375	Teamsters	1980	8	8	yes	yes
4	steel products	500	IAM	1978	11	40	yes	yes
5	mobile trucks	100		1980	6	11	yes	no
6	consumer plastics	125		1978	23	65	yes	no
7	pleasure boats	125		1978	23	42	yes	no
8	outdoor products	350		1979	20	32	yes	no
9	machine shop	300	UK	1977	8	15	NV	yes
10	metal stampings	150		1979	7	18	yes	yes
11	machine shop	85		1978	13	36	yes	yes
12	metal treating	40		1978	8	16	yes	yes
13	automotive	1550	Rubber	1980	3	8	yes	yes
14	electrical products	800		1981	8	16	yes	yes
15	photocopy products	500		1980	9	15	yes	yes
16	heavy machinery	1600	IUE	1979	12	52	yes	yes
17	foundry	275	Moulders	1979	11	52	sus	yes
18	auto parts	750	UAW	1979	15	32	yes	yes
19	trucks	250		1976	13	32	yes	yes
20	automotive	200		1978	7	16	NPC	yes
21	automotive	800	UAW	1980	3	8	yes	yes
22	electrical	250	UAW	1980	3	6	yes	yes
23	consumer	35		1979	10	30	yes	yes
24	consumer	120		1981	7	(3)	yes	yes
25	industrial	125		1979	8	42	NPR	yes
26	chemical	675	OCAW	1980	9	16	yes	yes
27	chemical	650		1980	6	12	yes	yes
28	appliances	400	IUE	1979	20	32	NPC	yes
29	electrical	75		1981	5	10	yes	yes
30	chemicals	375		1979	3	10	NV	no
31	paper	150	Paper Wkrs	1978	8	44	yes	yes
32	paper	1000	USW	1980	2	3	NL	no
33	chemical	450		1980	9	15	NV	yes
34	lumber	40	Woodwkrs	1980	10	18	yes	yes
35	clay products	65	USW	1981	30	30	yes	yes
36	lumber products	85	Woodwkrs	1981	24	34	yes	yes
37	concrete	20		1976	8	22	yes	no
38	plastics	250		1981	6	18	yes	yes
39	steel wire	250	USW	1981	11	26	yes	yes
40	machine shop	125		1981	8	26	yes	yes
41	machine shop	250		1981	9	16	yes	yes
42	sheet metal	50		1979	8	22	yes	yes
43	wood products	175		1979	4	22	yes	yes
44	furniture	300		1977	8	19	yes	no
45	upholstered furn.	200		1979	11	15	yes	no
46	furniture	35	Brick Clay	1978	12	70	yes	yes
47	upholstered furn.	100	Furn Wkrs	1979	12	16	yes	no
48	upholstered furn.	125		1979	9	20	yes	no
49	wood furniture	450		1981	16	31	yes	no
50	leather	55	AMC	1976	12	25	yes	yes

Table 1 (cont)   PRODUCTIVITY IMPROVEMENT OBTAINED IN IMPROSHARE® COMPANIES

Company	Manufacturing	Total Empl	Union	Year estab	% Gain (1) Start	% Gain After 1 yr	Still oper(2)	Prod team
51	fabric	100	Textile	1980	12	44	yes	yes
52	warehouse	50	Textile	1980	18	75	yes	yes
53	consumer	65		1977	18	35	yes	yes
54	consumer	250		1978	12	18	yes	yes
55	electrical	65		1978	35	55	yes	yes
56	metal stampings	90		1978	16	35	yes	yes
57	metal products	35		1978	10	25	yes	yes
58	metal furniture	550	Teamsters	1981	10	55(4)	yes	yes
59	chemical	300	Teamsters	1979	0	0	no	no
60	automotive parts	110	Independ.	1980	11	28	yes	no
61	consumer	140		1978	12	22	yes	yes
62	trucking	75		1975	5	0	no	no
63	consumer	550	IUE	1979	6	8	no	no
64	fabric	15	UAW	1981	15	(3)	yes	no
65	machine shop	65	UE	1981	9	(3)	yes	yes
66	electrical	155		1981	21	(3)	yes	yes
67	electrical	160	IUE	1981	19	(3)	yes	yes
68	oil equipment	600		1977	6	0	NL	no
69	machine shop	110	Teamsters	1980	8	10	yes	yes
70	sheet metal	85	Teamsters	1978	30	58	NPC	yes
71	plastics	220	Teamsters	1977	11	8	no	no
72	construction	200		1978	7	0	no	no

Notes:

(1) first 6 weeks
(2) as of 2/10/82
(3) less than one year — Total plants 5
(4) expected year end — 1

Plans not operating:

		Total plants
Sus	suspended, will restart	1
N	No, discontinued	5
NPC	No, plant closed	3
NPR	No, plant rebuilt	1
NV	No, volume reduced	3
NL	No, labor problems	2

Summary increase percent

	12 wks	1 yr
Average all companies	9.4	22.2
Average companies continued plan	10.3	24.4

Author

Mitchell Fein is a consulting IE primarily concerned with solving the people problems that arise between employees and management to retard productivity. Fein developed the IMPROSHARE® Plan. Since 1937 he has served over 500 companies in a broad range of industries, concentrating on developing ways of increasing productivity. He has written numerous articles for technical and trade journals and lectures extensively to management groups in the US and abroad. He is a Fellow of AIIE, was Adj. Assoc. Prof. of IE at New York University, serves on arbitration panel of the American Arbitration Assn., is a PE in New York and New Jersey, and is a CMC.

II. Direct Timing and Performance Rating

Direct observation timing—time study—traditionally has been one of the most commonly used techniques to set production standards. Even though predetermined time systems have been growing in popularity in recent years, direct time measurement is still used by many organizations. Direct observation timing may be defined as a technique to determine the time required by a qualified and well-trained person working at a normal (average) pace to perform a specified task under standard environmental conditions. The Industrial Engineering Terminology Standard Z94.12 defines time study as "a work measurement technique consisting of careful time measurement of the task with a time measuring instrument, adjusted for any observed variance from normal effort or pace and to allow adequate time for such items as foreign elements, unavoidable or machine delays, rest to overcome fatigue, and personal needs. Learning or progress effects may also be considered. If the task is of sufficient length, it is normally broken down into short, relatively homogenous work elements, each of which is treated separately as well as in combination with the rest."

It is important that the conditions specified in the above definitions be adhered to in order for the time study to be valid. First, the individual in the study must be qualified and well-trained. He or she must be instructed in the proper method and have had sufficient time to practice the job. Second, the worker should be performing at a rate or effort level that would be expected of a conscientious, self-paced employee when working neither fast nor slow and giving due consideration to the physical and mental requirements of the specific job. Third, the task must have a well-defined beginning and end. Fourth, the task should not be measured at times when atypical environmental conditions are present. Finally, it is important that all personnel, including the operator, the manager, and the bargaining unit representative, be informed of the study, thus allowing all parties to prepare for it properly. If any of these prerequisites are not met, the results will probably not be representative of the true time required for the task.

TIMING EQUIPMENT

The equipment used for direct observation timing ranges from a simple mechanical or electronic stopwatch to a video camera linked to a computer. The traditional apparatus is a decimal stopwatch (or stopwatches) mounted on a time study board to which the time sheet is clamped. There are two commonly used methods for the actual timing of elements: snapback and continuous. In the snapback method, at the end of each element the observer reads the watch and simultaneously resets, or snaps back, the watch to zero. In the continuous method, the watch runs continuously, and the observer notes and records the time at the end of each element. Each methods has advantages, but the continuous method is usually preferred. The snapback method saves the time required to compute elemental times, but takes more time overall because of the need to reset the timing device repeatedly to zero.

Today, microprocessor input devices are increasingly used to record time study observation data, thereby minimizing recording problems. Most of these devices can be linked to a computer, thus eliminating arithmetic calculations and associated errors. Whatever method and equipment are used, observers must be well trained in their application.

PROCEDURE FOR DIRECT OBSERVATION TIMING

The following steps are typical of a time study:

1) determine the operation and the operator to be studied;

2) verify the method for correctness;

3) notify all involved parties and obtain their cooperation;

4) divide the cycle (operation) into proper elements;

5) determine the number of cycles needed to ensure accuracy and confidence level;

6) observe and record the element times;

7) rate the operator's performance;

8) apply allowances;

9) compute the time standard and complete required documentation supporting the study.

The operation to be studied must be specified. It is then decided which operator will be studied. This decision should be made by the time study analyst and the foreman or manager, in order to ensure that a "qualified and well-trained" worker is selected. If possible, a worker representing the normal or average is selected to minimize the rating difficulty.

The operator and bargaining unit representative are then notified. This assures that the worker has a chance to voice an opinion as to any difficulties associated with the job. The bargaining unit official also has a chance to express any concerns or to bring in the bargaining unit's time study person. At this time, all parties involved review the operation to ensure that the proper method is in place and is being followed, and that there are no outside factors hindering the performance of the operator. All problems must be resolved before data recording begins.

A detailed study of the operation is performed to determine exactly how to divide the cycle into elements. The elements should be chosen by several criteria. First, the elements should be as short as possible but of sufficient duration to be accurately timed. Any machine-paced time should be separated. If possible, select elements so that a characteristic sound or motion can be used as a terminal point. Finally, separate any constant times from those that are variable. A complete description of each element should then be recorded and the study form or computer input prepared.

The actual recording of times can now proceed. The analyst should notify the worker ahead of time as to when the times will be taken. The analyst should also have the proper attitude and posture while the timing is being conducted. The worker should not alter his or her pace or routine, nor should there be any unnecessary interruptions of the work; and the analyst should not interfere or converse with the operator, but should maintain the proper position to observe the entire task. The analyst should record sufficient cycles to ensure a fair sampling of normal conditions. Enough data should be collected so that the results obtain the desired (designed) statistical significance. (This requirement is discussed at length in Mitch Fein's article "How 'Reliability,' 'Precision' and 'Accuracy' Refer to Use of Work Measurement Data," which appears in Chapter I of this volume.) During the timing, the worker must be rated (leveled) and compared with normal or average for that location. The importance of this rating cannot be overemphasized; it should be performed only by trained personnel.

Now the results of the study can be computed by summing the rated or leveled elemental times and dividing by the number of observations to obtain the normal time. This value is then increased to include allowances, and becomes the standard time. (Allowances and the impact of fatigue are discussed in Chapter III.)

A final note—in using any work measurement technique, including time study, it is important to consider the cost/benefit ratio of the particular method. (The concept of cost-effective technique selection is covered in Chapter I in the articles "A Procedure for an Economic Comparison of Work Measurement Techniques, Part I: The Model" and "Part II: An Application" by Keith Adams and Timothy McGrath.)

PERFORMANCE RATING

Performance rating is the procedure to adjust the observed time values so they coincide with the times that would be required by a "normal" or "average" worker. The Industrial Engineering Terminology Standard Z94.12 defines performance rating as "(1) The process whereby an analyst evaluates observed operator performance in terms of a concept of normal performance. (2) The performance rating factor."

The concept of normal performance is not standardized and varies from organization to organization. A number of factors influence the established level of normal performance within a specific organization, such as geographic location, products or services being produced, processes utilized, quality requirements, and the workforce in place. In general, the worker should perform at a rate that is reasonable for that employee population, be qualified and well-trained for the job, and use all abilities to best advantage. The established level of normal performance within a worker population at a specific location typically remains fairly constant. Raising or lowering this level usually requires considerable time and effort. Exceptions can result if a large turnover in labor force occurs within a short period of time, if an incentive program is introduced, or if any other major event influences the total worker population.

Some of the terms associated with performance rating are speed/effort rating, pace rating, objective rating, multi-factor leveling, and synthetic rating. In speed/effort rating, the operator's speed is compared to a normal speed commonly attained by average workers within a work group population, and the difference determines the rating factor. This procedure makes no allowance for variations in methods or worker technique which may affect the actual output level.

Pace rating employs an additional concept. This method recognizes that all jobs are not equal in terms of tempo, and so the speed is judged in relation to the normal speed for that type of work. The analyst establishes a normal pace for the job through viewing a series of films of similar operations.

Objective rating is similar to pace rating but divides the activity into two steps. First, the pace of the operation is judged in comparison to a normal pace for all jobs. Then each job rating is adjusted to compensate for the difficulty of the operation.

Several rating methods utilize multi-factor leveling systems. Four factors are commonly used to obtain the final rating: skill, effort, conditions, and consistency. The analyst determines numerical values for each, with skill and effort usually weighted most heavily.

Synthetic rating removes the need for subjective judgment by the analyst. This is achieved by obtaining selected elemental times and comparing their values to known standards such as MTM for those elements, and deriving the rating factor as a ratio of the two.

SUMMARY OF ARTICLES

The lead article in this chapter, "Industrial Worker Pace Variability—A Study with Real Time and Posterior Analysis," by S.Y. Nof, S.D. Ansell, Haim Gershoni, and Ayala Cohen, reports that the pace of manual workers assigned to operator-controlled tasks varies during the working day. Particularly, the authors noted a wave-like pace variability in their experiment. To study this phenomenon, they programmed and installed a real-time minicomputer system to record the cycle time of six factory workers and simultaneously display on an oscillograph the moving average of each worker's output. Hourly pace varied significantly around the workers' daily mean, and the wave-like variation pattern occurred with both low and high frequencies. Commonly proposed factors of variability such as boredom were also observed. Recovery from micro-errors is proposed as the cause of the high frequencies found. However, the pace variability had no consistent pattern per worker or per day. This research study supports the need for consistent rating practice during any direct timing operation.

Rodney K. Shultz and Thomas H. Campbell report in "The Worker's Perception of a Change in Work Rate" that many workers are able to notice small inaccuracies in standard times. The experiment utilized both short- (0.5 minutes) and long-cycle (2.31 minutes) assembly tasks. Each task session consisted of a base work period, a rest period, and a second comparison work period; multiple assembly tasks were performed in each work period. In general, the subjects detected a change when the work rate between periods was either increased or decreased. The duration of the task, the length of the rest period, and the magnitude of the difference between the rates for the two work periods did not affect the perceptions of the work rates. However, the study did show that slower work pace rates were overrated and faster work rates were underrated. The errors were largest when the rates were significantly different from 100 percent normal. This study showed that a worker performing a task may be able to determine work rates better than some time study analysts. This fact would explain worker sensitivity to inconsistencies in performance rating by analysts. The results also confirm the long-existing practice of selecting a worker who closely represents normal or average for time study participation.

In "An Analysis of Performance Rating," Robert W. Wygant presents a study reviewing performance rating data. The results confirm the view that most time study analysts—even those new to the discipline—tend to rate conservatively, that is, to underrate high levels of performance and overrate low levels. The author attributes the tendency to a failure to minimize subjectivity in performance rating, to biases built into the teaching of the method, and to factors in the work environment that reinforce those biases.

Proper training can improve performance rating techniques. In "New Way to Learn Pace Rating" Vincent G. Reuter explains a method for improving training films in the field. After showing half of a normal pace scene, the film is stopped, and the students rate the scene. The true rating is then given to the students, and the other half of the scene is shown. This method has produced impressive results.

"Performance Rating for Service Jobs" by Robert A. Brown and William F. Sowder describes an experiment in which the observational methods of Berne's Transactional Analysis (TA) are investigated to see if they can be used to create profiles of personnel involved in service activities. The study showed that trained observers can create profiles reliably; that very short observation periods, in some cases of a minute or less, are sufficient to apply the principles of work sampling; and that exposure to TA concepts can alter the measured behavior of a subject. The authors recommend applications for personnel selection, training, and placement test validation.

Industrial Worker Pace Variability--A Study with Real Time and Posterior Analyses

S. Y. NOF
SENIOR MEMBER, AIIE
School of Industrial Engineering
Purdue University
West Lafayette, Indiana 47907

S. D. ANSELL
Research Coordinating Unit
Wisconsin Board of Vocational, Technical,
and Adult Education
Madison, Wisconsin

H. GERSHONI
Industrial and Management Engineering
Technion–Israel Institute of Technology
Haifa
Israel

A. COHEN
Industrial and Management Engineering
Technion–Israel Institute of Technology
Haifa
Israel

Abstract: Evidence has been reported that the pace of manual workers varies during the working day. Particularly, a wavelike pace variability was noted. In an effort to study this phenomenon a real-time minicomputer system was programmed and installed to record the cycle time of six factory workers and simultaneously display on an oscillograph the moving average of each worker's output. An observer monitored the experiment to determine if work pace variations truly occurred and if so, to establish their nature and identify their cause through observation of the workplace. The computer system proved to be an excellent means of conducting such an experiment. Hourly pace varied around the worker's daily mean with ranges of between 14% to 64%. The wavelike variation pattern was indicated both with low frequencies of between 20 to 80 work cycles and with high frequencies of between 2 to 7 work cycles. Commonly proposed factors of variability such as boredom were observed. Recovery from micro errors is proposed as the cause of the high frequencies found. However, the pace variability had no consistent pattern per worker or per day.

■ During the course of several research projects carried out at the Technion – Israel Institute of Technology (Gershoni [7], [8], [9]), it was noted that the pace of workers performing a repetitive task seemed to vary non-randomly. It would appear that the pace of manual workers slowly oscillates between two extremes, as the data from one of the laboratory studies show in Fig. 1. Here it can be seen that the moving average of 100 observations plotted throughout the day clearly illustrates a distinct wavelike tendency. It is this type of wavelike performance pattern with which this study is concerned.

Production variability in repetitive unpaced work was noted by early observers (Wyatt and Fraser [27], Wyatt and Langdon [29], Davis and Josselyn [4]); they noticed a decrease in productivity during the work period towards the end of the day. This phenomenon was attributed to fatigue and boredom. Actually, these researchers found that the cycle time became longer during the day, and that small pauses in work appeared more often later in the day. A similar increase in time per unit was reported by Bartlett (Murrell [16], p. 382). However, he also noted some significantly shorter cycle-times relative to the usual worker's performance, namely, periods of fast

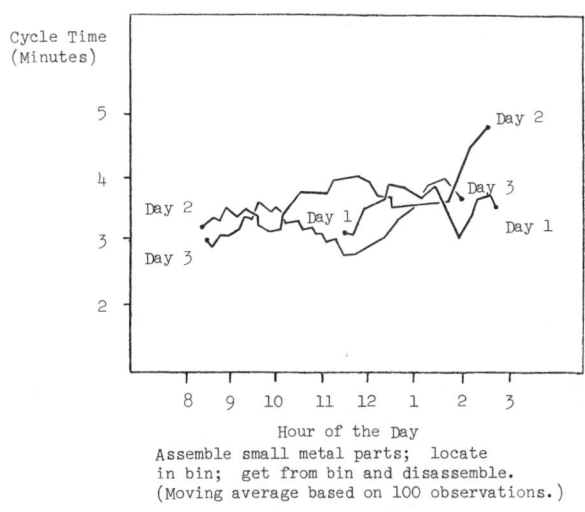

Fig. 1. Moving average for worker performing manual task in a laboratory.

Paper received June 1975; revised August 1977, April 1978.

work followed by rest. Contradictive findings (e.g., by Wyatt and Fraser [28]) indicated reduced time per unit towards the end of the day. Frequency distributions of the cycle times for repetitive unpaced work (Murrell [15]; Dudley [6]) showed that in every case the resultant distribution for skilled and motivated workers was positively skewed toward the shorter times, with the majority of cycles not substantially longer than the minimum.

In repetitive elemental motions a variation in the cycle time was reported by Schmidtke and Stier [22]. The standard deviation, or coefficient of variation of the cycle time in repetitive work, was likewise noted to vary (Wyatt and Langdon [29], Gershoni [8]). This variation was suggested as an indicator which could be used to schedule rest periods at the end of active periods (Murrell [16]). The wavelike pattern which was discussed above had been noted as periodicities in the performance of a repetitive task in an experiment by Murrell and Forsaith [17]. It was found that the output fluctuated on a 45 minute cycle, while irregular long cycle times ran on a 75 minute cycle. Such periodicities may indicate that the cycle times are not independent of shift time. Sury [24] tested serial dependence with an autocorrelation technique using a lag of one observation in a sequence of unpaced repetitive cycle times. In all cases a low and non-significant coefficient of correlation was found except for a single case, where the correlation was found to be significant.

Objectives

In an effort to more fully understand the phenomenon of wavelike variation of the worker's cycle time, this research was undertaken with the following objectives:

a) Development of a device for the automatic monitoring of the work performance of several workers. The purpose of this monitor was to record cycle time and to indicate to the observer variation in output pace at the time of occurrence.

b) Establish whether in an actual factory environment any wavelike patterns of work pace occurred.

c) If the pace variation did occur, determine if the pattern of variation correlated with any of the expected causal factors.

Six workers were monitored in order to provide an adequate sample and also to investigate possible interactions or different kinds of behaviors.

The Experiment

Plant and Workers

The experiment was conducted during seven consecutive work days in a shirt factory with women workers performing four tasks as shown in Table 1. Each worker had been performing her task between four and six months. They all had a similar background and education, were from the same age group, and were single (except for worker 4).

The workers worked 460 minutes a day, starting at 7:00 a.m., with a 20 minute coffee break at 9:00 a.m., and a lunch period of 30 minutes at 1:00 p.m. The work day ended at 3:30 p.m.

The workers worked under an incentive plan. They were paid a fixed salary plus productivity premium directly related to their performance above the norm (1% bonus per 1% increase in good production).

The work pace was controlled by the worker herself. In the sewing tasks a foot lever was used to regulate the speed of the sewing machine, and in the pressing operation the manual elements dominated the work cycle.

All four tasks were similar in their complexity. The elements of each task were unchanged during the experiment. After batches of production had been completed they were inspected by the supervisor, and defective items were later corrected by an available worker, and sometimes by the supervisor. However, those operations were not included in the experiment, and defective items were not monitored.

Because of absenteeism, some workers missed one or two days of the experiment. Consequently, data are missing in some cases from the tables and figures. Where data are missing it is clearly indicated.

Table 1: Operators, tasks, and overall results.							
				Experiment Results			
Task	Standard Time (Min.)	Worker	Age (Years)	Mean Time (Min.)	Coefficient of variation (%)	Skewness	Kurtosis
A Collar turning and pressing	.40	1	16	.57	38	2.4	14
B Collar stitching	.58	2	19	.56	41	2.4	9
		4	25	.58	34	2.2	9
C Pocket sewing	.55	3	17	.93	34	-0.2	3
		5	16	.83	33	1.5	6
D Button sewing	.58	6	16	.60	43	1.8	7

Apparatus

This system was designed to provide an observer with real-time information so that he could, on-the-spot, discover and link the causes with the fluctuations in workers' pace. Further, it collected the data required for subsequent analysis.

The apparatus was based on an Elbit 100 minicomputer (Fig. 2). Inputs from each worker signalled the completion of each cycle. The worker accomplished this by pressing a switch plate located on the work table (Fig. 3). Within a very short time this signaling motion became quite automatic to the worker.

A real-time clock within the computer fed timed impulses into the computer which, in essence, allowed the computer to record in its memory the elapsed time since the completion of the previous cycle for each worker.

In order to provide information to the observer on the performance of the workers in a useful format, the computer was programmed to compute for each worker the moving average cycle time for the ten most recent cycles and to plot this on an oscillograph. A ten-cycle period was chosen for this display to give adequate smoothing of the curve with a relatively short time delay. It was found in other experiments that the moving average is a clear indicator of the pace wavelike variation when it occurs. At the end of each cycle, a paper tape was punched with each individual cycle time, time of day, and the worker's identification.

Fig. 2. The minicomputer system with teletype, oscillograph, and a high speed punch as used in this field study.

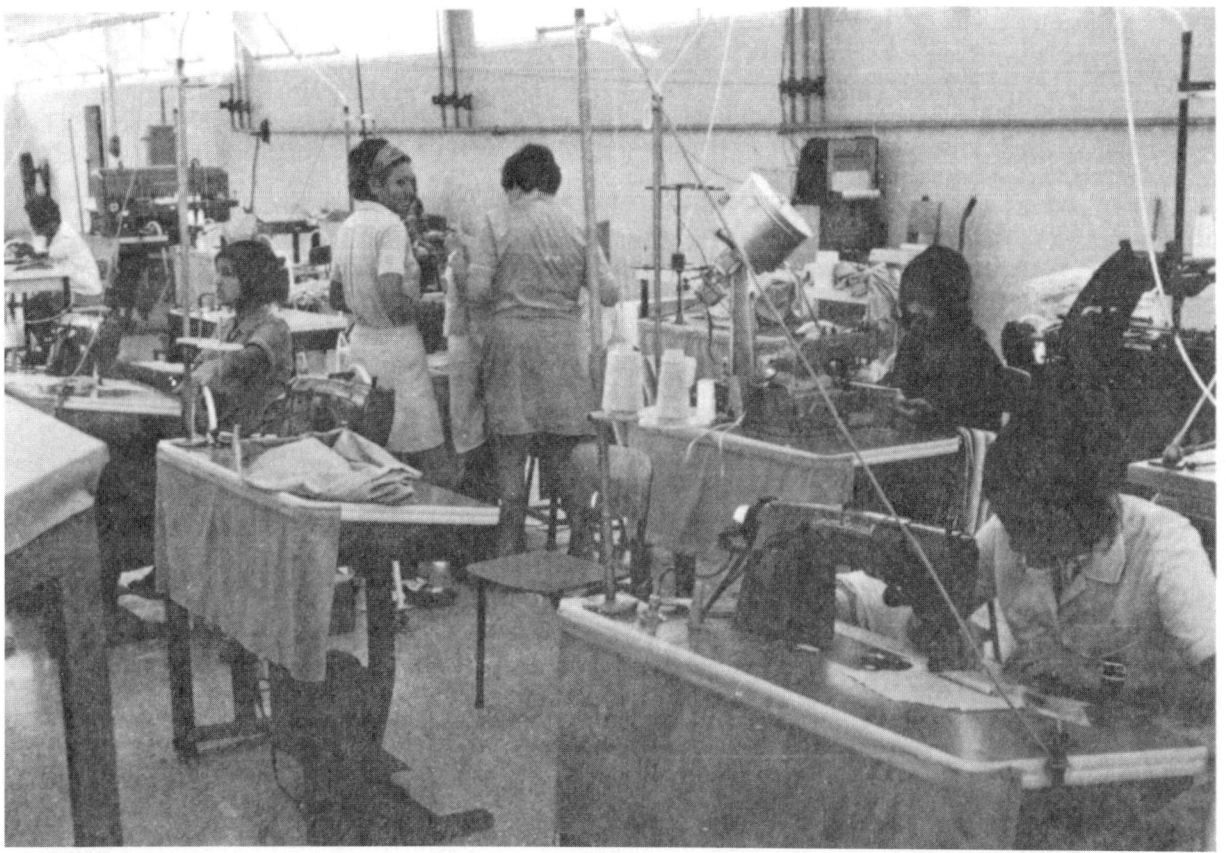

Fig. 3. Workers at workplace. Note switch plates used to signal the beginning of a new cycle to the computer.

Results

Automatic Monitoring of Workers

Use of the minicomputer was completely successful. The observer was able to follow the progress of six workers without having to personally record any data or make any calculations. The oscillograph automatically presented a graphic display of the current moving average cycle time, illustrating peaks and valleys of output (Fig. 4). Hence the observer was not only able to devote full attention to the worker's behavior but also knew from the oscillograph when wave peaks occurred for each worker and could accordingly direct his search for causes.

Since the displays of the moving average of cycle times for all six workers were displayed simultaneously on different channels of the oscillograph, common factors which might affect the performance of all the workers could be immediately discerned.

An interesting phenomenon was noted during the experiment. As the graphic display showed the production of the workers, they could obtain feedback of their personal productivity (i.e., output rate) relative to the other workers, as soon as they looked at the oscillograph. This could reduce the feedback period (usually a day) to a much shorter lag. Indeed, immediately as a break started (at 9:00 a.m., at 11 a.m., and at the end of the day) the workers would typically rush to the display to study the performance. Potentially, this could create a strong peer pressure to come to a level of production accepted by the group. However, at least during the experimental period, effects were quite the opposite. The capability to compare personal production with relatively short feedback delay generated a constructive competition, as a result of which significant increases in worker productivity were noted during the experiment. (Increases of 50% to 100%, and in one case of 400%, compared to the pre-experimental period were noted by the factory manager; however, there is no documentation to substantiate this.) This finding leads to questions such as what are the consequences of longer or shorter feedback periods, and whether productivity improved due to the availability of faster feedback, due to the introduction of change (i.e., the experimental system), or due to a change in the production method. Such questions, however, were beyond the scope of this study.

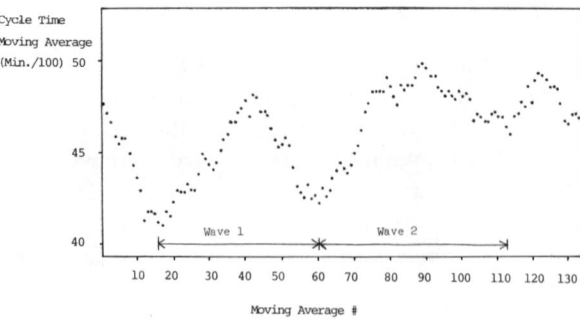

Fig. 5. Moving averages of cycle time for one worker (based on twenty (20) observations).

Pattern of Pace Variation Occurring in a Factory Environment

Wavelike Pace Variation

In the course of this experiment wavelike variations were not only clearly seen in the real-time display (see for example Fig. 4), but were more dramatically shown in the posterior analyses. Figure 5 illustrates the "waves" of a worker based on the moving average of twenty successive observations. (A general comment about the basis for the moving average: using 50 and 100 observations as the basis for the moving average resulted in over-smoothing of the observed data. On the other hand, using 3, 5, or 7 observations for the moving average basis yielded insufficient smoothing effects. The basis of 20 observations in this analysis yielded enough smoothing without losing the typical variations. Note that a basis of 100 observations was useful in Fig. 1, but the cycle time there is about 6 minutes, compared to the less than 1 minute cycle times in this experiment.)

Two types of wave patterns were actually found in this experiment: high frequency waves and low frequency waves.

We define low frequency waves here as those waves that are visible from the data, such as in Fig. 5. High frequency waves were indicated by spectral analysis, as described below, but they are not easily visible.

Studying the moving averages of each worker it was found that the cycle time mean fluctuates during the day between peaks and valleys. These fluctuations are actually increase or decrease trends of performance, and the worker does not abruptly change her mean cycle time.

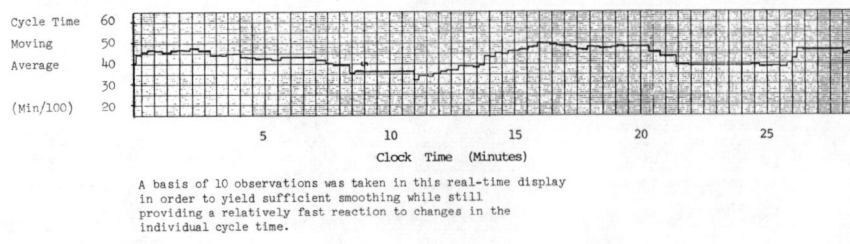

Fig. 4. Information displayed to the observer for one of six workers.

To illustrate the low frequency wave pattern, Fig. 5 shows a wave of 45 observations (from 15 to 60), followed by a 55 observation wave (from 60 to 115).

Although the low frequency waves are non-periodic, namely, consecutive wave periods are not identical, they were found to be in the range of 20 to 80 work cycles. These wave periods (of increase followed by decrease and again increase trends) correspond in this experiment to periods of 12 to 75 minutes, compared to 45 to 75 minute periods found by Murrell and Forsaith [17] for the actile period.

In order to determine whether a periodic wave really exists in the data and is not only apparent graphically, a spectral analysis test was applied. The theoretical background of spectral analysis can be found in various sources, e.g., Lewis [12]. In essence, the spectral analysis identifies the spectrum of frequencies in the data, and indicates which frequencies are more dominant.

The spectral analysis identified dominant high frequencies, namely, short cyclic periods in the data for each worker in each day. In the example of Fig. 6, worker 3 in day 3 has a dominant frequency of 2.3 radians, or about 2.7 work cycles per wave cycle (identified by the highest peak in the diagram). In day 5 worker 3 has a dominant frequency of 2.1 radians, or about 3 work cycles per wave cycle.

The frequencies identified by the spectral analysis were in the range of 2.2 to 6.7 work cycles per wave cycle for all workers. Thus, we may conclude that superimposed on the low frequency, non-periodic waves apparent in Fig. 5 there might also be a high frequency wave.

No uniformity among workers was found regarding the low or the high frequency waves, i.e., the workers did not have the same frequencies during the same time of the day. Moreover, the worker's frequency was not consistent from day to day. For example, Table 2 lists the high frequencies found for worker 2 during the experiment.

Table 2: Cycle time waves for worker 2.	
Day	Work cycles per wave cycle
Tue	3.2, 4.6
Wed	2.3, 5.4
Thu	3.0
Fri	2.5
Sun	2.2, 2.7
Mon	3.5
Tue	3.3

A comment about the spectral analysis: applying the spectral analysis on the cycle time data resulted in high values of the spectrum for the low frequencies because of the high serial correlation of the data. As a result, there was not sufficient information about the spectrum of the high frequencies. In order to suppress this effect of the serial correlation filtering by first differences was used (i.e., instead of the data themselves the differences between consecutive observations were computed and then used as the input to the spectral analysis).

While high frequencies in the spectrum obtained were significant, the low frequencies were not. The reason might be that in the filtering process the low frequencies were reduced or that not enough low frequency waves of the same period occurred in one day's data.

In an effort to study the peaks and valleys between which the pace wave fluctuated, extreme hourly cycle time averages were compared with the daily mean (see Table 3). For example, worker 2 on Sunday performed the task with a daily cycle time mean of .61 minutes. However, during the seventh hour of her work her mean was just .50 minutes, or her output rate 18% higher than the daily mean. Similarly, during the sixth hour of her work her mean cycle time was .68 minutes, or her output rate 11% less than the daily mean.

Using analysis of variance to test whether the maximum and minimum hourly means are statistically equal we found: in six cases, for workers 1, 2, 4, and 5, the maximum and minimum hourly means may be considered equal. In all other 21 cases (we ignored the first Tuesday's data to allow for the initial learning period with the experimental system) we can say that the two daily extreme means are not equal, with significance level of .05 in 10 cases and .01 in 11 cases.

Thus, it would appear that not only did the wavelike variation occur, as seen in Fig. 5, but that the peaks and valleys were of considerable difference. Some periods of stable cycle duration did occur occasionally, but not with any regularity. When stability did appear, it occurred within the general pattern of wavelike variability.

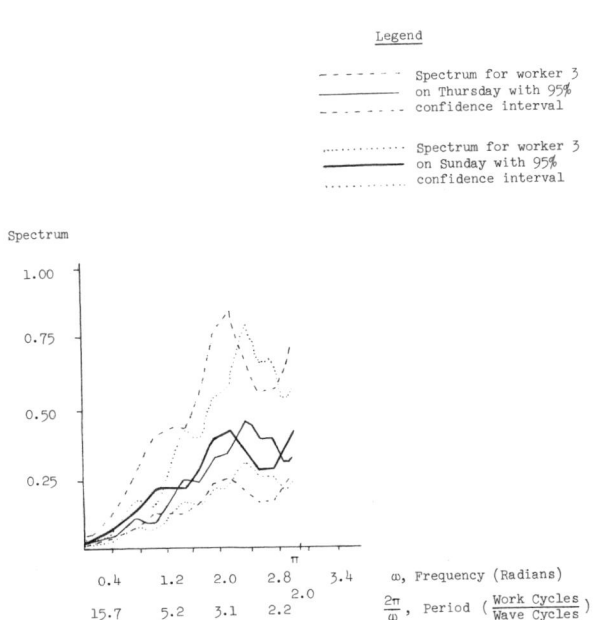

Fig. 6. Results from spectral analyses.

Other Cycle Time Statistics

The overall results of the experiment in terms of the mean cycle time, the coefficient of variation (the standard deviation divided by the mean), the skewness and the kurtosis

Worker	Day	Mean (Min)	Coef. of Variation (%)	Min. Mean (Min)	Hour of the Day	Max. Mean (Min)	Hour of the Day	Difference from Daily Mean (%)	±2 Standard Deviations	±3 Standard Deviations
				Extreme Hourly Cycle Time Means					% Outside Control Limits	
1	Tu*	.77	56	.61	11:00	1.51	7:00 9:00	-21/+96	7	4
	W	.56	29	.47	12:00	.59	10:00	-16/+5	3	0
	Th	.53	30	.46	2:00	.59	7:00	-13/+11	4	2
	Su	.60	38	.52	9:00	.70	12:00	-13/+16	4	2
2	Tu*	.71	54	.52	1:00	.97	12:00	-27/+37	8	2
	W	.54	40	.49	8:00	.58	11:00	-9/+7	7	4
	Th	.52	46	.47	11:00	.66	7:00	-10/+27	8	3
	F	.53	33	.47	10:00	.60	12:00	-11/+13	5	3
	Su	.61	32	.50	1:00	.68	12:00	-18/+11	5	3
	M	.59	40	.47	11:00	.68	2:00	-20/+15	5	2
3	Tu*	.85	50	.70	2:00	1.23	9:00	-17/+45	5	1
	W	.57	50	.45	11:00	.78	8:00	-21/+37	6	0
	Th	1.03	20	.91	11:00	1.11	7:00	-12/+8	6	2
	Su	1.14	18	1.06	10:00	1.36	1:00	-7/+19	2	0
	M	1.05	15	.99	8:00	1.13	11:00	-6/+8	5	0
	Tu	1.08	17	1.02	12:00	1.20	9:00	-6/+11	4	2
4	Tu*	.72	41	.62	12:00	.76	11:00	-14/+5	8	4
	W	.66	28	.61	2:00	.72	8:00	-8/+9	6	3
	Th	.57	25	.53	10:00	.66	2:00	-7/+15	5	3
	F	.80	26	.76	8:00	.87	9:00	-5/+9	10	3
	M	.46	43	.43	11:00	.51	9:00	-7/+11	7	3
	Tu	.56	38	.50	9:00	.74	8:00	-12/+32	6	4
5	Tu*	1.04	45	.70	12:00	1.30	9:00	-32/+25	6	0
	W	.83	33	.71	11:00	1.04	8:00	-14/+25	3	0
	Th	.72	25	.64	9:00	.85	8:00	-11/+18	6	2
	Su	.90	27	.81	2:00	1.22	8:00	-10/+35	5	0
	M	.91	29	.82	2:00	1.07	9:00	-10/+17	4	0
	Tu	.79	27	.74	8:00	.99	7:00	-6/+25	5	1
6	Tu*	.79	66	.60	10:00	1.21	9:00	-24/+53	8	0
	W	.59	37	.50	12:00	.68	9:00	-15/+15	6	2
	Th	.53	39	.50	8:00 10:00 1:00 2:00	.70	12:00	-6/+32	6	3
	Su	.68	41	.48	1:00	.92	10:00	-29/+35	8	0
	M	.67	40	.57	8:00	.79	7:00	-15/+18	7	0

*The first Tuesday includes the learning period.

were summarized in Table 1. The main purpose of the following discussion is to compare results of this experiment with some previous findings.

The actual mean cycle time of workers on tasks B and D were very close to the standard time set by the factory management. On the other hand, the workers on tasks A and C missed their standard time by about +40% to +70%. Since there does not seem to be any apparent reason for this, we have to conclude that it relates to the workers' personal performance, or to the accuracy of the standard.

The properties of the cycle time frequency distributions usually agree with previous studies, as described in the introduction. Except for worker 3, the distributions are positively skewed (see Fig. 7 and Table 1) with the kurtosis between 6 and 14. Worker 3 displays a distribution with skewness of -0.2 and kurtosis of 3, which are close to the normal distribution properties. Comparing to Dudley [6], this may imply that either this worker is not motivated, or possibly that she did not reach the level of skill acquired by her peers with the same experience.

The change in the mean cycle time and the coefficient of variation from day to day during the experiment can be seen in Table 3 and in Fig. 8. All workers improved their performance on the second day of the experiment, and ex-

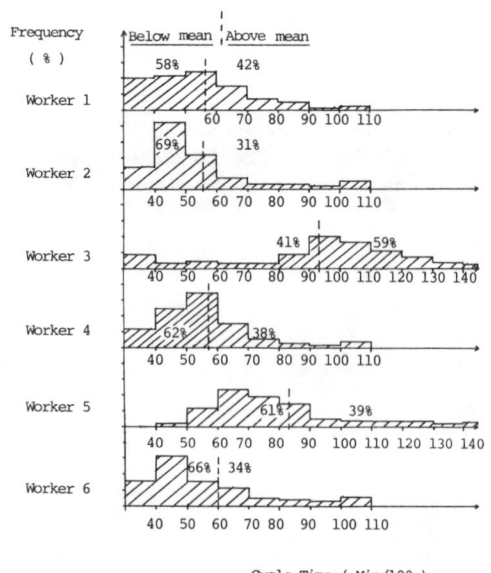

Fig. 7. Histograms of operator's cycle time.

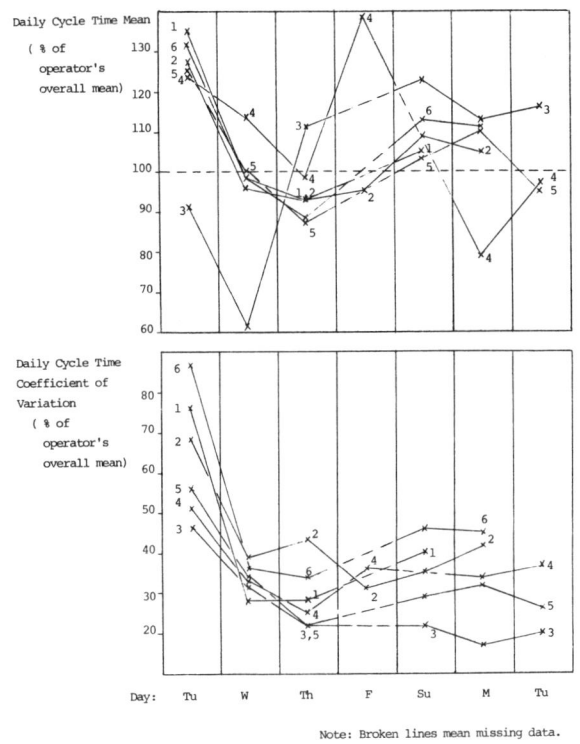

Fig. 8. Operator's daily performance.

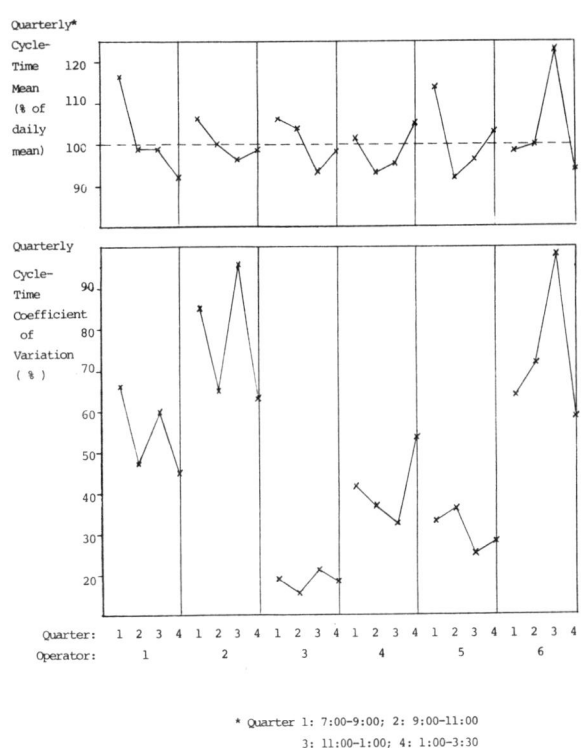

Fig. 9. Operator's quarterly performance on Thursday.

cept for worker 3, continued to improve on Thursday. All workers started the new week (on Sunday) with performance worse than in the last previous work day (Thursday or Friday), and improved in the following days. Testing whether all daily means of each worker are equal, it was found that the null hypothesis is rejected, and we may assume that the daily worker's cycle time means are different with significance level of .05 for four workers and of .01 for the other two workers.

Studying the deviation of extreme cycles from the worker's daily mean we found that 2% to 3% of the cycles were outside the three standard deviations control limits (see Table 3), and about 5% outside the two standard deviations control limits. Worker 3 again behaved differently, with only .5% outside the three standard deviations control limits, and 1% outside the two standard deviations control limits. These results seem to agree with the general positive skewness, because the meaningful control limit is the +2 or the +3 standard deviations, and the long positive tail corresponds with percentages which are higher than in the normal distribution.

With respect to the change of the cycle time mean and coefficient of variation during the day (see Figures 9 and 10) the results in this experiment seem to confirm previous findings which are discussed in the introduction. Both increase and decrease in the cycle time mean and coefficient of variation were indicated towards the end of the day. No uniformity was found among the workers each day, and no consistency was found in the performance of each individual worker regarding the pattern of change from start to end of the day.

Using analysis of variance to test whether the cycle time means in the quarters of the day per worker (as displayed in Fig. 8) are equal we found: in twelve cases, for all workers, the quarterly cycle time means can be considered equal. In all other 15 cases, the quarterly means are different with significance level of .05 in one case and .01 in 14 cases.

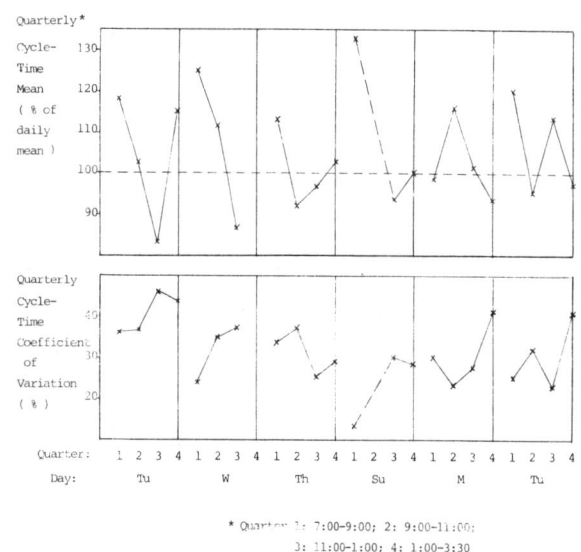

Fig. 10. Operator 5 quarterly performance

Note that above we found more significant cases of differences between extreme hourly means than here for the quarterly means. A possible explanation is that the low frequency wavelike pace variation had an upper limit of 75 minutes, while a quarter of the work day is about 120 minutes. Therefore, peaks and valleys which were apparent within the hour became averaged when a quarterly mean was computed.

Correlation of the Wavelike Variation and Observable Factors

One of the major objectives of this research was to attempt to correlate the pace wavelike variability pattern found with likely external factors. Unfortunately, the worker behaviors seemed to be random.

The most common factors listed by the observer for lower pace work, or breaks between cycles, were: conversation, apparent boredom, and day-dreaming. However, such factors were random and could not be significantly correlated with either the low or high wave frequencies.

Fatigue is often considered a major factor of pace variation (Davis and Josselyn [4]). Fatigue could be measured in this experiment by the cumulative time of work spent by the worker during each day. However, no significant correlation was found between the pace variation and the time of day. In the previous section we discussed the daily, quarterly, and hourly variation patterns of the cycle mean, which showed no significant consistency. In addition, a further check was made to see if the output during the day varied according to a clear pattern. Means for every 20 minutes and for each one-third day were also computed and although there was variation, this variation did not have any consistent pattern for any of the workers. Nor was a clear relationship found between the coffee and lunch breaks and the peaks and valleys of production. Consequently, we may conclude that the wave patterns seem to be a product of the worker herself and reflect her individual reaction to a combination of external and internal factors.

Discussion

Explanation of the Wavelike Variation

Inasmuch as the wavelike pace variation could not be significantly correlated to any observable factors, nor did it have any consistency, it may be explained by reasons related to the worker and to the task performed.

For the explanation, the work pattern can be separated into two components:

(i) low frequency periodicities; and
(ii) high frequency periodicities.

"Low" and "high" here correspond to the low and high frequency waves discussed before.

The low frequency periodicities are generated by the slow variation of the cycle time within a range of values. In this experiment, for example, ranges of 14% to 64% around the daily mean were noted. These slow variations may be accounted for by the compensation of physiological fatigue and psychological efforts by some rest and recovery during the work cycle. Further, day-dreaming and variable levels of concentration may contribute to such low frequency patterns. Recovery from micro-errors may also influence this variability as discussed in the next paragraph.

The high frequency periodicities which were found result from the variations of the cycle time around the local cycle time mean. It is suggested that this variability is the outcome of micro-errors in the performance, and is caused by the need to correct those errors. Micro-errors are defined as mistakes in micro-motions of the task, for example, inaccurate positioning of a pocket on a shirt, which requires repositioning.

Hancock [11] and Thomas et al. [26] presented combined manual and decision-making tasks as critical path networks, in which the cycle time is determined by the critical path of decision and manual operations. Although the tasks in this experiment were not the common combined manual and decision-making tasks, such as inspection and sorting, they and practically all tasks involve some decision-making aspects. Decisions and information processing are required whenever the worker is distracted from the ideal path of the task's motions. Figure 11 shows how one micro-error would increase the cycle time of a purely manual task by time components associated with the micro-error. These components include the time to identify and correct the micro-error, and an additional component of time to recover from the disturbance. Some non-error manual motions may

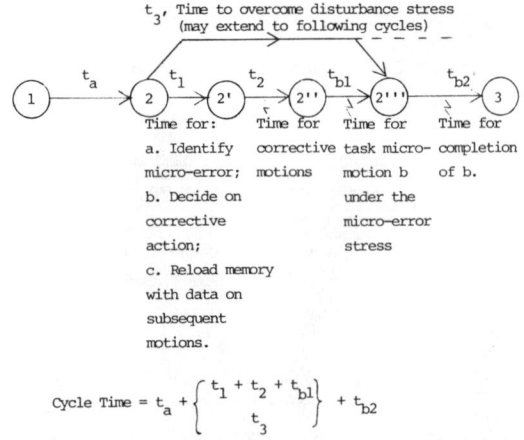

Fig. 11. The task cycle time with and without a micro-error.

take more time due to this stress either in the same cycle or in subsequent cycles. After the recovery (node $2'''$ in Fig. 11) the task may proceed normally, unless more micro-errors occur.

The time to overcome the psychological stress due to the micro-error may be part of the critical path of the cycle time. Thus, if the worker has difficulties in overcoming this stress, a greater variability in the cycle time may be expected. Moreover, since recovery from errors is more difficult for older people, it may be expected that they will avoid errors, and consequently their cycle time will show less variability in the high frequencies.

Another important factor in determining the effect of micro-errors on the high frequency variability of the cycle time is the number of micro-errors which occur during the cycle. This number may be related to the complexity of the task, and the worker's experience and fallibility. The last two hypotheses could not be checked here, because all workers are young and all tasks are similar in their complexity. Further investigations are required for the study of the micro-errors' possible effects on the cycle time.

Implications for Time Study

A point of practical significance which arises from this study is the effect pace variation patterns may have on work measurement practice. Work pace was found in the experiment to vary in the same worker (with high and low frequency periodicities) in the same day and from day to day, and also from worker to worker. No consistent pattern of variation was found (e.g., pace could not be found to be always higher towards the end of the day, nor lower).

In a standard industrial time study, when a cycle-to-cycle variation in a worker's performance is found, the time study man may set tolerance limits for his standard measured time. The number of cycles to be measured so as to yield an acceptable standard is then calculated. Because output may vary from worker to worker depending on ability and effort, the time study technician will rate or level the performance to allow for such variation.

If the time study was carried out at an extreme period during which there was little cycle-to-cycle variation, it is reasonable that the averaging would compensate for random variation and for high frequency periodicities. However, it is possible that the pace rating will not fully compensate for the low frequency pace variation which may occur after the study.

Numerous studies (e.g., Anson [2], Sury [23], Mansoor [14]) of the accuracy of various rating or leveling procedures show that they are rather imprecise because of their subjective nature. It would appear that in cases where pace variation might be significant, several spaced time studies during the same day or several days will improve the accuracy of the result. This way the time study technician will have a better chance of studying the worker during highs and lows of his performance, and will be able to reduce the inaccuracy of his result.

Conclusions

A practical minicomputer system was developed and successfully tested for the study of human work performance. This system is easily expandable to monitor more than six workers and can be used to investigate other types of operations in industry.

A major difference between this study and previous efforts to determine courses of cycle time fluctuations is that here the variations can be identified as they occur as well as in subsequent analyses.

Wavelike patterns of the cycle time were identified in this study with dominant high frequency (by spectral analysis), and with low frequency (by visual inspection). These patterns were not constant from day to day and could not be explained by any observable, external factor. Low frequency periodicities which were detected could be a result of fatigue, and high frequency periodicities may be explained by micro-error recovery time.

Significant differences which can occur between peaks and valleys in the work pattern may have serious implications for time study if not fully compensated for by pace leveling. No constant pace variability patterns were found in this experiment, but since it seems that the pattern of the variability is non-random it is certainly worthwhile to compare cycle time variability in a variety of tasks and with different groups of workers, in an effort to establish the relationship between cyclic patterns, the worker, and the task. The use of a method similar to the method described above in an industrial environment is recommended.

Acknowledgment

The authors acknowledge helpful comments regarding this research by Dr. W. M. Hancock and Dr. A. L. Sweet.

References

[1] Ansell, S. D. and Smith, K. U., "Application of a Hybrid Computer System to Human Factors Research," A.S.M.E. Human Factors Conference Paper, 8 pp. (1966).
[2] Anson, C. J., "Accuracy of Time Study Rating," *Engineering*, **177**, 301-304 (1954).
[3] Chorafas, D. N., *Control Systems Functions and Programming Approaches,* Academic Press, New York, Vols. A and B (1966).
[4] Davis, L. E. and Josselyn, P. D., "How Fatigue Affects Productivity," *Personnel,* **30**, 54-59 (1953).
[5] Davis, L. E. and Schlegel, C., "Effects of Auditory Visual Rhythmic Disturbances on Work Measurement Performance Rating," *J.I.E.,* **VII**, 3, 210-216 (1960).
[6] Dudley, N. A., "Work Time Distributions," *Int. J. Prod. Res.,* **2**, 2, 137-144 (1963).
[7] Gershoni, H. J., "Variations in Work Measurement Techniques" *Work Study Management,* **12**, 7, 422-429 (1968).
[8] Gershoni, H. J., "The Influence of Motivation and Micro-Method on Learning Manual Tasks," *Technion Research Report* (1969).
[9] Gershoni, H. J., "Stability of Workers' Micro-method," Technion Research Report (1970).

[10] Granger, C.W.J. and Hatanaka, M., *Spectral Analysis of Economic Time-Series,* Princeton University Press, Princeton, N.J., Chap 3 (1969).

[11] Hancock, W. M., "The Design of Inspection Operations Using the Concepts of Information Theory," in Univ. of Mich. Sum. Conf. Notes–*New Developments in the Prediction of Human Performance on Industrial Operations,* Ann Arbor, Michigan (July 25-29, 1966).

[12] Lewis, P.A.W., Statistical Analysis of Series of Events, *IBM Systems Journal,* **5**, 4, 202-225 (1966).

[13] MacGuire, R. W., "Equivalence of Rating Motion Picture Films and Actual Operations," *J.I.E.,* **IX**, 1, 10-13 (1958).

[14] Mansoor, E. M., "An Investigation into Certain Aspects of Rating Practice," *J.I.E.,* **17**, 2, 184-190 (1967).

[15] Murrell, K.F.H., "Operator Variability and Its Industrial Consequences," *Int. J. Prod. Res.,* **1**, 2, 39-55 (1962).

[16] Murrell, K.F.H., *Ergonomics,* Chapman and Hall, London, Chap. 17 (1965).

[17] Murrell, K.F.H. and Forsaith, B., "Laboratory Studies of Repetitive Work," *Int. J. Prod. Res.,* **2**, 4, 247 (1963).

[18] Netter, M. A., Jr., "Critical Path Analysis of Repetitive Man-Machine System Operation," MTM Res. Rep. No. 117, MTM Asso., Fair Lawn, New Jersey (1969).

[19] Pike, J. W., "Speed and Effort Rating," *T and MS,* **13**, 10-16 (1964).

[20] Sadosky, T. L., "Prediction of Cycle Time for Combined Manual and Decision Tasks," PhD Dissertation, University Microfilm, Ann Arbor, Michigan (1968).

[21] Saueressig, R.,"Hybrid Computer Feedback Analysis of Aging Function in Work Behaviour," Unpublished PhD Dissertation, University of Wisconsin, (1969).

[22] Schmidtke, H. and Stier, F., "An Experimental Evaluation of the Validity of Predetermined Elemental Time Systems," *J.I.E.,* **12**, 182-204 (1961).

[23] Sury, R. J., "Comparative Study of Performance Rating Systems," *Int. J. Prod. Res.,* **1**, 2, 23-38 (1962).

[24] Sury, R. J., "An Industrial Study of Paced and Unpaced Operator Performance in A Single Stage Work Task," *Int. J. Prod. Res.,* **3**, 2, 91-102 (1964).

[25] Thomas, C., "Special Report on a Test of a Group of Time Study Men," *Adv. Mgt.,* **18**, 13-18 (1953).

[26] Thomas, M. U., Hancock, W. M. and Chaffin, D. B., "Performance of a Combined Manual and Decision Task with Discrete Uncertainty," *Int. J. Prod. Res.,* **12**, 3, 409-425 (1974).

[27] Wyatt, S. and Fraser, J. A., *Studies in Repetitive Work,* I.F.R.B. Report No. 32, H.M.S.O., London (1925).

[28] Wyatt, S. and Fraser, J. A., The Effect of Monotony in Work, I.F.R.B. Report No. 56, H.M.S.O., London (1929).

[29] Wyatt, S. and Langdon, J. N., The Machine and the Worker, I.F.R.B. Report No. 82, H.M.S.O., London (1938).

Dr. Shimon Y. Nof is an Assistant Professor of Industrial Engineering at Purdue University. Formerly, he worked in the R & D Division of Manufacturing Data Systems, Inc. in Ann Arbor, Michigan, and was an Assistant Professor of Production Management in the University of Michigan, Dearborn. His research interests include production systems design, industrial information systems, and computer analysis of manufacturing systems. Dr. Nof holds a BSc and MSc in Industrial and Management Engineering from the Technion, Israel Institute of Technology, and a PhD in Industrial and Operations Engineering from the University of Michigan, Ann Arbor. He is a member of ACM, APICS, and AIIE.

Haim Gershoni is an Associate Professor in Industrial and Management Engineering at the Technion, Israel Institute of Technology. Formerly, he was General Manager of Klil-Israel Non-Ferrous Industries, and Chief Industrial Engineer at two American companies: Richmon Brothers and Joseph H. Cohen and Sons. His research interests include human performance, time and motion study, and work analysis. M₁. Gershoni holds BS and MS degrees from the Masschusetts Institute of Technology.

Dr. Sherman D. Ansell is a Research Consultant for the Wisconsin Board of Vocational, Technical and Adult Education. From the University of Wisconsin he received a BS in Mechanical Engineering, MS in Industrial Psychology and PhD in Industrial Relations. He was a Senior Lecturer in Industrial and Management Engineering at the Technion from 1969 to 1971.

Dr. Ayala Cohen is a Senior Lecturer in Statistics at the Faculty of Industrial and Management Engineering, The Technion, Haifa, Israel. She obtained the PhD degree in Statistics in 1967 from the Johns Hopkins University, Baltimore, Maryland. Her interests are in applied statistics and she published both in the basic statistical journals and in sociological, medical, technological journals where the research was jointly performed with nonstatisticians.

The Worker's Perception of a Change in Work Rate

Rodney K. Schutz
AT&T Bell Laboratories

Thomas H. Campbell
Western Publishing Company

ABSTRACT

Two bench assembly tasks having cycle times of .5 minutes and 2.3 minutes were evaluated. Each test session consisted of a work period, a rest period and a second work period. At the end of the second work period the subjects were asked to estimate the percentage change in work rate between the two work periods. The work rate during the first work period was paced at 85, 100 or 115 percent of the MTM normal time. The work rate during the second work period was changed by -6, -3, 0, +3 or +6 percent. Rest periods of .5 minutes and 10 minutes were evaluated. Eight, fully trained subjects were used. The subjects ability to evaluate changes in work rate was not effected by a) the task cycle time, b) the duration of the rest period or c) by the work rate during the first work period. The subjects could consistently detect a 6 percent change in work rate. In estimating the magnitude of the change, the subjects' absolute estimation error varied between 3 and 5 percent. This implies that workers in industry may be able to detect and accurately evaluate errors in time studies.

INTRODUCTION

Can workers really detect inconsistencies or inaccuracies in time standards? The present study provides insight and tentative answers to this question.

In 1951 Lifson[4]. investigated the accuracy of pace rating during time study. Five workers performed four jobs at each of five work rates. Six expert time study engineers rated the work performance. Some of the conclusions from the study were:

1. slow work rates are overrated and fast work rates are underrated

2. the rating error increases when the observed pace is significantly faster or slower than normal

3. a few workers can self-rate the work pace more accurately than the best time study engineer, but workers in general are much less consistent than time study engineers.

4. the general standard error for rating was 14.5 percent, that is across all jobs, all workers, all work rates and for all time study engineers.

5. the most accurate and the least accurate time study engineer had a standard error of 11.1 percent and 14.7 percent, respectively.

6. When the same time study engineer rated one worker, on one job, at one work rate, several times, the average standard error was 7.2 percent.

A study conducted by the University of Birmingham in 1954 involved 572 qualified time study engineers rating films of 9 jobs at each of five work rates [6]. It was found that 95 percent of the calculated normal times fell within 24.5 percent of the mean normal time.

An earlier study by the Committee on Rating of Time Studies evaluated 599 qualified time study engineers on 24 jobs at each of five different work rates [2]. Over 50 percent of the time study engineers had an absolute error of more than 10 percent.

Lehrer investigated the ratings made by 31 time study engineers of four jobs at various work rates [3]. It was found that 58 percent of the ratings had an absolute error of more than 10 percent. The absolute error was less than 5 percent in 19 percent of the cases.

Mundel and Keim conducted a study in which 50 time study engineers rated 57 jobs at various paces [5]. It was found that 54 percent of the ratings had an absolute error of more than 10 percent.

In summary, many industrial practitioners claim that they can accurately rate the pace at which a job is performed. However, these claims have not been emperically substantiated. Conversely, the above laboratory studies yield very consistent results and indicate that the average absolute error in rating is at least 10 percent.

METHODS AND PROCEDURES

Each test session consisted of a work period (referred to as the base period), a rest period, and a second work period (referred to as the comparison period). At the end of the comparison period, the subject was asked to estimate the percentage change in work rate between the base period and the comparison period. The work rate during the base period was set at 85, 100 or 115 percent of the MTM normal time. The work rate during the comparison period was set at 94, 97, 100, 103 or 106 percent of base period work rate, that is differences in work rate of -6, -3, 0, +3 and +6 percent were evaluated. Test session where the comparison period work rate was -3 or +3 percent were performed twice by each subject. All other test sessions were performed once by each subject. Two bench assembly tasks were used in the study - a short cycle task of .49 minutes and a long cycle task of 2.31 minutes. The subjects alternated between assembly tasks (e.g. the short cycle task was only performed during even numbered test sessions). The order of testing was randomized with reguard to all other factors. Each subject performed 42 test sessions; two tasks, three base rates and seven comparion rates (two of the five comparison work rates were replicated).

Eight male, college students were used as subjects (age 22-27 years). The subjects were randomly divided into two, four-man groups. The first group rested 10 minutes between work periods. The second group was not allowed a rest period and always began the second work period within 30 seconds after the completion of the first work period.

Assembly Tasks

The assembly tasks were simple psycho-motor tasks. They were designed to minimize the effects of learning. None of the components of the task were thought to require any special skills. A detail MTM analysis of the tasks as well as photographs and dimensioned layouts are given in Campbell [1]. The subjects were required to practice the tasks prior to testing. On the average, the subjects required three hours of practice to become proficient at both tasks.

The short cycle task consisted of the following:

a. Assemble a cardboard box with fold over corners (computer card box, 2000 card size)

b. Put together pairs of snaplock beads and place them in the cardboard box (Fisher Price, 1 1/2 inch baby beads). Each pair of beads is gotten from a separate parts bin.

c. Close box and place a rubber band around box.

d. Use a rubber stamp and ink pad to stamp a computer card.

e. Staple card to box.

The task was performed 15 times per work period and the average duration of the work periods was 7.5 minutes.

The long cycle task consisted of the following:

a. Construct a paper box (.72 minutes)
  1. Fold premarked sheet of paper
  2. Use razor knife to cut corners
  3. Use stapler to fasten corners

b. Assemble a toy tractor using 27 snap-together blocks (Lego toy tractor). The parts were separated into 18 bins. The finished tractor is placed in the box (1.59 minutes).

Ten cycles were performed per work period and the average duration of the work periods was 23.1 minutes.

The cardboard or paper used in making the boxes was held by the analyst and was issued to the subjects at prescribed time intervals. This procedure caused the subjects to perform the tasks at the desired work rate.

Subject Instruction

The basic experimental procedures were explained to the subjects prior to testing. They were told that they would be asked to perform twenty-one work rate comparisons on each of two tasks. They were informed that the work rate for the second work period could be equal, faster or slower than the work rate for the first work period. They were told that rate changes less than 1 percent would not be presented.

After completing a test session the subject was asked the following questions. Do you feel that you were forced to work at a faster, slower or equal rate during the second work period? If faster or slower, by how much in terms of a percentage?

In some cases, more than one test session was performed by a subject in one day, with a maximum of three tests per day. A break of at least two hours was allowed between test sessions.

RESULTS

The following Analysis of Variance (ANOVA) model was used in analyzing the data:

$$y_{ijklmn} = \mu + T_i + B_j + C_k + R_l + S_{m(l)} + \text{(interactions)} + e_{ijklmn}$$

This basic model was evaluated four times, each time using a different dependent variable (the Y's). The dependent variables considered were:

- SR - the subject's response or estimate of the direction and magnitude of the change in rate

- EE - the subject's estimation error, the subject's estimate of the direction and magnitude minus the actual direction and magnitude of the change in rate, i.e. SR - Actual.

- EM - the subject's error magnitude, the subject's absolute error in estimating the change in rate, i.e. |EE|.

- DP - the subject's detection probability, the probability of correctly detecting the direction of change in the work rate, that is DP was set equal to one or zero as follows:

Actual Change	Subject's Response		
	increase	same	decrease
increase	1	0	0
none	0	1	0
decrease	0	0	1

For example, if the subject felt that there was a two percent increase in work rate (+2) when the actual work rate had increased six percent (+6), the dependent variables would be SR = +2, EE = -4, EM = +4, and DP = 1.

The parameters in the model are:

μ - general mean

T_i - assembly task, I=2, short or long cycle

B_j - base period work rate, J=3, 85, 100 or 115 percent of MTM normal time

C_k - comparison period work rate, K=5, 94, 97, 100, 103 or 106 percent of base period work rate

R_l - rest period, L=2, zero or 10 minutes

$S_{m(l)}$ - subject effect, nested within R_l, M = 4

e_n - residual error, N=1 or 2

A total of 336 work rate comparisons were evaluated in the study (42 test conditions on each of eight subjects). The raw data is given in Campbell [1]. The data is plotted in Figure 1.

Significance Tests

The ANOVA tables for the four models are sumarized in Table 1. All of the factors in the models were considered to be fixed effects when calculating F ratios. The factors that were found to be statistically significant at an α level of 1 percent and 5 percent are denoted by asterisks (*) and daggers (†), respectively. Since the degrees of freedom for residual error is very large (i.e. 258), a 5 percent α level is a nominal degree of statistical significance. The power of the tests is very high, and the models should be able to detect extremely small differences in the levels of any given factor.

As was to be expected, most of the models indicated a statistically significant subject effect. Moreover, several of the subject interaction terms were significant at an α level of .05.

Differences in work rate during the comparison period were statistically significant in all four models. This means that the subjects response is altered when the change in work rate between periods is increased or decreased.

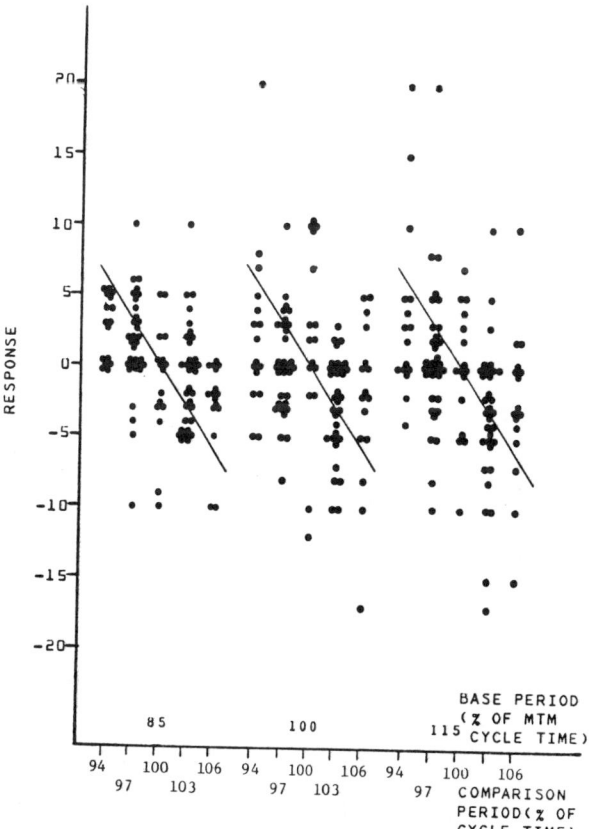

Figure 1 Raw Data

For the error magnitude model, the duration of the rest periods and the differences in work rate during the base period seemed to effect performance.

None of the other factors in any of the models were statistically significant. This general lack of statistical significance is the most important conclusion to be noted from Table 1. The absolute duration of the two assembly tasks, one-half minute versus two and one-third minutes, did not influence the subjects' perceptions of work rate. Similarly, differences in work rate during the base period, a very fast 85% rate versus a very slow 115% rate, did not effect rate perception. Finally, increasing the duration of the rest period between work periods to 10 minutes did not make it more difficult for the subjects to detect changes in work rate.

Within Subject Variability

As can be seen from Figure 1, the subjects were very erratic in estimating changes in work rate. This is to be expected and is consistant with the results obtained in previous studies concerning time estimation, speed estimation and size estimation. Within subject variability is also referred to as inherent variability or pure error variability. It is due to differences observed under test-retest conditions. A different response will be given when the same subject performs the same test on two different days. Since the test sessions where the work rate was 97 or 103 percent were replicated, the within subject variability can be estimated. The standard deviations for test-retest variability are given in the first column of Table 2 under the heading 'pure error'. These estimates may slightly overestimate the actual within subject variability, particularly for the detection probability model. Not all test conditions were replicated, only changes in work rate of +3% and -3%. The subjects had much more difficulty in detecting a 3 percent, in comparison to a 6 percent, change in work rate.

The second and third columns of Table 2 are the standard deviations for the models and the raw data, respectively. The standard deviations in the second column are usually called standard errors of estimations. Referring to Table 1, these standard deviations are calculated as the square root of the residual mean squares. For example, the residual mean square for the SR model (i.e. the variance of the model) is 20.0. The square root is 4.47 and is given as the first entry of the second column of Table 2 under the heading 'model error'. Any prediction or estimate of SR that is made using the SR model will have a standard deviation of 4.47. If the standard deviation for a model is very small then very precise predictions can be made. Of course the standard deviation for a model can never be less than the pure error standard deviation.

Table 1 ANOVA TABLES
(* is 1% and † is 5% significance)

Source	df	Subject Response SS	MS	F	Error Estimation SS	MS	F	Error Magnitude SS	MS	F	Detection Prob. SS	MS	F
T_i, task; short-long	1	6.6	6.6	.33	6.6	6.6	.33	11.1	11.1	1.23	.027	.027	.11
B_j, base period work rate	2	18.8	9.4	.47	18.8	9.4	.47	74.8	37.4	4.14†	.399	.199	.84
C_k, comparison period rate	4	1199.9	300.0	15.01*	1547.9	387.0	19.36*	159.0	39.7	4.40*	4.196	1.049	4.42*
R_l, rest time	1	10.4	10.4	.52	10.4	10.4	.52	161.6	161.6	17.87*	.860	.860	1.00
$S_{m(1)}$, subject	6	413.9	68.8	3.44*	412.9	68.8	3.44*	184.1	30.7	3.40*	2.565	.428	1.80
TB_{ij}	2	.8	.4	.02	.8	.4	.02	1.1	.5	.06	.232	.116	.49
TC_{ik}	4	29.7	7.4	.37	29.7	7.4	.37	23.5	5.9	.65	.952	.238	1.00
TR_{il}	1	.4	.4	.02	.4	.4	.02	6.6	6.6	.73	.074	.074	.31
BC_{jk}	8	227.3	28.4	1.42	227.3	28.4	1.42	86.2	10.8	1.19	1.268	.158	.67
BR_{jl}	2	8.1	4.1	.20	8.2	4.1	.20	45.7	22.9	2.53	.184	.092	.39
CR_{kl}	4	109.2	27.3	1.37	109.2	27.3	1.37	1.00	.3	.03	1.036	.259	1.09
$TS_{im(1)}$	6	308.9	51.5	2.58†	308.9	51.5	2.58†	50.8	8.5	.94	1.828	.305	1.28
$BS_{jm(1)}$	12	502.6	41.9	2.09†	502.6	41.9	2.09†	91.5	7.6	.84	1.417	.118	.50
$CS_{km(1)}$	24	387.2	16.1	.81	387.2	16.1	.81	259.5	10.8	1.20	7.601	.317	1.33
e_{ijklm}, residual	258	5157.6	20.0		5157.6	20.0		2332.2	9.0		61.29	.238	
total	335	8380.2			8728.2			3488.5			83.93		

The variability of any prediction can never be less than the inherent, test-retest variability of the data.

Referring again to the ANOVA results in Table 1, the last entry in each SS column is the total sum of squares. The variance of the raw data can be estimated by dividing these total sums of squares by their degrees of freedom. The corresponding standard deviations are given in the third column of Table 2 under the heading 'total error'. For example the variance of the SR data is 25.02 (i.e. 8380.2/335), and the standard deviation is 5.00. The variability in the data is due to a) inherent variability (within subject, test-retest differences) and b) external variability (differences in test conditions).

Table 2
VARIABILITY ESTIMATES

Model	Standard Deviations		
	Pure Error	Model Error	Total Error
SR, Subject Response	4.13	4.47	5.00
EE, Error Estimation	4.12	4.47	5.10
EM, Error Magnitude	3.11	3.01	3.23
DP, Detection Probability	.52	.49	.50

Recall that the pure error variability was calculated from a bias subset of the data. As a result the entries in the first column of Table 2 tend to overestimate the within subject variability. This is particularly apparent for the EM and DP models because the standard deviation for pure error is larger than the standard deviation for the model. Nevertheless, several general conclusions can be derived from Table 2.

In reviewing Table 2, it can be seen that the different test conditions imposed by the analyst in conducting the experiment, had a very minor influence on the subjects' ability to predict changes in work rate. The variability in the raw data (third column of Table 2) is only slightly larger than the within subject variability (first column of Table 2).

For the SR model approximately 80 percent of the variability in the data is due to pure error (i.e. 4.13/5.00 = .83). Conversely, only 20 percent of the variability in the data is due to the differences in rest time, in work rates or in the assembly task. The same conclusion was reached in the previous discussion of the ANOVA tables; most of the factors in the model were not statistically significant.

The results from Table 2 also show that the present ANOVA models give a reasonably good fit to the data. Although the subjects' perception of a change in work rate was only slightly influenced by the differences in the test conditions, the models were able to account for the majority of these differences when they were present. For the SR data, the ANOVA model was able to account for approximately 60 percent of the test variability (i.e. (5.00 - 4.47)/(5.00 - 4.13) = .61).

Parameter Estimates

Unbias, minimum variance, least squares estimates of the parameters in the model were calculated. The parameters for the main effects in the models are shown graphically in Figures 2 through 5.

DISCUSSION

Four different models were evaluated because different points of view can be taken in interpreting the data. The subject response model, SR, is the most general with the raw data being used as the dependent variable. The dependent variable in the error estimation model, EE, is the difference between the perceived change in work rate and the actual change in work rate. The results from the SR and EE models are similar, but the EE model more clearly indicates any tendency the subjects may have to overestimate or underestimate. In the error magnitude model, EM, the absolute value of the estimation error is the dependent variable. The EM model gives a better indication of how accurately a subject can esitmate the degree of change. In the SR and EE models, overestimations (+) and underestimations (-) tend to cancel each other. The detection probability model, DP, describes the subjects' ability to recognize whether a change in rate has or has not occured. Irregardless of whether a subject can accurately estimate the magnitude of the change, did the subject realize that the work rate had been increased or decreased.

Assembly Task

The absolute cycle time of the assembly task did not influence the subjects' rate perception ability. Prior to collecting the data, it was thought that changes in work rate may be more easily recognized for long cycle tasks. A given percentage change in work rate represents a larger absolute time difference for a long cycle task. If the work rate during the comparison period is 3 percent faster, then the cycle time during the comparison period will be reduced by approximately 1 second for the short cycle task and by approximately 4 seconds for the long cycle task. This suggests that when performing the long cycle task the subjects may be more apt to at least recognize that a change has occured. However, this

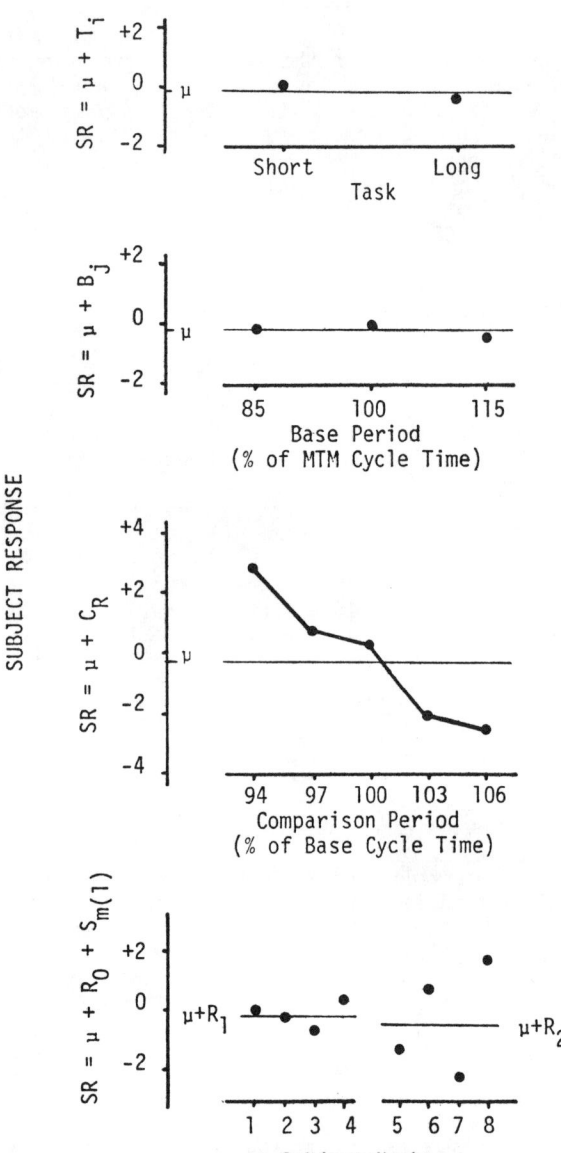

Figure 2 Subject Response

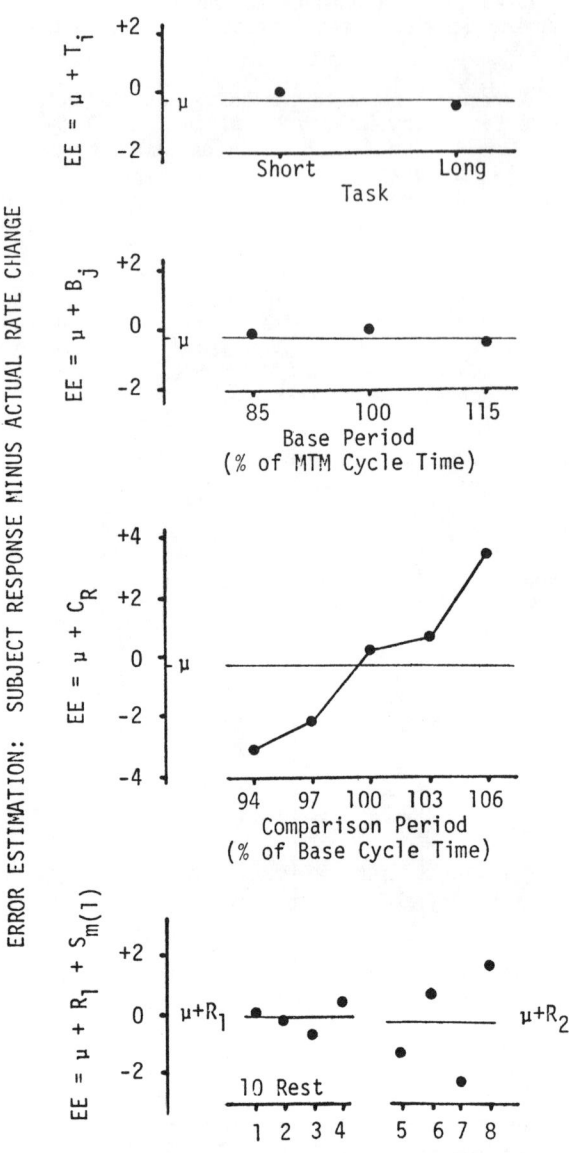

Figure 3 Error Estimation

original hypothesis was not substantiated by the data.

With regard to the EM model, the present data indicates that the absolute error remains constant as the cycle time is increased. Previous research on time perception has produced conflicting results with regard to whether the absolute error remains constant, increases or decreases as the length of the time interval to be estimated is increased.

The fact that the number of cycles performed during a work period was different for the short cycle task and the long cycle task is not thought to have influenced the results. Extensive pilot studies were conducted before the present data was collected. In those pilot studies it was noted that the subjects' rate perception accuracy did not improve as more cycles were performed. As few as three cycles of work were sufficient in many cases.

Base Period Work Rate

Differences in the work rate during the base period had at best a nominal influence on performance. It can be seen in Figure 5, that the

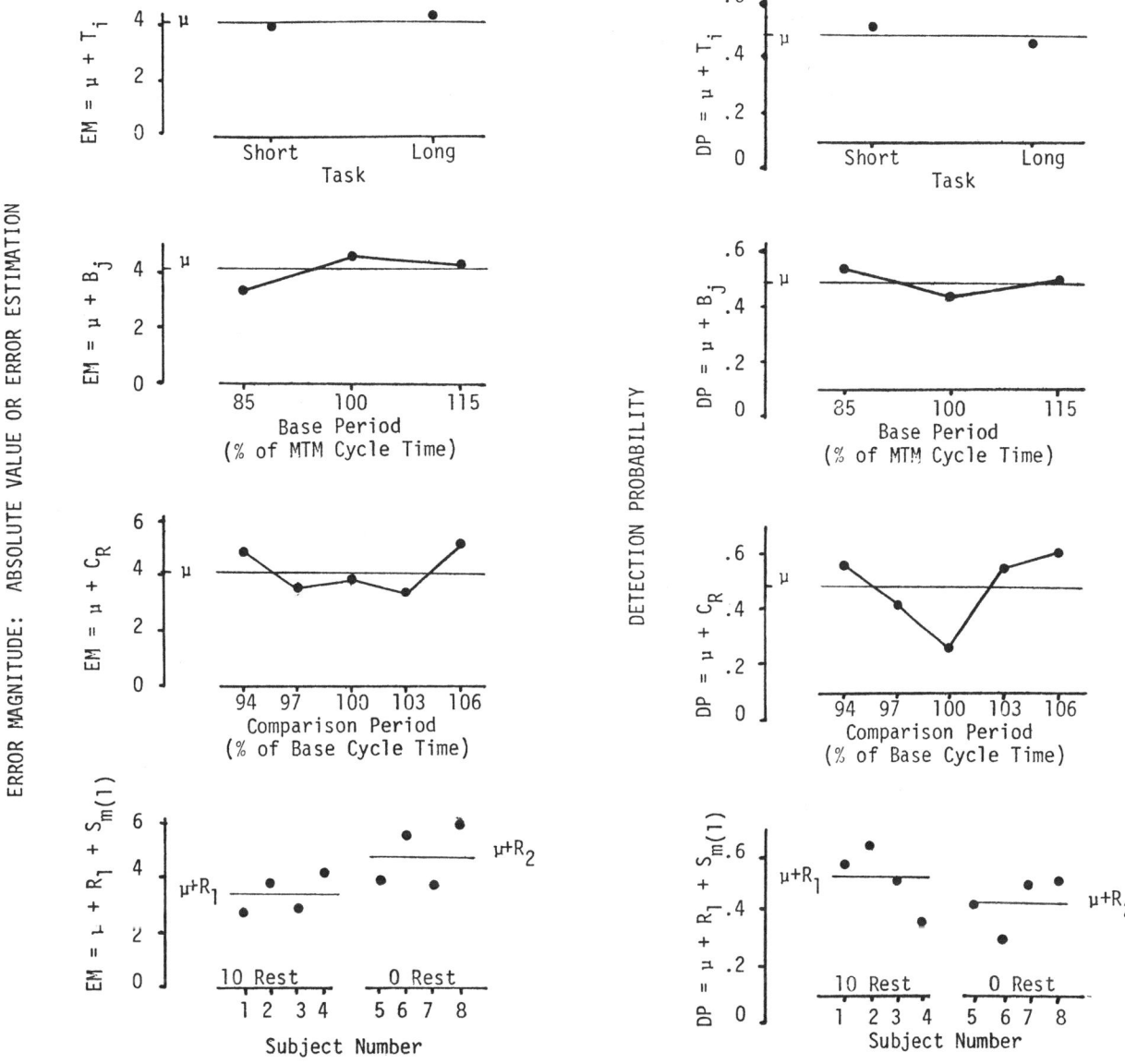

Figure 4 Error Magnitude

Figure 5 Detection Probability

subjects were slightly more proficient in detecting a change in rate if the base rate was faster or slower than normal. Consistent with this observation, Figure 4 shows that the absolute estimation error was slightly smaller when the work rate during the base period was 85 percent or 115 percent of the normal rate. These observations are in conflict with Lifson's data [4]. Lifson found that the work rate was more accurately estimated if the work rate was close to the normal rate. This same conclusion is alluded to in many time study books. Thus, pending further data the differences in work rate shown in Figures 4 and 5 are probably negligible.

Comparison Period Work Rate

From Figures 2 and 3 it can be seen that the subjects tend to underestimate the magnitude of a change in work rate. If the work rate is increased 6 percent then the subjects on the average feel that there has been a 3.5 percent increase (see Figure 3). Similarly, if the work rate is decreased 6 percent, the subjects perceive the change to be about -3 percent. The subjects tend to be conservative in estimating the magnitude of a change in work rate. In those cases where the work rate was not changed, the subjects average error was approximately zero. The subjects do not seem to have a bias to expect the work rate to either increase or decrease.

From Figure 4 it can be seen that the absolute estimation error increases when large differences in work rate are considered. When the work rate is changed by 6 percent, the absolute estimation error is approximately 5 percent. If the work rate is not changed or only changed by 3 percent, the absolute estimation error is 3 percent.

From Figure 5 it can be seen that the ability to recognize that the work rate has increased or decreased improves when a large percentage change in work rate has occured. The probability of detecting a 6 percent change in work rate is approximately .60. This indicates that the subjects have very good rate perception ability. A 6 percent change in the short cycle task represents a change of less than 2 seconds. Figure 5 also shows that the subjects tend to feel that the work rate has been changed when in fact it has not. When the work rate was not altered, the subjects thought a change had occured 75 percent of the time.

Subject Differences

As can be seen in Figures 2 through 5 there are marked subject differences. This is not surprising and the size of the differences are similar to what has been reported in other studies.

In terms of detection probability, subject #2 could detect a change in work rate twice as often as subject #6 (see Figure 5). In considering the absolute estimation error, subjects #1 and #3 were twice as accurate as subjects #6 and #8 (see Figure 4). These differences are thought to be typical between subject differences. All of the subjects seemed well motivated and performed the tests in a conscientious manner.

Rest Period

Subject differences are nested within the differences in rest time. This nesting was unavoidable. At present each subject participated in 35 hours of laboratory testing. The number of hours of laboratory time would be almost doubled if a non-nested design were used. The use of a small number of subjects in conjunction with the nested design, makes it impossible to reach any conclusions concerning the influence of rest time on performance. In fact the results in Figures 4 and 5 are exactly opposite to what is to be expected. The absolute estimation error decreases and the detection probability increases if there is a 10 minute break between work periods. Intuitively, it would seem that perception and recall would be enhanced if the subjects began the second work period immediately after completing the first work period. The results in Figures 4 and 5 are probably due to differences in the two groups of subjects. One or more of the subjects in the group that was given a ten minute rest period probably have better than average rate perception ability. Conversely, one or more of the subjects in the zero rest time group could have had slightly worse than average ability in judging work rates.

CONCLUSIONS

The present subjects were able to recognize extremely small changes in work rates. It seems that an industrial worker (actually performing a task) may be more accurate than a time study engineer (outside observer) in estimating work rates. All of the studies cited in the introduction section of this paper found that the average error in rating for time study engineers was at least 10 percent. In extrapolating the data in Figure 5 it would seem that a change in work rate of 8 to 10 percent will almost always be recognized (detection probability of .9). Given that a change in work rate is detected, the subjects can not accurately estimate the degree of change (see Figure 4) and they tend to underestimate the magnitude of the change (see Figure 3).

When the work rate was not changed, the subjects invariably felt that it had changed. This occured over 70 percent of the time (see Figure 5). This percentage must be interpreted with caution in that the present results are very dependent upon the instructions given to the subjects prior to testing. In the present experiment, the pre-test instructions implied that the subjects were expected to recognize changes in work rate as small as 1 percent. These instructions may have been to stringent.

Although the subjects had a compulsion to respond even though the work rate had not been changed, their actual estimation error was relatively small in these cases. The subjects usually stated that the change in work rate was 5 percent or less (see Figure 4).

These results imply that workers in industry may be very prone to complain about inconsistencies in work rates. They will almost always recognize the errors made by time study engineers. Moreover, the workers are apt to feel that a 5 percent error exists between two jobs when in fact the jobs have been very consistently and accurately rated. An industrial environment may actually exaggerate these conditions. The present subjects were evaluated in an emotionally sterial environment. Workers in industry may be more impulsive in evaluating work conditions. If the work being performed is monotonous and boring or if there is a wage incentive system, the workers perceptions may be bias.

BIBLIOGRAPHY

1. Campbell, T. H., "A Worker's Perception of Changes in his Rate of Work", MS Thesis, Georgia Institute of Technology, June 1978.

2. Committee on the Rating of Time Studies, A Fair Day's Work, Society for the Advancement of Management (New York), 1954.

3. Lehrer, R. N., "Development and Evaluation of Pace Scale for Time Study Rating", Ph.D. Dissertation, Purdue University, 1949.

4. Lifson, K. A., "A Psychological Approach to Pace Rating", Ph.D. Dissertation, Purdue University, 1951.

5. Mundel, M. E. and Keim, A. J., Proceeding of Second Annual Purdue Motion and Time Study Work Sessions, Purdue University, April 1945.

6. University of Birmingham (United Kingdom), Work Measurement Research Unit, Work Measurement: Some Recent Studies, MacMillan, London, 1968.

AN ANALYSIS OF PERFORMANCE RATING

Robert M. Wygant
Associate Professor
Industrial Engineering
Western Michigan University

It is widely recognized that time study analysts tend to overrate low levels of performance and underrate high levels of performance. This study reviewed performance rating data from 48 subjects. The results of this analysis show that even new analysts that have had no previous exposure to time study tend to follow this same pattern of conservative rating. This tendency is partially explained by the methods in which performance rating is taught. Another contributing factor is the degree of subjective judgment that is involved in performance rating. If these judgments are to be linear, then the function must increase logarithmically.

INTRODUCTION

Over 80 years ago Frederick W. Taylor introduced work measurement as part of his ideas on scientific management. He recommended that each worker should be assigned a given task to be performed in a standard amount of time. Taylor described a "fair day's work" as the amount of work which a first-class man can do under favorable circumstances [29]. Since this first introduction, the use of work measurement as a management tool has grown and expanded into widespread usage throughout industry. A survey by Industrial Engineering and Patton Consultants reported that 89 percent of the responses reported that they are using work measurement [24].

The most common technique used to measure work is time study. This method is based on measuring the time to complete a task using a stop watch or similar timing device. An important element of any time study is determining the "normal" time for the operation that is being observed. This requires that a subjective judgment of the work pace be made by the time study analyst while observing the task that is being studied. This subjective judgment may be referred to as leveling, effort rating, pace rating, speed rating, or performance rating [36]. For more details on time study procedures see any of the current books on time study [4][13][14][19][22].

The subjective element of time study has received considerable attention relative to the accuracy and consistency of performance rating. Over the years industrial engineers have claimed that with proper training an analyst can judge the performance of an operator consistently within plus or minus 5 percent of the true pace. Labor organizations have been the harshest critics of this subjective judgment. In 1961 the AFL-CIO adopted a resolution opposing all work measurement techniques [10]. Gomberg has questioned the validity of time study and in particular the idea of a "normal" worker. He suggests that the concept of the "normal" worker has been at the bottom of much of the conflict between management and labor.

In reviewing performance rating Murrell [20] points out that when subjective judgments are made of physical functions, the function must increase logarithmically if the judgment is linear. He uses the Weber Fraction, $dI/I = K$, as the basis for this reasoning.

A research study by the MTM Association [17] indicates that one of the major causes of differences in output is the change in methods. Seymour [27] shows that although higher speeds of performance are traditionally associated with greater effort, there is no proof of this.

Two work measurement experiments on the rating of performance levels of the S.A.M. films were conducted by C.L.M. Kerkhoven [15]. He suggests that the performance levels of the S.A.M. films M (shovel sand) and N (stack cartons) are open to question since the "calibrated" ratings have taken place by estimation rather than by measurement.

In another study Gershoni [9] concluded that time study analysts from both the United Kingdom and Israel were unable to detect partially trained workers or variations in the micro-method. He concluded that "due to the fact that our observers could not detect level differences, the estimated performance could vary enormously. For example, it was found that the United Kingdom observations might have varied between 0.18 and 0.44 minutes" for the same task.

A recent study of a case-piling operation was made by Wilson and Wygant [35]. Four workers were studied while lifting two different size cases at three different controlled paces. Using Time Study, the analysts were unable to establish a performance level that accurately predicted the true controlled pace.

METHOD

Forty-eight subjects were involved in this analysis. They were divided into three groups of 16. The first two groups (Group 1 and Group 2) were all university students with no industrial work experience. The third group (Group 3) was made up of individuals with industrial engineering experience ranging from one to eleven years. Each of the groups were given a lecture which described the various rating systems and their application. Following the classroom sessions they were shown a film "Industry's Perennial Problem" (Tampa Management Institute) which attempts to develop the concept of "normal" or 100% work pace through demonstrations of walking and card dealing.

Following the introductory film the subjects participated in a practice session in which they rated 15 different industrial operations. After a discussion of the individual results the subjects were asked to think about the concept of "normal" before returning to the next session.

Two days after the practice films, all three groups were asked to rate 30 different industrial scenes. The subjects in Group 1 were not advised of the correct or true rate until they had completed the analysis of the entire film. Groups 2 and 3 were given the true rating immediately after they had recorded their observed rate for each operation.

RESULTS

The means and standard deviations for 480 observations from each group are shown in Table 1. A paired-comparisons T-test indicates that there is no significant difference between the group means at the .001 level.

True Rating	Group 1 x		Group 2 x		Group 3 x	
60	65.3	11.6	59.7	7.6	59.1	9.5
70	93.4	7.0	92.5	10.6	78.4	13.0
85	88.4	8.5	85.6	13.9	79.0	10.4
90	97.7	12.1	94.4	8.7	88.3	8.9
100	96.7	9.6	96.8	9.0	94.4	8.0
110	104.8	8.9	102.7	13.5	100.1	9.6
120	119.5	11.2	113.3	14.3	111.9	9.3
130	125.9	9.5	123.6	15.2	116.9	11.8
140	130.0	8.0	124.4	8.5	116.3	9.6
150	129.4	13.8	125.5	16.4	130.9	15.0
160	140.9	12.9	143.8	18.8	130.3	17.7

Table 1. True Rating vs Actual Rating

The data indicates that there is little relationship between the standard deviation of observed values and the true value. Group 1 had an average standard deviation of 10.4 over the range of values that were studied, with a low of 7.0 for the true values of 70 and a high of 13.8 for the true values of 150. Groups 2 and 3 showed similar tendencies with average standard deviations of 11.9 and 10.5.

Some of the large deviations from the true values can be attributed to inexperience of the analysts. However, it is interesting to note that Group 3, which averaged over 4 years of industrial experience, were no more consistent in rating than the student groups.

Assuming a linear relationship exists between the true values and the observed values over the range of 60 to 160, the linear regression model gives:

Observed value = 32.1 + 0.661 x True value
Correlation coefficient = .82

A plot of the combined data is shown in Figure 1. These results verify what has been acknowledged for many years -- analysts tend to overrate low levels of performance and underrate high levels of performance [20][22][33]. This trend is often referred to as conservative rating.

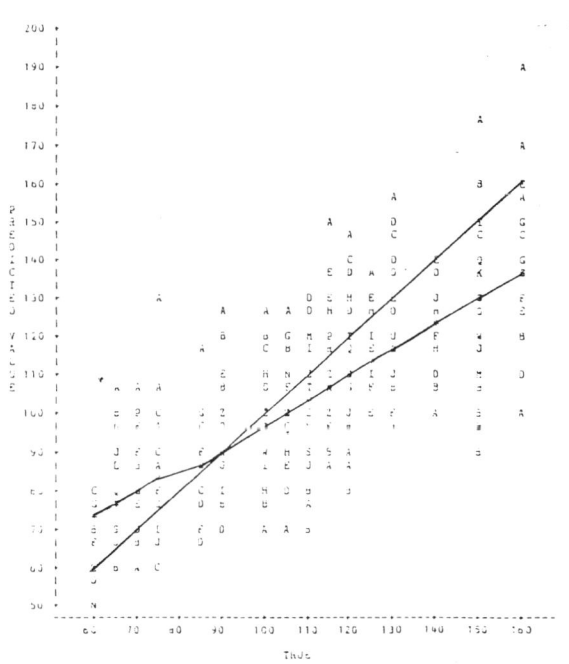

Fig. 1 - Observed Rates vs True Rate

Data that has been transformed into a log/log relationship as proposed by Murrell results in the following regression line:

log observed value = 0.626 + 0.628 x log true value
Coorelation coefficient = 0.82

This data is shown in Figure 2.

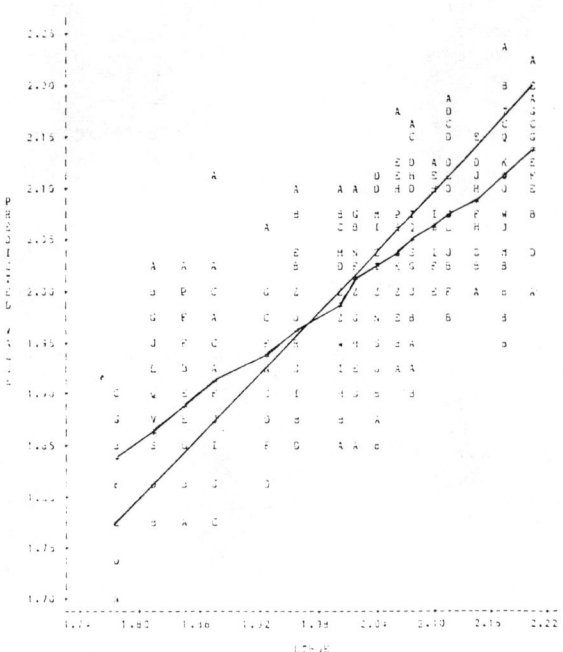

Fig. 2 - Log Observed Rate vs Log Actual Rate

Whitmore has suggested that if the observer plots his ratings against observed times that the readings should describe part of a hyperbola [34]. He recommends plotting the reciprocal of the ratings. The reciprocals are shown in Figure 3. The regression analysis for the combined data are:

Reciprocal of observed value = .003 + .69 x Reciprocal of true value

Correlation coefficient = .81

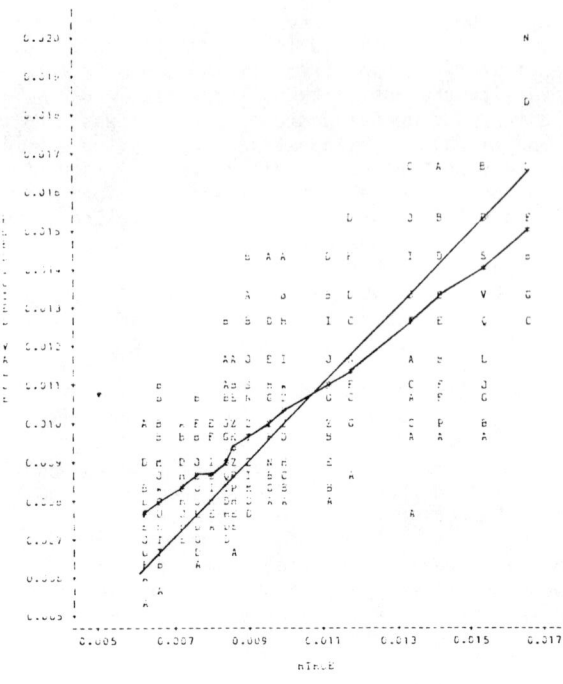

Fig. 3 1/Observed Rate vs 1/True Rate

To check the effect that extreme values have, the data was modified to include only the true values between 80 and 130. The resulting equation is:

Modified observed value = 19.040 + .781 x true value

Correlation coefficient = .84

The relationship between the true rating and observed rating at various levels is illustrated in Table 2. There appears to be little difference between the linear and log/log data. Using the reciprocal method indicates a steeper slope than the other two values and is higher than the true values of all levels. It is somewhat surprising that eliminating the extreme values did not improve the overall results to any appreciable extent.

Combined Data

True	Linear	Log/Log	Recip	Modified
60	71.8	69.0	62.9	
80	85.0	83.9	83.8	81.5
100	98.2	97.7	104.8	97.1
120	111.4	110.7	125.6	112.8
140	124.6	122.9	146.6	

Table 2 - Relationship between True and Observed

CONCLUSIONS

The trend that time study analysts overrate low levels of performance and underestimate high levels of performance appears to hold true for new students who are just being introduced to work measurement techniques. Part of this tendency may be explained by the theory of subjective judgments which shows that for the judgment of a physical function to be linear, the function must increase logarithmically.

I believe a second contributing factor to the conservative or flat tendency in rating is related to the method in which performance rating is taught. In almost all cases the students are shown that there are limits to human performance which make it more difficult to perform as the operator pace moves further from the "norm" or 100% performance. Since a new analyst has been cautioned concerning extreme values, it is only natural that the individual rater will attempt to avoid these extremes by overrating low performance and underrating high performance. Once out in the actual working environment this will be reinforced by labor to raise low values and management to limit high values.

It is obvious from this study, as well as many previous projects, that there exist wide variations in performance rating between individuals. Much of the criticism that has been directed toward performance rating has been justified. Subjective rating still relies heavily on the experience and skill of the analysts.

Since work measurement is being used by the vast majority of manufacturing companies, continued research in this area should be encouraged to arrive at an equitable method to define and judge the concept of "normal." Ayoub and Manuel suggest that "it may be found possible to performance rate on an individual basis by first measuring the individual's resting and next his working ventilation rate" [2].

Another area that I believe will help to eliminate some of the variations between analysts is the application of predetermined time systems. Gershoni [9] has indicated that the use of such a system to identify the micro-method would eliminate a large source of the error.

If we are to improve on the methods for developing accurate estimates of the time required to perform tasks in the working environment, we first have to recognize that there is a problem - ignoring it won't make it go away!

REFERENCES

[1] Anderson, Clifton A., "Performance Rating," Industrial Engineering Handbook, 3rd ed, New York: McGraw-Hill Book Co. (1971).

[2] Ayoub, M.M. and Robert R. Manuel, "A Physiological Investigation of Performance Rating for Repetitive Type Sedentary Work," J of Industrial Engineering, July, 1966, 17-7, 366-376.

[3] Barnes, Ralph M., "What Has Been Done to Improve Rating Operator Performance," National Time and Motion Study Clinic, Nov., 1945.

[4] Barnes, Ralph M., Motion and Time Study, 7th ed., New York: John Wiley & Sons. (1980).

[5] Cohen, Leonard and Strauss, Leonard, "Time Study and the Fundamental Nature of Manual Skill," J of Consulting Psy, May, 1946.

[6] Conrad, Robert B., "The Effect of Interruption on Standard Time," AIIE Transactions, 1970, 2-2, 150-156.

[7] Donath, Nir and Globerson, Shlomo, "The Use of Multiple Regression Analysis and Linear Programming for Work-Measurement Applications," J of MTM, 4-4, 16-21.

[8] Fein, Mitchell, "Work Measurement: Concepts of Normal Pace," Industrial Engineering, Sept., 1972, 34-39.

[9] Gershoni, Haim, "An Analysis of Time Study Based on Studies Made in the United Kingdom and Israel," AIIE Transactions, Sept., 1969, 1-3, 244-251.

[10] Gomberg, William, A Trade Union Analysis of Time Study, 2nd ed., New York: Prentice-Hall, Inc. (1955).

[11] Gottlieb, Bertram, "Unions and Industrial Engineering: Policies and Practices," J of Industrial Engineering, Sept.-Oct., 1964, 249-253.

[12] Honeycutt, John M., Jr., "Comments On An Experimental Evaluation of the Validity of Predetermined Elemental Time Systems," J of Industrial Engineering, May-June, 1962, 13-3, 171-179.

[13] Jay, Tony A., Time Study, Poole, UK: Blandford Press, Ltd. (1981).

[14] Karger, Delmar W. and Franklin H. Bayha, Engineered Work Measurement, 3rd ed., New York: Industrial Press, Inc. (1977).

[15] Kerkhoven, C.L.M., "The Rating of Performance Levels of the S.A.M. Films," J of Industrial Engineering, July-Aug., 1963, 14-4, 170-174.

[16] Konz, Stephan, Work Design, Columbus, OH: Grid, Inc. (1979).

[17] Lang, Andrew M., An MTM Analysis of Performance Rating, Pittsburgh: The MTM Association for Standards. (1952).

[18] Lowry, S.M., H.B. Maynard, and G.J. Stegmerten, Time and Motion Study, 3rd ed., New York: McGraw-Hill Book Co. (1940).

[19] Mundel, Marvin E., Motion and Time Study, 5th ed., Englewood Cliffs, N.J.: Prentice-Hall, Inc. (1978).

[20] Murrell, Hywell, "Performance Rating as a Subjective Judgement," Applied Ergonomics, 1974, 5-4, 201-208.

[21] Nadler, Gerald, "Critical Analysis of Motion Time System," National Time and Motion Study Clinic, Nov., 1952.

[22] Niebel, Benjamin W., Motion and Time Study, 6th ed., Homewood, IL: Richard D. Irwin, Inc. (1976).

[23] Presgrave, Ralph W., The Dynamics of Time Study, 2nd ed., New York: McGraw-Hill Book Co. (1945).

[24] Rice, Robert S., "Survey of Work Measurement and Wage Incentives," Industrial Engineering, July, 1977, 9-7, 18-31.

[25] Salvendy, Gavriel and Seymour, W. Douglas, Prediction and Development of Industrial Work Performance, New York: John Wiley & Sons. (1973).

[26] Schmidtke, Heinz and Stier, Fritz, "An Experimental Evaluation of the Validity of Predetermined Elemental Time Systems," J of Industrial Engineering, May-June, 1981, 12-3, 182-204.

[27] Seymour, W. Douglas, Industrial Skills, London: Sir Issac Pitman, Inc. (1966).

[28] Sheridan, Thomas B. and Ferrell, William R., Man-Machine Systems, Cambridge, MA: The MIT Press. (1974).

[29] Taylor, Frederick W., Scientific Management, New York: Harper & Row. (1911).

[30] Thompson, David A., "Time Study Sample Size-The Effect of Effort Rating Variation," J of Industrial Engineering, Mar.-Apr., 1961, 12-2, 122-125.

[31] Wechsler, David, The Range of Human Capacities, 2nd ed., Baltimore: Williams & Wilkens. (1952).

[32] Welford, A.T., Fundamentals of Skill, London: Methuen & Co. (1968).

[33] Welford, A.T., Skilled Performance, Glenview, IL: Scott, Foresman & Co. (1976).

[34] Whitmore, Dennis A., Work Measurement, London: William Heinemann, Ltd. (1975).

[35] Wilson, Jeffrey L. and Robert M. Wygant, Study on Heart Rate To Determine Work Load, Kalamazoo, MI: Unpublished report, Western Michigan Univ. (1983).

[36] Williams, Robert L., "Industrial Engineering Terminology Manual," J of Industrial Engineering, Nov.-Dec., 1965.

[37] _____, A Fair Day's Work, New York: Society for Advancement of Management. (1954).

[38] _____, "The MTM Performance Level," Pittsburgh: Maynard Research Council. (1962).

[39] _____, "An Introduction to the Leveling Procedure," Pittsburgh: Maynard Research Council. (1962).

New way to learn pace rating

A few simple cuts and splices in pace-rating films permits adaptation to programmed instruction method. Trainees say new procedure makes learning the rating process much faster and attains better results.

VINCENT G. REUTER, Arizona State University, Tempe, AZ

Before using the programmed instruction method of teaching pace rating (part of a motion and time study course) students were frustrated by lack of feedback. We were (and still are) using the SAM (Society for the Advancement of Management) rating films.

The students were to rate five work scenes of three different operations, with each scene shown at a different pace and each scene *randomly placed within the film*. The students' constant complaint was that they were expected to make their ratings without having seen films representing normal pace (100 percent) and without feedback of true ratings during the rating process.

Traditional procedure

Four of the eight SAM Rating Series films were used in the course. Each film consisted of eighteen scenes. The first three scenes detailed the method used for each operation and were not to be rated. The remaining fifteen scenes represented five scenes for each of the three operations A, B, and C, projected in the following order: A-1, B-1, C-1, A-2, B-2, C-2, etc. Each scene was shown for approximately 12-15 seconds, followed by a short break to permit the observer to record his rating. Upon completion of the entire film, the instructor provided the actual ratings for recording and comparison with the observed ratings.

In attempting to evaluate the accuracy with which the observer rated the operator pace, three calculations were made:

1. Absolute error, the arithmetic sum of the deviations between the true and observed ratings and then obtaining the average amount of deviation from the true ratings. This measures the magnitude of the average error but does not tell whether the observer rated high or low.

2. Systematic error, the algebraic

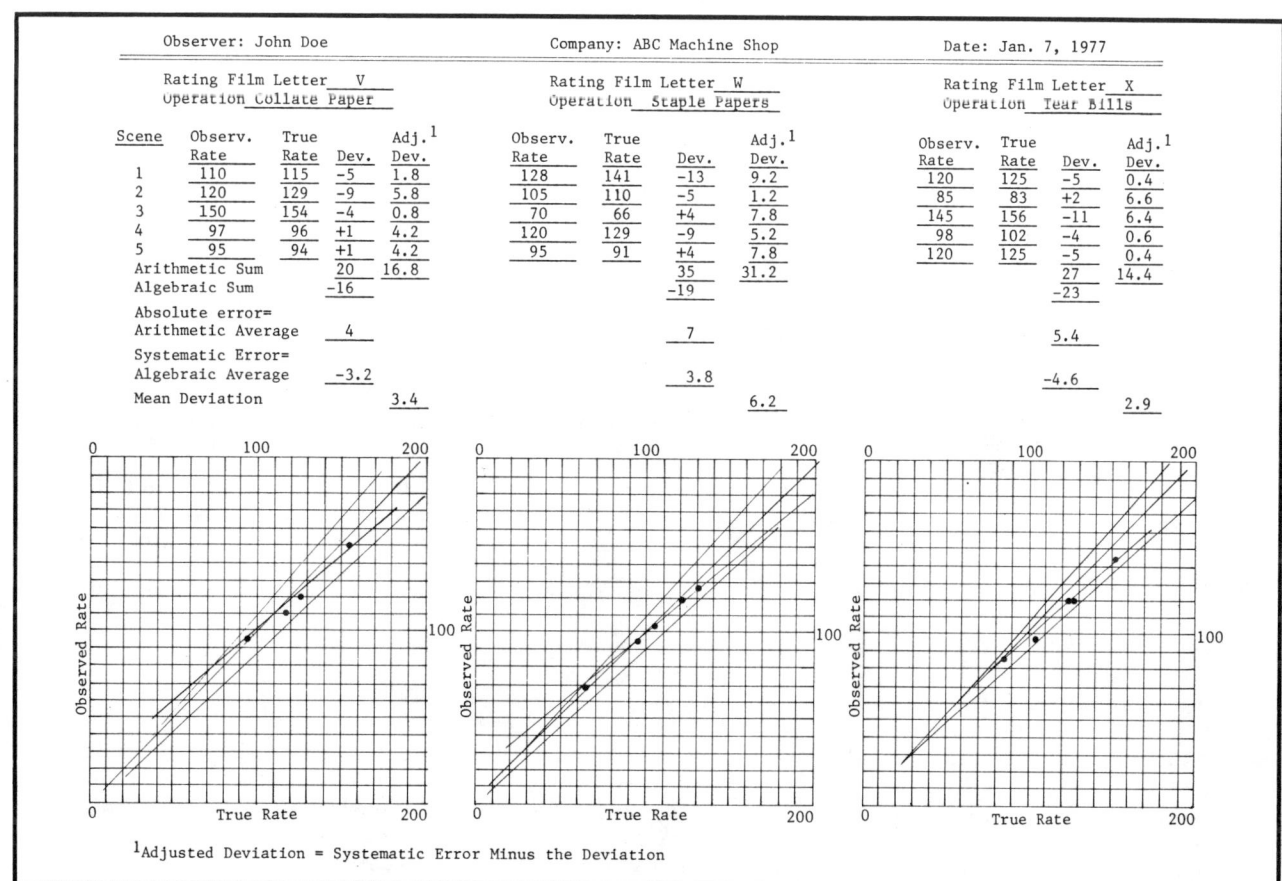

Figure 1. *Form and data used to measure results of observer's pace rating of scenes in SAM films.*

sum of the deviations and is the average amount of algebraic deviation of each person's rating from the true ratings. It tells how high or low the person rates on the average.

3. Mean deviation, the average deviation of each person's errors from his own systematic error. It measures the consistency with which the observer rates around (plus or minus) his own systematic error.

Figure 1 represents a typical observer's rating sheet after completing his observations, calculations, and trend lines.

The problem

Assume that we are to compare two raters. Both raters have the same systematic error of −10 but rater A has a mean deviation of ± 2 and rater B has a mean deviation of ± 8. Both raters have the same constant error or bias from zero error, but rater A is more consistent and has a much smaller variable error as demonstrated by the comparative mean deviations. Rater A will definitely be expected to achieve greater improvement in rating than rater B. Rater A seems to recognize variation in pace or tempo; therefore he needs only to adjust his scale of rating upward and he will be on target. B, on the other hand, must first learn to differentiate more accurately between different tempos or he will not score well, even after adjusting his sights upward. Furthermore, the rater can more rapidly adjust his sights toward the concept of normal pace if he is provided feedback on each scene as he observes his ratings. Feedback can be provided, and it can be done in such a way as to facilitate rapid orientation.

Revised procedure

Programmed learning is designed to provide early feedback to the learner so that he may take steps to correct his errors and to reinforce his correct actions. Programmed learning was adapted to the rating procedure by special preparation of two of the four films utilized in the course. Films 1 and 6 were cut apart and respliced so that the scene rated closest to normal, 100 percent, was sequenced first for each of the three operations in both films. The revised sequences are shown in Table I.

Using the programmed learning approach, films 1 and 6 were shown during the first rating session. After stopping the film approximately halfway through each scene, the student observers were asked to rate each scene, based upon their concept of normal. Then the students were given the true rating and shown the second half of the scene for reinforcement of the true pace rating. During the second rating session, film number 6 was shown again, using the programmed learning approach, to further develop and reinforce the proper concept of normal pace. Next, film number 8, still in its original sequence format, Table I, was shown to test the students' newly learned perception of normal pace and its variations. Subsequently, films 8 and 4 were used repeatedly to provide additional tests of the students' rating ability.

Results

The students have ceased to complain and they readily attain a quick understanding of the rating concept. They unanimously agree that the revised procedure makes learning the rating process much faster, more interesting, and attains better results. This programmed learning technique could be used by industrial firms who use the SAM films for training in pace rating. **IE**

Film 1		Film 6		Film 8	
Deal Cards		Tape Boxes		Collate Papers	
A3*	102	P4*	97	V1	115
A1	115	P1	126	V2	129
A2	144	P2	124	V3	154
A4	137	P3	137	V4	96
A5	89	P5	93	V5	94
Transport Marbles		Seal Cartons		Staple Papers	
B5*	102	Q4*	103	W1	141
B1	87	Q1	126	W2	110
B2	137	Q2	109	W3	66
B3	123	Q3	126	W4	129
B4	109	Q5	133	W5	91
Toss Blocks		Pack Cans		Tear Bills	
C2*	103	R2*	90	X1	125
C1	115	R1	160	X2	83
C3	127	R3	155	X3	156
C4	153	R4	121	X4	102
C5	88	R5	131	X5	125

Table I. Sequences of scenes in SAM rating films. Those asterisked in films 1 and 6 were changed to first position for programmed instruction. Film number 8 was not changed.

Dr. Vincent G. Reuter is Associate Professor of Management at Arizona State University. Prior to entering university teaching, Dr. Reuter's work experience included Assistant Division Manager, Tool Coordination Division, Wright Aeronautical Division, Curtiss-Wright Corporation, plus line and staff positions with both Republic Aviation Corporation and Honeywell. He also served as an industrial engineer with consulting firms.

He holds an AA degree in engineering, BSC and MA degrees in commerce, and the PhD degree in management from the University of Iowa. He is a Lieutenant Colonel (retired) in the Transportation Corps of the U. S. Army Reserve.

Performance Rating for Service Jobs

ROBERT A. BROWN
The University of Alabama in Huntsville
Huntsville, Alabama 35807

WILLIAM F. SOWDER
McDonnell Douglas Corporation
Huntsville, Alabama 35801

Abstract: An experiment is described in which the observational methods of Berne's Transactional Analysis are investigated to see if they can be used to create profiles of personnel involved in service activities. It is shown that: i.) Trained observers can create profiles reliably; ii.) very short observation periods, in some cases of the order of a minute or less, are sufficient allowing the principles of Work Sampling to be applied; and iii.) exposure to TA concepts can alter the measured behavior of a subject. Applications to the areas of personnel selection, training, and placement test validation are pointed out.

■ The measurement of industrial production is nearly a century old. The measurement of industrial quality is approximately a half-century old. Thus the process of measurement in industry with respect to both production and quality is well established. It is also well established that in our present economy more than half the jobs are in service industries, not manufacturing.

We also find the concepts of production and productivity measurement quite common in the service industries. For example, the number of units of medication prepared is a common measure of production in nursing; similar output measures have been constructed for the processing of insurance policies, in police work, etc. Similarly, in the service industries some attempts have been made to measure quality. Medication errors in nursing, for example, are one common measure of quality of care delivered.

There is one aspect of service, however, which to date has defied measurement. This aspect is exemplified by the sales clerk's "rapport with the customer," the nurses' ability to deliver "tender loving care (TLC)", or a receptionist's ability to "put the customer at ease." All these desirable qualities share the common property of being subjective in their definition. Everyone has had the experience of remembering a pleasant encounter with a service organization and, contrariwise, those experiences which for some vague reason seem to be less satisfying than one would desire.

Actually the same problem arises and has been solved in the industrial context. In the development of any incentive system, it is necessary to apply a subjective concept of the speed or pace of work to an observed performance in order to develop a work standard. This function is familiar to all industrial engineers as rating. A suitably trained industrial engineer can reliably and consistently measure human work pace. Why, then, can he not also measure the subjective performance of a person involved in one of the service industries?

The answer to this question seems to be that there is no ready-made classification scheme for categorizing observed behavior in a way which is related to job performance in the service industries. Let us for the moment assume that such a system exists. We could then examine the performance of successful secretaries, receptionists, nurses, salespersons, etc., and construct a profile of desirable behavior based upon the analysis of those who are agreed to be good at their jobs. Then, potential employees to fill such positions could be examined for their possession of the requisite profiles to best suit the individual to the job. Examinations might take the form of role-playing in a simple scenario, or further research might be able to demonstrate a relation between

---

Paper based on a presentation to AIIE 1978 Spring Annual Conference and published in the *Proceedings*. Paper was handled by Organization Planning/Engineering Economics Department.

desirable profiles and certain written psychological tests. Periodic on-the-job evaluations could be performed to ensure that an individual's performance was not deteriorating with time. Should such a deterioration be noted, a retraining program could be initiated in which the change in the individual's profile would document the successful achievement of the retraining program's goals.

All of the above is speculation, of course. It does, however, point out the desirability of investigating the fundamental question of whether such on-the-job performance, similar to but different from pace rating, can be measured. A search by the authors disclosed that Transactional Analysis (TA) is a system, first devised for clinical purposes, of classifying human behavior. The fact that it depends for its derivation on certain assumed "internal states" of the individual really has no bearing on the present application. For the present we only wish to enquire whether the method can be used objectively to classify human behavior; the question of whether these classifications are optimally related to job performance is necessarily left to further research.

Transactional Analysis Concepts

In the mid 1950's, Dr. Eric Berne put forth his principles and concepts of Transactional Analysis. The basic theories of TA may be categorized into (i) personality aspects, and (ii) social interaction aspects. The personality aspects are concerned with those things that involve our understanding of ourselves. The social interaction aspects deal with those things involved when we relate to others, that is, the dynamics of the human interaction when two or more people get together. The original objectives of TA were to provide means or techniques whereby individuals could increase their human awareness and bring their physical, mental, and emotional attributes into an appropriate balance.

Under Berne's theory of social interaction, when two or more individuals get together social interactions or actions and responses in the various participants take place. These actions and responses within TA take the form of transactions, games, and scripts. The transaction is the unit of TA. The theory holds that a person's behavior is best understood if examined in terms of transactions. If two people encounter each other one will speak or otherwise give an indication of acknowledging the presence of the other. This is called a *transactional stimulus*. The other person will then do something related to the stimulus called the *transactional response*.

There are three basic types of transactions. However, before we discuss the transactions it is necessary to explain a little bit about the theory of personality because the transactions are in part named for the presumed personality states associated with the transactions.

According to Berne, an individual can be in a PARENT (P) state, an ADULT (A) state, or a CHILD (C) state. The PARENT state is one in which the subject represents a figure of authority; consequently his verbalizations will involve words like "should," "ought," and "must." In the second or ADULT personality state, the individual assumes the role of a reasoning, decision-making person. The third ego state is that of the CHILD; in this state the individual is most concerned about feelings, fun, social behavior, etc. These ego states are distinguishable by physical manifestations and by the content of verbal utterances. Certain gestures, postures, mannerisms, facial expressions, tone of voice, and choice of words are typically associated with each of these ego states. There is also a second-order analysis possible in examining ego states which further subdivides the PARENT and CHILD into two sub-states each.

Returning now to our explanation of transactions, an example of a simple transaction between two ADULT ego states is the following: stimulus—"What time is it?"; response—"Four-thirty." This type of transaction is called a complementary transaction. An example of a crossed transaction would be a stimulus from a manager's P-state to a worker's C-state: "Let's get to work." The response from the worker might be from *his* P-state to the *manager's* C-state: "Don't you boss me." There are numerous combinations of complementary and crossed transactions possible. There is also an ulterior transaction, but it is not necessary for our purposes to understand these completely. It suffices to know that they are categories used by a TA observer to classify a subject's behavior.

A third aspect of observable behavior is the way in which a person structures his time. It has been postulated that an individual structures time in some or all of the following six ways:

Withdrawal	Rituals
Activities	Pastime
Games	Intimacy

Some of these time structures, e.g., games, are very complex in their description and definition, but again precise definitions need not concern us here.

The foregoing three classification schemes provide us with three different profiles which may be constructed to represent an individual's observed behavior. Hypothetical examples of these are shown in Figures 1, 2, and 3. The Egogram shows the percentage of observed behavior in which the subject adopted each of five second-level divisions of ego state. The Timeogram is a similar percentage breakdown of the individual's time structuring behavior. Finally, the Transactogram shows a percentage breakdown of the transactions between the subject and other parties involved in the transactions.

Hypotheses

The foregoing brief introduction leads us to the research hypotheses for this study.

Hypothesis 1

Ego states, time-structuring and transactions of an individual cannot be reliably and consistently identified by persons knowledgeable in TA concepts.

Hypothesis 2

The length of time a person has been associated with TA applications is not significant in identifying ego states, time-structuring or transaction types.

Hypothesis 3

People familiar with TA concepts cannot identify ego states, time-structuring and types of transactions with consistency in observation times less than 15 minutes.

Hypothesis 4

Analysis of personality state profiles at different phases of TA development do not lead to measurable personality changes.

Methodology

It will be noted that each of the hypotheses above is stated in the form of a null hypothesis. This is done in order to provide the proper form for the statistical design of an experiment.

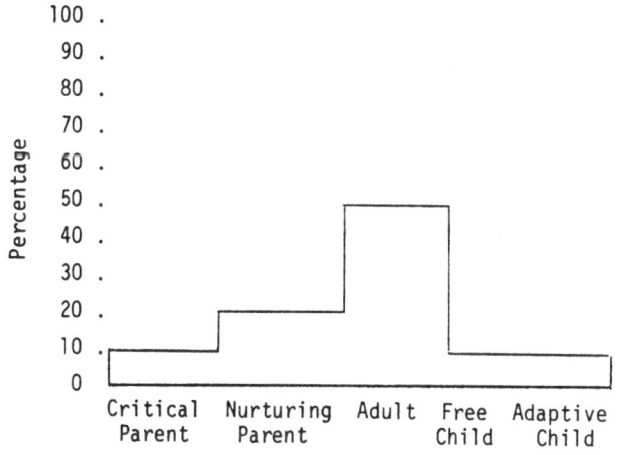

Fig. 1. Egogram

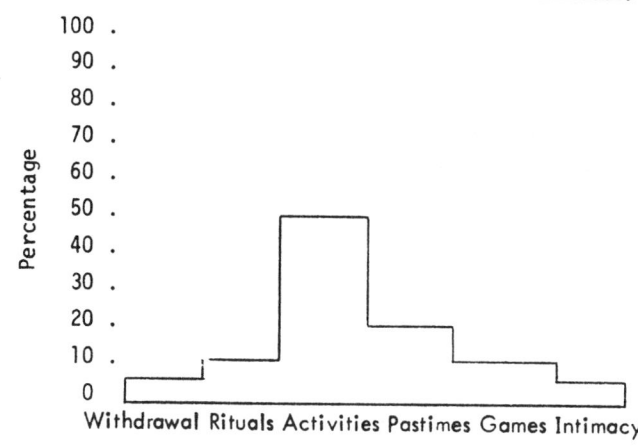

Fig. 2. Timeogram

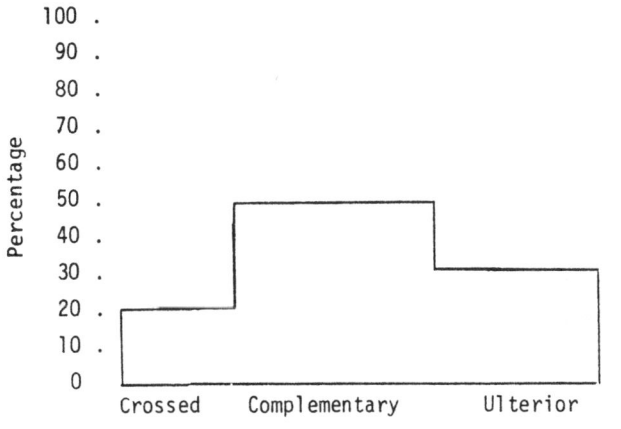

Fig. 3. Transactogram

A nested factorial experiment was designed for each personality categorizing scheme—ego states, time structuring, and type of transaction. A summary of the factors and their levels of evaluation is shown in Table 1. "Observers" was a nested factor, nested within the experience groups. A total of 18 experimental conditions was therefore required.

To accomplish this experiment, an individual engaged in a business office environment where the social interaction was among the internal workers in the office rather than with the external public was selected. The particular subject was a girl secretary in an office with six highly technically trained engineers. Her job responsibilities included general typing, preparation of technical reports generated by the engineers, filing, answering the telephone and taking messages, making travel arrangements for personnel she supported and keeping the engineers' expense and travel records. The cadre of engineers included a manager and five support type, scientifically oriented people. The subject was 27 years years of age and her education included two years of college, where she majored in English.

The experiment was conducted only during normal working or business hours from 8 am to 5 pm. The working environment included social interactions among the internal office workers only; that is, no effects from external persons or customers were included. No effects of the subject's environment away from work were recorded. Therefore, this experiment and its results apply only to a single subject and on-the-job environments. This is in keeping with the philosophy that the experimentation emphasize observer-related aspects rather than subject-related aspects to demonstrate the experimental methodology.

The subject was observed with a portable Sony vidicon camera and recording unit. Each video tape was restricted to approximately 15 minutes recording time. The relationship of the subject to the camera is shown in Fig. 4. Although the subject was aware that she was being observed by a camera, she did not know when the actual recording was

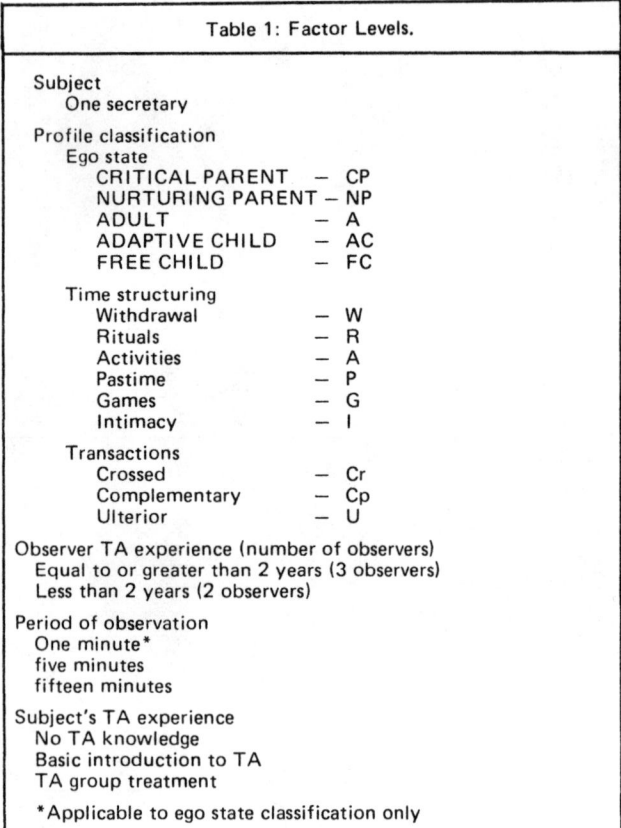

Table 1: Factor Levels.

Subject
  One secretary
Profile classification
  Ego state
    CRITICAL PARENT   — CP
    NURTURING PARENT — NP
    ADULT   — A
    ADAPTIVE CHILD   — AC
    FREE CHILD   — FC

  Time structuring
    Withdrawal   — W
    Rituals   — R
    Activities   — A
    Pastime   — P
    Games   — G
    Intimacy   — I

  Transactions
    Crossed   — Cr
    Complementary — Cp
    Ulterior   — U

Observer TA experience (number of observers)
  Equal to or greater than 2 years (3 observers)
  Less than 2 years (2 observers)
Period of observation
  One minute*
  five minutes
  fifteen minutes
Subject's TA experience
  No TA knowledge
  Basic introduction to TA
  TA group treatment

  *Applicable to ego state classification only

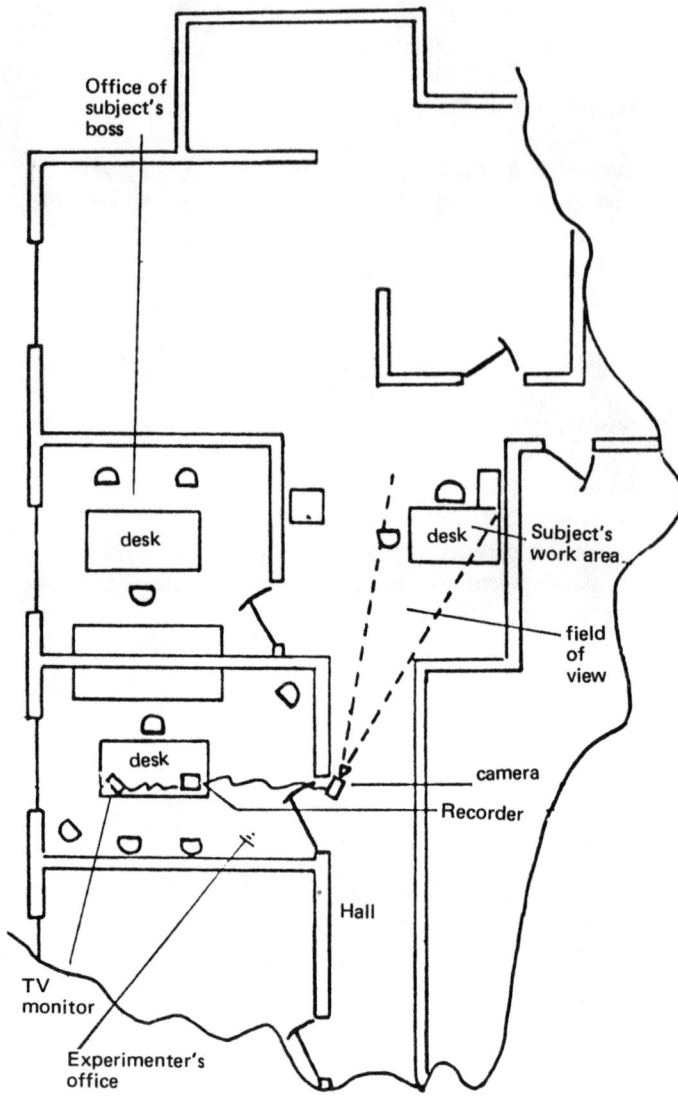

Fig. 4. Layout of observation equipment.

accomplished. Recording was initiated by the experimenter with the aid of a TV monitor which was in the experimenter's office and out of sight of the subject.

Material was recorded on the tapes in segments since the secretary often left her desk and was out of sight and sound of the recording equipment. The recorder was turned off during these times when the subject was away from her desk. It is possible that these recording procedures could have resulted in favorable observation times which might result in a bias in the experimental results. Although this effect is possible, it was not the intent of the experimenter to choose favorable recording times. Future experiments should use a recording technique using random numbers to select the recording times. Also a person independent of the experimenter could be used to turn on the recorder resulting in a "double blind" experimental recording technique.

Data were recorded by each observer as they observed each video tape recording. The percentage of time that the subject was in each of the elements of the various personality profile areas was recorded as raw data. The recording techniques for each 15 minute tape are as follows:

Ego States – Record in one minute segments the percentage of time spent in CRITICAL PARENT, NURTURING PARENT, ADULT, ADAPTIVE CHILD or FREE CHILD. The tape was stopped at the end of each one minute segment and the observers recorded data. Fifteen segments of data were obtained from each tape.

Time-structuring– Record in five minute segments the percentage of time spent in Withdrawal, Rituals, Activities, Pastime, Games or Intimacy. The tape was stopped at the end of each five minute segment and the observers recorded data. Three segments of data were obtained for each tape.

Transactions–Record in five minute segments the percentage of time spent in Complementary, Crossed or Ulterior type transactions. Three segments of data were obtained for each tape.

Time structuring and transactions data were collected in five minute segments because of the difficulty the observers had in assessing the percentage of time in each element for smaller increments of time. That is, if the tape were stopped for time intervals less than 5 minutes, the observers lost continuity within the data that precluded them from making good observations.

Data were recorded in this manner for each phase of the subject's TA knowledge—namely no knowledge, basic TA introduction and TA group treatment. Thus, the observers were aware of what phase the subject had completed. This could detract from the randomness of the data analysis. However, it is unlikely that the observers were influenced by this sequencing of tape observation since they were not aware of the experimental results until all the data had been collected and analyzed. Also it was not clear what to expect from the subject subsequent to completion of each phase. Thus, it was not easy to anticipate the changes in the subject's personality profiles.

To make the variances independent of the proportions for ANOVA a different scale of measurement was needed. It was decided that a standard data transformation for each piece of raw data of

$$y = 2 \arcsin \sqrt{h},$$

where h is the proportion (percentage/100), was appropriate. Based on a test of homogeneity of variances, the data transformation function was adequate.

Because it is possible due to recording techniques and data analysis sequencing procedures that the experiment accidentally selected "favorable" or "good times" for observation purposes, preliminary data analyses were conducted to evaluate experimental repeatability. In this type of experimentation with a human subject under natural environmental conditions, pure experimental replication is impossible. However, repeatability analyses of the personality profiles were undertaken in lieu of replication. For ego states where the subject had no knowledge of TA, two different video tapes recorded at different times were analyzed to obtain a cursory measure of repeatability. This preliminary analysis indicated that repeatability existed. Therefore, this factor was not further included in this experiment.

Results

The ANOVA tables for ego states, time structuring, and transactions are shown, respectively, in Tables 2, 3, and 4. Since the results are largely the same for all three dependent variables, only the results for ego states will be discussed here.

We note first that TA groups based upon observer experience do make a difference at the 1% level of significance. Thus more experienced observers can be expected to do a better job of categorizing behavior.

We also find that the main factor "Categories" is significant at the 1% level. Thus categories of classification do show meaningful distinctions for all three behavioral classification schemes.

The absence of significance for the "Observer" factor is interpreted to mean that the observers were relatively homogeneous within the "Experience" groups. Expressed

Table 2: Nested Factorial Analysis of Variance Table for Ego States.								
Source of Variation	Degree of Freedom	Sum of Squares	Mean Square	F	$F_{.01}$	S	$F_{.05}$	S
Main Effects								
G - TA Groups	1	2.9308	2.9308	8.40	6.63	X	3.84	X
O - Observers	3	2.1909	0.7303	2.09	3.78		2.60	
P - Period	2	0.1914	0.0957	0.27	4.61		3.00	
S - Subject status	2	0.2061	0.1031	0.30	4.61		3.00	
C - Categories	4	35.6344	8.9086	25.50	3.32	X	2.37	X
Two Factor Interactions								
GS	2	0.0228	0.0114	0.03	4.61		3.00	
GC	4	2.3875	0.5969	1.71	3.32		2.37	
GP	2	0.0178	0.0089	0.03	4.61		3.00	
OS	6	0.2075	0.0346	0.10	2.80		2.10	
OC	12	6.0528	0.5044	1.44	2.18		1.75	
OP	6	0.0703	0.0117	0.03	2.80		2.10	
SC	8	8.0832	1.0104	2.90	2.51	X	1.94	X
SP	4	0.0462	0.0116	0.03	3.32		2.37	
CP	8	0.3262	0.0408	0.12	2.51		1.94	
Three Factor Interactions								
GPS	4	0.0457	0.0114	0.03	3.32		2.37	
GSC	8	3.5784	0.4473	1.28	2.51		1.94	
GCP	8	2.8385	0.3548	1.02	2.51		1.94	
OSC	24	18.9673	0.7903	2.26	1.88	X	1.57	X
OCP	24	7.9776	0.3324	0.95	1.88		1.57	
OPS	12	0.5426	0.0452	0.13	2.18		1.75	
SCP	16	10.9749	0.6859	1.96	2.08		1.69	X
Four Factor Interactions								
GSCP	16	6.3663	0.3979	1.14	2.08		1.69	
OSCP	48	0.0488	0.0010	0.003	1.65		1.41	
Error	225	78.6059	0.3494					
Total	449	188.3139						

Table 3: Nested Factorial Analysis of Variance Table for Time-structuring.									
Source of Variation	Degree of Freedom	Sum of Squares	Mean Square	F	$F_{.01}$	S	$F_{.05}$	S	
Main Effects									
G - TA Groups	1	1.5786	1.5786	5.31	6.77		3.89	X	
O - Observers	3	0.0137	0.0046	0.02	3.90		2.65		
P - Period	1	0.0010	0.0010	0.003	6.77		3.89		
S - Subject status	2	0.0838	0.0419	0.14	4.73		3.04		
C - Categories	5	32.4626	6.4925	21.83	3.12	X	2.26	X	
Two Factor Interactions									
GS	2	0.0016	0.0008	0.003	4.73		3.04		
GC	5	1.2387	0.2477	0.83	3.12		2.26		
GP	1	0.0004	0.0004	0.001	6.77		3.89		
OS	6	0.0984	0.0164	0.06	2.90		2.14		
OC	15	2.4411	0.1627	0.55	2.14		1.72		
OP	3	0.0178	0.0059	0.02	3.90		2.65		
SC	10	7.1656	0.7166	2.41	2.42		1.88	X	
SP	2	0.0017	0.0009	0.003	4.73		3.04		
CP	5	1.1115	0.2223	0.75	3.12		2.26		
Three Factor Interactions									
GPS	2	0.0137	0.0069	0.02	4.73		3.04		
GSC	10	1.7078	0.1708	0.57	2.42		1.88		
GCP	5	1.2811	0.2562	0.86	3.12		2.26		
OSC	30	6.5875	0.2196	0.74	1.81		1.52		
OCP	15	2.8330	0.1889	0.64	2.11		1.72		
OPS	6	0.1702	0.0284	0.10	2.90		2.14		
SCP	10	9.8155	0.9816	3.30	2.42	X	1.88	X	
Four Factor Interactions									
GSCP	10	2.3274	0.2327	0.78	2.42		1.88		
OSCP	30	4.6912	0.1564	0.53	1.81		1.52		
Error	180	53.5293	0.2974						
Total	359	129.1735							

Table 4: Nested Factorial Analysis of Variance Table for Transactions.									
Source of Variation	Degree of Freedom	Sum of Squares	Mean Square	F	$F_{.01}$	S	$F_{.05}$	S	
Main Effects									
G - TA Groups	1	3.2057	3.2057	6.37	6.78		3.89	X	
O - Observers	3	0.0992	0.0331	0.07	3.88		2.65		
P - Period	1	0.0002	0.0002	0.0004	6.78		3.89		
S - Subject status	2	0.0185	0.0093	0.02	4.73		3.04		
C - Categories	2	27.9686	13.9843	27.79	4.73	X	3.04	X	
Two Factor Interactions									
GS	2	0.0185	0.0093	0.02	4.73		3.04		
GC	2	0.3580	0.1790	0.36	4.73		3.04		
GP	1	0.0002	0.0002	0.0004	6.78		3.89		
OS	6	0.1373	0.0229	0.05	2.91		2.14		
OC	6	1.5396	0.2566	0.51	2.91		2.14		
OP	3	0.0081	0.0027	0.01	3.88		2.65		
SC	4	1.6826	0.4207	0.84	3.44		2.42		
SP	2	0.0013	0.0007	0.001	4.73		3.04		
CP	2	0.8268	0.4134	0.82	4.73		3.04		
Three Factor Interactions									
GPS	2	0.0201	0.0100	0.02	4.73		3.04		
GSC	4	1.2506	0.3127	0.62	3.44		2.42		
GCP	2	0.4632	0.2316	0.46	4.73		3.04		
OSC	12	7.4992	0.6249	1.24	2.29		1.80		
OCP	6	1.6630	0.2772	0.55	2.91		2.14		
OPS	6	0.1545	0.0258	0.05	2.91		2.14		
SCP	4	1.8186	0.4547	0.90	3.44		2.42		
Four Factor Interactions									
GSCP	4	1.8707	0.4677	0.9295	3.44		2.42		
OSCP	12	35.1394	2.9283	5.8194	2.29	X	1.80	X	
Error	90	45.2867	0.5032						
Total	179	131.0306							

another way, differences in experience between observers explained essentially all of the observer-to-observer differences.

Again, the absence of significance for the period of time over which observations were recorded indicates that the time period factor is not important in determining the reliability of the observed results. Therefore observation intervals as short as five minutes (one minute in the case of ego states) are sufficiently long to permit adequate classification of behavior.

The absence of significance for the "Subject" main effect requires some explanation. One would presume that absence of significance for this factor demonstrates that the subject's behavior was not influenced by exposure to TA concepts. A little reflection shows, however, that this is an artifact of the way in which the data are collected. The original data were obtained as percentages; the subject while under observation had to be doing *something*. Therefore, when averaged across classification categories the subject's mean performance for every state of TA knowledge is simply 100% divided by the number of classification categories. Significance for the degree of exposure to TA concepts is strongly demonstrated in the two-factor Subject-Category interaction for ego states, and somewhat less strongly for time structuring, indicating a differential effect on how behavior was classified depending upon the subject's state of TA knowledge. Table 5 summarizes the original hypotheses and the experimental results related to them. No attempt has been made to explain the three-factor and higher order interactions.

Conclusions and Recommendations

Based on the results of this experiment involving the testing of Transactional Analysis concepts in a business office environment with a 27 year old girl secretary as the subject, it is concluded that:

(i) Berne's concept of ego states, time-structuring and transactions can be reliably identified by people familiar with TA concepts.

(ii) It is desirable that people who perform identification of ego states, time-structuring and transactions have sufficient training and experience in TA concepts and be thoroughly familiar with identification clues and diagnosis techniques.

(iii) The amount of time needed for identification of ego states, time-structuring and transactions appears not to be significant. Fifteen minutes or less may be adequate.

(iv) The construction of personality profiles (egograms, timeograms, and transactogram) appears to indicate some measure of variations or changes in a person's personality development.

The subject's behavior in this experiment appears to be normal for a business office environment with an ADULT ego state, activities and pastiming time usage, complementary transactions being the dominant elements of the personality profiles.

Overall, it appears that the use of ego state classifications is the most sensitive and reliable of the classification schemes. This is probably also the most easily taught and understood of the basic TA concepts. Thus it seems that a ready-made behavioral classification scheme is available which can be used for further research on the classification of behavioral interactions in industrial human relations.

The present research is considered to be only a pilot study. There is a definite need to repeat and replicate the current experiment, especially through the use of more than one subject. Further research is needed to determine whether the transactional analysis classification scheme is the best one for industrial applications. Certain techniques of cluster analysis might be applied to a replicated experiment to determine, possibly through the use of Principle Components Analysis, whether a modification of Berne's TA classification methods would be more appropriate.

Following the development of the best classification system, research needs to be undertaken to determine desirable profiles for different types of work. These would undoubtedly be a function of the society, possibly regions of the country, etc. Furthermore such research might address

	Table 5: Summary of Test of Hypotheses.	
Hypothesis	Thesis	Results
1	Ego state, time-structuring and transactions cannot be identified by people knowledgeable in TA concepts	**Must be rejected.** — These characteristics can be consistently identified.
2	The length of time an observer has been in TA is not significant in identifying ego states, time-structuring and transactions.	**Must be rejected.** — The experience level of an observer in TA appears to be important in identifying these personality characteristics.
3	People familiar with TA concepts cannot identify ego states, time-structuring and transactions in observation times less than 15 minutes.	**Must be rejected.** — Identification of these personality characteristics can be accomplished in 15 minutes or less; there is no distinguishable difference between one and 15 minutes for ego states, and 5 and 15 minutes for time-structuring and transactions.
4	Analysis of personality state profiles at different phases of TA development do not lead to measurements of personality changes.	**Must be rejected.** — Variations in personality state profiles are detectable by trained observers indicating that these profiles can be used to measure personality changes.

the possibility that more than one profile could be equally effective doing a particular job. This, in a supervisor, could be termed research on alternate styles of management.

References

[1] Berne, Eric, *Games People Play*, New York: Grove Press, (1964).
[2] Berne, Eric, *Transactional Analysis in Psychotherapy*, New York: Grove Press (1961).
[3] Berne, Eric, *What Do You Say After You Say Hello*, New York: Grove Press (1972).
[4] Harris, T. A., *I'm OK – You're OK*, New York: Harper and Row (1969).
[5] James, Muriel and Jongeward, Dorothy, *Born to Win*, Reading, Massachusetts: Addison-Wesley (1971).
[6] Sowder, William F., *Experimentation in Transactional Analysis*, thesis, The University of Alabama in Huntsville, Huntsville, Alabama, 1976 (also available through University Microfilms, Inc.).
[7] Steiner, Claude, *Games Alcoholics Play,* New York: Grove Press (1971).
[8] Thomson, George, "The Identification of Ego States," *Transactional Analysis Journal*, **2**, **4**, 196-211 (October 1972).

Dr. Robert A. Brown received a BS degree from the U.S. Naval Academy at Annapolis, Maryland and MS and PhD in the Department of Industrial Engineering at Ohio State. After teaching for five years at Ohio State, Dr. Brown moved to The University of Alabama in Huntsville, where he is presently the Chairman of the Department of Industrial and Systems Engineering. He has served as the Chairman of the Huntsville Section, IEEE and as a member of its board. Dr. Brown also has wide consulting experience in aerospace and manufacturing industries. He is also a member of ORSA, AIIE, and ASEE, on whose Engineering Manpower Committee he currently serves.

Mr. William F. Sowder received his Bachelor's of Aerospace Engineering degree from Ohio State University in 1954 and a MSE degree from The University of Alabama in Huntsville in 1976. Mr. Sowder has been employed by McDonnell Douglas Corporation for the past 18 years and is currently serving as Group Engineer in the Systems Development Department.

III. Allowances and the Impact of Fatigue

Following the determination of normal time by any work measurement technique, an additional adjustment must be performed in order to arrive at the standard time. This adjustment adds time for personal delays, unavoidable delays (interruptions), and any slowdowns caused by fatigue. In direct observation timing, for example, the readings are taken over a relatively short period, and abnormal readings, unavoidable delays, and time for personal needs are removed from the study in determining the normal time.

For work measurement, the Industrial Engineering Terminology Standard Z94.12 defines allowance as "a time value or percentage of time by which the normal time is increased, or the amount of non-productive time applied, to compensate for justifiable causes or policy requirements which necessitate performance time not directly measured for each element or task. Usually includes: irregular elements, incentive opportunity on machine controlled time, minor unavoidable delays, rest time to overcome fatigue, and time for personal needs."

Allowances must be derived through sound engineering practices. Otherwise, the standard time may be incorrect, and the entire work measurement process will have proved pointless. Companies assigning an arbitrary percentage for allowances may overestimate, thus inflating the cost of a product. Conversely, they may underestimate allowance times, with the result that workers cannot meet production standards, or can meet them only at increased safety or health risk. It is also important to note that allowances may not be applied uniformly to the total cycle time. Different values may be applicable to machine-controlled time and human-effort time. Allowances applicable to the cycle time are usually expressed as a percentage of the total.

METHODOLOGY FOR DEVELOPING ALLOWANCES

There are two methods commonly used to develop allowance data. One is continuous observation, which requires the observer to study actual operations over a substantial period of time. The observer records the duration of and reason for each lost time interval. Upon establishing a reasonably representative sample, the observer summarizes the findings to determine the percentage allowance for each applicable lost-time event. The data obtained in this fashion, like those for any time study, must be adjusted to the level of normal performance. Because of the long period of observation required, this method is exceptionally tedious, not only for the analyst but also for the operators. Another disadvantage is the tendency to take too small a sample, which may result in biased results.

The most efficient method of establishing the allowance percentages is through properly designed work sampling studies. This method involves recording a number of random observations, thus requiring only part-time or intermittent attention. No timing device is used; the observer is required only to record what each operator is doing at the appropriate sampling time. The number of lost-time intervals recorded, divided by the total number of observations, will closely approximate the actual percentage of time required by the operator for allowances.

PERSONAL ALLOWANCES

Allowances for personal delays may be defined as the time provided for cessations in the work necessary for maintaining the general well-being of the employee. In most work environments, this usually includes trips to the drinking fountain and rest room. The general working conditions and type of work will influence the time necessary for personal delays. For example, operations involving heavy work performed at high temperatures, such as that done in the pressroom of a rubber-molding department or in a hot-forge shop, would require greater allowance for personal needs than would light assembly or office work performed in temperature-controlled areas. Detailed continuous observation studies have demonstrated that 5 percent allowance for personal time, or 24 minutes in eight hours, is common for typical shop conditions. This value is also recommended by the International Labor Office. The amount of time needed for personal delays will, of course, vary to some extent with the person as well as the type of work. While the 5 percent figure cited appears to be adequate for the majority of workers, some bargaining groups have included time for personal delay as a negotiated contract item. In general, this is not a recommended practice. All personal allowances should be based on properly engineered studies for a given worker population, the task, and work place conditions.

FATIGUE

Closely associated with the allowance for personal needs is the allowance for fatigue. Fatigue has many meanings: not feeling like working, feeling tired, a reduction in output due to becoming overheated, or the actual tiring of the body muscles. The Industrial Engineering Terminology Standard Z94.12 defines fatigue as "a psychological and physiological process that reduces the performance capacity and motivation of humans subjected to excessive or repeated work stresses." Fatigue allowance may be defined as the time added to the normal time to compensate for a lessening in the capacity for the ability to work. It is important to note that several human variables affect the need for fatigue allowances, including age, sex, physical conditioning, and the general health of the worker.

Fatigue may be psychological, environmental, or physical in nature. In extreme cases it can be a symptom of a disease. Psychological fatigue is characterized by tiredness or not feeling like working; lack of motivation may produce apparent psychological fatigue. Environmental fatigue results from the stress placed on the body by the environment. This stress is usually related to the body's ability or inability to maintain its normal temperature of 98.6 degrees F. because of heat produced during physical work. Physical fatigue may be characterized by a drop in blood sugar or a local muscular fatigue caused by accumulating lactic acid in an intensely working muscle group.

Psychological fatigue is by far the least studied of the fatigue types, but it is receiving more attention with the continuing shift of the population to knowledge work and away from physical work. Most physical work has some associated mental tasks which can add to the fatigue component. Studies have indicated that output reduces with time for continuous work and that periodic breaks can increase the total output.

Environmental fatigue, most commonly heat stress, is caused by the inability of the body to transfer excess heat. A minor rise in body temperature will usually cause no harm to the individual. The safe upper limit of body temperature is approximately 100.4 degrees F. Heat stress has been studied enough to allow the compilation of tables that show maximum allowable exposure time and recovery periods given certain parameters. The calculated time for a task can then be adjusted upward to allow relief so that the worker may receive the recommended time to recover.

Physical or physiological fatigue can arise through either exceeding the body's maintainable energy output or by requiring a single muscle group to be used repeatedly without time to relax. In general, the body can be expected to maintain a maximum work load of 3.0 to 5.5 kcal/min. over an eight-hour period; this usually decreases after eight hours. The value is based on a percentage of the energy expenditure at a person's aerobic capacity and on other human variables as mentioned above. The aerobic capacity should be measured using the same muscle group or groups as in the projected tasks. If measurement indicates that the actual work load is greater than a person's allowable work load, then a fatigue allowance or recovery period should be calculated and implemented.

Local muscular fatigue is a function of the percent of maximal tension required for a task. In general, the greater the percent of maximal producible tension a task requires, the shorter the length of time that the task can be performed. For any sustained contraction of more than a few seconds, this fatigue must be considered. Allowances for this type of fatigue are not normally recommended, however. Instead, the task should be redesigned to distribute the work so that a period of rest for the muscle group is built in. For example, a task could be alternated between hands so that each hand would have a period of rest after performing the task.

DELAY ALLOWANCES

Time should also be provided for any work slowdown or stoppage beyond the control of the worker. Examples include communicating with the foreman and waiting for material or tooling.

SUMMARY OF ARTICLES

William H. Bostion provides an overview of the subject in "Development of Personal, Fatigue and Delay (PF&D) Allowances," which presents the results of a MTM Association survey of member companies. The article lists specific considerations for each category of allowances.

"Methods for Estimating Physical Fatigue" by Arun Garg reviews four different approaches to the task: 1) output degradation, 2) motion and time study and predetermined motion time data, 3) psychological methodology, and 4) physiological criteria. Methods for determining metabolic expenditure rate are also reviewed. Standards based on motion and time study and psychophysical methodology are compared with those based on metabolic expenditure and heart rate. Comparison shows that neither the standards based on subjective estimates of work analysis nor the acceptable work loads based on psychophysical methodology agree with the physiological fatigue criteria. The article recommends that the physiological assessment of a job should be made to assist the work study engineer in determining relaxation allowances more accurately when moderate to heavy manual work is performed, especially under trying environmental conditions.

The article by Anil Mital and Richard L. Shell, "Determination of Rest Allowances for Repetitive Physical Activities That Continue for Extended Hours," presents a comprehensive model using job and worker profile inputs, such as metabolic energy expenditure rate required for the job, shift duration, age, sex, and body weight of the worker. The model predicts the total length of rest breaks as a percentage of shift duration. The modeling concept was verified by collecting data on male and female industrial workers. The model can be used to calculate recommended rest allowances for eight hour shifts or for extended shifts up to twelve hours. All modeling assumptions and concepts are described.

"Development of Design Data Base for Manual Lifting Activities for Extended Work Shifts," also by Mital and Shell, presents data useful in establishing comprehensive lifting guidelines. The data norms for long and extra-long shifts are provided for various lifting frequencies, heights of lift, and container sizes; the effects of selected task-related variables, worker age, and shift duration are also discussed. Somewhat surprisingly, the study also indicated that the age of the worker was not significant in terms of lifting capacity.

The study reported by Shell and O. Geoffrey Okogbaa in "The Effect of Mental Fatigue on Knowledge Worker Performance" used three measures—work output, heart rate, and brain wave activity—to compare the performance of knowledge workers with hourly rest breaks and with no rest breaks. The results show a clear advantage in output by the workers with scheduled breaks.

DEVELOPMENT OF PERSONAL, FATIGUE & DELAY (PF&D) ALLOWANCES

William H. Bostion, Supervisor
Industrial Engineering

Westinghouse Electric Corporation
Defense and Electronic Systems Center
Baltimore, MD 21203

GENERAL

Personal, Fatigue and Delay (PF&D) is the time allowed a worker to compensate for attending to personal needs, for fatigue, and for delay occurring due to conditions beyond their control. This time is an additive to the normal time required to accomplish a task. The inclusion of this allowance is common practice in the development of a labor standard.

Historically, these allowances have not received the same attention as the basic time standards, also work measurement texts have devoted proportionately few pages to this subject.

In 1976, the MTM Association decided to conduct a survey, of their membership only, to determine the Work Measurement allowance techniques used by the individual member companies. This project was undertaken by the Associations Research Committee. They prepared a survey form containing 14 questions regarding PF&D practices, which resulted in 186 useable responses. Some of the results were:

a. 99% employed PF&D allowances in their work standards.

b. 61% have an official breakdown of the allowances.

c. 56% stated that the company and workers both believe PF&D to be a gross measurement which is tolerated since part of the allowance cannot be measured with any precision.

d. 18% stated their PF&D allowance was subject to negotiations in their labor contact. Only 21% of these had significant disputes of the allowances between the union and the company.

e. 47% had more than one PF&D allowance.

f. The values for the allowance were established by the following techniques.

1. Work sampling, or a similar technique	22%
2. Time studies	7%
3. Historical value; source unknown	27%
4. Provided by management consultants	15%
5. Based on judgement	21%
6. Union negotiated	8%

g. Only 39% had standards used for wage incentive purposes.

h. The number of employes covered by the survey was:

No. of Employes	% of Response
1 thru 100	31%
101 thru 500	43%
501 and above	26%

The results of this survey probably reinforces some conceptions concerning allowance practices of industry, but also contained some surprises.

Present practices for computing PF&D allowances have resulted in varied interpretation of the factors being considered and the use of different techniques to establish them. Variances in application ranged from an allowance for each element within a standard to the adoption of a fixed or blanket allowance for all standards in an organization or activity. As a result of these different practices, standards for identical work are inconsistent and result in different measurement criteria for identical jobs or functions, as well as incomparable data at summary levels.

In order to minimize these variances, it is necessary to establish a standardized method of computing the PF&D allowances consistent with the organizations or activity.

The guidelines for developing allowances suggested here have been accepted and used extensively, for some time, throughout the Department of Defense and are established as a standardized method for development of PF&D allowances.

Where appropriate, a fixed PF&D allowance, based on the standardized method, may be developed one time for a specific function or for groups of personnel doing similar work under similar conditions. The fixed allowance applies to all standards in the function or group and precludes the need to individually compute the allowance for each standard. In work situations where these guidelines are not applicable, the fixed allowance should be developed through work measurement techniques such as time study or work sampling.

CONDITIONS FOR CONSIDERATION

The development and application of PF&D allowances requires that the various conditions under which a task is performed be examined and considered. To ensure that all conditions are considered, separate factors are provided for each of the three areas; Personal, Fatigue and Delay. The persons analyzing the conditions must be completely objective in establishing the allowances which correctly reflect the true situations inherent to the task.

For example, your operation may or may not have established break periods for the employe to stop work from time to time to attend to personal needs, (go to the restroom, get a drink of water, take a smoke, etc. . .) however, the fact remains that an allowance should be considered because these situations are inherent to the job.

ALLOWANCES FOR PERSONAL TIME

Consider the surroundings, work conditions, and job requirements which cause the employe to attend to personal needs such as:

a. Break periods as a result of management policy, union agreements or realized experience.

b. Working conditions
  1. normal
  2. Slightly disagreeable
  3. Extremely disagreeable

c. Time allowed by management, at the beginning of the shift to make ready and/or at the end of the shift to get/put away tools and equipment, clean up work area, or to don/remove special work clothing such as aprons, smocks, etc. Additional supplemental allowance would be required for super-clean room conditions when operators are required to utilize special clothing. This must be included each time the clothing is put on/taken off during their shift.

d. Other conditions, relative to your specific operations, may occur and must be considered such as paid lunch periods where the work period is 8 consecutive hours and the time is allowed at the expense of the company/government.

ALLOWANCES FOR FATIGUE (PHYSICAL & MENTAL)

a. Consider the average weight handled per person and only those elements of time that the operation is under load, the height that the load must be manually lifted, and the effective net weight for sliding or rolling objects to provide a realistic allowance. "Realistic" has been defined as an allowance acceptable to the worker, supervisor, and the analyst.

b. Consider the position which the employes must assume to perform the operation, such as; sitting, standing, walking, climbing and/or working in close, cramped positions.

c. Consider the degree of concentration necessary to perform the task and the amount of variety associated.

d. Consider the amount of light on the working surface in relation to the fineness of details upon which the operator works, the glare on the work surface, rapid changing or "hypnotic" effect on the work surface.

e. Consider the general noise of the work area as well as any annoying, sharp, or intermittent noises occuring during more than 50% of a work day. If ear plugs or ear muffs are worn, their sound deading effect must also be considered.

f. Consider the "Monotony" fatigue resulting from fast, highly repetitive operations. The cycle is the time elapsed from starting one element until the same element is started again.

g. Consider restrictive safety devices and clothing which are required by the job and which cause fatigue when worn. No allowance should be considered here unless it is necessary to remove the device(s) occastionally for relief.

ALLOWANCE FOR DELAY

a. Consider the job in relation to adjacent jobs — how long can any adjacent job be shut down before the task being analyzed is affected?

b. Consider delays inherent in the task, such as:
  1. Supervisory interruptions
  2. Moving from one work station to another
  3. Special delays, fall into two categories:
     a) Those which occur on a non-foreseeable basis (power failure, minor repairs to defective parts, waiting for job assignment.
     b) Those which occur on a time basis (daily, weekly, hourly).

The following are examples of the types of special delays which should be considered.

a. Obtain job information from the supervisor, inspector or production control.

b. Waiting for special tools already being used if waiting time cannot be eliminated.

c. Power failure of non-reportable duration.

d. Work interference.

e. Minor rework elements if not caused by operator error.

f. Extra work required due to hidden parts or material defects if minor.

g. Unsuccessful hunt for parts or material.

h. Machine breakdown of non-reportable duration.

APPLICATION OF ALLOWANCE(S)

Since the productive time in the work day is a variable inversely proportional to the amount of PF&D allowance, it is necessary that all factors are a constant expressed as a percentage of the total work day in order to provide a constant base. It is, therefore, necessary that all locally determined factors are similarly expressed.

The application of the allowances requires that the total percent of PF&D be determined by adding the percentage for the applicable factors, as developed for the specific areas analyzed, for the productive day before it can be applied. This is accomplished by dividing the total work day by the productive day expressed as a percent of the work day, i.e.,

$$\text{Allowance Factor} = \frac{100\%}{100\% - \text{allowance (\% of the work day)}}$$

EXAMPLE:

Assume that all factors total 15 percent allowance (which is 72 minutes of a 480 minute work day). Converting this allowance to a percentage of the productive day (408 minutes) results in an allowance of 17.6 percent.

$$\text{Allowance Factor} = \frac{480 \text{ mins}}{480 - 72 \text{ mins}} = \frac{480 \text{ min}}{408 \text{ min}} = 1.176$$

The final step in the application of the allowance is to multiply the normal time by the allowance factor. For example, the productive time to be 408 mins, the job standard would be:

$$408 \text{ minutes} \times 1.176 = 480 \text{ minutes}$$

DEVELOPMENT OF ALLOWANCES

Various techniques can be employed in the establishment of PF&D allowances. However, several basic rules should be applied to assure consistency of the data collected. For example, the rules should provide for:

a. Logical development and application to support conclusions.

b. Understanding of objectives by all individuals concerned with development and application

c. Consistency in application (training the analyst).

d. Supportable documentation for review by others-utilize forms as much as possible.

e. Review and approval procedures.

One technique would be to develop forms (ref attachment no. 1) which contain the applicable data and allowances associated with your requirements. Another would be to list the job conditions (ref attachment no. 2 & 3) and assign the applicable allowance percentage. The option of the technique employed is open-ended.

DETERMINATION OF ALLOWANCE(S)

The development of allowances over a deversified activity will more than likely result in a range of percentage allowances which could cause administration problems.

For example, ref attachment no. 4, an analysis of 37 areas resulted in 22 different allowance factors, which would be difficult to administer however, this can be consolidated by grouping them into a manageable number.

In this example we reviewed the elapsed hours, for a given period of time, and developed a percent of total hours. This provided the capability to evaluate the impact of consolidating the grouped areas.

The elapsed hours were multiplied by the allowance factor to obtain the allowance hours which when added together, would give the ability to establish an average allowance factor by dividing the allowance hours by the elapsed hours. This then reduced the previous 22 allowance factor to (5) five factors or would allow for a single overall allowance factors.

SUMMARY

The purpose of this article has been to demonstrate that the ability to establish meaningful PF&D allowances is not a major task and that by applying some very simple techniques these factors can be established using a standardized method. The PF&D factors will then have consistency of interpretation, application, and measurement criteria.

REFERENCE MATERIALS

DOD 5010.15.1-M Basic Volume, Appendix II

MTM Association-Work Measurement Allowance Survey.

BIOGRAPHICAL SKETCH

William H. Bostion, Supervisor of Industrial Engineering, Westinghouse Defense and Electronics Systems Center, Baltimore, Maryland. Mr. Bostion is currently responsible for the total Work Measurement activities, which includes the computerized predetermined time systems, performance reporting and coordination of the Industrial Engineering Training Program. He is a Senior Member of AIIE and Past President of the Baltimore Chapter. He currently serves as Program Chairman for the Aerospace Division. Prior to joining Westinghouse, he was Senior Engineer for National Industries for the Severely Handicapped, Washington, D.C., assisting sheltered workshops in costing technique, overhead burden development and development of manufacturing capabilities. His background includes industrial engineering, cost estimating and cost control with companies such as Bendix and Cutler Hammer.

Attachment #1

PF&D Allowance Survey - DOD Basic Manual
5010.15.1M

Analyzed By:_____ Date:_____

Area Analyzed _____

Also Applicable To _____

(Partial Listing)

		DWM Stop Allowance	Allowance Used
Personal	Basic Allowance (2 - 10 min Breaks/Day)	4.2	
	Normal Office Conditions	0.0	
	Normal Shop Conditions	1.0	
	Slightly Disagreeable Conditions	3.0	
	Extremely Disagreeable Conditions	6.0	
	Preparation & Clean-Up - 5 min/Day	1.0	
	10 min/Day	2.1	
	15 min/Day	3.1	
	20 min/Day	4.2	
	Super-Clean Room	4.0	
	Other (Specify)	—	
	Total Personal Allowance		

COMMENTS _____
_____
_____

	Effective Net Weight Handled	Percent of Time Under Load					DWM Stop Allowance	Allowance Used
		1-12	13-25	26-50	51-75	76-100		
Fatigue	1-10#	0	1	2	3	4	PERTABLE	
	11-20#	1	3	5	7	10		
	21-30#	2	4	9	13	17		
	31-40#	3	6	13	19	25		
	41-50#	5	9	17	25	34		
	51-60#	6	11	22	X	X		
	61-70#	7	14	28	X	X		
	71-80#	8	17	34	X	X		

ATTACHEMENT #2

Examples of Application

Unloading Boxes from Truck

a. Job Conditions - Crew is unloading boxes from a truck and placing them on a pallet and the following conditions are in effect.

    (1) The operation is performed at a warehouse ramp.
    (2) The boxes weigh 25 pounds each and the employee is under load 25% of the time. The boxes are being taken from stacks slightly higher than his waist and are placed on pallets resting on the truckbed.
    (3) The work is purely routine.
    (4) The employee walks approximately five feet with each box.
    (5) The cycle time (per box) is .500 minutes, actual under load elements equal .125 minutes (if per pallet the % may be somewhat less).
    (6) No restrictive safety devices are required.
    (7) A forklift operator is considered a part of the unloading crew.

b. Computation of Allowance

		Percent
(1)	Personal	
	Base	4.2
	Class B Slightly disagreeable, exposed to weather	3.0
(2)	Fatigue	
	Physical - 25 pounds handled 25% of the time (total under load element time, .125 divided by cycle time, .500 = 25%).	4.0
	Mental - Class A - work committed to habit	0.0
	Position - Class C (walking)	1.0
	Monotony - Class C (0.50 minutes)	2.0
(3)	Delays	
	Class A. Little coordination with adjacent jobs.	1.0
(4)	TOTAL ALLOWANCE	15.2

c. Allowance Factor

$$AF = \frac{100\%}{100\% - 15.2} = \frac{100\%}{84.8\%} = 1.179$$

d. Computation of Standard

If this operation is studied and the normal time is determined to be 0.500 minutes, the standard time would be computed as follows:
0.500 X 1.179 = 0.590 standard minutes. The number of decimal places used would depend on the time increments used in the manhour accounting system and the volume of production.

ATTACHMENT #3

AIRCRAFT INSTRUMENT ASSEMBLY

a. Job Conditions

An employee receives tray of parts and assembles small aircraft instrument. Completed instrument is delivered to outgoing window in clean room. Cycle time is 15 minutes.

    (1) Work is performed in "super" clean room.
    (2) No formal break periods have been established, but employees are free to attend to personal needs as necessary.
    (3) Instrument weighs less than one pound.
    (4) No clean up period at end of shift.
    (5) Employee performs work seated at work bench.
    (6) No restrictive devices are required.
    (7) Only occasional visual and mental concentration required.
    (8) Unavoidable delays have been established at 5% by seperate study.

b. Computation of Allowances   Percent

(1)	Personal	
	Basic	4.2
	"Super" clean room	4.0
(2)	Fatigue	
	Position-sitting	1.0
(3)	Unavoidable Delay	5.0
(4)	TOTAL ALLOWANCE	14.2

c. Allowance Factor

$$AF = \frac{100\%}{100\% - 14.2\%} = \frac{100\%}{85.8\%} = 1.166$$

d. Computation of Standard

Standard time is computed in the same manner as shown in the preceding example.

Attachment #4

Determination of Allowances

Area No.	① Factor	② Elapsed Hours	③ Percent of Total Hours	④ Allowance Hours (1 × 2)	Avg Factor (4 ÷ 2)
1	15.0	28220	1.68	4233	
2 - 6	15.3	111600	6.63	17075	
7 - 11	15.4	474180	28.16	73024	
Totals		614000	36.47	94332	15.4
12	16.4	13580	0.81	2227	
13	17.3	510480	30.32	88313	
14	17.4	28200	1.68	4907	
15 - 17	17.7	76900	4.57	13611	
18 & 19	17.8	36185	2.15	6441	
20 - 22	18.9	29655	1.76	5605	
Totals		695000	41.28	121104	17.4
23	21.3	11210	0.66	2388	
24	21.9	24704	1.48	5410	
25	23.2	4100	0.24	951	
26	24.3	94648	5.61	22999	
27	24.6	14108	0.84	3471	
28	24.9	2150	0.13	535	
Totals		150000	8.96	35754	23.7
29	25.0	20760	1.23	5190	
30	25.2	21260	1.26	5358	
31 & 32	26.2	60605	3.60	15879	
33 & 34	27.2	77550	4.61	21094	
35	28.7	23350	1.39	6701	
Totals		203525	12.09	54222	26.6
36	41.3	5920	0.35	2445	
37	46.7	14080	0.84	6575	
Totals		20000	1.19	9020	45.1
Grand Totals		1683425	99.99	314432	18.7

ATTACHMENT #5

Information For Ordering
DOD 5010.15.1-M Volumes From The
U.S. Government Printing Office

March 14, 1979

Address: Superintendent of Documents
Government Printing Office
Washington, D.C. 20402

Phone: (202) 783-3238 (Order Desk)

Title: DOD 5010.15.1-M. "Standardization of Work Measurement"

VOLUME	TITLE	COST	STOCK NO
Basic	General Guidance, Jan 77	$6.25	008-007-02850-2
	Change 1	$2.75	008-007-02929-1
I	Professional, Managerial Technical	Not available at this time	
II	Clerical & Sales Occupations. Dec 75	$1.90	008-007-02743
	Change 1	$ .80	008-007-02891-0
III	Service, Jun 75	$1.25	008-007-02721-2
	Change 1	$ .60	008-007-02892-8
IV	Farming, Fishery, Foresty & Related, Jun 75	$ .85	008-007-02720-4
	Change 1	$ .60	008-007-02930-4
V	Processing, Jun 75	$1.15	008-007-02719-1
	Change 1	$ .60	008-007-02931-2
VI	Machine Trades, Nov 74	$2.00	008-007-02652-6
	Change 1	$1.00	008-007-02893-6
VII	Bench Work, Feb 77	$2.85	008-007-02830-8
	Change 1	$1.00	008-007-02894-4
VIII	Structural Work, Jun 75	$1.25	008-007-02718-2
	Change 1	$ .80	008-007-02895-2
IX	Miscellaneous (Materials Handling, Packaging, Transportation), Jan 77	$3.50	008-007-02821-9
	Change 1	$1.00	008-007-02896-1
X	Universal, Apr 77	$3.25	008-007-02835-9
	Change 1	$ .90	008-007-02897-9

METHODS FOR ESTIMATING PHYSICAL FATIGUE

Arun Garg

The University of Wisconsin
Milwaukee, Wisconsin 53201

ABSTRACT

Four different methods of estimating physical fatigue are reviewed. These are (1) output degradation, (2) motion and time study and predetermined motion-time data, (3) psychophysical methodology, and (4) physiological criteria. New methods for determining metabolic energy expenditure rate are also reviewed. Standards based on motion and time study and psychophysical methodology are compared with those based on metabolic energy expenditure rate and heart rate. Comparison shows that neither the standards based on subjective estimates of work analyst nor the acceptable work loads based on psychophysical methodology are in agreement with the physiological fatigue criteria. Some general conclusions and recommendations are presented.

INTRODUCTION

Frederick W. Taylor, generally known as the father of scientific management, was responsible for the definite approach to work measurement. His findings that the greatest production results when each worker is given a definite task to be performed in a definite manner still holds true. Productivity depends upon worker's production efficiency (i.e. shorter movements, fewer redundant motions, better balance in man-machine times, etc.) and also upon fewer disruptions of the worker's productive efforts due to occupationally caused accident, ill health or fatigue. The techniques for improving worker's production efficiency have been a traditional concern of industrial engineers and have been extensively studied. The techniques for objectively predicting the effects of physical effort on worker's health and fatigue are much newer and are presented in this paper.

The term "fatigue" has various meanings depending upon the nature of the job. Therefore, a number of different techniques are necessary to study physiological, perceptual and psychological demands placed on individuals working in an industrial organization. For example, some monitoring or inspection jobs place heavy demands on the perceptual process. Some jobs require the operation of complex machinery where the operator may be placed under heavy decision - making stresses for a considerable period of time. In other jobs the worker is asked to perform work that results in a heavy physical work load. In the interest of brevity, this paper will confine itself to a discussion of those types of industrial jobs which are physically demanding. In particular, four different methods of estimating physical fatigue are reviewed and acceptable work loads based on these methods are also compared. These methods, as classified in the literature, are (1) output degradation, (2) motion and time study and predetermined motion-time data, (3) psychophysical methodology, and (4) physiological criteria.

In physical work, at least respiratory, circulatory, muscular and nervous systems are involved. Chaffin [9] defined physical fatigue as "reduction in a person's capacity to perform a given task which is related to the prior strain caused in various human subsystems by the stress of physical elements in the task." In other words, physical fatigue may be defined as a decrease in work capacity caused by work itself resulting in a decrease in power of muscle to contract, i.e. muscle tension. The primary cause of physical fatigue is muscular effort involved. The muscular effort may result in either inadequacy of supply of oxygen and foodstuffs to the muscles and/or inadequacy of muscle's enzymes and oxidizing capability [10]. Physical fatigue may also be due to an alteration of the physiochemical state, i.e. a breakdown of homeostasis. For example, loss of sodium chloride in a hot environment may lead to excessive physical fatigue and sickness.

Some of the effects of physical fatigue are localized muscle irritability, increased tremor, decreased precision positioning ability, sensations of discomfort, loss of coordination and strength, increased breathing and heart rate, chest pains, etc. [10]. These effects lead to a decrease in productivity and an increase in absenteeism [10,17].

Factors Affecting Physical Fatigue

Strenuous physical effort can be divided into two types of muscular activity: (1) dynamic muscular work such as cranking, walking, climbing, etc., and (2) static muscular work such as holding a tool, maintaining an awkward body position, etc. Most of the industrial operations are a combination of static and dynamic muscular efforts. In general, static work is more fatiguing than dynamic work [17]. During static muscular work the blood supply to the active muscles may be reduced. Sufficient blood flow to the muscles is necessary to supply the muscles with oxygen and foodstuffs and remove heat, carbon-dioxide, lactic acid and other metabolites. Fatigue develops when the applied force is 15% or more of the maximum force. Strength and working capacity decreases exponentially with exertion time and 50% of maximum force can last at most one minute [17]. Physical fatigue is measured differently for static and dynamic muscular work though the primary cause of physical fatigue is the same for both types of exertions. This paper discusses the measurement of physical fatigue for dynamic muscular work. Excellent discussions of static muscular efforts are given in Greenberg and Chaffin [18], Chaffin [10], Grandjean [17] and Rohmert [24].

The various factors which determine physical fatigue can be classified into three main categories. These are (1) human variables or worker characteristics, (2) task variables, and (3) environmental variables. Human variables include worker's age, sex, physical fitness, anthropometry, training, motivation, etc. These variables, in general, determine maximum aerobic power of the worker or the capacity to perform dynamic muscular work. The major task variables are weights and forces involved, frequency and speed of exertion, body posture and technique involved, duration of exertion, movements of whole body and body limbs involved, vertical heights at which a task is performed and nature of the task itself. For example, lowering a load is less fatiguing than lifting the same load [14]. Also, straight-back, bent-knee method of lifting is more fatiguing than bent-back, straight-knee method of lifting [14]. These task variables determine the loading of respiratory, cardiovascular and musculoskeletal systems of the body. A work analyst should consider the physical demands of the job against the capacity of the individual worker (maximum aerobic power) to avoid excessive strain on respiratory and cardiovascular systems. Then, there are environmental variables such as temperature, humidity, ventilation, chemical contaminats, light, noise, etc. which produce extra stress on the worker. For example, under extremes of hot and/or humid environment blood flow to skin increases to dissipate body heat. This reduces the blood flow (venous return) to the heart, and, therefore, cardiac output per beat decreases. To meet the oxygen requirements of the working muscles, cardiac output is maintained by (1) an increase in heart rate and (2) further reduction of blood flow to abdominal organs. Under extreme conditions this may lead to muscle fatigue, leg swelling, abdominal upsets (nausea), loss of consciousness, severe headache, visual disturbance, fainting, giddiness, and possible cardiac failure.

METHODS FOR MEASURING PHYSICAL FATIGUE

At present there are four different methods for measuring physical fatigue as classified in the literature. These are (1) output degredation, (2) motion and time study and predetermined motion-time data, (3) psychophysical methodology, and (4) work physiology. These methods are reviewed in the following sections.

Output Degradation as an Indicator of Fatigue

Lowry et al. [21] defined output degradation as an indicator of fatigue. Because of simplicity in concept and ease in measurement, fatigue was defined as that effect of work upon an individual's mind and body which tends to lower his rate or grade of quality of production, or both, from his optimum [2]. The problem with decrease in output as an indicator of fatigue is that it does not give enough information to allow a work analyst to easily eliminate the sources of fatigue. To safeguard worker's health and to maintain optimum efficiency, it is important to reduce sources of fatigue rather than relying upon fatigue allowances. Secondly, a decrease in output can also be due to loss of interest in the job, personal problems, lack of incentive, lack of sleep, drugs, illness, and employee's attitude toward his fellow workers or supervisor [5, 19].

Motion and Time Study and Predetermined Motion-Time Data

Conventionally, time standards for a job is set by determining a normal time through time study, predetermined motion-time data, elemental time data, or work sampling. Allowances are added to normal time for personal time, fatigue, unavoidable delays and other factors to set standard time. The advantage of such an approach is that it is simple and can be understood and applied with minimal formal training.

The major problem with such an approach is that both the rating of the job and relaxation allowances are primarily subjective. As stated by Barnes [5] "rating is a matter of judgement on the part of time study analyst, and unfortunately there is no way to establish a time standard for a job without having the judgement of the analyst enter into the process." Different rating systems have been developed but all of them rely on the subjective judgement of the analyst. The number of different factors that have a bearing on task difficulty is very large and it is practically impossible to estimate accurately the difficulty of a job by merely observing it [12]. According to Davis et al. [12] "situations in one department that superficially appear to be very similar in another department can prove to be very different... Preconceptions about the difficulty of the job or about the relative importance of the components of jobs are frequently inaccurate." For example, often too much emphasis is given to the object weight in determining fatigue allowances. However, lifting 10 kg of load eight times per minute is more fatiguing than lifting 20 kg of load four times per minute [14].

The term 'work measurement' itself may be a misnomer, because in fact it is the time which is measured rather than stress and strain on the body [28]. Work measurement gives the time required to perform a task; the time should be modified by some physiological response of the body to determine fatigue allowance.

In some cases, because of simplicity of this approach, the magnitude of fatigue allowances is negotiated in union contracts. If the product, process or work force physical capabilities change, new standards are needed. However, time and expense of such studies may be prohibitive.

Psychophysical Methodology

A modification of the in-plant observations of acceptable work load was devised by Snook and Irvine [26] to determine population capabilities for manual handling operations. Essentially, using this approach the worker is given control of one of the task variables (generally pace or weight of the load) to determine maximum acceptable work load. All other variables such as frequency, size of the load, height, distance, etc. are controlled. The worker determines the maximum acceptable work load based on his or her own feelings of exertion and fatigue. The arguments in favor of psychophysical methodology are that it is simple to use and only the individual worker can sense the various strains associated with manual handling operations. Only the individual worker can integrate the sensory inputs into one meaningful response. For example, Magora [22] and Dehlin et al. [13] have reported greater frequency of low-back injuries on those jobs which the workers believed to be harder.

Several studies have been made to determine maximum acceptable work loads for various manual materials handling operations such as lifting, lowering, pushing, pulling, carrying, etc. [4, 25]. Table I shows that there are large variations in manual materials handling capabilities in the population. Also, lifting capability is significantly affected by several task variables such as size of the load, frequency of lifting, height of lifting, etc.

One general point which needs to be recognized with the psychophysical approach is that it is not clear that the recommended work load limits are for "safe loads' [11]. For example, an individual worker may believe that he can sustain a work load, but his cardiovascular response may show an extreme amount of strain [11, 12]. Secondly, recommendations based on psychophysical methodology are for simple operations such as lifting, lowering, pulling etc. Most industrial jobs consist of several operations and the recommended maximum work loads for simple operations, though they provide useful guidelines, are difficult to apply to most industrial situations. Lastly, criteria are needed to relate individual worker's characteristics to acceptable work loads.

Physiological Criteria

During the performance of a work task, especially if it involves moderate to strenuous exertion, physiological changes take place within the body. Work physiology has, as its premise, that the measurement of certain of these changes provides indices of the level of stress imposed upon the worker. Alterations in work methods, performance levels, or certain environmental factors, etc. are reflected in the stress levels of the worker and may be evaluated. Interest in work physiology is increasing in this country, partly because a more objective method of measuring physical work is needed and also because better and more compact apparatus has become available to measure physiological variables in industrial settings [5].

So far the physiological measurements during manual materials handling operations are concerned, metabolic energy expenditures rate and heart rate are the most widely accepted and reliable indicators

Table I: Maximum Acceptable Weight of Lift (kg) for Male Workers Based on Psychophysical Methodology [25]

Lifting Origin	Lifting Distance (m)	Object Width (m)	Lifting Frequency (Lifts/Min)	% Population 90	75	50	25	10
Floor Level	0.76	0.75	1	15	18	22	26	30
			4.3	12	15	19	23	26
			6.6	11	14	17	20	24
			12.0	8	10	13	16	10
Floor Level	0.76	0.36	1	18	23	29	34	40
			4.3	15	20	25	30	35
			6.6	13	17	22	27	31
			12.0	10	13	17	20	24
Knuckle Height	0.76	0.75	1	14	17	21	26	29
			4.3	14	18	21	25	29
			6.6	12	16	19	22	25
			12.0	9	12	14	17	19
Knuckle Height	0.76	0.36	1	14	18	23	28	32
			4.3	15	19	23	26	30
			6.6	13	17	20	23	26
			12.0	10	13	15	18	20

of the level of physiological demands. It is generally accepted that for many different tasks with a hetrogeneous population, a safe, eight-hour average metabolic rate is about one-third of the population's maximum aerobic power [15]. This often equals to slightly above 5 kcal/minute, assuming an average, healthy male population [10]. Similarly, a heart rate of 115 beats per minute is taken as an acceptable limit for eight hours of work for a normal healthy male population [10]. Similarly, a heart rate of 115 beats per minute is taken as an acceptable limit for eight hours of work for a normal healthy young male [10]. Similar guidelines are available for female workers and for work duration other than eight hours [10]. Davis et al. [12], Aquilano [3], Krager and Bayha [20], Grandjean [17], Brouha [8] and many others have recommended the use of metabolic energy expenditure rate and/or heart rate to determine rest allowances. However, the application of work physiology to design of work place layout, work methods and to determine rest allowances has been limited to a few industries in this country. The primary reason is that there has been no easy way to determine metabolic rate or heart rate for a given job in industry. On the job measurement of oxygen utilization (metabolic rate) has been difficult due to interference of the measuring equipment with normal work methods. On the other hand, tabulated values are essentially for the 'average worker' and may be very inaccurate due to their overly simplistic job descriptors [14]. Garg [14] has shown that relatively minor changes in the job physical parameters that are commonly used to describe a person's manual activity result in significant changes in metabolic energy expenditure rate.

Most of the above problems have been overcome in the last few years. For example, advancements in the electronics have resulted in light-weight, compact equipment for the measurement of oxygen uptake and heart rate.

ESTIMATING METABOLIC RATE FOR A JOB

Garge et al. [15] recently developed a technique to estimate metabolic energy expenditure rate for a job based upon physical parameters of the job. Their approach is based on the assumption that the average metabolic energy expenditure rate is simply equal to sum of the energy demands of various operations (task elements) and the maintenance of body postures, averaged over time. Energy demands for various operations and maintenance of body postures are estimated from experimentally developed regression equations. Table II shows an application of their model for a steel work operation. The highly repetitive job is divided into eleven different elements and the estimated metabolic cost for each element is given in the last column. These metabolic costs sum to a total of 31.2 kcal. The postural component was estimated to be 2.18 kcal/min. The job was analyzed for a total of 5.27 minutes. Therefore, the average metabolic rate (M) for the job is:

$$M = \frac{2.18 \times 5.27 + 31.2}{5.27}$$

$$= 8.1 \text{ kcal/min}$$

This is a fairly high metabolic rate as compared to an eight-hour limit of 5.5 kcal/min [1]. The needed rest allowance can be calculated from the following equation [1]:

$$T_{rest} = \left(\frac{M - 1.5}{4}\right) \times 100$$

T_{rest} = Needed rest allowance (percentage of working time)

Simple calculation indicates that 65 minutes of rest should be provided for every 100 minutes of work. Workers on this job should be screened by the medical department and should be physically fit to

Table II: Partitioning of the Job into Tasks and Estimated Net Metabolic Costs of the Tasks [15]

Task #	Task	Technique	Number of Times Performed	Load (Kg)	Task Description or Vertical Movement of Work Piece (m) From	To	Time (Min)	Net Estimated Metabolic Cost (Kcal)
1.	Lift	Stoop	14	30.7	0.25	1.17	-	10.74
2.	Lift	Arm	4	30.7	1.17	1.52	-	1.29
3.	Lift	Arm	2	30.7	1.17	1.78	-	1.03
4.	Lift	Arm	1	30.7	1.17	2.03	-	0.71
5.	Carry	In Front	14	30.7	7.5	Steps	0.092	9.69
6.	Lateral Movement of Arms of 90°	Standing Both Hands	28	30.7	-	-	-	3.78
7.	Lower	Stoop	1	30.7	1.17	0.51	-	0.29
8.	Lower	Arm	4	30.7	1.17	0.81	-	0.54
9.	Lower	Arm	1	30.7	1.52	1.17	-	0.18
10.	Lower	Arm	2	30.7	1.78	1.17	-	0.58
11.	Walk	-	14	0	7.5	Steps	0.115	2.54

avoid any potential cardiovascular problems. One of the advantages of using Garg et al's [15] approach is that it relates the metabolic cost to useful parameters such as weights, frequencies, heights, etc. which are more easily understood by a work analyst. Secondly, this approach takes into account some of the worker's characteristics (gender and body weight) and, therefore, the relaxation allowances are assessed for a given job condition as well as the particular worker.

It is worth mentioning that both the frequency and the length of rest periods are significant factors in the recovery from the effects of fatigue [1, 10, 17]. In other words, in the above example, it is not only important to provide 65 minutes of rest for every 100 minutes of work but also when and how the allowed rest should be taken. In this regard, Chaffin [10] has developed a procedure to determin maximum working time for any single work interval. In general, short and frequent rest pauses are preferred over long and infrequent breaks [1, 17]. There is less cumulative fatigue with many short rest pauses than with the same amount of rest time alloted in infrequent, long breaks.

COMPARISON OF DIFFERENT PHYSICAL FATIGUE CRITERIA

Metabolic Rate versus Time Study

Aquilano [3] measured the metabolic energy expenditure rates for six jobs on six workers while working at 100 and 128 percent of normal pace as determined by three experts on predetermined motion-time data systems and stopwatch time study. At incentive pace, energy expenditure levels of workers ranged from 32 percent below the assumed standard level (4 kcal/min above the resting level) to 189 percent in excess of that level. Performance at standard pace ranged from 46 percent under to 125 percent over the accepted standard of metabolic energy expenditure rate. In other works, great differences in energy demands from reasonable levels and a large variance among tasks were observed. The major discrepency appeared in the rest allowances. Similarily, Wyndham et al. [29] reported that men working at a performance index value of 100 percent (i.e. well motivated in work-study sense) were liable to work at a rate which is about 70% of the maximum oxygen uptake rate. This would be excessive for the average worker.

The computation of rest allowances by traditional time-study techniques are not always consistent with the physiological needs of the workers [23]. Moores [23] has demonstrated that as the work rate increases the energy expenditure cost increases at an increasing rate, where as performance rating assessments increase, but at a decreasing rate. Moreover, there is a "flatness of rating", i.e. a tendency to overate low performances, and underestimate high ones.

The ratio of actual energy expenditure to a person's aerobic power represents the degree of utilization of man's capacity to perform work [7]. Since the aerobic power varies greatly among the individuals, it is important to consider worker's characteristics in determining rest allowances. Inter-individual coefficients of variation were calculated from Aquilano's data on six subjects and are given in Table III. Subjects employed by Aquilano [3] were well-trained workers, in excellent health and 21 to 25 years of age. It can be noted from Table III that the inter-subject variability ranges from 9.6 to 19.2 percent on six workers while each worker was working at the same pace. This suggests that worker characteristics can significantly affect rest allowances because, in general, the worker consuming greater amount of energy would appear more fatigued. To determine rest allowances, both the time study and the physiological measurements should be made on a large number of workers because different workers experience different amounts of fatigue while working at the same pace.

Heart Rate Recovery versus Traditional Methods

Dr. Brouha recommended measuring heart rate while the worker is recovering from work to determine a safe stress level throughout the working shift. He suggested constructing heart rate recovery curves by determining the pulse rate at intervals of 0.5 to 1.0, 1.5 to 2.0, 2.5 to 3.0 minutes following the cessation of work effort [8]. According to Brouha [8], a safe stress level throughout the working shift is sustained provided (i) the first recorded pulse rate is 110 beats per minute or less and (ii) the third recorded pulse rate is at least 10 beats per minute lower [8].

Table III: Inter-Individual Differences in Energy Expenditure for Load Lifting Tasks (From Aquilano's Data [3])

Task	Weight (kg)	Vertical Range (m)	Coefficient of Variation %		
			Slow pace	Standard Pace	Incentive Pace
1.	4.5	0 to 0.91	14.3	15.3	15.8
2.	11.4	0 to 0.91	16.0	7.9	9.6
3.	4.5	0 to 1.67	15.4	16.4	13.4
4.	11.4	0 to 1.67	15.0	11.3	9.9
5.	4.5	0.91 to 1.67	19.2	14.8	14.4
6.	11.4	0.91 to 1.67	14.1	10.6	11.8

Measurement of heart rate during work and rest can also be used to determine adequate rest periods. For example, Davis et al. [12] demonstrated the effects of two and seven minutes of rest breaks after each ten minutes of work on heart rate for a lifting task. The heart rate continued to increase during working periods with two minutes of rest, thus indicating that a two-minute rest period was inadequate. The heart rate during working periods did achieve a steady state with a seven-minute rest period.

In this study, four different operations were studied in a local foundry to compare the fatigue allowances based on traditional methods with those based on heart rate recovery curve. The fatigue allowances for the four foundry operations stand grind, chip and grind, cold knockoff, and bench coremake were 14, 12, 16 and 7 percent, respectively. These fatigue allowances were based on historically developed tables and in consultation with other foundaries.

The workers were allowed to work approximately for an hour and a half and then rest. The heart rates were measured immediately following the cessation of work effort, i.e., during the rest period. The heart rates during recovery from work effort for the four foundry operations studied are plotted in Figure 1. It can be noted from Figure 1, that three foundry operations (stand grind, chip and grind, and bench coremake) are well within the safe stress level criteria suggested by Brouha. However, the fourth operation (cold knockoff) will produce excessive physical fatigue. The heart rate at the start of recovery for the worker on cold knockoff was 180 beats per minute which was extremely high. After 12 minutes of rest, the heart rate was more than 95 beats per minute indicating that total recovery from work stress had not occurred.

Certainly, the fatigue allowance of 16 percent and/or sequence of working and resting schedule was not adequate to meet physiological fatigue criteria proposed by Brouha [8]. The worker on this job

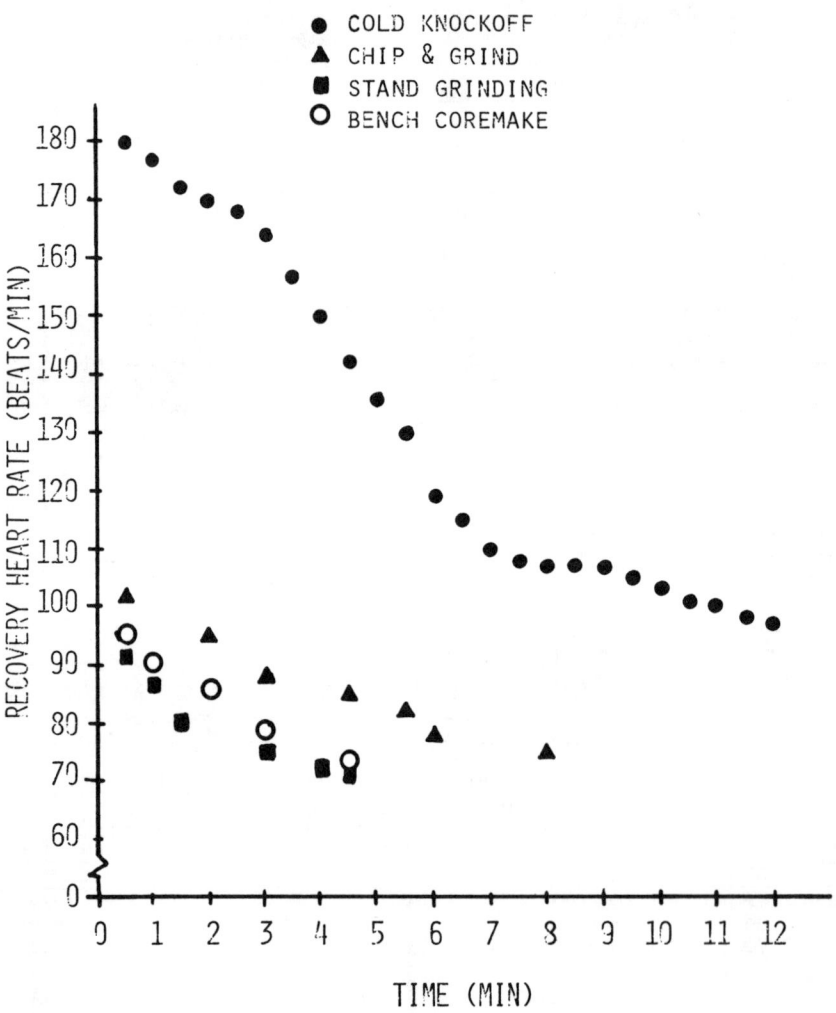

Fig. 1. Heart rate recovery curves for the four foundry operations studied.

would be sustaining an average heart rate which would be in excess of 112 beats per minute as proposed by Snook and Irvine [27]. The heart rate recovery curve shows that the fatigue allowance and the sequence of working and resting schedule should be reevaluated for this job to safeguard the worker's health and for meaningful work measurement. Further, the workers on this job should be carefully screened by the medical department to avoid any potential cardiovascular problems.

Psychophysical versus Physiological Measurements

Snook and Irvine [27] measured heart rates of nine subjects while performing lifting tasks at their chosen levels. The means and standard deviations of measured heart rates were 112 and 10.1 beats/min for leg work and 99 and 11.3 beats/min for arm work. In general, the average heart rate responses for leg work were within acceptable bounds for eight hours of work, but some workers demonstrated higher than expected heart rates for the task.

Recently, Garg and Saxena [16] measured metabolic rates on six male subjects while they were lifting loads from floor to 0.5 m height at 3, 6, 9 and 12 lifts per minute. The subjects were allowed to select the weight of the load according to their own feelings of exertion or fatigue. The subjects were not instructed to lift by any particular technique and were allowed to choose a lifting posture which they felt most comfortable. The measured metabolic rates are plotted in Figure 2. It can be noted from Figure 2 that the metabolic energy expenditure rates increase with an increase in lifting frequency. The means of the measured metabolic rates from the six subjects varied from 3.28 to 5.56 kcal/min as compared to accepted eight-hour criteria of 5.0 to 5.5 kcal/min [12, 1]. In other words, the subjects did not adjust the weight of the load to produce uniform metabolic rates at different frequencies. In fact, one of the subjects selected a workload which resulted in metabolic energy expenditure rate of 6.25 kcal/min which is approximately 20 percent more than the recommended value of 5.2 kcal/min [6]. Under monetary and social incentives a worker may choose a work pace which may be beyond the worker's physiologically safe level.

SUMMARY

Four different methods for measuring physical fatigue were reviewed. These methods were (1) output degradation, (2) motion and time study and predetermined motion-time data, (3) psychophysical methodology, and (4) physiological criteria. It is concluded that out of the four methods presented, only the physiological measures provide an objective scale to measure the physical difficulty of the job. Simple techniques are available to determine metabolic energy expenditure rate requirements of a given job.

Physiological studies of manual materials handling operations which were also evaluated in terms of stopwatch time studies and predetermined motion-time data showed significant difference between the phsyiologically measured and observed

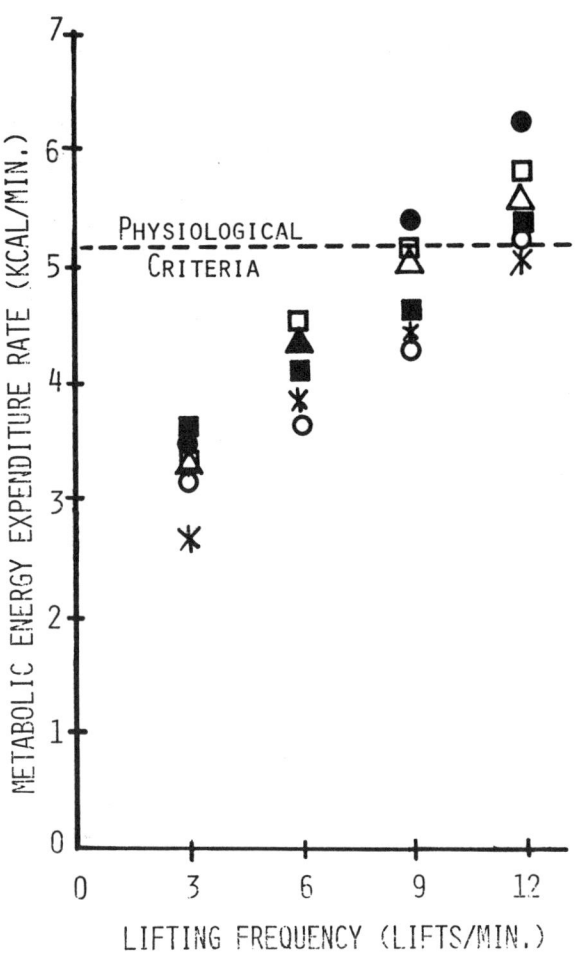

Fig. 2. Measured metabolic rates for lifting from floor to 0.5 m height at work loads determined by psychophysical methodology.

indices of effort. A comparison of physical fatigue based on psychophysical methodology and metabolic energy expenditure rate also showed that the two criteria were not in agreement.

It is recommended that the physiological assessment of a job should be made to assist the work study engineer in determining relaxation allowances more accurately when moderate to heavy manual work is performed, especially under trying environmental conditions. This could lead to improvements in methods, application of work measurement and production standards.

REFERENCES

[1] AIHA Technical Committee, "Ergonomics Guides," Amer. Ind. Hyg. Assoc.J., 32, 560-564 (1971).

[2] Alford, L.P. and Bangs, J.R., Production Handbook, The Roland Press, New York (1944).

[3] Aquilano, N.J., "A Physiological Evaluation of

Time Studies for Strenuour Work as Set by Stopwatch Time Study and Two Predetermined Motion Time Data Systems," *J. of I.E.*, 19, 9, 425-432 (1968).

[4] Ayoub, M.M., Bethea, N.J., Martz, H.F., Asfour, S.S., Nakken, G. and Mital, A., Determination and Modeling of Lifting Capacity, HEW (NIOSH) Report, Grant #5 R010H00545-2 (August 1978).

[5] Barnes, R.M., *Motion and Time Study*, John Wiley & Sons, New York (1968).

[6] Bink, B., "The Physical Working Capacity in Relation to Working Time and Age," *Ergonomics*, 5, 1, 25-28 (1962).

[7] Bonjer, F.H., "Actual Energy Expenditure in Relation to the Physical Working Capacity," *Ergonomics*, 5, 29-31 (1962).

[8] Brouha, L., *Physiology in Industry*, Pergamon Press, New York (1960).

[9] Chaffin, D.B., "Physical Fatigue: What it is - How it is Predicted," *J. of Methods-Time Measurement*, 14, 3, 20-28 (1969).

[10] Chaffin, D.B., Some Effects of Pgysical Exertion, Technical Report, Dept. of Industrial and Operations Engineering, The University of Michigan (1972).

[11] Chaffin, D.B., "What Basis Exists for Determining How Much We can Safely Lift?" *Proceedings of AIIE*, Washington (1975).

[12] Davis. H.L., Faulkner, T.W. and MIller, C.I., "Work Physiology," *Human Factors*, 11, 2, 157-166 (1969).

[13] Dehlin, O., Hedenrud, B., Horal, J., "Back Symptons in Nursing Aids in a Geriatric Hos-Pital," *Scandinavian J. of Rehabilitation Medicine*, 8, 47-53, (1976).

[14] Garg, A., A Metabolic Rate Prediction Model for Manual Material Handling Jobs, Ph. D. Dissertation, The University of Michigan (1976).

[15] Garg, A., Chaffin, D.B. and Herrin, G.D., "Prediction of Metabolic Rates for Manual Materials Handling Jobs," Amer. Ind. Hyg. Assoc. J., 39, 8, 661-674, (1978).

[16] Garg, A. and Saxena, U., "Effects of Lifting Frequency and Technique on Physical Fatigue with Special Reference to Psychophysical Methodology and Metabolic Rate," *Amer. Ind. Hyg. Assoc. J.*, (in press).

[17] Grandjean, E., *Fitting the Task to the Man*, Taylor and Francis Ltd., London (1969).

[18] Greenberg, L. and Chaffin, D.B., *Workers and Their Tools: A Guide to the Ergonomic Design of Hand Tools and Small Presses*, Pendell Publishing Co. (1977).

[19] Karpovish, P.V. and Sinning, W.E., Physiology of Muscular Activity, W.B. Saunders Company, Philadelphia (1971).

[20] Krager, D.W. and Bayha, F.H., *Engineered Work Measurement*, Industrial Press Inc., New York (1977)

[21] Lowry, S.M., Maynard, H.B. and Stegemerten, G.J., *Time and Motion Study, and Formulas for Wage Incentives*, McGraw-Hill, New York (1927).

[22] Magora, A., "Investigation of the Relation Between Low Back Pain and Occupation," *Industrail Medicine and Surgery*, 39, 504-510 (1970).

[23] Moores, B., "A Comparison of Work-Load Using Physiological and Time Assessments," *Ergonomics*, 14. 1. 61-69 (1970).

[24] Rohmert, W., "Problems in Determining Rest Allowances," *Applied Ergonomics*, 4, 2, 91-95 (June 1973).

[25] Snook, S.H., "The Design of Manual Handling Tasks," *The Ergonomic's Society Lecture*, Bedfordshire, England (April 1978).

[26] Snook, S.H. and Irvine, C.H., "Maximum Acceptable Weight of Lift," *Amer. Ind. Hyg. Assoc. J.*, 28, 4, 322-329 (1967).

[27] Snook, S.H. and Irvine, C.H., "Psychophysical Studies of Physiological Fatigue Criteria," *Human Factors*, 11, 3, 281-290 (1969).

[28] Whitmore, D.A., *Work Measurement*, William Heinmann Ltd., London (1975).

[29] Wyndham. C.H., Morrison, J.F., Williams, C.C., Heyns, A., Margo, E., Brown, A.N. and Astroup, J., "The Relationship between Energy Expenditure and Performance Index in the Task of Shovelling Sand," *Ergonomics*, 9, 5, 371-378 (1966).

BIOGRAPHICAL SKETCH

Dr. Arun Garg received his BSME from the Indian Institute of Technology in 1969, MME from the Villanova University in 1970, PE and PhD degrees in Industrial and Operations engineering from The University of Michigan in 1973 and 1976, respectively. He is presently Assistant Professor in the Systems-Design Department at The University of Wisconsin-Milwaukee, Milwaukee, Wisconsin. His areas of interest are ergonomics and work-physiology considerations in the design of contemporary man-machine systems and bio-mechanical modeling for manual materials handling jobs. He is a member of AIIE, Human Factors Society and Tau Beta Pi.

DETERMINATION OF REST ALLOWANCES FOR REPETITIVE PHYSICAL ACTIVITIES
THAT CONTINUE FOR EXTENDED HOURS

Anil Mital and Richard L. Shell
Department of Mechanical and Industrial Engineering
University of Cincinnati
Cincinnati, Ohio 45221

ABSTRACT

This paper presents a comprehensive model for determining rest allowances for physical activities. Using job and worker profile inputs, such as metabolic energy expenditure rate required by the job, shift duration, age, sex, and body weight of the worker, etc., the model predicts the total length of rest breaks as a percent of shift duration. The modeling concept was verified by collecting data on five male and five female industrial workers. The widely accepted belief that if a person is operating at a pace which requires a metabolic energy expenditure rate equivalent to a third of his/her aerobic capacity (approximately 5 kcal/min.) can operate for eight-hours without undue accumulation of fatigue was not validated. The modeling assumptions and concepts are also described.

INTRODUCTION

Rest allowances are generally provided so that workers may recover from fatigue that is generated as a result of producing work. Lack of proper rest allowances frequently lead to tiring of the muscles and decline in quality and quantity of output. When muscles get tired, the resulting fatigue is called the physiological muscle fatigue. Traditionally, physiological fatigue allowances are provided when, and if, the energy demands of the job, for eight hours of work, exceeds one-third (33%) of the aerobic capacity of a person [1-7]. What this percentage should be for ten or twelve hour shifts is presently not known. Appropriateness of this methodology for determining rest allowances is also questionable since it does not take into consideration the total energy available for producing work.

Acceptable level of metabolic energy expenditure rate (which does not lead to excessive fatigue), based on total energy consumption, is also widely used for determining rest allowances [1,3]. There is, however, some conflict as to what is the total energy consumption per day. While one source suggests a value of 4200 kcal [1], the other proposes 4800 kcal [3]. Surprisingly, both sources [1 and 3] recommend an occupational energy expenditure rate limit of 5 kcal/min. as the limit of acceptable level of job demand. Obviously, both sources differ in their assessment of basal and leisure metabolism. Regardless of which one is correct, for most industrial workers these limits are unrealistically high [8]. Recent nutritional studies also indicate that the assumption that humans have plenty of food reserves to meet the demand may not be correct [9]. Given excess food supply and adequate oxygen consumption, still only a certain amount of calories will be consumed. Additional calories (excess food) will increase the weight of the individual by adding to the body fat. Inadequate supply of food, on the other hand, will first result in the reduction of fat and then reduction in the muscle mass.

The above discussion raises the obvious question, "Which method is appropriate for estimating the duration of rest breaks?", since both have been recommended in the published literature [1,2]. The purpose of this work was to seek an answer to this paradox and to develop a comprehensive procedure for estimating rest allowances which are unique to the individual performing a physical task. For this purpose, an experiment was conducted on industrial subjects.

EXPERIMENTAL DETAILS

Five male and five female workers, ranging in age from 24 years to 55 years, were recruited from industries located in the greater Cincinnati area. The subjects were experienced in industrial palletizing and stacking tasks. Table 1 shows the basic descriptors of the subject population. The aerobic capacity of the subjects was estimated [10] and it was assumed that they reflected the capabilities of the average industrial work force rather than large sized highly active professionals such as lumber jacks, construction workers, etc.

Table 1: Subject Descriptors

	Males (N=5)		Females (N=5)	
	$\bar{x}$	R	$\bar{x}$	R
Age (years)	32	24-45	36.2	24-55
Height (centimeters)	166.2	155.7-171.5	163.4	154.1-176.7
Body weight (pounds)	165.25		160.07	120.6-190.8
Experience (years)	7	1-20	5	3-10
Aerobic Capacity (Kcal/min*)	16.9	14.5-18.5	11.0	9.7-11.6

Note: $\bar{x}$ = Mean; R = Range
*Estimated from [10].

The experiment simulated palletizing/stacking tasks which were performed by these subjects under controlled laboratory conditions. Each subject was asked to stack containers of different sizes from one height to another height. Containers varied in size from twelve inches to twenty-four inches while the height ranged from approximately thirty-two inches to sixty inches. Four different paces were included in the experiment; containers were stacked at a rate of either 1, 4, 8, or 12 per minute. Each subject performed only one combination of container size, height, and pace (out of 36 possible) selected at random. He or she was allowed to control one element of the work rate (weight) and were asked to set it up such that the final work rate would be the maximum rate that could be sustained continually, with breaks for lunch, supper, and other personal needs, for eight and twelve hours. It took each subject approximately 20 to 45 minutes to set the maximum acceptable work rate. On the second day, the subjects were asked to perform with this maximum acceptable work rate for eight and eventually twelve hours. Subjects were given a fifteen minute break after two hours, a lunch break of 30 minutes after four hours, another fifteen minute break after six hours, a thirty minute supper break after eight hours, and finally another fifteen minute break after ten hours. No two subjects performed on any given day. Physiological responses of the workers (heart rate and metabolic energy expenditure rate) were monitored continually during the work on both days.

A special palletizing equipment was constructed to simulate the task in the laboratory. An MRM-1 Oxygen Consumption Computer and a Quinton Heart Rate Meter were used to record physiological responses of the subjects.

RESULTS

The results of this experiment indicated that the subjects initially (during the 45 minute estimation period on the first day) selected a work rate, that they thought they could sustain for eight and twelve hours, corresponding to approximately one-third of their aerobic capacity. As has been indicated in the literature [1], [12], this should be the case if individuals work at their peak but at levels which avoid undue fatigue. However, on the second day, as the day progressed, subjects made reductions in the maximum acceptable work rate (Figure 1). These adjustments in the work rate

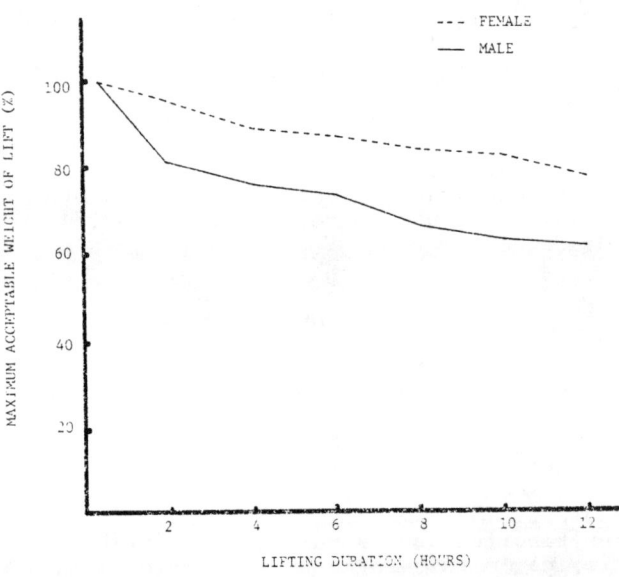

Figure 1. Decline in Weight with Time

continued all day long. The metabolic energy expenditure rate of the subjects also declined continually with time. Figure 2 shows the change in metabolic energy expenditure rate with time both for males and females. Heart rates of the individuals, on the other hand, did not change significantly with time ($p \leq .10$). Figure 3 shows the variation in heart rate of males and females with time. The decline in heart rate due to reduced work load was compensated by increasing fatigue, thereby causing almost no change in heart rate with time. The decrease in metabolic energy expenditure rate, on the other hand, was not expected.

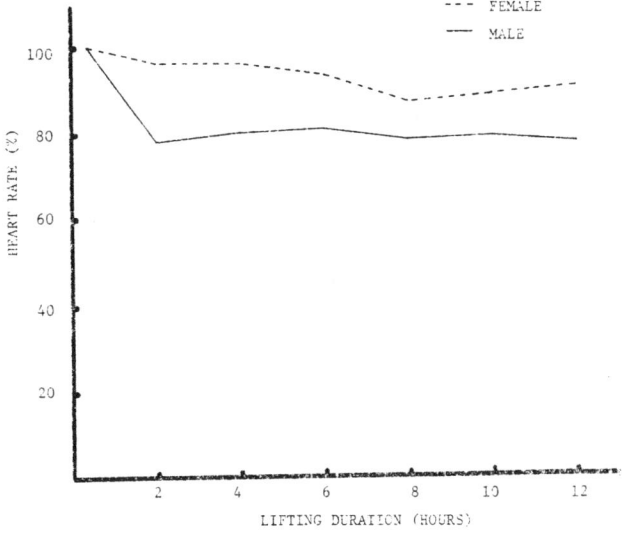

Figure 2. Change in Heart Rate with Time

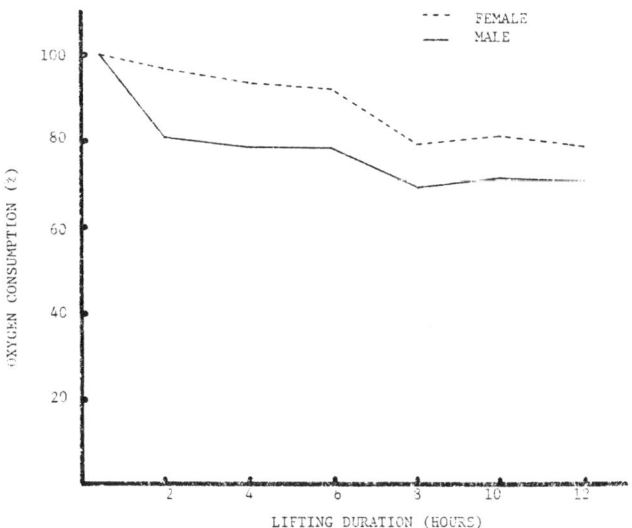

Figure 3. Decline in Oxygen Uptake with Time

Such subjects earlier had selected a work-rate which elicited an energy expenditure rate equivalent to approximately one-third of their aerobic capacity; it was expected that work-rate could be sustained throughout the working day. It did not happen. Work-rates corresponding to one-third of aerobic capacity could not be sustained. The actual work-rates sustained for eight hours corresponded to approximately 28 to 29% of subject's aerobic capacity; for twelve hours this number was 23 to 24%. Since the aerobic capacity approach could not explain the observed pattern, an explanation based on total energy available to produce work was sought.

Using the basis of total available energy, and the fact that it has limits, a comprehensive model structure was developed to determine physiological fatigue rest allowances unique to the worker. The subsequent sections describe the modeling content and its validation.

THE MODEL

Concept

The modeling concept used is not new and has been proposed earlier [1,3]. This proposed model also determines rest allowances based on the total energy available to produce work. There are, however, major differences between the proposed procedure and earlier procedures. The total daily energy requirement of an individual is based on newer data [8] and is adjusted for age. It is also adjusted for energy required for food ingestion. The calculation of basal and leisure metabolism is based on actual hours these activities are performed rather than standard assumption that each lasts for eight hours. The work (shift) duration is also accounted for.

Assumption

The major assumption made in the development of this proposed model is that there is a limit to the amount of energy available to produce work. If the job requires more energy expenditure capacity of the individual, excess energy cannot be supplied by additional food intake. Recent nutritional studies indicate [9] that excess food in such cases only adds to the body weight and is not burned to meet greater job demands. Inadequate food supply, on the other hand, leads to a reduction in the body fat followed by a reduction in the body muscle mass.

Model Structure

Figure 4 shows the model flow chart. The inputs required are:

(i) Worker Sex (M or F),
(ii) Worker Age (Years),
(iii) Body Weight (Pounds),
(iv) Number of Hours of Sleep Per Day (Hours),
(v) Shift Duration (Hours),
(vi) Number of Tasks Performed During the Shift,
(vii) Time Duration for Each Task (Hours),
(viii) Metabolic Energy Requirement for Each Task (Kcal), and
(ix) An Estimate of Worker's Physiological Condition (High, Average, or Low).

Two additional pieces of information are needed to increase the accuracy of the procedure. In the absence of these, the recommended rest allowances will be very

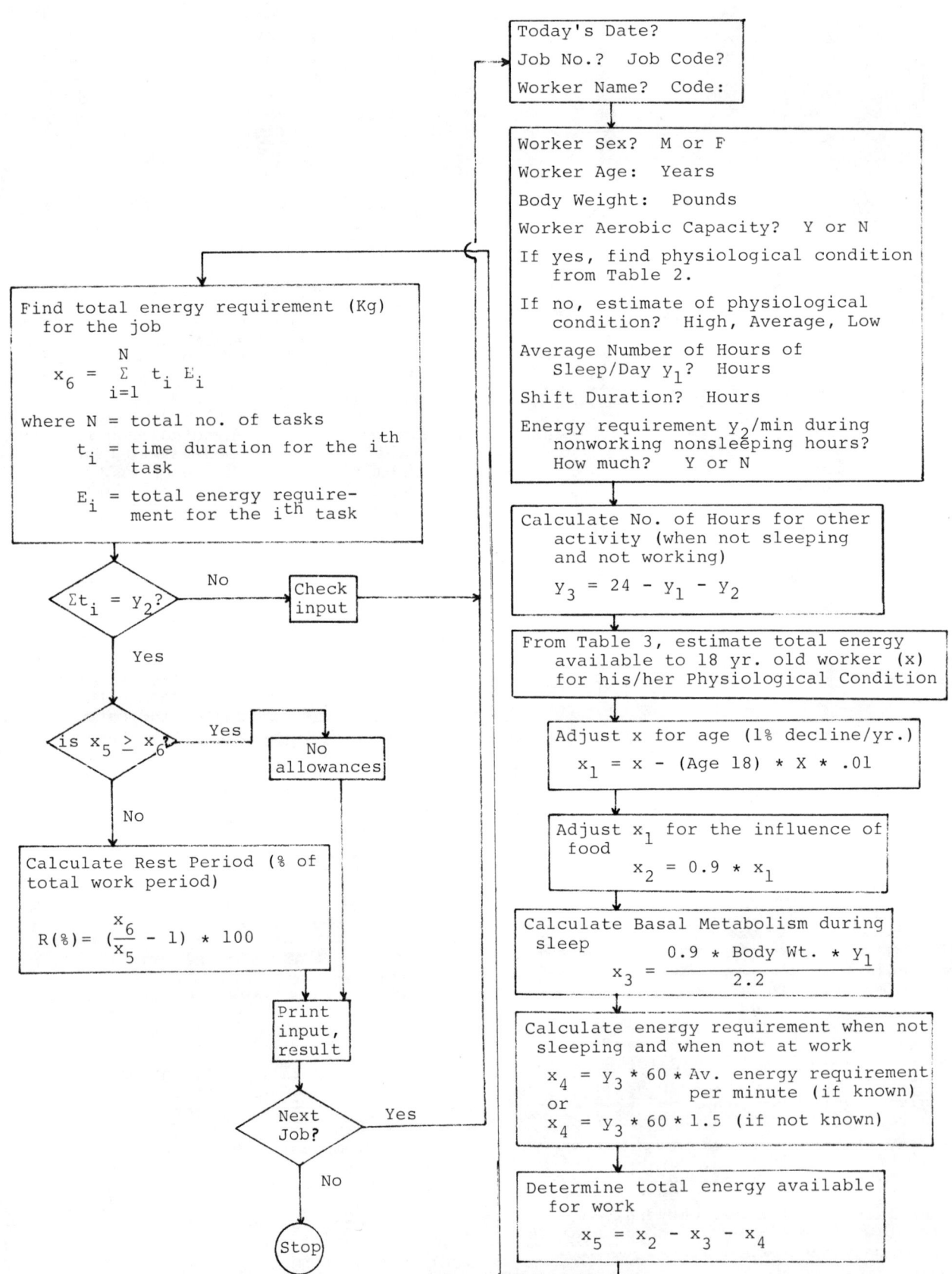

Figure 4: The Model Flow Chart

conservative. This optional information is:

(i) Worker Aerobic Capacity (ML/MIN/KG)

and (ii) Average Energy Requirement/Minute When Not Working or Sleeping (Kcal).

As shown in Figure 4, if the aerobic capacity of the worker is known then his/her physiological condition is calculated from Table 2. This table gives a relationship between worker sex, age, aerobic capacity, and physiological condition and is reproduced here from the original reference [13]. Five different categories (low, fair, average, good, and high) are used to describe the physiological condition. In case the aerobic capacity of the worker is not known, a qualitative estimate (low, average, or high) should be provided. Once the general physiological condition of the worker is known, the total amount of energy available to an eighteen year old per day is determined from Table 3. Next, this total amount is adjusted for age. A decline of one percent per year of age above 18 years is used as the adjustment factor. This decline was determined from the nutritional requirement data that have been reported [8]. Further adjustments are made to this new value for the energy required in digesting the food. Approximately ten percent of the total energy is expended in this process [14].

Once the adjustments have been made for age and the influence of food, basal metabolism and energy expenditure during the leisure period are calculated. A value of 0.9 Kcal/Kg of body weight per hour of sleep (corrected for metabolic savings in sleep) is used for determining the basal metabolism. For determining the energy requirement during leisure time either the actual value (in Kcal/min.) is used or a default value of 1.5 Kcal/min. is used (this value represents energy expenditure during a typical leisure time sedentary task).

The basal and leisure metabolism are subtracted from the adjusted total energy to determine the amount available for producing work. Next the total energy requirement for the job is determined as follows:

$$\text{Total Energy Requirement for the Job (Kcal)} = \sum_{i=1}^{N} t_i E_i$$

where t_i = time (in hours) for the ith task and E_i = its energy requirement (in Kcal). The last step is the determination of the rest duration. The rest period as a percent of total work period (shift duration) is given by:

Table 2: The Range of Expected Maximal Aerobic Capacity, By Sex and Age*

Age (years)	Men Maximal Oxygen Uptake (ML/KG/MIN)				
	Low	Fair	Average	Good	High
20-29	<25	25-33	34-42	43-52	53*
30-39	<23	23-30	31-33	39-49	49*
40-49	<20	20-26	27-35	36-44	45*
50-59	<18	18-24	25-33	34-42	43*
60-69	<16	16-22	23-30	31-40	41*
Women					
20-29	<24	24-30	31-37	38-48	49*
30-39	<20	20-27	28-33	34-44	45*
40-49	<17	17-23	24-30	31-41	42*
50-59	<15	15-20	21-27	28-37	38*
60-69	<13	13-17	18-23	24-34	35*

*Reproduced from [13].

Table 3: Total Energy Available Per Day (Kcal) to An 18 Year Old*

Sex	Aerobic Capacity	Physical Condition				
		Low	Fair	Average	Good	High
Male	Known	2100	2575	3050	3225	4000
	Not known	2100	--	3050	--	4000
Female	Known	2100	2325	2550	2775	3000
	Not known	2100	--	2550	--	3000

*Source [8,9].

$$\text{Rest}(\%) = \left[\frac{\text{Total Energy Requirement for the Job}}{\text{Adjusted, Total Energy Available for Work}} - 1\right] * 100$$

The above procedure is applicable to any shift duration and any combination of physical tasks.

MODEL VALIDATION

The model proposed in the previous section was validated by comparing its predicted performance with the actual data obtained in the experiment described earlier. As shown in Figure 2, the metabolic energy expenditure rate of the subjects declined continuously with time. On the average, oxygen uptake of males declined by approximately 2.6% per hour. It meant that males could sustain an energy expenditure rate of 3.96 Kcal/min for 8-hours; for 12-hours, the metabolic rate was 3.44 Kcal/min. For females the corresponding numbers were 2.97 and 2.70 Kcal/minute. If in fact subjects performed at these levels of physical demand, they would not require any rest allowances. Two examples, one male and one female, are given.

In each example, the total energy available for producing work was calculated. The procedure described in Figure 3 was used for this purpose. The metabolic energy requirement for the job was calculated based on the shift duration and the metabolic energy expenditure rate that could be sustained for that duration by either males or females (3.96 and 3.44 Kcal/min for males for 8- and 12-hours, respectively and 2.97 and 2.70 Kcal/min for females for 8- and 12-hours, respectively; for shift durations between 8- and 12-hours, linear interpolations between the two values may be used).

Example 1

The first subject involved was a male with the following particulars:

Age = 31 years
Body weight = 144.5 pounds
Aerobic Capacity = 51.82 ML/Kg/Min
Hours of sleep/day = 8
Shift duration (Hours) = 8 (Case 1) and 12 (Case 2)
Energy Requirement During Leisure Time (Kcal/min) = 1.4
Energy Required for the Job (Kcal/hour) = 237.6 (= 60 min x 3.96 Kcal/min) - Case 1
= 206 (= 60 min x 3.44 Kcal/min) - Case 2

Two different cases were examined: Case 1 - 8-hour shift; Case 2 - 12-hour shift.

Table 4 shows the program output for the two cases. In both cases, no rest allowance is needed. The total energy requirement for the job is quite close to the total energy available for producing work. The differences in the two numbers are primarily due to:

(i) Energy expenditure during leisure time was estimated, not measured.
(ii) A linear decline in oxygen uptake was assumed (2.6%/hour).
(iii) Hours of sleep were estimated of the actual decline shown in Figure 2.

It should be kept in mind that the above analysis was based on the experimental procedure which allowed subjects to work only 7 out of 8 hours and 10.25 out of 12 hours.

Example 2

The second subject was a female with the following particulars:

Age = 24 years
Body weight = 120.6 pounds
Aerobic Capacity = 42.37 ML/Kg/Min
Hours of sleep/day = 8
Shift duration (hours) = 8 (Case 1) and 12 (Case 2)
Energy Requirement During Leisure Time (Kcal/Min) = 1.25
Energy Required for the Job (Kcal/hour) = 178.2 (= 60 min x 2.97 Kcal/min) - Case 1
= 162 (= 60 min x 2.7 Kcal/min) - Case 2

Table 5 shows the program output for both the cases. As before, the differences between the available and required metabolic energies are small. The reasons for these differences have been outlined earlier.

FINAL REMARKS

This paper proposed a new method for determining rest allowances for physical activities that may continue for eight hours or for extended shifts up to 12 hours. The method was validated by collecting experimental data. The closeness between the estimated available energy and total required energy indicates that the proposed method is indeed quite reliable.

ACKNOWLEDGEMENT

This work was supported by the National Institute for Occupational Safety and Health Grant Nos. 1-R01-OH-01429-01 & -02. Acknowledgement is also

Table 4. EXAMPLE 1 - OUTPUT

PROGRAM OUTPUT

```
DATE :          2/5/84
JOB NUMBER :    EXAMPLE 1
JOB CODE :      CASE 1
WORKER NAME :                   MR. X
SEX         :                   M
AGE (YEARS) :                   31
WEIGHT IN POUNDS :              144.5
AEROBIC CAPACITY IN ML/KG/MIN :         51.82
PHYSIOLOGICAL CONDITION COMPUTED FROM THE TABLE :       5
HOURS OF SLEEP PER DAY :        8
SHIFT DURATION (HOURS) :                        8
ENERGY AVAILABLE FOR 18 YEAR OLD FROM THE TABLE (X) :   4000
ENERGY REQUIRED WHEN NOT SLEEPING AND NOT WORKING :     672
TOTAL ENERGY AVAILABLE FOR WORK :       1987.09

TOTAL NUMBER OF JOBS :  2
TIME DURATION OF JOB            ENERGY REQUIRED FOR JOB
    7                               237.6
    1                               84

TOTAL ENERGY REQUIREMENT FOR THE JOB (KCAL) :           1747.2
REST PERIOD AS PERCENT OF TOTAL WORK PERIOD :           0
```

PROGRAM OUTPUT

```
DATE :          2/5/84
JOB NUMBER :    EXAMPLE 1
JOB CODE :      CASE 2
WORKER NAME :                   MR. X
SEX         :                   M
AGE (YEARS) :                   31
WEIGHT IN POUNDS :              144.5
AEROBIC CAPACITY IN ML/KG/MIN :         51.82
PHYSIOLOGICAL CONDITION COMPUTED FROM THE TABLE :       5
HOURS OF SLEEP PER DAY :        8
SHIFT DURATION (HOURS) :                        12
ENERGY AVAILABLE FOR 18 YEAR OLD FROM THE TABLE (X) :   4000
ENERGY REQUIRED WHEN NOT SLEEPING AND NOT WORKING :     336
TOTAL ENERGY AVAILABLE FOR WORK :       2323.09

TOTAL NUMBER OF JOBS :  2
TIME DURATION OF JOB            ENERGY REQUIRED FOR JOB
   10.25                            206
    1.75                            84

TOTAL ENERGY REQUIREMENT FOR THE JOB (KCAL) :           2258.
REST PERIOD AS PERCENT OF TOTAL WORK PERIOD :           0
```

Table 5. EXAMPLE 2 - OUTPUT

PROGRAM OUTPUT

```
DATE :          2/5/84
JOB NUMBER :    EXAMPLE 2
JOB CODE :      CASE 1
WORKER NAME :                   MS. Y
SEX             :               F
AGE (YEARS)          :          24
WEIGHT IN POUNDS     :          120.6
AEROBIC CAPACITY IN ML/KG/MIN :         42.37
PHYSIOLOGICAL CONDITION COMPUTED FROM THE TABLE :       4
HOURS OF SLEEP PER DAY :    8
SHIFT DURATION (HOURS)      :           8
ENERGY AVAILABLE FOR 18 YEAR OLD FROM THE TABLE (X) :   277
ENERGY REQUIRED WHEN NOT SLEEPING AND NOT WORKING :     600
TOTAL ENERGY AVAILABLE FOR WORK :       1352.96

TOTAL NUMBER OF JOBS :          2
TIME DURATION OF JOB            ENERGY REQUIRED FOR JOB
    7                               178.2
    1                               75

TOTAL ENERGY REQUIREMENT FOR THE JOB (KCAL) :           13
REST PERIOD AS PERCENT OF TOTAL WORK PERIOD :           0
```

PROGRAM OUTPUT

```
DATE :          2/5/84
JOB NUMBER :    EXAMPLE 2
JOB CODE :      CASE 2
WORKER NAME :                   MS. Y
SEX             :               F
AGE (YEARS)          :          24
WEIGHT IN POUNDS     :          120.6
AEROBIC CAPACITY IN ML/KG/MIN :         42.37
PHYSIOLOGICAL CONDITION COMPUTED FROM THE TABLE :
HOURS OF SLEEP PER DAY :    8
SHIFT DURATION (HOURS)      :           12
ENERGY AVAILABLE FOR 18 YEAR OLD FROM THE TABLE (X) :
ENERGY REQUIRED WHEN NOT SLEEPING AND NOT WORKING :
TOTAL ENERGY AVAILABLE FOR WORK :       1652.96

TOTAL NUMBER OF JOBS :          2
TIME DURATION OF JOB            ENERGY REQUIRED FOR JOB
   10.25                            162
    1.75                            75

TOTAL ENERGY REQUIREMENT FOR THE JOB (KCAL) :
REST PERIOD AS PERCENT OF TOTAL WORK PERIOD :
```

extended to Mr. Jagtar S. Chaudhry, a research assistant in the graduate industrial engineering program at the University of Cincinnati, for writing the BASIC program for calculating the results in Tables 4 and 5.

REFERENCES

[1] American Industrial Hygiene Association: Ergonomics Guide to Assessment of Metabolic and Cardiac Costs of Physical Work, *American Industrial Hygiene Association Journal*, 32, 560-564, 1971.

[2] Krager, D.W. and W.M. Hancock: *Advanced Work Measurement*, Industrial Press, Inc., New York, New York, 1982.

[3] Grandjean, E.: *Fitting the Task to the Man*, Taylor & Francis Ltd., London, 1981.

[4] Lehmann, G.: *Pratische Arbeitsphysiologie*. 2. Auflage, Theime Verlag, Stuttgart, 1962.

[5] Monod, H.: La depense Ernegetique Chez l'homme. In *Physiologie du Travail* (Editor: J. Scherrer), Masson, Paris, 1967.

[6] Murrell, K.F.H.: *Ergonomics - Man in His Working Environment*. Chapman and Hall, London, 1965.

[7] Spitzer, H.: Physiologische Grundlagen Fur Den Erholungszuschlag Bei Schwerarbeit. REFA-Nachrichten, Heft 2, Darmstadt, 1951.

[8] Recommended Dietary Allowances. Food and Nutrition Board, National Academy of Sciences, National Research Council, Washington, D.C., 1980.

[9] Bozian, Richard C.: Personal Communications. Professor of Medicine and Director Division of Nutrition, University of Cincinnati School of Medicine, Cincinnati, Ohio, April, 1983.

[10] Chaffin, D.B.: Some Effects of Physical Exertion. Department of Industrial Engineering, University of Michigan, Ann Arbor, Michigan, July 1972.

[11] Astrand, P.O. and Rodahl: *A Textbook of Work Physiology*. McGraw-Hill, New York, New York, 1977.

[12] Mital, A., Asfour, S.S., and M.M. Ayoub: Physiological Approach - Work-Rate Recommendations and Container Configuration for Manual Lifting and Lowering Activities and Comparison with the Psychophysical Approach. *Journal of Human Ergology*, 11 (2), 143-156, 1982.

[13] American Heart Association: *Exercise Testing and Training of Apparently Healthy Individuals*. A Handbook for Physicians, 1972.

[14] Darden, E.: *Nutrition and Athletic Performance*. The Athletic Press, Pasadena, California, 1976.

BIOGRAPHICAL SKETCH

Dr. Anil Mital is an assistant professor of industrial engineering and the human factors engineering graduate coordinator at the University of Cincinnati. He holds a bachelor's degree in mechanical engineering (with highest honors) from Allahabad University, India. He also holds a M.S., from Kansas State University, and a Ph.D., from Texas Tech University, in industrial engineering. His research interests are in the areas of applied ergonomics and modeling of man-machine systems. He is a member of Pi Tau Sigma, Alpha Pi Mu, Tau Beta Pi, Phi Kappa Phi, Omicron Delta Kappa and Sigma Xi. He is presently serving on the editorial boards of *Human Factors Journal* and the International Foundation on Production Research. His professional society memberships include IIE (Senior), AIHA, and HFS. He has written numerous technical articles and reports. In 1984 he was awarded the Young Engineer of the Year Award by the Engineers and Scientists of Cincinnati. He was awarded the Junior Morrow Research Chair for the 1982-83 academic year by the University of Cincinnati College of Engineering.

Richard L. Shell, Ph.D., P.E., is Professor and Director of Industrial Engineering in the Department of Mechanical and Industrial Engineering at the University of Cincinnati. Dr. Shell's past business experience has included engineering and management positions with Bourns, Ampex, and IBM. During the past several years, he has served as an engineering and management consultant for government and private industry. He is presently serving as a board of directors member for several corporations.

Dr. Shell has been active in the Institute for over 20 years and has held offices at the chapter and regional levels. In 1978, he received the Phil Carroll Achievement Award from the Work Measurement and Methods Engineering Division, and is presently Director for the Division.

DEVELOPMENT OF DESIGN DATA BASE FOR MANUAL LIFTING ACTIVITIES FOR EXTENDED WORK-SHIFTS

Anil Mital and Richard L. Shell
Ergonomics Research Laboratory
Department of Mechanical and Industrial Engineering
University of Cincinnati
Cincinnati, Ohio 45221

ABSTRACT

This paper presents data for designing jobs which involve lifting objects manually for extended work-shifts. Thirty-seven male and thirty-seven female industrial workers provided the necessary data. Using the psychophysical approach, maximum weights acceptable to them for continuous lifting were determined. The data norms, for long and extra-long shifts, are provided for various lifting frequencies, heights of lift, and container sizes. The effects of these task related variables, worker age, and shift duration are also discussed.

INTRODUCTION

Overexertion injuries caused by manual materials handling have been of serious concern to ergonomists for over two decades. Even though this has been of major concern to industry and government, the number and severity of overexertion injuries, especially those caused by manual lifting, have continued to increase over the years. Case studies and injury statistics published in the literature demonstrate this very well [1,2]. The direct cost of compensation, lost work days, medical care, etc., has been estimated to be approximately fifteen billion dollars annually [3]; the indirect costs may be as much as four times this number. It is critical, therefore, to develop ways to alleviate this problem. Essentially safeguards should be built in the 'job-worker' sub-system. The ways to do this include: (i) development of screening procedures to ensure proper job-worker match, (ii) worker training to enhance individual's lifting capabilities, (iii) designing manual lifting jobs which are within the capabilities of a certain group of population, and (iv) selection and installation of mechanical aids to alleviate undesirable lifting stresses.

This paper concentrates on the third alternative. That is, development of population lifting capacity profiles to aid in manual lifting job design. Specifically, the effect of lifting duration on what people are willing to lift was investigated. All previous maximum acceptable weight of lift data have been developed while assuming an eight-hour shift. Overtime and extended work-shifts, ten or twelve hours per day, are frequent in industry. Under these circumstances, when the total working duration per day exceeds eight hours, worker's risk of personal injury also increases. There are several reasons; the most important being the additional accumulation of fatigue (chronic effect). It was hypothesized in this study that, if given the choice, with increasing shift duration workers will reduce the weight of lift to avoid excessive fatigue and, therefore, their lifting capabilities for long and extra-long shifts (10- and 12-hour shifts) will be lower than the regular 8-hour shift lifting capabilities. To date no lifting capability data have been generated for extended work-shifts.

The specific objectives of this investigation were to: (i) develop design data bases for male and female industrial workers for 10- and 12-hour work-shifts and (ii) to investigate the effects of several task related variables, such as work force, container size, and lifting height, on the lifting capability of an individual.

TESTING THE HYPOTHESIS

The hypothesis, stated earlier, that 'with increasing shift duration workers will reduce the weight of lift, if given the choice, to avoid excessive fatigue' was subjected to an experimental verification prior to collecting lifting capability data. Ten industrial male and female subjects participated in the verification process which was conducted over a period of two days, for each subject. On the first day, each subject determined the maximum weight he/she was willing to lift in a given size container, at a given pace, across a specified height level. The psychophysical approach was employed and the subjects were asked to assume they were working on a 12-hour shift. The maximum weights were estimated based on an experimental trial period of 40-45 minutes. The heart rate and oxygen uptake at the estimated weight were also measured.

On the second day, each subject verified the weight estimated on the first day by actually lifting the container for twelve hours at the specified pace. Responses (weight, heart rate, and oxygen uptake) were recorded every two hours.

During the estimation and verification processes, subjects wore normal work clothes. The climatic conditions were controlled and only one subject performed at any given time.

The data analysis indicated that individuals cannot accurately estimate the acceptable weight of lift for 12-hours based on a trial period of 40-45 minutes. In order to determine the actual weights, the estimates must be adjusted according to

(i) the maximum acceptable weight of lift decreases with lifting duration; the rate of decline is 3.4% per hour for males and 2.0% per hour for females.

(ii) the heart rate decreases by about 1.9% per hour for males and 0.8% per hour for females, and

(iii) the metabolic energy expenditure rate decreases by 2.6% per hour for males and 1.9% per hour for females.

Thus, the stated hypothesis was proved correct and workers did choose lower weights for longer shifts.

DEVELOPMENT OF DESIGN DATA BASES

In order to develop lifting capability norms for 10- and 12- hour shifts, an experiment was conducted. The experimental details are given in the following subsections.

Subjects

Thirty-seven male and thirty-seven female subjects, recruited from local industries, participated in the experiment. Each subject had at least six months of experience in manually lifting goods. Individuals with a history of back problems or those who were on medication or had any other physical ailment, were screened. Thus, only healthy and experienced subjects were chosen. Once selected, their isometric strengths and anthropometric measurements were recorded. These physical characteristics of the sample population were comparable with those reported by other studies [4,5].

Experimental Design

An incomplete balanced block factorial design was used. Subjects were blocks. Each subject performed nine out of thirty-six possible treatment combinations (4 frequencies - 1, 4, 8 and 12 lifts/min. x 3 container sizes - 30.48, 45.72, and 60.96 cm in the sagittal plane x 3 lifting heights - floor to knuckle, knuckle to shoulder, and shoulder to reach). There were a total of 37 male and 37 female subjects. Subjects were stratified by age in three age groups - up to 29 years, 30 to 39 years, and above 39 years. Each group had approximately the same number of subjects. This permitted the representation of a very wide population cross-section. One of the thirty-six treatments was randomly selected and repeated as treatment number thirty-seven for the purpose of balancing the design.

The maximum acceptable weight of lift and heart rate and oxygen uptake at that weight were the response variables. Other variables, such as technique, climatic conditions, motivation, etc., were controlled.

Equipment

A special lifting equipment, which allowed automatic lowering of containers, was constructed. It consisted of two adjustable shelves, the upper one capable of moving downwards under the weight of the container. The two shelves were adjusted, each time as dictated by the treatment combination. Once adjusted, the lower shelf would become immobile while the upper one would retain its ability to move down. The design allowed containers to be lowered automatically to the original point of lift and, thus, permitting it to be lifted again through the lifting height set originally at the beginning of each treatment.

Experimental Procedure

As mentioned earlier, each subject performed nine out of thirty-six possible treatment combinations as specified by the experimental design. All nine treatments were performed in a random order; the experimental procedure being the same for each combination. All treatments, for a particular subject, were performed on one day with no rest break between the treatments. Subjects were asked to work as hard as possible, but without getting tired or exhausted.

The psychophysical methodology was used to determine the maximum acceptable weight of lift. For each treatment combination, subjects were randomly started with either a very heavy or very light load in the container and were allowed to make adjustments to it by removing or adding some weight to it. Assuming a 12-hour shift, subjects lifted the container over the specified height level at a pace set with the help of a metronome. Adjustments were made continually, while lifting, until the weight in the container approached the maximum weight acceptable to the particular individual for 12-hours of lifting. For each treatment combination, this weight and the heart rate and oxygen consumption at this weight were recorded. The entire process took approximately 40-45 minutes per combination, at the end which equipment were adjusted for the next treatment.

Data Adjustments and Norms

Once all the pyschophysical data were collected, data adjustments were made. Multipliers, based on observed trends (see section 'Testing the Hypothesis'), were used to arrive at the maximum weights acceptable to male and female industrial workers for long and extra-long work-shifts (10- and 12-hour shifts). Figures 1-6 show the lifting capability of Industrial workers for various container sizes, frequencies, and height levels for 10- and 12-hour long shifts. The corresponding heart rate and oxygen uptake values are given in Tables 1 and 2 for 10- and 12-hour shifts, respectively.

EFFECTS OF AGE AND TASK VARIABLES

Effect of Age

Three age categories were included in this investigation: (i) up to 29 years of age, (ii) 30 to 39 years of age, and (iii) 40 years and above. No age related differences were found. The lifting capability of a younger worker was not different from that of an older worker ($p \geq .10$). Apparently, experience on the job compensates for any muscular weakness or decline in physical capabilities.

Effects of Shift Duration

As stated earlier, the lifting capability decreased with shift duration. The lifting capability of males for 12-hour shifts was 6.8% lower compared to 10-hour shifts. The corresponding reduction for females was 4%. The decline in heart rate, when the shift duration increased from 10 to 12 hours, was 4.5 bpm for males and 2 bpm for females ($p < .05$). The corresponding decline in oxygen uptake was 0.3 lit/min for males and 0.15 lit/min for females. Small differences in heart rate indicate almost constant circulatory burden (reduction in lifting capability was compensated by increased fatigue build up).

Effects of Container Size

Figures 1 and 4 show the effect of container size on lifting capabilities of males and females for 10- and 12-hour shifts, respectively. As the container-size increased, the lifting capability decreased. Bigger containers have longer moment arms which is the main reason for this decline. The physiological responses, however, were not affected ($p \geq .10$). The reduction in lifting capability, when the container size increased from 30.48 cm to 60.96 cm, was 8.25% for males and 4.9% for females.

Effects of Frequency

Frequency effects are plotted in Figures 2 and 5. With increased frequency, the lifting capability decreased while the physiological costs increased. This decrease and increase was significant at the one-percent level. The lifting capability for males decreased by almost 30% when the frequency increased from 1 lift per minute to 12 lifts per minute. For females, the corresponding decline was about 24%. The increase in heart rate and oxygen uptake for males was about 30% and 100%, respectively. For females, these increases were 28% and 100%, respectively.

The reduction in lifting capability and increases in physiological costs were expected. With frequency, work-rates generally increase, causing these changes.

Effects of Lifting Height

Figures 3 and 6 show the lifting capability of males and females for the three height levels for 10- and 12-hour shifts. More weight was lifted from the floor and knuckle heights compared to the shoulder to reach height level. Metabolic energy expenditure rate and heart rate were also highest for the floor to knuckle height and decreased with height. These responses were expected. The floor level lifts involve larger muscle mass and can generate more force. This causes greater lifting capability. Since larger muscles are involved and more weight is lifted, including part of body weight, physiological costs are higher. With height, the lifting capability decreased by up to 12% for males and 16% for females. Corresponding changes in physiological responses are shown in Tables 1 and 2.

CONCLUDING REMARKS

This paper presents lifting capability data of the industrial population. The effects of important task variables (container-size, frequency, and lifting

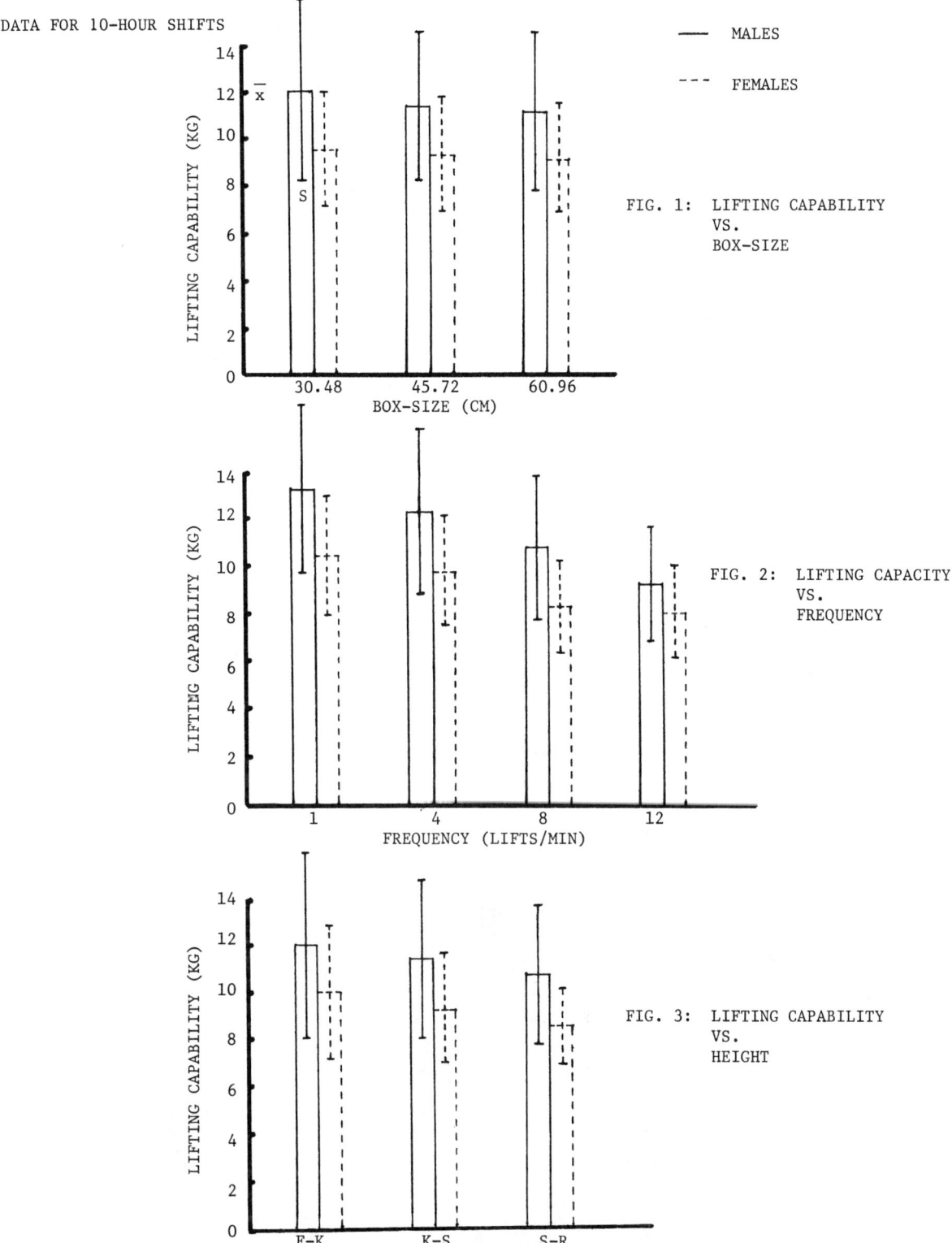

FIG. 1: LIFTING CAPABILITY VS. BOX-SIZE

FIG. 2: LIFTING CAPACITY VS. FREQUENCY

FIG. 3: LIFTING CAPABILITY VS. HEIGHT

DATA FOR 12-HOUR SHIFTS

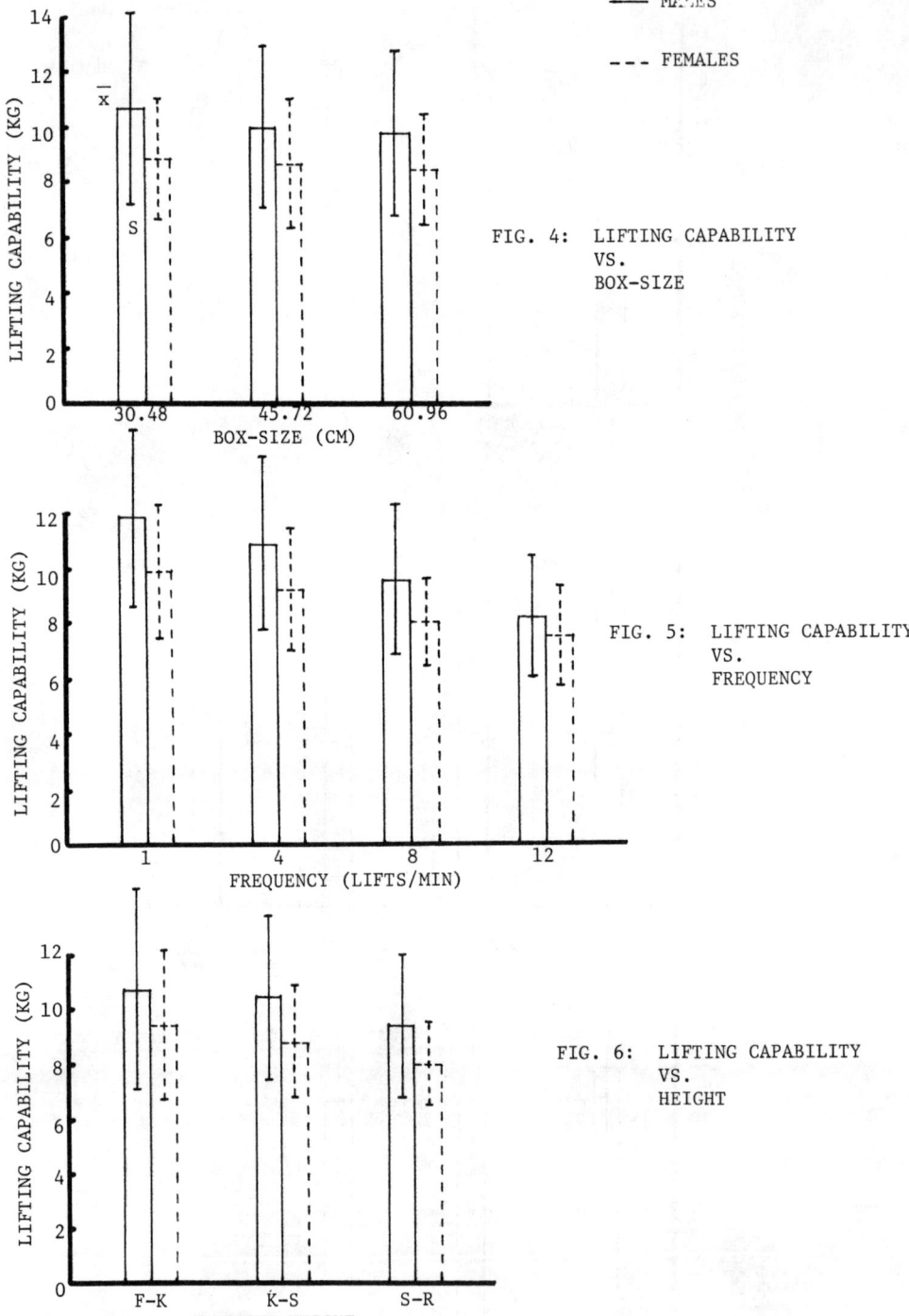

FIG. 4: LIFTING CAPABILITY VS. BOX-SIZE

FIG. 5: LIFTING CAPABILITY VS. FREQUENCY

FIG. 6: LIFTING CAPABILITY VS. HEIGHT

Table 1: Means ($\bar{x}$) and Standard Deviations (s) of Physiological Responses for 10-Hour Shift

		Males		Females	
Variable	Response	$\bar{x}$	s	$\bar{x}$	s
Box-Size (inches)					
12	Heart Rate (bpm)	94.83	17.87	104.30	15.94
	Oxygen Uptake (l.min^{-1})	0.65	0.29	0.43	0.22
18	Heart Rate	94.84	16.23	103.41	16.50
	Oxygen Uptake	0.66	0.28	0.43	0.22
24	Heart Rate	93.94	18.93	99.79	16.58
	Oxygen Uptake	0.67	0.27	0.43	0.23
Frequency (lifts/min)					
1	Heart Rate	80.62	13.31	89.79	10.03
	Oxygen Uptake	0.38	0.17	0.27	0.15
4	Heart Rate	93.30	15.27	98.97	13.91
	Oxygen Uptake	0.65	0.22	0.41	0.21
8	Heart Rate	100.54	15.96	106.20	13.95
	Oxygen Uptake	0.78	0.25	0.51	0.22
12	Heart Rate	103.84	16.83	114.70	16.14
	Oxygen Uptake	0.83	0.24	0.53	0.21
Height of lift					
Floor to knuckle	Heart Rate	97.61	19.13	105.45	17.27
	Oxygen Uptake	0.78	0.30	0.53	0.23
Knuckle to shoulder	Heart Rate	97.02	16.19	103.49	15.38
	Oxygen Uptake	0.62	0.25	0.41	0.20
Shoulder to reach	Heart Rate	88.74	16.13	98.61	16.07
	Oxygen Uptake	0.56	0.24	0.35	0.20

Table 2: Means ($\bar{x}$) and Standard Deviations (s) of Physiological Responses for 12-Hour Shift

		Males		Females	
Variable	Response	$\bar{x}$	s	$\bar{x}$	s
Box-Size (inches					
12	Heart Rate (pbm)	90.07	16.85	102.22	15.45
	Oxygen Uptake (l.min^{-1})	0.59	0.27	0.40	0.21
18	Heart Rate	90.27	15.77	101.74	16.00
	Oxygen Uptake	0.60	0.27	0.40	0.21
24	Heart Rate	89.48	17.78	97.97	16.02
	Oxygen Uptake	0.61	0.25	0.40	0.22
Frequency (lifts/min)					
1	Heart Rate	76.60	12.55	88.16	9.68
	Oxygen Uptake	0.34	0.16	0.25	0.15
4	Heart Rate	88.75	14.66	97.33	13.82
	Oxygen Uptake	0.60	0.21	0.38	0.20
8	Heart Rate	95.62	14.95	104.20	13.11
	Oxygen Uptake	0.71	0.24	0.47	0.21
12	Heart Rate	98.92	15.89	112.56	15.55
	Oxygen Uptake	0.76	0.23	0.50	0.20
Height of lift					
Floor to knuckle	Heart Rate	93.06	18.27	103.34	16.72
	Oxygen Uptake	0.71	0.28	0.50	0.22
Knuckle to shoulder	Heart Rate	92.03	15.17	101.86	14.98
	Oxygen Uptake	0.57	0.23	0.38	0.19
Shoulder to reach	Heart Rate	84.47	15.37	96.76	15.36
	Oxygen Uptake	0.51	0.23	0.32	0.19

height) on lifting capability are also discussed. At the present time, no other data are available to aid ergonomists in designing manual lifting jobs which are to be performed for long and extra-long shifts. These data, therefore, should be of considerable help in establishing comprehensive lifting guidelines. Physiological response measures should also aid in estimating fatigue. The average metabolic energy expenditure rates of males and females, at the 10-hour lifting capability, correspond to approximately 26% of their aerobic capacity. The values for 12-hour lifting capability were 23% and 24%, respectively.

The results also clearly show that age of the worker is not a relevant factor as far as his/her lifting capability is concerned. Older and experienced workers are physically as capable as their younger, but relatively less experienced, counterparts.

ACKNOWLEDGMENT

This work was supported by the National Institute for Occupational Safety and Health, Department of Health and Human Services, through Grants 1-R01-OH-01429-01 & 02.

REFERENCES

[1] Work Practices Guide for Manual Lifting, 1981, National Institute for Occupational Safety and Health, Cincinnati, Ohio.

[2] Ayoub, M.A., 1982, The Manual Lifting Problem: The Illusive Solution. Journal of Occupational Accidents, 21.

[3] Taber, M., 1982, Reconstructing the Scene: Back Injury. Occupational Health and Safety, 51.

[4] Ayoub, M.M., Bethea, N.J., Dievanayagam, S., Asfour, S.S., Bakken, G.M., Liles, D., Mital, A., and Sherief, M., 1978, Determination and Modeling of Lifting Capacity. Final Report, DHHS (NIOSH) Grant No. 5-R01-OH-00545-02.

[5] Chaffin, D.B., Herrin, G.D., Keyserling, W.M., and Foulke, J.A., 1977, Pre-employment Strength Testing. Final Report, DHHS (NIOSH) Contract No. CDC-99-74-62.

BIOGRAPHICAL SKETCH

Anil Mital, Ph.D. is an associate professor of industrial engineering and the human factors engineering graduate coordinator at the University of Cincinnati. His research interests are in the areas of applied ergonomics and modeling of man-machine systems. He is a member of Pi Tau Sigma, Alpha Pi Mu, Tau Beta Pi, Phi Kappa Phi, Omicron Delta Kappa and Sigma Xi. He is presently serving on the editorial board of Human Factors Journal and on the board of the International Foundation on Production Research. His professional society memberships include IIE (Senior), AIHA, and HFS. He has written numerous articles and reports. He was awarded the Junior Morrow Research Chair, for the 1982-83 academic year, by the University of Cincinnati College of Engineering. In 1984 he was awarded the Young Engineer of the Year Award by the Engineers and Scientists of Cincinnati. He also received the 1984 Sigma Xi Distinguished Research Award. Presently he is editing Advances in Ergonomics/Human Factors I which will be published by North Holland in December, 1984.

Richard L. Shell, Ph.D., P.E., is Professor and Director of Industrial Engineering in the Department of Mechanical and Industrial Engineering at the University of Cincinnati. Dr. Shell's past business experience has included engineering and management positions with Bourns, Ampex, and IBM. During the past several years, he has served as an engineering and management consultant for government and private industry. He is presently serving as a board of directors member for several corporations. Dr. Shell has been active in the Institute for over 20 years and has held offices at the chapter and regional levels. In 1978, he received the Phil Carroll Achievement Award from the Work Measurement and Methods Engineering Division, and is presently Director for the Division.

THE EFFECT OF MENTAL FATIGUE ON KNOWLEDGE WORKER PERFORMANCE

Richard L. Shell, Ph.D., P.E.
and
O. Geoffrey Okogbaa, Ph.D.
Department of Mechanical and Industrial Engineering
University of Cincinnati
Cincinnati, Ohio 45221

ABSTRACT

The measurement and quantification of mental work output and fatigue has always been a major challenge and concern to many industrial engineers and managers. The problem has been compounded by current technological advancements and automation of human-machine systems. This paper examines the impact of such advances on the knowledge worker and presents the results of one laboratory experiment conducted to simulate typical knowledge worker functions. The experiment was carried out under two types of working conditions, namely with rest breaks and without rest breaks. The duration of the rest break was ten minutes for every hour of work. Three performance measures; work output, heart rate and brain waves, were obtained. The results showed performance deterioration due to mental fatigue and significant work output gains with rest breaks.

INTRODUCTION

Researchers engaged in the enhancement of human output through better human-machine interactions have long been interested in the phenomena of fatigue in the workplace. Actual documentation of fatigue research, however, dates only as far back as World War I. Since then, the investigations have passed through three different eras of which the third is still ongoing.

The first period of interest was in Britain during World War I. The research at this time focused primarily on the effects of fatigue on productivity in the munitions industry. It was believed then that fatigue effects were compounded by shift changes, illumination and ventilation, workplace design and plant layout. The criterion for existence of fatigue or the lack thereof was based on total output of manufactured items and thus any ensuing reduction in output was ascribed to fatigue (Cameron 1973).

The second era of interest in fatigue was demonstrated in the periods during and immediately following the second World War. The major research effort at this time was on military and civilian aviation. The focus of some of the research was on ways to establish appropriate standards of operation of aircrafts to avoid fatigue. The general conclusion of most of the investigations was that output performance alone should not be the exclusive area of interest in describing fatigue. The researchers also advanced the notion that fatigue had three distinct facets, namely:

(1) Subjective feelings
(2) Impairment due to oxygen debt in the tissues
(3) Work output decrement.

The third era of interest in fatigue started in the early fifties and is still active at this time. The research focused on two different but related areas. One of these areas was fatigue during car driving and the resultant effects on accident rates (Brown, 1967). Fatigue research in driving assumes that safety is the ultimate goal and as a result the number of accidents is used to measure the effects of fatigue. The other area of research is the effect of fatigue on Air Traffic Controllers (ATC) and the resultant health changes (Rose et al, 1978). The emphasis in this area of research is on the chronic and cumulative effects of fatigue and a concern with the long-term wellbeing of the workers. While it has not been shown that fatigue leads to accidents, it has been recognized that the long-term effects on the health and on-the-job satisfaction of ATC crews are legitimate areas of concern.

Most of these investigations have been and are concerned with fatigue as it affects physical and paraphysical work. Little has been reported, however, regarding the phenomena of fatigue and its effects on the knowledge worker, especially during those activities which can be classified as mostly mental.

Except in a few cases, no work activity can be dichotomized as strictly mental or physical. While physical activities make demands upon certain levels of the central nervous system, mental activities are also accompanied by some activity of the muscles. Mental fatigue may sometimes also result because of the contraction of vocal muscles during the process of thinking or because of other muscular activity such as facial expression associated with mental work. In those purely mental operations such as reading, mental arithmetic and the solution of verbal intelligence problems, the functional impairment of neuronic cells account for mental fatigue evident in work output decrement and in the feeling of weariness.

Evidence suggests that fatigue (whether physical or mental) may be reduced but not eliminated. The mechanization and automation of most material handling tasks mean that manual work in industry will continue to decrease. Consequently, there will be less muscular tiredness due to the stressing of muscles as industries become more automated. For mental work, however, the effect of mechanization and automation has had exactly the opposite effect on mental fatigue.

OCCUPATIONAL TRENDS FOR THE KNOWLEDGE WORKER

The current industrial climate is characterized by the continued increase in the efficacy of the technology of production in manufacturing and service organizations. A major implication of this increase in automation and sophistication of human-machine systems is the change in the function and nature of human work. This change in function and nature has underscored the importance of task performance of certain mental functions such as information assimilation and decision making. In addition, the rapid technological advances in digital computers in particular and electronics in general and the enhanced utilization of automated work systems mean increased mental demands and attendant mental fatigue on workers and more specifically knowledge workers. This trend in rate of development and improvement suggests a future dominated by knowledge workers.

Growth rates for knowledge workers - professionals, technical and management jobs - and those for blue collar occupations - craft, laborer, and farm occupation - have grown and differed considerably. Once a small proportion of the labor force, knowledge workers now represent about half the total. The number of knowledge workers has grown while the blue collar work force has grown only slowly and farm workers have declined. Over two-thirds of the nation's work force is currently employed in industries that provide such services as health care, trade, education, government and banking, among others. Professional and technical workers will grow from 14.2 million in 1978 to 16.9 million in 1990 (source: US BLS, 1981), reference Figures 1 and 2. Managers and administrators will grow from 10.1 million (1978) to 12.2 million (1990), the reason for growth being the increased dependence on trained management specialists especially in highly technical areas of operation. In contrast, employment in the goods producing industries rose only 9 percent between 1965 and 1968. In 1980, the total manufacturing labor force numbered about 23 million. Significant gains in productivity resulting from automated production, improved machinery and technological breakthroughs permitted large increases in output without additional workers.

The military has not been left behind in the technological breakthroughs. While the impact of new and future advances in technology is uncertain, current advances have transformed the occupation needs of the armed forces. Whereas in earlier eras the armed forces consisted mainly of combat personnel engaged in purely military duties, today the forces include large numbers of specialists and technicians whose responsibility is to support the combat mission. The evolution of thermonuclear weapons, the developments in aircraft systems, the advances in computer based command, control and communication and the introduction of space technology have imposed great requirements on the armed forces for skilled manpower. Knowledge workers now make up 46 percent of the total trained enlisted personnel of the United States forces as opposed to 28 percent in 1945 (Binkin, 1980), well in agreement with the growth of the white collar employment in the economy as a whole. Available data show that in the military the change occurred mainly in the technical fields which now require computer specialists, teletype and electronic instrument technicians, radio system operators, health technicians, meteorologists and a host of support staff whose tasks are similar to those performed in the civilian sector. Blue collar enlisted workers now constitute about 54 percent of the military labor force, compared with 72 percent in 1945, a change due mainly to the sharp decrease in the percentage of ground combat soldiers during the years following World War II. On the whole, while military occupational needs are uncertain and subject to a variety of unpredictable circumstances, it would be safe to assume that the armed forces' needs for knowledge workers will continue to grow as will the cost of training.

The attendant implication of higher levels of automation and other improve-

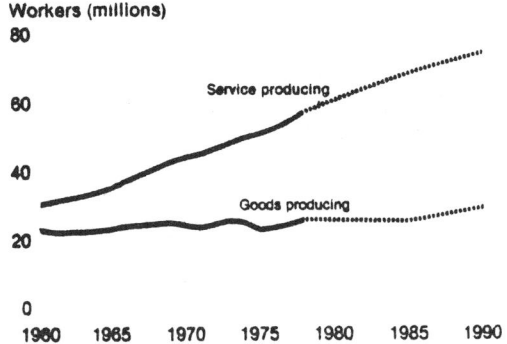

Figure 1. Projected growth of workers in goods and service industries.

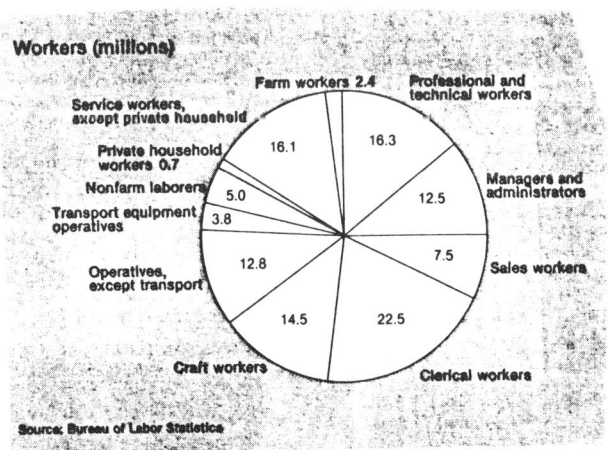

Figure 2. Projected distribution of employment by occupation in 1990.

ments in technology is that more and more workers in both the civilian and military sectors would be required to engage in occupations in which their mental abilities rather than their physical capability is called into play. The major problem of worker fatigue in the near future will not be so much physical as mental. Until now, the understanding of the phenomena of mental fatigue is not only rudimentary but minimal at best.

INVESTIGATING THE PROBLEM

Researchers at the University of Cincinnati have conducted ongoing experiments in which typical knowledge worker functions were simulated in the laboratory. The aim was to investigate the possible relationship between worker output performance and mental fatigue. Such a relationship would provide the impetus to determine limits, standards, and allowances for mental activities similar to the design rules and rest allowances for manual (physiological) work tasks. The mental task experiments were chosen to resemble those encountered by knowledge workers in the course of their daily work. The tasks chosen were (1) abstract in nature such as finding solutions to arithmetic-logical problems (Bennett, 1976; Sjoberg, 1977, Leplat, 1978; Beetchereva, 1981) and (2) cognitive in nature such as the reading of standardized texts (Thorndike, 1917; Carver, 1976).

Subjects, Sample Size and Design

In one investigation, fourteen subjects were selected randomly from a group of male University student volunteers. It should be noted that since all participants were volunteers, this project was one of two compulsory assignments needed to fulfill the requirement of a course in which all the subjects were enrolled. As a result, the problem of low motivation which is sometimes associated with volunteer subjects was minimized. All subjects for the study were junior level engineering students with an average age of 21.8 years and standard deviation of 1.2 years. The group was thus considered homogenuous in terms of age and ability. Standardized tests for both reading and arithmetic-logical problems from such examinations as the SAT, ACT and GRE were modified and used.

A symmetric design was used in the procedure. Two treatment groups were used for each of the two task types with randomization of treatment. Two additional groups were used for a combined task assignment also with randomization of treatment. Subjects were habituated to the experimental setting by participating in a pre-test experiment lasting twenty minutes. The procedure for the cognitive process and the abstract process tasks were as follows:

Reading   The two groups (Rest/No Rest) were required to read standardized texts over a five hour period. The only difference between the groups was that one had occasional pauses for scheduled rest of 10 minutes for every 50 minutes of continuous work. Various rest break durations ranging from 5 to 15 minutes have been suggested and used by various authors (Sjoberg, 1980; Bennet et al, 1976; Zwaga, 1973). The National Institute of Occupational Safety and Health recommends 15 minutes for two hours of continuous VDT work for moderate work load and 15 minutes for one hour of continuous work for high work load. The other group performed the same task but with no scheduled rest. To control for skimming, tests of comprehension were scheduled during each work segment.

Arithmetic-Logical Problems   This type of task is one that induces high information load. Standardized tests were administered and were such that over 95 percent of the population chosen would complete them without difficulty. However, the problems were challenging enough that some thought was needed before a response was made. As in the reading task, one group worked continuously without rest breaks while the other group had scheduled rest.

Combined Task   The two groups which engaged in the combined task, spent half of the time on the reading assignment and the other half on the problem solving assignment. One of the groups rested between work while the other group did not.

EEG (Electroencephalogram) - Brain Waves

Bitemporal recordings were made over the skull of each subject with the reference electrode on the left or right mastoid depending on whether the subject was left or right handed. The signals were appropriately amplified and then fed to an FM recorder for storage. It was then routed to a Fast Fourier Transform Analyzer for processing to yield the power spectra and the corresponding time and frequency components.

ECG - Heart Rate

Recordings of ECG signals were initiated using a four-channel physiograph with the reference electrode on the abdomen and the active electrodes on the lower and upper left corners of the chest. The raw amplified signal was recorded and from this the number of heart beats/minute were obtained.

RESPONSE AND INDEPENDENT VARIABLES

Independent Variables

The variables of interest were

(1) Task Difficulty - The difficulty level for both task types were constant throughout each experimental session.

(2) Task Length - The length of the tasks were longer than the standard length for each task type. This was to ensure that no subject completed a task before the end of a segment or the experimental session.

Response Variables

The variables included

(1) EEG waves - Alpha, Beta, Theta components and the theta/alpha ratio.

(2) Heart Rate (beats/minute).

(3) Reading rate (words/min) and reading comprehension (words/min).

(4) Number of problems attempted (both reading and arithmetic-logical problems).

(5) Number of correct responses (both tasks).

(6) Number of incorrect responses (both tasks).

RESULTS

Task Output

For a given task, the differences between individual performance were not significant ($\alpha = 0.05$). As a result, it was assumed that the subjects were from identical populations in which case it was correct to talk about a single measure for a task such as average output rather than individual output measures.

Figure 3 indicates typical trends for the cognitive process task (Reading Comprehension) for both groups (Rest, No Rest). The trend shows that output performance decreases with time. After some re-expression and transformation, the ensuing data was analyzed and the results indicated a higher minimum output value for all subjects who had scheduled rest breaks. This suggests that the minimum attainable output performance for subjects without rest breaks was smaller than those with rest breaks. The trend for abstract process task (arithmetic-logical problems) and the combined task also indicated output decrement as a function of time (Figures 4 and 5).

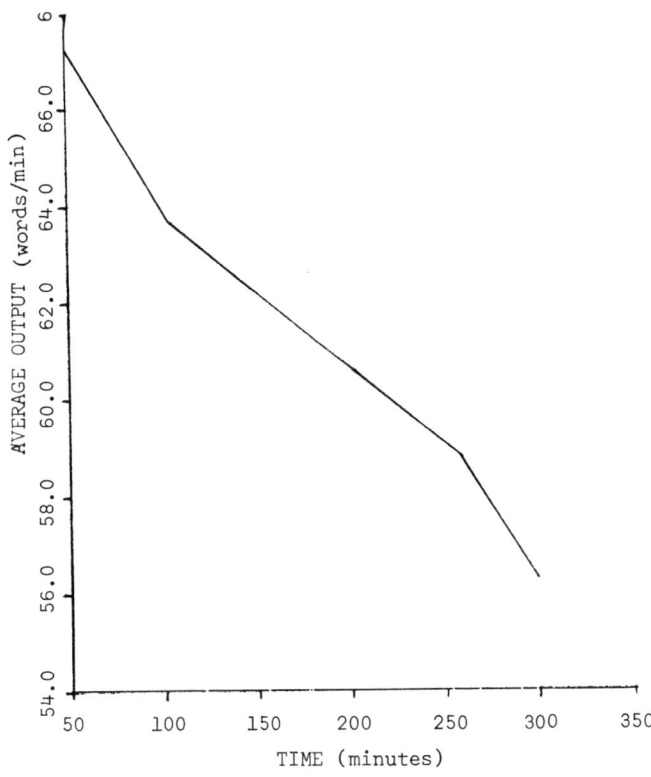

Figure 3. Raw Output for Reading Task Without Rest

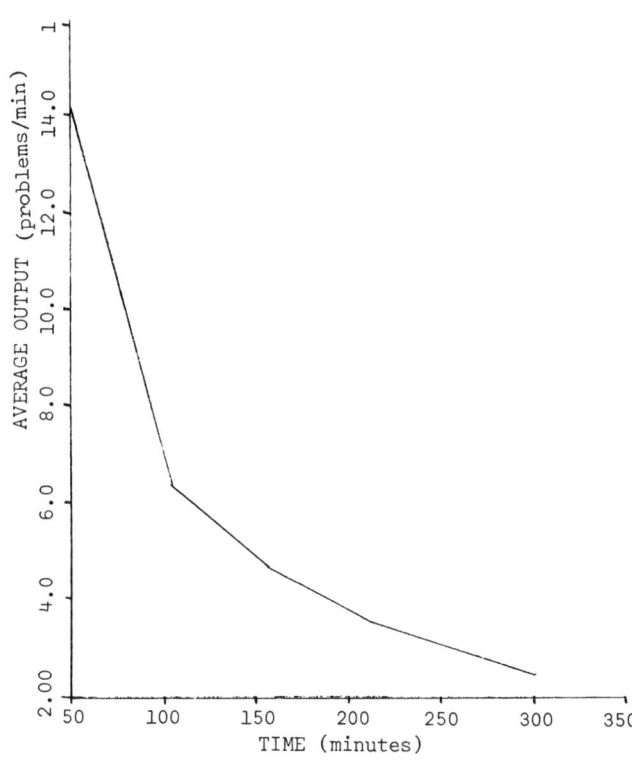

Figure 5. Raw Output for Reading/Arithmetic Tasks Without Rest

Task Condition Differences (Rest, No Rest)

The minimum mental work output for the Rest subgroup was higher than the No Rest classification. The periodic output from the tasks were transformed into 'Areas under the output performance curve'. Using these areas, performance indices were developed, namely:

(1) Mental Output Index (MOI)

$$\text{where MOI} = \frac{\Sigma_i \text{Area}}{\text{Maximum Possible Area}}$$

(2) Mental Fatigue Index (MFI)

where $MFI = 1 - MOI$

The MFI index was lower for those tasks with rest breaks. Not only were the MFI values lower for the rest subgroup but the actual differences between the mean mental output index (MOI) for the two subgroups was significant. The actual numerical difference between the two subgroups was 6 percent. This represents a 6 percent increase in productive output for the subgroup with rest breaks (Table 1).

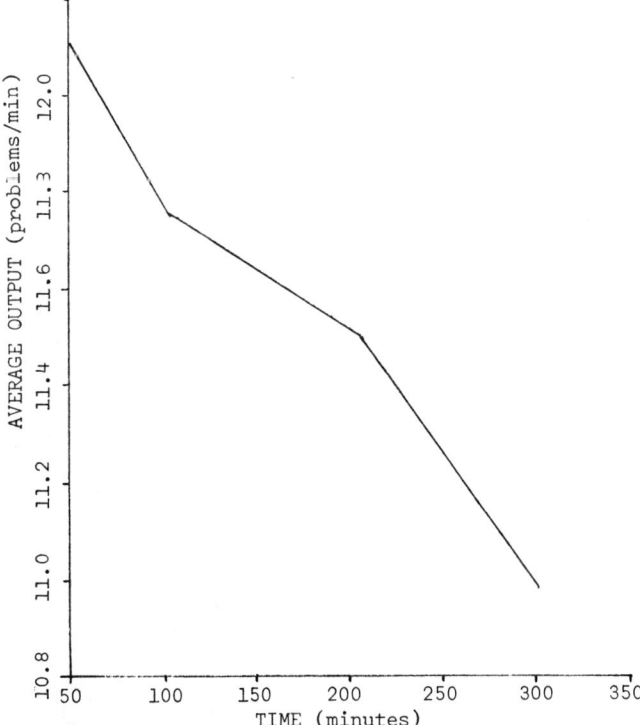

Figure 4. Raw Output for Arithmetic Task Without Rest

Task Differences

After appropriate transformation of the raw data, the resulting cumulative average for the arithmetic (abstract) task was higher than the same measure for the reading (cognitive) task. The differences between the tasks were also statistically significant. The Mental Output Index (MOI) was consistently higher for the arithmetic task. Such a comparison would be meaningless without data transformation because of the different magnitudes and units of measurement for the tasks.

Heart Rate

The mean and standard deviation of heart beats for each subject was recorded. The mean values ranged from 48 beats to 82 beats/minute. Compared with the accepted normal average of 75 beats/minute, most subjects averaged much lower beats.

TABLE 1

DIFFERENCES IN % MOI FOR REST VERSUS NO REST SUBGROUP

Sample Size n	No Rest (X_1)	Rest (X_2)	$X_2 - X_1$
1	38.5	41.4	2.9
2	40.0	40.4	0.4
3	39.9	44.1	4.2
4	40.2	42.4	2.2
5	40.0	44.4	4.4
6	41.4	42.5	1.1

1. $H_o: \sigma_1^2 = \sigma_2^2$, $H_1: \sigma_1^2 \neq \sigma_2^2$

   $S_2^2 = 2.359$, $S_1^2 = 0.852$

   $F = 2.77$, $F_{0.01,5,5} = 11$;

   $F_{0.05,5,5} = 5.05$

   ∴ Cannot reject H_o = variances are the same.

2. $H_o: \mu_1 = \mu_2$, $H_1: \mu_1 \neq \mu_2$

   $\bar{x}_1 = 40$, $\bar{x}_2 = 42.533$

   $t = 3.16$

   $t_{0.01,10} = 2.764$, $t_{.05,10} = 1.812$

   ∴ Reject H_o = mean values are different

A test of randomness indicated that the sequences of heart beat was random, suggesting that heart rate may not be affected by mental fatigue. But as indicated, the average beats for most of the subjects were consistently below the normal rate of 75/minute.

Brain Waves

There was a significant correlation between alpha and beta components and between alpha and theta components for those tasks with no rest breaks. The correlation between these components were not as high for the tasks with rest breaks. This is an indication that alpha components were dominant during tasks with no rest breaks and corroborates the view that a high alpha rhythm is synonymous with reduced readiness to react to a stimuli.

A test of randomness for the sequences of power spectral density values for the alpha, beta, theta, and theta/alpha components showed that the sequences were not random especially for 'No Rest' task classification.

CONCLUSIONS AND GENERAL REMARKS

Conclusions

The results of the experiment support the notion of a time-dependent mental work output decrement due to mental fatigue.

Mental work output and fatigue differences between the two task types (cognitive, abstract), and the working conditions (Rest, No Rest), were all significant. Additionally, mental work output for tasks with scheduled rest breaks were superior (6 percent more) to the tasks with no scheduled rest breaks. No significant differences were observed for the combined task.

The sequences of heart rates were random for all subjects with the mean heart rate consistently below the norm. Heart rate variability may be useful in explaining the apparent discrepancy.

Alpha components of the brain waves dominated during tasks with no rest breaks. Since the presence of alpha components indicate a reduced ability to respond to a stimuli, fatigue effects are likely minimized by scheduled rest breaks.

General Remarks

The average age of the participants ($\bar{x} = 21.8$, $s = 1.2$) is not typical of the general knowledge worker population. What is important, however, is to realize that while actual numeric values would probably be different, the trend for mental work output and fatigue will likely be the same across worker populations and ages.

In a general model of mental fatigue, or any human model for that matter, various variables interact very strongly, i.e., psychological, physiological, social and work related variables. Such variables can affect one's interest and motivation in a given task and consequently influence output and possible fatigue.

REFERENCES

Amosov, N.M.: Modeling of Thinking and the Mind. Spartan Books, MacMillan Co., Ltd., New York, NY, 1965.

Anon, 1946: The Role of Fatigue in Pilot Performance, U.S. Civil Aeronautics Administration Report No. 61.

Aquilano, N.: "Why IE's Can't Measure Fatigue," Journal of Industrial Engineering, March 1970.

Bennett, C.A.: "Counteracting Psychological Fatigue Effects by Stimulus Change," Proceedings of the 18th Human Factors Society Meeting, Huntsville, Alabama, 1976.

Beechtereva, N.P.: "Neurophysiological Correlates of Mental Processes in Man," Psychophysiology Today and Tomorrow (Beechtereva, N.P., editor), Pergamon Press, New York, NY, 1981.

Binkin, M., Kyriakopoulos, I.: Manning the Modern Military, Brookings Institution, Washington, D.C., U.S.A., 1979.

Boepple, E., Kelly, L.: "How to Measure Thinking," Journal of Industrial Engineering, July 1971.

Brown, C.C.: Methods in Psychophysiology, The Williams and Wilkins Company, Baltimore, Maryland, 1967.

Bureau of Labor Statistics (USBLS), Department of Labor, 1981.

Cameron, C.: "A Theory of Fatigue," Ergonomics, 1973, Vol. 16, No. 5

Carver, R.P.: "Optimal Rate of Reading Prose," Reading Research Quarterly No. XVIII, Fall, 1982.

Carver, R.P.: "Toward a Theory of Reading and Rauding," Reading Research Quarterly, 1977-1978, 13:8-63.

Charnock, D.M., Manenica, I.: "Spectral Analysis of R-R Interval Under Different Work Conditions," Ergonomics, 1978, Vol. 21, No. 2.

Farr, R.C.: Reading: What Can Be Measured? International Reading Association, Research Fund, Newark, Delaware, 1969.

Fiscal Year 1982 - The Department of Defense Annual Report.

Fiscal Year 1980 - The Department of Defense Annual Report.

Fry, E.: "Reading Rate in 1908," Journal of Reading, May 1970.

Leplat, J.: "Factors Determining Work-Load," Ergonomics, 1978, Vol. 21, No. 3.

Rose, L.: "The Reading Process and Some Research Implications," Journal of Reading, October 1969.

Rose, R.M.; Jenkins, C.D.; Hurst, M.W.: Air Traffic Controllers Health Change Study, Report No. FAA-AM-78-39, Contract No. FA73WA-3211, Federal Aviation Administration, Washington, D.C., 1978.

Sjoberg, H.: Physical Fitness and Mental Performance During and After Mental Work," Ergonomics, 1980, Vol. 23, No. 10.

Thorndike, E.L.: "Reading as Reasoning: A Study of Mistakes in Paragraph Reading". Reading Research Quarterly, No. 4, 1971.

Zwaga, H.J.: "Psychophysiological Relations to Mental Tasks: Effort or Stress," Ergonomics, 1973, Vol. 16, No. 1.

BIOGRAPHICAL SKETCHES

Richard L. Shell, Ph.D., P.E., is Professor and Director of Industrial Engineering in the Department of Mechanical and Industrial Engineering at the University of Cincinnati. Dr. Shell's past business experience has included engineering and management positions with Bourns, Ampex, and IBM. During the past several years, he has served as an engineering and management consultant for government and private industry. He is presently serving as a board of directors member for several corporations. During the past decade, Dr. Shell has authored over fifty publications including What Every Engineer Should Know About Human Resources Management, Marcel Dekker, December 1980, co-authored with Dr. Desmond Martin. Dr. Shell has been active in the Institute for over 20 years and has held offices at the chapter and regional levels. In 1978, he received the Phil Carroll Achievement Award from the Work Measurement and Methods

Engineering Division, and is presently Director Elect for the Division.

O. Geoffrey Okogbaa, Ph.D., is an Assistant Professor of Industrial Engineering in the Mechanical and Industrial Engineering Department at the University of Cincinnati. He received the BSISE (1977), MS (1979) from the Ohio State University and Ph.D. (1983) from the University of Cincinnati. Dr. Okogbaa's research interests are in the areas of modeling of human-machine systems and applied statistics. He is a member of Alpha Pi Mu, Texnikoi, and an active member of the Cincinnati chapter of IIE.

IV. Work Sampling

INTRODUCTION AND MATHEMATICAL BASIS

Work sampling is a technique used to investigate the proportions of total time devoted to various activities that comprise a job or work situation. The Industrial Engineering Terminology Standard Z94.12 defines work sampling as "an application of random sampling techniques to the study of work activities so that the proportions of time devoted to different elements of work can be estimated with a given degree of statistical validity." The technique can be applied to humans, machines, or any observable state or condition of an operation. The underlying assumption of work sampling is that the sampling percentage of any observed state of nature estimates the actual time spent in that condition.

The theory of work sampling is based on the fundamental laws of probability. If at a given instant an event can either be present or absent, then:

$$(p + q)^n = 1$$

where:
p = probability of a single occurrence
q = (1−p) the probability of an absence of occurrence
n = number of observations

This expression is expanded according to the binomial theorem; the first term gives the probability that x = o, the second term that x = 1, and so forth. The distribution of these probabilities is known as the binomial distribution. In most sampling studies each condition is taken at a time, and all the other conditions are then considered as nonoccurrence of this one event.

In the typical industrial situation, p is unknown to the analyst. The best estimate of p ($\bar{p}$) may be computed as x/n where x is the number of observations for that occurrence and n is the total observations during the trial period. If the number of random observations is small, the analyst must be concerned with the accuracy of $\bar{p}$. The larger the number of observations, the more accurate $\bar{p}$ becomes. When n is sufficiently large, the normal distribution can be used as a satisfactory approximation of the binomial distribution.

METHODOLOGY

The following steps are recommended for conducting a work sampling study:

1) Define the problem; state the objective or purpose, and describe in detail each element to be measured.

2) Obtain the approval of management before starting the study. Also, the workers to be studied must understand the purpose of the study, and their cooperation is needed.

3) Define the desired confidence level and allowable error.

4) Conduct a preliminary study to estimate the percentage of the smallest activity ($\bar{p}$). This estimation can be based on historical data or sampling for a short period of time, for example, one or two days.

5) Design the study by determining the number of observations to be made; the number of observers needed; the number of days or shifts needed for study; and a detailed plan of taking observations.

6) Conduct the observations according to the plan.

7) Analyze and summarize the data.
8) Check the accuracy of the data at the end of each observation day.
9) Report the conclusions and recommendations at the end of the study period.

ADVANTAGES AND DISADVANTAGES

Advantages of work sampling include the following:

1) Activities which are impractical or costly to measure by time study or predetermined time systems can be easily and cost-effectively measured by work sampling.
2) A simultaneous work sampling study of several operators or machines can be made by a single observer.
3) Work sampling requires fewer analyst hours and costs less than continuous time study or the application of predetermined time systems.
4) Work sampling results may be interrupted at any time during the study without affecting the results.
5) Work sampling studies are less fatiguing and less tedious to make on the part of the observer than time study or predetermined time systems.
6) In work sampling the degree of accuracy and/or error limits can be changed by varying the number of observations.
7) Work sampling is impersonal and group oriented, and because there is no direct timing workers usually feel positive about the study.
8) Work sampling is often the only suitable work measurement technique to be applied to professional or knowledge workers.

Some of the disadvantages associated with work sampling include the following:

1) Work sampling is usually uneconomical for studying a single operator or machine, or for studying operators or machines located over widely dispersed areas. Other work measurement techniques, e.g., time study or predetermined time systems, are preferred for repetitive operations.
2) Time study provides a more detailed breakdown of activities and delays.
3) Management and workers may not understand statistically based work sampling as readily as they do direct observation time study.

STUDY DESIGN

The number of observations (n) depends on the level of confidence and accuracy desired. The larger the sample size, the more accurate the results become. By approximating the percentage occurrence of the element being sought ($\bar{p}$), and by determining the allowable error and confidence levels, n can be computed as follows:

$$n = \frac{\bar{p}(1-\bar{p})}{\sigma_{\bar{p}}^2}$$

$\sigma_{\bar{p}}$ = standard deviation or standard error of a percentage

$\bar{p}$ = an estimate of the true percentage occurrence of element being sought, expressed as a decimal n = total number of observations upon which $\bar{p}$ is based

The control chart (from statistical quality control theory) is a device which helps in judging whether or not a process is within defined control limits. By plotting the daily and the accumulative results of a sampling study, it can be checked to see if any data points fall outside the designed control limits. If this does occur, it may result from a chance variation or from some assignable cause. Usually, assignable causes are relatively large variations that are attributable to such reasons as differences among workers, pieces of equipment, and materials. When the process is out of control due to any cause, the problem must be corrected before taking additional samples. All data collected during the out-of-control period should be excluded from the study. Control limits for work sampling can be calculated by using the formula commonly employed in statistical quality control applications.

SUMMARY OF ARTICLES

In "Past-Present-Future of Work Sampling," Chester L. Brisley provides an excellent summary of the origin and development of the technique. He begins the history in 1927, when L.H.C. Tippett developed the "snap-reading" method, which involved taking snapshots at random intervals to determine the incidence and duration of loom stoppages in a weaving factory. He then follows the evolution of the term to Robert Lee Morrow's "ratio delay" in 1935 and to Brisley's own introduction of "work sampling" in 1952. Brisley covers Wallace J. Richardson's use of work sampling in cost reduction (1959), Merle D. Schmid's introduction of "work measurement sampling" to set work standards (1959), and Harold O. Davidson's argument for the superiority of fixed interval over random sampling (1960). After summarizing other major developments of the succeeding years, including the huge growth in the role of the computer, he notes that engineers today can create a work sampling program for almost any situation by combining the many techniques available to them. The article concludes with a provocative look at the future of the field.

The articles "Selection of Work Sampling Observation Times: Part I—Stratified Sampling" by Joe Moder and "Selection of Work Sampling Times: Part II—Restricted Random Sampling" by Joe Moder and Henry Kahn discuss and compare five methods of random sampling: 1) simple, 2) systematic, 3) restricted, 4) stratified continuous, and 5) stratified noncontinuous. The authors give particular weight to statistical efficiency and stress the advantages of stratification. They also recommend variance analysis to account for the presence of several variation sources, in addition to sampling error.

"Preparatory Computer Assistance for a Work Sampling Study" by David M. Rhyne and Douglas K. Freeman (edited by Gary E. Whitehouse) presents a BASIC program that will generate the essential randomness of a work sampling study. The program features a menu selection with the following output:

- the required number of random work sampling observations;
- observation records and forms for the work sampling study;
- the listing of a matrix of random observation times;
- analysis of the work sampling results.

(Several other *Industrial Engineering* articles have provided microcomputer programs to assist work sampling studies. Foremost are "Work Sampling Observation Generator" by Gary E. Whitehouse and Donald Washburn (March 1981); "Work Sample Size Analyzer," also by Whitehouse and Washburn (April 1981); and "Program Calculates Results of Work Measurement Sampling" by Geoffrey R. McNall (July 1984). Written in BASIC, these three programs were assembled and tested on a TRS-80; the first two were also tested on an Apple II. All will also run on other microcomputers programmable in BASIC, including the IBM PC and compatibles.)

The final article in this chapter, "Work/Activity Sampling—Contemporary Design Analysis Methodology and Applications: Part II—Work Sampling Calculations Revisited" by Elinor S. Pape, defines a procedure for computing work sampling interval estimates of proportions and sample sizes. The procedure incorporates realistic assumptions about the work situation and about the manner in which data is collected. It uses traditional work sampling calculations as a starting point and progresses to alternate formulas that allow for correlations between members of a work crew and for random variations from day to day (sampling period to sampling period). Pape develops mathematical formulas for three variance estimators and recommends guidelines for application.

PAST - PRESENT - FUTURE OF WORK SAMPLING

Chester L. Brisley, Ph.D.
Department of Engineering
University of Wisconsin-Extension
Milwaukee, Wisconsin

ABSTRACT

Snap readings, ratio delay, group timing technique, work measurement sampling, activity sampling are snyonomous terms applied to what we shall refer to as "work sampling." Almost fifty years have elapsed since work sampling was discovered and developed by L.H.C. Tippett, who named the technique "snap-reading" after the snap-shot concept of clicking a camera at random and photographing the scene desired.

The technique has developed some refinements and is being used in the armed services, government, industry, stores, educational institutions, hospitals, construction, and every segment of our society. It is used to study clerical activities, employee and machine utilization, cost reduction, indirect labor, management, engineers, hospital personnel, construction workers, allocation of overhead, and many variable conditions relating to machines and people. This paper suggests the importance of extracting more information from work sampling in the future.

L. H. C. TIPPETT 1927 TO 1952

Work sampling had a very excellent birth by an English statistician, L. H. C. Tippett in the May, 1953, issue of Time and Motion Study published in London, England.[1] In his own words he reviews his experience:

"Round about 1927 I was making surveys in weaving sheds to discover the causes and durations of loom stoppages with a view to estimating how much of the productive capacity was lost for various causes. At first I used the obvious method of timing looms with a stop watch. This caused no difficulty from the operators because I was timing the looms and not the weavers and no one thought of my activities as having any connection with time study as conventionally understood. But the work was tedious; and as it was practical to record only two, or three, or four looms at a time, I had to move about the shed and observe many looms in turn before a reasonably reliable average could be determined.

"One day a weaving manager remarked: 'I can tell at a glance whether the weaving in the shed is good. If most of the weavers are bent over their looms mending warp breaks, weaving is bad; if the weavers are mostly watching running looms, weaving is good.' In a moment "the penny dropped". It became clear that a snapshot of the state of the looms in a shed taken at any instant was in some way an indication of the rate of production in a short interval surrounding that instant and of the losses in output due to various causes.

"After a little thought I decided that the proportion (or percentage) of looms snapped as running was equal to the proportion (or percentage) of time the looms ran, on the average, during that short interval, and thus estimated directly the loom running efficiency. Likewise, the percentage of looms snapped as being stopped for any given cause estimated directly the percentage of the time looms were stopped for that cause, on the average, during a short time surrounding the instant of observation.

"Thus was started the ratio delay technique or, what I still think is a better term, the snap-reading method. An observer progresses round a mill and as he comes to each loom he takes a snap-reading of its state, whether working or stopped, and if stopped, the cause of the stop. In this way he collects several thousand snap-readings, classifies them, and estimates the various percentages." As we can see, Tippett employed basically the same approach that we use today.

WORK SAMPLING 1952 - 1959

What have we learned since Tippett's original contribution and what is the future of work sampling? This is the purpose of this paper.

Wallace J. Richardson/Robert Lee Morrow

In preparation for this assignment, I reviewed Professor Wallace J. Richardson's article from Factory, September, 1959, "Work Sampling Today."[2] Seven years earlier, in 1952,[3] I had the privilege of writing an article in which Factory introduced the term "work sampling," converting the name from the Ratio Delay technique which name had been coined by Professor Robert Lee Morrow in 1935. As noted by Tippett, the name had originally been the "snap reading method." The 1952 article in Factory with the dice demonstration and the new name attracted attention; and, as shown in Richardson's article, great progress was made using this technique during the period of seven years from 1952 to 1959.

Professor Morrow, in the second edition of his book Motion Economy and Work Measurement,

1957,[4] could not bring himself to change his choice of names completely; but he compromised and wrote Chapter 23, "The Ratio Delay Study," and Chapter 24, "Work Sampling". In the ratio delay chapter, he reviews the effort of Taylor, Merrick, Barth and Tippett in the establishment of delay and variation allowances to be added to time studies. He proceeds to use the term "Ratio Delay Study" to describe Tippett's snap reading method as a technique for establishing allowances. In his next chapter, he reviews the new term:

"Work sampling is the term now generally applied to what heretofore was called the ratio delay study. The reason for the change in terminology is that the ratio delay technique is no longer confined to delays. Work sampling is a more descriptive term because it covers the more general application and varied uses of this technique." He then proceeded to describe its various uses in conjunction with work simplification, cost reduction, and elevator traffic studies.

Richardson, in his 1952 to 1959 review of the progress of work sampling, cited "how it's growing." He said: "Personal knowledge and experience support these five facts:

1. About two-thirds of all large manufacturing firms use work sampling.
2. Many smaller companies use it.
3. Service industries, such as warehousing, are picking it up.
4. Growth of use is rapid.
5. There has been a big shift in the literature. Six years ago the emphasis was on describing the technique. Today most articles about it describe new uses or new refinements in application."

Cost Reduction

Professor Richardson has had great success in employing work sampling as one of the first steps in a cost reduction program.[5] He has followed the pattern of using the supervisor in indirect labor areas as an observer.

He suggests that we make work sampling one of the first steps in a cost reduction program; and, consequently, management will have a much better chance of success. He also recommends if management will pay strict attention to measuring work units of output, they are able to produce a situation in which the most sensible course open to the supervisor, from any point of view, is to give as valid a study as is possible. This is true because the first work sampling study will be used as a "bench mark" from which to direct effort toward improvement and also to develop some broad "productive time per unit of output" measures. Therefore, these two factors tend to balance one another in that any distortion of the results can lead to conclusions which will work some hardship on the supervisor. The most successful approach, in the experience of Richardson and others, is being patient and truthful and not cutting the throat of the supervisor with the results of the work sampling study.

Work sampling is currently being used in industry, hospitals, offices, warehouses, construction, stores, banks, and every segment of society. Many universities and professional societies have conducted educational sessions. More and more institutions and industries are using this broad-gaged tool of work measurement for cost reduction and manpower budgeting.

FIXED INTERVAL WORK SAMPLING

What have we learned since 1959? Harold O. Davidson did some in-depth analyses when he presented "Work Sampling - Eleven Fallacies" in the September-October, 1960, issue of the Journal of Industrial Engineering.[6] His conclusion, after evaluating eleven supposedly factual principles of work sampling relating to accuracy, is that, while the work sampling technique has a surface appearance of exceeding simplicity, we cannot expect to develop skill in anything without digging below its surface. He compared random sampling with systematic sampling and decided it is entirely possible that systematic or fixed interval sampling may yield a more random error than random sampling. He also deduces that systematic sampling with intervals smaller than the activity duration is fundamentally the same process as timing with a stop watch. Others have supported Davidson's views, [7], [8], [9], [10]; therefore, as the years have passed, less emphasis is being placed on random sampling. This change of emphasis has come about largely because the utilization of observers on a random basis is a source of inefficiency. Likewise, managers of industrial engineers have intuitively, even before research was done on the subject, established randomness by varying the pattern of paths through the area being studied. The trend toward Systematic Work Sampling was recognized by the Work Measurement and Methods Engineering division of the AIIE in 1969 when a collection of the Journal of Industrial Engineering articles was published under the title, "Fixed Interval Work Sampling." [11]

WORK MEASUREMENT SAMPLING

Another approach making use of work sampling was introduced by Merle D. Schmid in 1959 which he calls "Work Measurement Sampling."[12] Schmid distinguishes the two terms as follows:

Work Measurement Sampling differs from work sampling in that the primary objective is to estimate job standards rather than percent of time spent on various classes of activity. Collection of data by sample observations differs in work measurement sampling from work sampling mainly by a procedure that preserves the sequence of and identifies the time of day for each observation in the new technique. Professor Schmid presents his approach in a book he has written on the subject published by the University of Dayton.[13]

The statistical procedure on which Work Measurement Sampling is based requires that, each time a job is observed, four points in time be identified. In Figure 1, these four points are referred to as O_1, O_2, O_3, and O_4; i.e., the four observations that straddle the beginning and end of Job ABCD.

Consider the time axis in Figure 1 with a series of random observations (X points) on it. At each point in time represented by the "X points" on the time axis an observation has been made of a man working at a given time to determine what he is working on. From the viewpoint of Job ABCD, let us consider that he is either working on Job ABCD or he is not working on Job ABCD. We observe that:

1. At point O_1 he is not working on Job ABCD.
2. At point O_2 and each subsequent point to and including O_3 he is working on Job ABCD.
3. At O_4 he is not working on Job ABCD.

It is logical to conclude that Job ABCD is at least $O_3 - O_2$ long and it is not longer than $O_4 - O_1$.

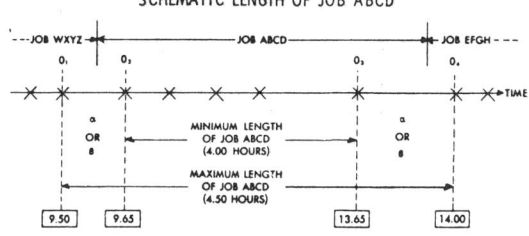

Figure 1

If from the example, time of day values (in continental decimal hours) is placed on the points O_1, O_2, O_3, and O_4, it is reasonable, using the arbitrary time values posted, that Job ABCD has a maximum length of 4.50 and a minimum length of 4.00 hours. Therefore, a good estimate of the length of Job ABCD would be the median of all possible lengths of Job ABCD that could have occurred under the restrictions of the known values O_1, O_2, O_3, and O_4. This of course would be:

$$\frac{(O_3 - O_2) + (O_4 - O_1)}{2} = \frac{4.00 + 4.50}{2} = 4.25 \text{ hours}$$

The Manufacturing Development Staff of General Motors Corporation was instrumental in the early development and testing of Work Measurement Sampling. Also, the Application Development Section of IBM Corporation gave Professor Schmid help in the development and refining of the computer programming of Work Measurement Sampling.

WORK SAMPLING AND THE COMPUTER

Since 1959, the computer has played a considerable role in the use of work sampling by the fact that the clerical drudgery of summarizing and analyzing data is reduced. One of the current outstanding programs using the computer is that of Chase Manhattan Bank developed by Arnold J. Taylor, Vice President, Production Planning and Control, Deposit Services Group.[14] The system uses the acronym MAST, Mechanized Activity Sampling Technique.

Taylor's technique is to automate all of the clerical and mathematical calculations associated with the recording of observations, computation of element percentages, performance ratings, and accuracies, as well as the preparation and maintenance of control charts.

A portable punch, the IBM 3000 Information Recorder I, is used by the observer as he makes his rounds of observations. A mylar overlay assists the observer in proper entry of the data. The specific benefits of MAST include:

1. Increased efficiency of Engineers and Analysts through elimination of tedious clerical routines.
2. Rapid analysis and professional presentation of Work Sampling data.
3. Reductions in the costs of Work Sampling studies.
4. Accurate computations of data.
5. Reduction of errors by engineers and analysts through diagnostic routines included in the system.

E. E. Gesler, Jr., of the DuPont Company[15] also had his samplers punch out the proper category of activity in a porta-punch card. These cards are run through a reproducer which allows the cards to run smoothly through a computer. In auditing a maintenance incentive program in which 1000 maintenance people are involved, the computer report shows the distribution of time in minutes for the average mechanic in each foreman group and each area. The purpose of these studies was to assess results of corrective action.

Gesler cites that the list of factors affecting productivity in maintenance is long with such items as:

1. Availability of material and tools
2. Use of tool belts and tool carts
3. Degree of cooperation-operations/engineering
4. Disciplinary practices
5. Employment practices
6. Transporation facilities
7. Communication facilities
8. Scheduling effectiveness
9. Weather shutdown
10. Vacation periods
11. Clarity of work orders/blue prints, etc.

AUDITING STANDARDS

Work Sampling is being used increasingly by many consultants and by industry to audit work measurement studies.

Frank DiGiovianni -- Clifford N. Sellie

Frank DiGiovianni of Standards International, Inc., suggests the following approach in auditing standards:[16]

Prepare a check list for and conduct a Work Sampling study to cover the whole spectrum of activity of the key jobs to determine the areas of greatest problems. The list should be prepared so that the study will result in reliable judgments as to the percentage of time the workers utilize for:

1. Manual work
2. Machine work
3. Production delays
4. Waiting for material
5. Waiting for instruction
6. Time to change shifts and/or punch out
7. Personal time
8. "Goof-off" time
9. Operator not in department
10. Miscellaneous delays.

"It may be germane to point out," he says, "that excessive allowance or downtime amounting to 10 minutes per day are equivalent to an additional week each year." Clifford N. Sellie, President of Standards International, is also an advocate, as is Wallace Richardson, of getting the Foreman on your side by having him make a work sampling along with you.

Harold F. Allard

Albert Ramond and Associates, Management and Industrial Consultants, specializing in work measurement and wage incentives, follows a practice of auditing by work sampling. Harold F. Allard, Senior Vice President of Ramond, has been a strong supporter of Work Sampling for years.

He looks upon work sampling as a valuable maintenance management aid,[17] as well as an auditing tool. He follows the practice of using Fixed Interval Work Sampling with groups of employees that do not have to move about. Downtime and other delays, as well as working time, are established in frequent fixed interval observations of five or six minutes. Random Work Sampling is used to audit those who are quite mobile.

In maintenance, he separates activities for skilled crafts into: Direct Work, Indirect Work, Travel, Waiting, and Nonproductive time. He has found that travel, waiting, and nonproductive time require the greatest attention and warrant a continuous audit. He suggests making one tour of observations a day and adding the results once a month. In this manner current productivity can be compared with past productivity.

Thomas G. Kay

In the June, 1972, issue of Industrial Engineering, Thomas G. Kay[18] indicated that, since 1964, the Link-Belt Chain Division of FMC Corporation has operated under a standard hour incentive system. Standards are audited by the Industrial Engineering Department to determine their accuracy.

Work sampling has been adopted as the general method of auditing standards as a means of covering large areas and conserving time. Workers are observed, and their activity, as well as pace, are recorded and summarized, and the end results compared with workers' earnings during the study time to see if earnings and work effort are compatible. Cycle times can be derived, delay allowance and personal time summarized, and then compared with the standards. After summaries are reviewed, determination can be made on conditions that are out of line with the original standards.

Elements sampled are usually "working," "idle," and "absent from work area." However, elements may differ from study to study, according to the types of jobs and the way in which standards are set.

The preliminaries to the study are completed by reviewing the standards involved, identifying the workers involved, selecting a fixed route to follow through the sampling area, and selecting the entry points on the route.

The advantages as a result of using this "selected fixed route with random entry points" were:

1. A large number of observations were collected in a shorter time span, while maintaining randomness between readings.
2. Workers appeared more relaxed and less apprehensive in the absence of a stopwatch, which is usually associated with a time study.
3. Work Sampling was easier for the observer since he did not have to keep track of time.
4. Workers were not able to tell when and from where the observer was coming; and, therefore, they were less likely to bias the study.
5. The observer can interrupt the study at any time without bringing about bias in the study. This is especially valuable when an observer wishes to discuss a work element with an individual, or check jobs with the time keeper.
6. Having a fixed path assures observation of all work areas on every cycle, giving more consistency to results.

Charles Travis

Another work sampling audit that brought about a correction in time standards was presented by Charles Travis[19] of the Hibriten Chair Company in Industrial Engineering, September, 1970.

In the upholstering department, the workers were paid on a piece rate basis for upholstering various types of chairs. The supervisors were receiving complaints from the upholsterers that they couldn't "make the rate." A rated work-sampling study was made to see if the cause could be determined quickly. The study showed a high average pace rating for the group with a very low percent idle. This analysis showed they were working at an above average pace rating. Also, they were working effectively during the day, as indicated by the low percent idle; and they were not abusing the allowed rest and personal time allowance. Therefore, it was concluded that, with this performance, most of the upholsterers should have been earning about standard in weekly pay. Therefore, as a result of the rated work-sampling analysis, the earnings records were checked; and it was found that the group was earning below standard. The cause of the complaints was then justified as some tight standards existed. This was also confirmed when the time standards were audited and resulted in them being corrected.

THE FUTURE OF WORK SAMPLING

In approaching the future of Work Sampling, I was impressed with the words of Lester R. Bittel, who presented "The Towne Lecture" at the 1973 ASME Winter Annual Meeting, published in Mechanical Engineering.[20] He advocates that "true motivation exists in the very nature of the work itself." The elements he identifies that operate as opportunities and freedoms rather than as restrictions and repressions are:

1. Free access to all the information that is needed to do the work well.
2. An opportunity to set one's own production goals.
3. An opportunity to achieve something that is worthwhile to others and meaningful to one's self.
4. Recognition by others of one's achievements.
5. An opportunity to interact with others, in particular, with one's supervisor without fear of recrimination.
6. An opportunity to exert self-control and to exercise self-discipline.

Work Sampling is an excellent technique whereby the behavior of the individuals as well as the group may be studied and analyzed by the very people involved in the study. Through work sampling, problems similar to those cited earlier by Gesler can be ascertained; such as:

1. Disagreements among crafts with respect to who should do the work.
2. Some personnel carrying paging units while others who could use such equipment do not have these units.
3. Problems because of equipment exposed to cold weather.
4. Lack of suitable shop facilities.
5. Noise level and communications problems as a result.
6. Failure of Receiving Inspection allowing defective fixtures to pass and the crew being held up because of this condition.

and many more detailed observations as the observers make their rounds.

The employee becomes his own general manager of his particular work place. This requires a rapid feedback of information by management to the employee's question, "How am I doing?" The employee must know very soon after he has completed work, information concerning the quality, quantity, costs, and other pertinent data as to the "score of the game" he is playing. He must be assured that his work center will be supplied with its necessary tools and materials. Instructions, blue prints, and specifications must be presented with great clarity. Therefore, work sampling will be used more and more to assist various work groups to set higher and more realistic goals of accomplishment. It is a tool that communicates needed information of the current situation.

Undoubtedly, in looking at the future, work sampling will be employed in many ways in conjunction with multiple linear regression, simulation, and other statistical and mathematical techniques. We employ whatever technique will help us solve the problem.

New mechanical devices are continually being developed to help us in observing behavioral patterns. The tape recorder, in conjunction with a computer as suggested in the October, 1969, issue of Industrial Engineering,[21] may be a tool that can assist the Industrial Engineer. Or perhaps the Video Tape Recorder may be employed in a similar manner, adding visual information as well as other data.

Our objective must be to strive to extract more information from a Work Sampling Study. A colleague of mine, Professor Russell W. Fenske, Assistant Vice Chancellor of the University of Wisconsin-Milwaukee, has been interested in this subject ever since an article was published on this subject by him.[22] He emphasizes that if conditions are fixed and the proportions of time spent on different activities have stabilized, they will be representative of the actual operations. However, if conditions have been changing, if supervision has changed, or if employees are being motivated to a greater or lesser extent, the final proportions may not be representative. This fact may not be detected in a conventional work sampling analysis, and incorrect conclusions may be drawn concerning the efficiency of operations or unrealistic standards may be established.

He points out the fact that, if conditions were changing during the study, for instance, a gradual reduction in setup time, the data would appear as in Figure 2. It should be noted that since production plus setup plus personal time equals 100 percent, a decrease in setup must result in an increase in one or more of the other factors; in this specific case, the increase is in production. This will be true of changes in the proportions for any work sampling study. Therefore, it is necessary to brainstorm with the supervision or the individuals being work sampled, why the data is not leveling out but, instead, is continuously changing. These conditions may be reflected in work sampling studies that are quite continuous over time because of the learning curve theory, or for some other reason. In the future, more effort should be devoted to isolating various deviations from normal with respect to our data, rather than including them in a broad average whereby the deviate behavior of our productivity is not evaluated. Therefore, we predict that Markov Chains, multiple regression, and other management science concepts, along with work sampling, may be employed to a much greater extent in squeezing more information from the data.

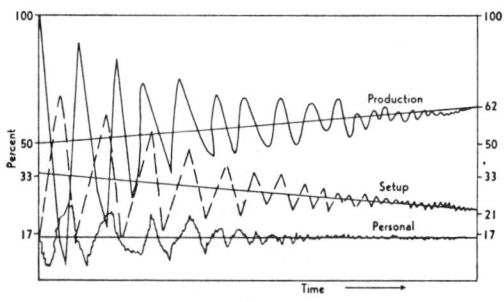

Figure 2

CONCLUSION

Changes are occuring continuously, and work sampling can be employed to determine these changes. Work must be measured with respect to input and output. All factors relating to work must be assessed as to machine utilization, energy usage, people utilization (not only in man-hours but in terms of dedication), transportation to and from the work place, communications to and from the employee, etc. Work Sampling is one tool that can be of great assistance in these behavioral science analyses.

REFERENCES

[1] Tippett, L. H. C., "The Ratio Delay Technique," Time and Motion Study, May, 1953, pp. 10-19, published by Sawell Publications Ltd., 4 Ludgate Circus, London, E.C. 4.

[2] Richardson, Wallace J., "Work Sampling Today," Factory, September, 1959.

[3] Brisley, C. L., "How You Can Put Work Sampling to Work," Factory Management and Maintenance, July, 1952, pp. 83-89.

[4] Morrow, Robert Lee, Motion Economy and Work Measurement, Second Edition, Chpts. 23 and 24.

[5] Richardson, Wallace J., "Work Sampling and Indirect Cost Reduction," Industrial Management Society, Industrial Engineering and Management Clinic, November 3-5, 1971, pp. 21-24.

[6] Davidson, Harold O., "Work Sampling--Eleven Falacies," The Journal of Industrial Engineering, September-October, 1960.

[7] Haines, I. Landis, "Work Sampling by Fixed Interval Study," The Journal of Industrial Engineering, July-August, 1958, Vol.IX, pp. 266-268.

[8] Davidson, H. O.; Hines, W. W.; and Newberry, T.C., "The Error of Estimate in Systematic Activity Sampling," The Journal of Industrial Engineering, July-August, 1960, Vol. XI, N.4, pp. 290-292.

[9] Flowerdew, A.D.J. and Malin, P.W., "Systematic Work Sampling," The Journal of Industrial Engineering, July-August, 1963, Vol. XIV, N. 4, pp 201-207.

[10] Jones, Ned Gene, and Ghare, P.M., "Confidence Intervals for Systematic Activity Sampling," The Journal of Industrial Engineering, May-June, 1964, Vol XV, N.3, pp 141-147.

[11] The above four articles are reprinted in the WMME (Work Measurement and Methods Engineering) Publication Number 1, "Fixed Interval Work Sampling," American Institute of Industrial Engineers.

[12] Schmid, Merle D., "Work Measurement Sampling" Proceedings, Industrial Management Society, Industrial Engineering and Management Clinic--1959.

[13] Schmid, Merle D., Work Measurement Sampling, University of Dayton, Dayton, Ohio, 1965, pp. 9-13.

[14] Taylor, Arnold J., MAST, Mechanized Activity Sampling Technique, System Users Manual, Operations Department, Deposit Services Group, Production Planning and Control, The Chase Manhattan Bank, New York, New York, November 12, 1973.

[15] Gesler, E.E., Jr., "Work Sampling and Cost Control," Chemical Engineering Progress, June, 1967, pp. 29-35.

[16] DiGiovanni, Frank, "Auditing Standards to Boost Morale While Lowering Costs," Industrial Management Society, Industrial Engineering and Management Clinic, November 3-5, 1971, pp. 33-37.

[17] Allard, Harold F., "Work Sampling: Valuable Maintenance Management Aid," Plant Engineer, September 19, 1968.

[18] Kay, Thomas G., "Timeless Work Sampling," Industrial Engineering, June, 1972, pp. 30-33.

[19] Travis, Charles, "Trouble-shoot with Rated Work Sampling," Industrial Engineering, September, 1970, pp. 19-22.

[20] Bittel, Lester R., "From Work Measurement to Work Measurement, From Wage Incentives to Work Itself," Mechanical Engineering, April, 1974, pp. 29-32.

[21] Samanta, D., "How a Tape Recorder Can Improve Work Measurement," Industrial Engineering, October, 1969, pp. 24-27.

[22] Fenske, Russell W., "Extracting More Information from a Work Sampling Study," Journal of Industrial Engineering, July, 1967, pp. xvii-xix.

Selection of Work Sampling Observation Times: Part I—Stratified Sampling

JOSEPH J. MODER
SENIOR MEMBER, AIIE
Departments of Management Science and Industrial Engineering
University of Miami
Coral Gables, Florida 33124

Abstract: Various methods of selecting observation times in work sampling studies are presented, including simple random, systematic, and stratified sampling, and a new method called restricted random sampling. The attributes of these sampling methods are evaluated, particularly statistical efficiency, and the important advantages of stratification are demonstrated. Finally, the analysis of variance is recommended for statistical analysis of work sampling data to account for the presence of several sources of variation, in addition to sampling error.

■ The sampling of work was first practiced by Frederick W. Taylor [7] in the 1890's by making continuous time studies of what workers did over a period of a day or more. It was some 30 years later that work sampling of a different form was originated by L.H.C. Tippett [10]. He collected data by taking snap readings on the state of activity of machines in a textile mill, using the binomial distribution as a model to *approximate* the accuracy of his statistical estimates. The word *approximate* is important here because Tippett, who was well aware of what he was doing, did not use simple random sampling (SRS), even though it it an underlying assumption in the use of the binomial distribution as a model for the sampling error in this application. This point has escaped many work sampling practitioners who even today insist upon the use of SRS. The main purpose of this paper is to develop the theme that SRS should almost never be used in work sampling because it is awkward to carry out, and more importantly, it can be quite inefficient from a statistical point of view.

In a stop-watch production study, one or two workers are observed continuously, usually for one full work day. On the other hand, in a work sampling study *many* workers are observed less intensely over a much longer period, typically a calendar month. For this reason work sampling observations are taken in "rounds" that may include many workers and require an appreciable amount of time to carry out, as much as an hour or more. Thus, SRS presents a serious data collection problem, since it requires each minute in the work day to have an equal chance of getting into the sample. *What do you do when two or more sample times are closer than the time required to make a round of observations?* W. J. Richardson [8] in a recent (1976) text on the practical aspects of work sampling prescribes the use of SRS only, and suggests the above problem be handled by either enlisting the aid of a second sampler, or by having the first sampler double back after completing his first round. It might be asked, what do you do when three observers are needed simultaneously? Although Richardson's procedures seem to remedy the situation, they leave much to be desired. It is shown below that time stratification and other restrictions on SRS are more expedient solutions to this problem. In the next section, a review and analysis is given for a variety of work sampling methods that are described in the current literature, and after this, a series of recommended procedures will be developed.

Alternative Work Sampling Methods

M. E. Mundel [6], in his recent (1978) work measurement textbook presents three different options to the problem of selecting work sampling observation times.

Received October 1978; revised October 1979. Paper was handled by Work Measurement/Methods/Ergonomics/Health Services Department.

Method 1 – "The first hour of the working day may be identified by the numeral 1, the second hour by the numeral 2, and so on. A table of random numbers is used to obtain a series of three-digit figures, the first digit representing the hour of the working day, and the next two digits, the minutes. Numbers representing hours not in the working day, or impossible minute values, are discarded. A sufficient number is obtained to give the required observation times for each day of the study."

Method 2 – "A second method is to stratify the observations by hours. To do this, we divide the number of observations to be made per day by the number of working hours. Following this, we again use the random number table to give us two-digit figures to represent the minutes for each hour.... If more than one observation per hour was required, the values for each hour need rearrangement for actual use. Each day should have a different list of time values."

Method 3 – "If a great number of observations is required, an observer may go continuously through the area where the observations are to be made, randomizing his route but constantly making observations."

Based on previous work by the author (1969), as well as private communication (1978) with Arnold J. Taylor of Bausch & Lomb, who has developed a very elaborate computerized work sampling procedure, a fourth method will be added to the above list.

Method 4 – Use SRS with the replacement of sample times that are too "close together," i.e. a sample time is rejected and replaced if its addition to the already selected sample times would not allow sufficient time for the completion of a round of observations.

To complete this review, it is noted that a combination of Methods 2 and 4 is recommended by George L. Smith [9] in his recent (1978) work measurement text.

Analysis of Sampling Methods

The usual convention of work sampling will be followed in the analysis of these four sampling methods. Time is recognized as a continuous variable, but for sampling purposes it will only be specified to the level of minutes. Randomization within the selected minute is left to the variation in the walking time of the observer.

Method (1) above is SRS and is not recommended for work sampling studies because of the data collection problems discussed above. It will be used in this paper as a benchmark for the evaluation of the statistical efficiency of other preferred methods.

Methods (2) and (3) above are both variations of the use of Stratified Random Sampling (StRS), although this fact has evidently escaped theoretical attention. First, consider Method (3) wherein the observer collects data continuously, and suppose that each round of observations requires 30 minutes of a 8-hour workday to complete. Thus, there would be 16 separate time strata to be sampled in each day of the study. At the completion of each round of observations a new 30 minute time stratum is begun by randomizing the route. That is, the starting point of each round is determined by randomly selecting a worker, or work station, along the observation route. This, in effect, amounts to the use of SRS in the selection of the observation time for *each* worker in *each* 30 minute time stratum, since each minute in each stratum is equally likely as the observation time for each worker. This combination of events renders Method (3) an application of proportional StRS.

Method (2) is also a form of StRS, however, two modifications of this method are proposed; one to correct a defect which it has, and one to improve its statistical efficiency. To explain the defect, suppose each time stratum has a duration of only 6 minutes, and a round of observations requires 2 minutes to complete. Thus, the possible samples in each stratum might be considered to be minutes 1-2, 2-3, 3-4, 4-5, 5-6. However, you will note that minutes 2, 3, 4 and 5 each appear in two samples, while 1 and 6 appear in only one sample. A modification to eliminate this sampling bias consists of the addition of 6-1 to the list of possible samples. This means that the selection of time 6 would require breaking the round time into two portions, collecting the first half in 6 and the second half in 1. In effect, this means that the "2nd half" of the round is collected before the "1st half" of the round. This will be referred to as the *Loop Principle* where the time units in a stratum are visualized as forming a continuous loop as shown in Fig. 1. A pragmatic engineer may (creatively) consider this nit-picking and decide to finish a round started in minute 6, during minute 1 of the *next* stratum, rather than the same stratum. This procedure (described by Smith

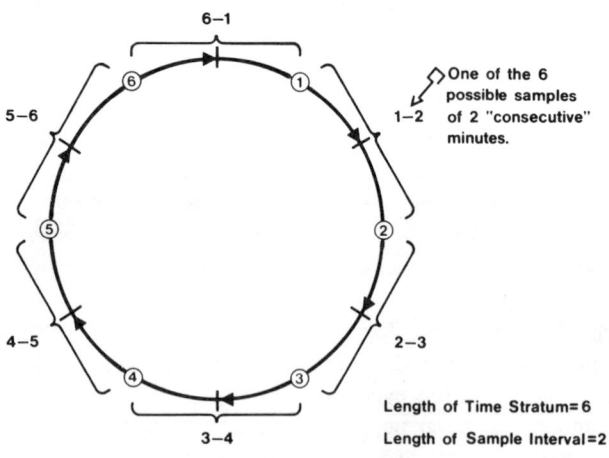

Fig. 1. Illustration of the Loop Principle in randomly sampling an interval of time in a time stratum.

[8]) lowers the probability of the early or late parts of a "*day*" getting into the sample. This may not be trivial since these are often atypical periods from a worker activity point of view. A study known to the author produced an estimate of percent machine down time which was rejected by the manager as being unrealistically low. After considerable discussion it was determined that the technician collecting the observations started work one-half hour after the

factory work shift began, and stopped sampling a similar period before the end of the shift. The correction of this biased sampling procedure eventually produced a useful work sampling estimate.

The above ad hoc procedure of letting a round of observations spill over into the next time stratum not only violates the definition of StRS, but it also opens the door to further complications if, for example, minute 1 is selected as the observation time for the next stratum. This might also be "fixed up," but a better solution to this problem will be presented when sampling Method (4) is discussed below.

The second (optional) modification to Method (2) to improve its statistical efficiency is given in the following redefinition; it carries the stratification process to the limit by reducing the size of the strata to the point where only one random observation time is selected in each stratum.

Method 2* — Stratify the observations by *time*, that is, divide the number of minutes in the work day by the number of rounds of observations to be collected per day to obtain the duration of a time stratum. Then, select one observation time in each stratum using SRS, with the proviso that the Loop Principle will be followed, i.e. an unfinished round started late in the time stratum will be completed at the beginning of that *same* time stratum.

Method (4) is one which has been studied by the author for some time [see reference 4]. It will be called Restricted Random Sampling (RRS) because it is essentially SRS with the Restriction that no samples will be closer than some minimum spacing required to complete a round of observations. This form of sampling is considered in detail in Part II of this series on work sampling.

Statistical Advantages of StRS

Methods (2*) and (3) are both forms of StRS; they are recommended for work sampling over SRS not only because they are much simpler to draw and to administer, but also because they are more efficient, statistically. To show this, consider StRS based on Methods (2*) or (3), where the estimators will pertain to an individual worker in the round of observations.

Let:

N = number of strata and also the number of observations,

π_i = parameter for stratum i denoting the fraction of time in which the worker is engaged in the activity category of interest,

$\pi = \sum_{i=1}^{N} \pi_i/N$ = parameter for the entire study period (N strata),

x_i = the value of the random variable corresponding to the observation in stratum i denoting whether the worker was in the activity state of interest ($x_i = 1$), or not in this state ($x_i = 0$), for $i = 1, 2, \ldots, N$, and

$V(X)$ = variance of X_i, assumed to be independent random variables.

Then,
$$\hat{\pi}_{StRS} = \frac{1}{N} \sum_{i=1}^{N} x_i \quad (1)$$

$$V(\hat{\pi}_{StRS}) = \frac{1}{N^2} \sum_{i=1}^{N} V(x_i) = \frac{1}{N^2} \sum_{i=1}^{N} \pi_i(1-\pi_i) \quad (2)$$

$$V(\hat{\pi}_{SRS}) = \frac{1}{N^2} \sum_{i=1}^{N} V(x_i) = \frac{1}{N^2} \sum_{i=1}^{N} \pi(1-\pi) = \frac{\pi(1-\pi)}{N} \quad (3)$$

Letting $\pi_i = \pi + s_i$, so that $s_i = \pi_i - \pi$, where $E(s_i) = 0$ and $E(s_i^2) = V(s)$, then

$$V(\hat{\pi}_{StRS}) = \frac{1}{N^2} \sum_{i=1}^{N} (\pi+s_i)(1-\pi-s_i) = \frac{\pi(1-\pi)}{N} - \frac{V(s)}{N}. \quad (4)$$

This last expression indicates that the effectiveness of stratification in reducing the variance of the estimate of π depends on N, and on $V(s)$ which measures the variance among the strata π_i's. It is instructive to examine Table 1 which shows this effect for 5 hypothetical cases where $V(\hat{\pi}_{StRS})$ varies from its maximum value of $V(\hat{\pi}_{SRS})$, to a minimum value of zero. It will be shown in the next section that actual worker activity distributions corresponding to cases 2, 3, and 4 in Table 1 are not uncommon.

	Table 1: Effect of $V(s)$ on $V(\hat{\pi}_{StRS})$, for five hypothetical cases where $\pi=.5$ and $N=4$.									
	Case 1		Case 2		Case 3		Case 4		Case 5	
i	π_i	s_i	π_i	s_i	π_i	s_i	π_i	s_i	π_i	s_i
1	.5	0	.25	-.25	.2	-.3	0	-.5	0	-.5
2	.5	0	.50	0	.4	-.1	.25	-.25	0	-.5
3	.5	0	.50	0	.6	+.1	.75	.25	1	.5
4	.5	0	.75	.25	.8	+.3	1.00	.5	1	.5
Average	.5	0	.5	0	.5	0	.5	0	.5	0
$V(s)$	0		.03125		.0500		.15625		.25	
$V(\hat{\pi}_{SRS})$	.0625		.0625		.0625		.0625		.0625	
$V(\hat{\pi}_{StRS})$	.0625		.0547		.0500		.0234		zero	
% Reduction	0%		12.5%		20.0%		62.5%		100%	
$V(\hat{\pi}_{StRS})/V(\hat{\pi}_{SRS})$	1.00		0.88		0.80		0.37		0.00	

Study of Actual Worker Activity

The effects of StRS predicted by Eq. (4), and illustrated in Table 1 for hypothetical activity distributions, were calcu-

Table 2: Effects of strata length on the relative efficiency ratios for 7 workers in a steel products factory.

Type of Worker	$V(\hat{\pi}_{StRS})/V(\hat{\pi}_{SRS})$							$V(\hat{\pi}_{SyRS})/V(\hat{\pi}_{SRS})$			$V(\hat{\pi}_{StRS})/V(\hat{\pi}_{SyRS})$			Average
	Strata Length in Minutes													
	7.5	15	30	60	120	240	480	7.5	15	30	7.5	15	30	
Floater 1	.57	.74	.83	.87	.89	.92	1.00	.27	.44	.53	2.11	1.68	1.57	1.73*
Floater 2	.48	.61	.71	.78	.95	.99	1.00	.59	.81	.36	0.75	0.75	1.97	1.02*
Lay-out Man	.44	.63	.76	.86	.91	.91	1.00	.19	.34	.71	2.32	1.85	1.07	1.48*
Crane Op. 1	.48	.68	.75	.88	.93	.97	1.00	.66	.64	1.00	0.73	1.06	0.75	0.83*
Crane Op. 2	.37	.55	.62	.91	.95	.98	1.00	.28	.45	.36	1.32	1.22	1.72	1.41*
Crane Op. 2	.36	.48	.65	.82	.84	.88	1.00	.19	.81	.62	1.89	0.59	1.05	0.92*
Timekeeper	.19	.39	.71	.93	.95	.98	1.00	.06	.20	.25	3.17	1.95	2.84	2.53*
Timekeeper	.24	.42	.73	.84	.97	.98	1.00	.04	.41	.36	6.00	1.02	2.03	1.72*
Mill Weigher	.46	.59	.66	.90	.96	1.00	1.00	.16	.61	.62	2.88	0.97	1.06	1.23*
Mill Weigher	.54	.62	.79	.82	.90	.98	1.00	.49	.65	.75	1.10	0.95	1.05	1.03*
Ave.	.41	.57	.72	.86	.92	.96	1.00	.29	.54	.56	1.41*	1.07*	1.30*	1.22*

*These entries are the ratio of averages, not the average of the ratios

lated from actual all-day time study data collected on production workers in a steel mill, and for a maintenance pipefitter. These tasks were chosen because they are non-repetitive and are typical of those covered in work sampling studies.

Hines and Moder [1, 2] studied 7 workers employed in the manufacture and fabrication of steel products. These workers, listed in Table 2, were studied on the basis of all-day (480 minute) time study data, where the activity category of interest was the *idle* worker state. *Floaters* engage in an irregular type of work, assisting *lay-out men* who mark steel components for welding. The two *crane operators* used an overhead crane to move heavy steel components and assemblies. The mill and warehouse *timekeeper* performed basic timekeeping functions, while the *mill weigher* weighed products and prepared identification tags and production reports. The last three workers were each studied for two days, with the results for each day shown separately.

The purpose of the original study of these data was to evaluate the effectiveness of systematic sampling (SyRS), which can be defined as the selection of a random minute in the first stratum of the "day," with subsequent observations taken systematically at intervals equal to the length of the strata. The data was reanalyzed here to evaluate the use of StRS, with strata lengths from 7.5 to 480 minutes, the last value being equivalent to SRS. The results of this study are presented in Table 2. The relative efficiency ratios $V(\hat{\pi}_{StRS})/V(\hat{\pi}_{SRS})$ were computed by applying Eqs. (2) and (3) to the all-day time study data given in the original study [2]. Since we are dealing with variances, a ratio of say 0.5 means that a StRS of Size N would be equivalent to a SRS of size 2N.

The second study involved a maintenance pipefitter for whom 7 days of all-day time study data was available. The relative efficiency ratios were again computed for strata lengths of 30 to 480 minutes, and for 4 different activity categories. The results of this study, presented in Table 3, are similar to those in Table 2.

Table 3: Effects of strata length on the relative efficiency ratio of $[V(\hat{\pi}_{StRS})/V(\hat{\pi}_{SRS})]$ for a maintenance pipefitter.

Activity category	Strata length in minutes				
	30	60	120	240	480
Production	.40	.75	.94	.97	1.00
Travel	.52	.80	.91	.98	1.00
Delays	.48	.72	.87	.96	1.00
Personal Time	.20	.60	.80	.92	1.00
Average	.40	.72	.88	.96	1.00

The average results of these two studies are plotted in Fig. 2 to point up the effect of strata length on the relative efficiency ratios. The shapes of the two curves are quite similar, with the knee of the curve occurring at strata lengths of approximately 50 and 90 minutes for the steel-mill workers and the pipefitter, respectively. Thus, strata lengths below one hour appear to be very effective in reducing the error variance, and hence the necessary sample size, for the types of workers included in these studies. Figure 2 indicates that

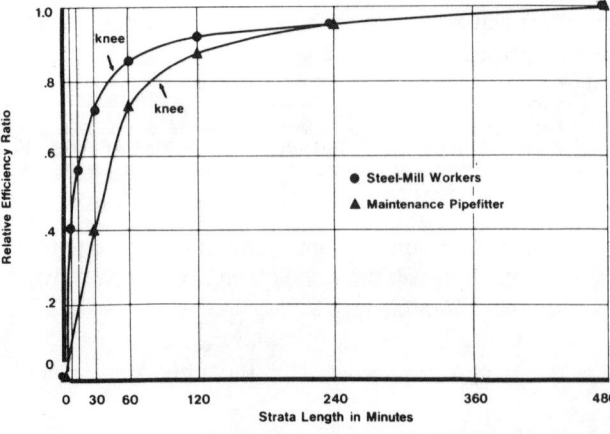

Fig. 2. Relative Efficiency ratio $V(\hat{\pi}_{StRS})/V(\hat{\pi}_{SRS})$ vs. Strata length for two classes of workers.

a reduction of as much as 50% in sample size can be achieved by using StRS in place of SRS. Unfortunately, one cannot carry this reduction in strata length too far because of its relationship with total sample size (number of rounds of observations) and study duration. This relationship is given in Eq. (5) for the case where one observation is taken in each stratum. All three variables must be considered jointly in designing a work sampling study.

(Study Duration) = (Total Sample Size) × (Strata Length) (5)

Comparison of Alternative Sampling Procedures

This paper has so far concentrated on SRS and StRS, with brief references to RRS and SyRS. In addition to these 4 basic procedures, a variation of StRS will be considered, giving a total of 5 sampling procedures. Each procedure will be defined below, and comparisons will then be made to serve as the basis for a recommendation on sampling procedures to be adopted in various work sampling situations. The following nomenclature will be used in these definitions. It is assumed that numerical values for these variables are chosen so that they all take on integer values.

T = minutes in the work "day" being sampled (the term day may actually be a half-shift if sampling is interrupted for the lunch period),

N_D = number of work "days" in the study,

N_T = total number of rounds of observations in the study,

$N = N_T/N_D$ = number of rounds of observations per "day,"

T/N = average number of minutes between rounds of observations,

M = minutes required to complete a round of observations, $M \leq T/N$ with equality holding for continuous sampling, Method (4) below, and inequality holding when sampling observations are part time.

1. SRS: Simple Random Sampling is defined as the selection of N minutes from the population of T minutes without replacement, such that each combination of N of the T minutes has the same chance of being selected. This is equivalent to requiring that each minute have an equal chance of being selected for the sample. If two or more sample times are closer than M minutes (and this will happen more often than you would normally expect) then it is assumed that additional observers will be recruited to assist in the data collection process. (The Loop Principle is normally not considered a part of SRS, but technically it should be).

2. SyRS: Systematic (Random) Sampling is based on a division of the work "day" into N *equal* time intervals; SyRS is defined as the drawing of a minute at random from the first sampling interval $(1, 2, \ldots, T/N)$ for the start of the first round of observations in the "day," with subsequent rounds of observations starting systematically at intervals of T/N minutes. If the first round starts at time $R(1 \leq R \leq T/N)$, then the remaining rounds during the "day" will start at $(R+T/N)$, $(R+2T/N)$, ..., $[R+(n-1)T/N]$. Also, if the start of the *final* round of observations in the "day" does not allow sufficient time to complete the round, then according to the Loop Principle (Fig. 1), an appropriate portion of the round must be conducted at the start of the "day."

3. RRS: Restricted Random Sampling is defined as the use of SRS in the selection of the sample of size N from the population of T minutes with the restriction that sample times will be rejected and replaced if they are less than M minutes from sample times already drawn. Again, the Loop Principle as described in SyRS applies to the final sample in the "day," but the randomization of the routes, described for StCRS below, is not required here and it is assumed that it will not be used.

4. StCRS: Stratified Continuous Random Sampling is defined as the continuous collection of data, where $M = T/N$, with the proviso that each round of observations will be based on a randomly selected route which can be accomplished in several ways depending on the circumstances dictated by the system being studied. A random route may be obtained by visualizing the worker (or work station) locations as forming a loop, numbering them from 1 to K, and then selecting a random integer from $(1, 2, \ldots, K)$ to determine the starting point of the round of observations. For example, if $K = 8$ and if 6 is drawn for the random starting point, then the route would be 6,7,8,1,2,3,4,5. If it is deemed necessary, the direction of movement of the observer along the route can also be randomized so that a coin toss of heads might call for the above route, while tails would call for the reverse route 6,5,4,3,2,1,8,7. Finally, if additional variation is deemed necessary, several different route structures can be defined and randomly sampled. For example, the above routes may form a loop through the factory. Another route such as 1,2,3,6,5,4,7,8, might follow a figure eight pattern through the factory. This latter refinement is utilized when it is desirable to further randomize the direction of approach of the observer at each work station. For this sampling procedure, the time to complete a round of observations (M) must include the average time required to deadhead to the random starting point of each route.

5. StNCRS: Stratified Non-Continuous Random Sampling is defined for the case where observation is non-continuous, i.e., $M < T/N$; it is based on the selection of a random starting time for the *single* round of observations taken in each time strata of T/N minutes, with the proviso that the Loop Principle as described in SyRS applies in *each* time stratum. If it is deemed necessary, the observation route can be randomized as described for StCRS.

Table 4: Rankings of the attributes of each of the six proposed sampling procedures.

Sampling procedure	Ease of drawing the sample	Ease of collecting the data	Variation among the observation times	Statistical efficiency of procedure	Total Score†	Final Rank
SRS Method 1	Good (2)	Poor (5)	Excellent (1)	Poor (5)	40	5
SyRS	Excellent (1)	Good (2)	Poor (5)	Excellent (1)	26	1
RRS Method 4	Fair (3)	Good (2)	Fair to Good (4)	Good (3)	33	3
StCRS Method 3	Poor to fair (4)	Good (2)	Good (2.5)	Good (3)	29.5	2
StNCRS Method 2*	Poor (5)	Fair (4)	Good (2.5)	Good (3)	36.5	4
Attribute Importance Weight (W_j)	1	3	3	4	--	--

†Total Score = $\sum_{j=1}^{4} W_j r_j$, e.g. for SyRS, TS = (1 × 1 + 3 × 2 + 3 × 5 + 4 × 1) = 26.

The attributes of each of these 5 sampling procedures are compared in Table 4, which also includes a reference to the corresponding sampling Methods 1, 2*, 3 and 4, described above in the section on Alternative Work Sampling Methods. The entries in Table 4 include an adjective descriptor along with an ordinal rank, where 1 is best and 5 is worst. A weighted sum of these ranks is then referred to as the "Total Score" for each method. The ranks and weights given in Table 4 are based on the judgement of the author, considering the typical conditions under which work sampling studies are made. These numerical assignments are not considered critical to the final conclusions reached in this study; they are presented here to illustrate a methodology which might be used to evaluate these sampling procedures under specific study conditions.

Considering the first attribute of Table 4, the ease of drawing the sample varies from the simplest for SyRS, which requires only one randomization for each "day" of the study, i.e. the start time of the first round of observations, to the most complex for StNCRS, which requires the selection of random starting times and the application of the Loop Principle, in each time stratum. The other methods are judged to have intermediate complexity in the order listed in Table 4. This aspect of work sampling is given a low importance weight of 1 (last row of Table 4) because it represents a relatively small portion of the total resources required in the work sampling study, and also because it is readily amenable to computerization.

The second attribute, ease of data collection, is of more importance and has been assigned a weight of 3. Here it is judged to be simplest for SyRS, RRS and StCRS, intermediate for StNCRS because of the frequent application of the Loop Principle, and poorest for SRS because it frequently requires the complication of recruiting additional observers, or compromising the sampling times by "quickly doubling back over the route."

The third attribute, variation among the observation times, is considered to be as important as the second. Here, SyRS ranks last because the workers will know almost exactly when they will be observed after the first random time observation is taken, and the variance among subsequent observation times is zero. The best performance on this attribute is for SRS, and it is its only redeeming feature. It has the maximum variation of the times between observations of all 5 procedures being considered. In fact, it can be shown (Part II of this series) that these times have an exponential distribution whose standard deviation is equal to the mean. However, it is not felt that this degree of variation is necessary to secure unbiased performance from the workers being studied. This point is not to be taken lightly, and the author does not mean to wish it away. However, if a work sampling study is to continue for the typical period of a month or more, and the proper ground work is laid, the variation in observation times afforded by the other sampling methods, except SyRS and RRS where M is close to T/N, should be adequate to avoid worker performance bias.

The last attribute in Table 4 is considered to be the most important, it is given a weight of 4 because data collection, which is directly proportional to the statistical efficiency of the sampling method, represents the largest cost factor in most work sampling studies. On this attribute, SyRS again wins out. Some evidence for this conclusion is given in Table 2 where the superiority of SyRS over SRS and StRS is shown. The middle portion of Table 2 shows that SyRS is *always* equal to or superior to SRS, while it is only superior to StRS *on the average*. There are some combinations of worker type and strata length where StRS was superior. The statistical efficiency of RRS will be evaluated in Part II of this series, where it will be shown to be comparable to StRS.

Recommended Sampling Procedures

The last columns of Table 4 give the total score and rank for each of the sampling procedures. Although this ranking is based on the subjective judgement of the author, it is felt to be indicative of the *relative* merits of these 5 procedures. The SyRS procedure ranks first, and its use is recommended where operator performance bias does not preclude its adoption. Where this is not the case, any of the other procedures are potential candidates. The StCRS procedure is the recommended second choice whenever it can be arranged to use continuous observation and random routes. When non-continuous observation must be used, a choice between StNCRS and RRS must be made. If the physical area covered by the study is quite large, it may be desirable to avoid the use of random routes because of the dead heading that it entails (movement to the randomized starting point in the route) as well as the splitting of the observation rounds to accommodate the Loop Principle. In this case, the use of RRS is preferred over StNCRS. These recommendations are summarized in Fig. 3 which presents the above statements in the form of a decision tree.

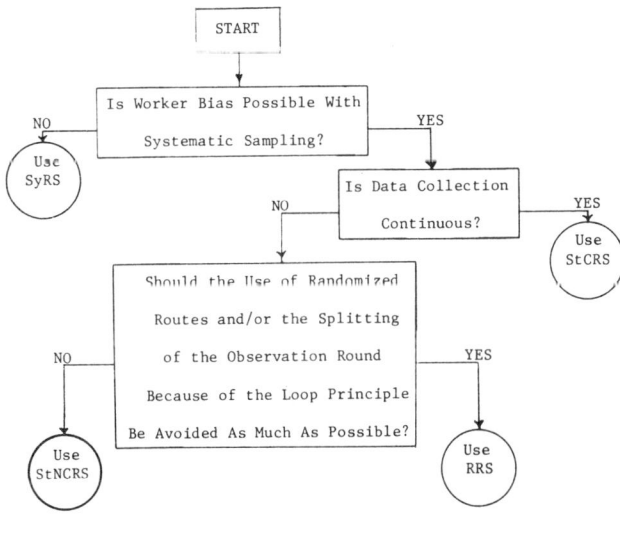

Fig. 3. Decision tree diagram for the selection of a sampling procedure for work sampling studies.

Work Sampling Study Data Analysis

In addition to the selection of a sampling method and specific observation times, one other statistical aspect of conducting a work sampling study needs consideration. That is, the choice of a sample size and the computation of confidence intervals for the parameters being estimated. The classical approach to sample size selection is to assume SRS is used (regardless of how the sample is actually drawn), and apply Eq. (3) together with a guesstimate of the parameter π. This procedure has some utility if the computed sample size is used as an approximate value to be modified by good judgement. The next step in the classical approach is to substitute the parameter estimate ($\hat{\pi}$) itself in place of π in Eq (3), to compute the error variance of $\hat{\pi}$. The latter procedure is considered to be inadequate for the following reasons:

1. If the recommendations in Fig. 3 are followed, some form of sampling other than SRS will be used, and it will generally have a lower sampling error variance than SRS. Thus, Eq. (3) will tend to *overestimate* the true error variance.

2. Work sampling data is often pooled over several workers in close proximity performing similar tasks, and for this reason the activities observed are frequently not independent physically or statistically. As a result, Eq. (3) will tend to underestimate the true sampling error variance (see [5]).

3. The objective of most work sampling studies is to estimate parameters that apply to the system in the future, not just to the historical period covered by the data collection. In this case the *random* component of the day-to-day variation in the parameters should be considered. This factor also results in Eq. (3) *underestimating* the true error variance.

To account for the composite of these sources of variation, the method of replicated estimates is suggested. Employing this procedure, the entire measurement study is broken up into a number of short periods (at least six and preferably more), and independent parameter estimates are computed for each period. Then the random variation among the period estimates is used to establish the accuracy of the pooled estimate. This procedure requires, of course, *that the period-to-period variation is purely random, free of systematic sources of variation associated with the periods*. To insure this, the use of the analysis of variance is recommended.

The natural time period to use in work sampling is the day, and because there may be a systematic variation in worker productivity for the days from Monday through Friday, the analysis of variance procedure is useful to isolate the fixed from the random effects. Consider a typical study in which sampling takes place for each of the I days of the week, and extends over a period of J *full* weeks. An appropriate mixed random and fixed effects linear model for this study can be written as:

$$p_{ij} = \pi + D_i + d_{ij} + e_{ij} ; \quad i=1,\ldots,I \text{ and } j=1,\ldots,J$$

where

p_{ij} = sample estimate for ith day of the jth week, giving the fraction of the observations made during the day in which the worker(s) was engaged in the activity category in question,

π = population mean for the process under study for the

past as well as the indefinite future, during which the process remains essentially unchanged,

D_i = fixed day effect, e.g. there may be differences between Monday and Friday, and the other days of the week, which repeat systematically from week-to-week; it is assumed that $D_1 + D_2 + \ldots + D_I = 0$, by definition, so that the use of full weeks in data collection will automatically balance out any fixed day effects.

d_{ij} = the random component of day-to-day variation, i.e. the random variation in the process mean for the ith day of the jth week, from the expected value of $(\pi + D_i = \pi_i)$ for the ith day of every week; it is assumed that the d_{ij}'s are independent random variables with $Ed_{ij} = 0$, $Ed_{ij}^2 = \sigma_d^2$ for all ij, and $E(d_{ij}d_{i*j*}) = 0$ if $i \neq i*$ and/or $j \neq j*$,

e_{ij} = random sampling error associated with the sample collected on the ith day of the jth week; it is assumed that the e_{ij}'s are independent random variables with $Ee_{ij} = 0$, $E(e_{ij}e_{i*j*}) = 0$ if $i \neq i*$ and/or $j \neq j*$, and $Ee_{ij}d_{i*j*} = 0$ for all ij and $i*j*$; however, it is recognized that $Ee_{ij}^2 \neq $ constant (σ_e^2) for all ij, as assumed in analysis of variance procedures, since the *sampling error variance* for a binomially distributed attribute variable depends on the quantity $(\pi + D_i + d_{ij})$. For this reason, it may be desirable to replace values of p_{ij} by a suitable function of them so that the new variables are more nearly homoscedastic. Since σ_e^2 should dominate σ_d^2 in most applications, a transformation, suitable for binomially distributed variables and suggested by Johnson [3], is given as follows:

$$\arcsin \sqrt{(x+3/8)/(N+3/4)},$$

where x is the number of occurrences of the activity category in question out of the N observations made on a given day. Because of the nature of the random term $(d_{ij} + e_{ij})$, as well as the fact that the e_{ij}'s will not be normally distributed, the statistical procedures described below which incorporates the Student t variable, should only be considered as approximate procedures.

In this type of study the fixed day effects as well as their estimates sum to zero, and the variance of the estimates of π and $\pi + D_i = \pi_i$ are shown in Table 5. An approximate confidence interval for π, the parameter of primary interest in work sampling studies, can then be constructed as given in Eq. (6). Also, if the null hypothesis of no fixed day effects ($H_o: D_i = 0$ or $\pi_i = \pi$ for all i) is rejected, then Eq. (7) can be used to construct approximate confidence intervals on the π_i's. If the D_i's are relatively large, then Eq. (7) may be quite useful in assessing and improving the productivity of a system

$$\Pr\{\bar{p}_{..} - t_{\alpha/2}\hat{\sigma}_r/\sqrt{IJ} < \pi < \bar{p}_{..} + t_{\alpha/2}\hat{\sigma}_r/\sqrt{IJ}\} = 1-\alpha \quad (6)$$

Table 5: ANOVA table for typical work sampling study.

Source	D.F.	S.S.	Approximate $E(MS)$
Days	$I-1$	$\sum_i\sum_j(\bar{p}_{i.}-\bar{p}_{..})^2$	$\sigma_e^2 + \sigma_d^2 + J\Sigma_i D_i^2/(I-1)$
Replications	$I(J-1)$	$\sum_i\sum_j(\bar{p}_{ij}-\bar{p}_{i.})^2$	$\sigma_e^2 + \sigma_d^2 = \sigma_r^2$
Total	$IJ-1$	$\sum_i\sum_j(\bar{p}_{ij}-\bar{p}_{..})^2$	

$\bar{p}_{i.} = \sum_j p_{ij}/J$, $\bar{p}_{..} = \sum_i\sum_j p_{ij}/IJ$, and $\Sigma\bar{p}_{i.} = I\bar{p}_{..}$

Parameter	Estimator	Variance
Population Mean π	$\hat{\pi} = \bar{p}_{..}$	$\hat{\sigma}_r^2/IJ$
Day Mean $\pi_i = \pi + D_i$	$\hat{\pi}_i = \hat{\pi} + \hat{D}_i = \bar{p}_{i.}$, and $\Sigma_i\hat{D}_i = 0$	$\hat{\sigma}_r^2/J$
Residual Variance	$\hat{\sigma}_r^2 = \sum_i\sum_j(\bar{p}_{ij}-\bar{p}_{i.})^2/I(J-1)$	

$$\Pr\{\bar{p}_{i.} - t_{\alpha/2}\hat{\sigma}_r/\sqrt{J} < \pi_i < \bar{p}_{i.} + t_{\alpha/2}\hat{\sigma}_r/\sqrt{J}\} = 1-\alpha \quad (7)$$

under study. In the above equations, $t_{\alpha/2}$ denotes the Student t value with $I(J-1)$ degrees of freedom. It should also be noted that separate estimates of σ_e^2 and σ_d^2 could be made if the entire study is replicated, say K times, as suggested by the model given in Eq. (8). That is, K sets of data are collected independently and simultaneously during the period of the study.

$$P_{ijk} = \pi + D_i + d_{ij} + e_{ijk}. \quad (8)$$

Conclusions

The thesis developed in this paper is that SRS should not be used in work sampling studies. As a matter of fact, it probably has never been used because the usual practice of taking the same number of observations in each "day" of the study renders the procedure a form of StRS, with "days" forming the strata. However, this does not carry stratification far enough to reap the statistical efficiency benefits that result when strata lengths of an hour or less are utilized (see Fig. 2.).

SyRS is an excellent sampling method for work sampling studies, but unfortunately, in many applications, worker performance bias would become a serious problem if it were used. For this reason, StCRS is recommended as a (close) second choice in preferred sampling methods. Because of its very desirable properties, continuous observation should be utilized whenever possible. When noncontinuous observation must be adopted, then the use of RRS or StNCRS is recommended, depending on the conditions under which the study is being made. Finally, the limitations of the binomial distribution, as a statistical model in work sampling are emphasized. While the binomial model has limited utility in establishing the sample size for a study, it should be replaced by the analysis of variance in the treatment of the data collected in a work sampling study to assess errors of estimate and construct confidence intervals on parameter estimates.

References

[1] Hines, W. W. and J. J. Moder, "Recent Advances in Systematic Activity Sampling," *The Journal of Industrial Engineering,* **16** 5, 295-302 (Sept.-Oct., 1965).

[2] Hines, W. W., "The Relationship Between the Properties of Certain Sample Statistics and the Structure of Activity in Systematic Activity Sampling," PhD dissertation, Georgia Institute of Technology, Atlanta, Georgia (March 1964).

[3] Johnson, N. L. and F. C. Leone, "Statistics and Experimental Design in Engineering and the Physical Sciences," John Wiley & Sons, Inc., 54-56 (1964).

[4] Kahn, H. D., "Restricted Random Activity Sampling," MS thesis, Department of Industrial Engineering and Systems Analysis, University of Miami, (1969).

[5] Moder, J. J. and W. J. Halladay, "Work Sampling Applied to Long Cycle Operations Performed by a Variable Labor Force," *The Journal of Industrial Engineering,* **7** 4, 164-168 (July-Aug. 1956).

[6] Mundel, M. E., "Motion and Time Study: Improving Productivity," Prentice-Hall, Inc., 5th Edition, 104-105 (1978).

[7] Person, H. W., "The Genius of Frederick W. Taylor," *Advanced Management,* 10:2-11 (Jan.-Mar. 1945).

[8] Richardson, W. J., "Cost Improvement, Work Sampling, and Short Interval Scheduling," Reston Pub. Co., Inc., Reston Virginia (1976).

[9] Smith, G. L. Jr., *Work Measurement: A Systems Approach*, Grid Pub. Inc., p. 20 (1978).

[10] Tippett, L.H.C., "Statistical Methods in Textile Research, Uses of Binomial and Poisson Distributions: A Snap-Reading Method of Making Time Studies of Machines and Operatives in Factory Surveys," *Journal of Textile Institute Transactions,* **26**, (February 1935).

Dr. Joseph J. Moder is Professor of Management Science and Industrial Engineering at the University of Miami. He received his BS degree from Washington University, his MS and PhD degrees in Chemical Engineering from Northwestern University, did Post Doctoral work in Statistics and Operations Research at Iowa State University and Stanford University, and was a Visiting Professor of Engineering Production at the University of Birmingham. Professor Moder is a member of AIIE, ORSA, TIMS, AIDS, ASA and PMI, and is a registered engineer in Georgia; his research interests are in the theory and applications of activity sampling, and network based project management techniques.

Selection of Work Sampling Observation Times: Part II—Restricted Random Sampling

JOSEPH J. MODER
SENIOR MEMBER, AIIE
Departments of Management Science and Industrial Engineering
University of Miami
Coral Gables, Florida 33124

HENRY D. KAHN
Environmental Protection Agency
Washington, D.C. 20460

Abstract: This paper considers the problem of sampling from a finite population defined on the interval $(1, 2, \ldots, T)$, using a restricted form of random sampling in which the spacing between consecutive observations must be at least some minimum value, M. For this type of restricted random sample, of size N, the value M can be arbitrarily specified in the range from zero, corresponding to simple random sampling with replacement, to a maximum of T/N, corresponding to systematic sampling. An efficient procedure is developed for generating a sample for specified values of N, M and T. It is then shown, by computer simulation, that values of M in the range from 60 to 90 percent of T/N retain most of the advantages of systematic sampling, while avoiding most of the disadvantages for some special processes. Also, the error variance associated with stratified sampling is comparable to restricted random sampling with a minimum spacing of about 70 percent, and both methods are almost as efficient as systematic sampling.

■ This paper considers the task of sampling from a finite population defined on the interval $(1, 2, \ldots, T)$, using a restricted form of random sampling in which the spacing between consecutive samples must be at least some minimum value, called M. In work sampling applications, M will be set equal to or greater than the time required to obtain a round of work sampling observations, with equality yielding the maximum variability between consecutive sample observation times.

This sampling procedure, called Restricted Random Sampling (RRS), was shown in Part I of this series [9] to be useful in work sampling when the two most desirable forms of sampling cannot be used. That is, when potential worker bias rules out the use of Systematic Random Sampling (SyRS), and when sampling is not continuous, ruling out the use of Stratified Continuous Random Sampling (StCRS). A formal definition of RRS is as follows:

DEFINITION: "A RRS of size N on the interval $(1, 2, \ldots, T)$ is one in which each sample time is drawn at random with constant probability over the interval, rejecting a sample time if it is less than a distance M from its nearest neighbor, and where 1 and T are considered to be adjacent so that the interval $(1, 2, \ldots, T)$ forms a closed loop." (See Fig. 1.)

The need for the closed loop in this definition was explained in Part I of this series where it was referred to as the Loop Principle. It is required in order to achieve the specification that each minute in the interval $(1, 2, \ldots, T)$ has an equal probability of entering the sample. The range of the minimum allowable spacing, M, is $(0, T/N)$, from which it can be seen that (SRS) Simple Random Sampling (with replacement) and SyRS are special cases of RRS, with $M = 0$ and T/N, respectively. A method of sampling with intermediate values of M will be described in the next section.

The Exponential Sampling Procedure

The basis for this sampling procedure is a theorem in

Received October 1978; revised October 1979. This investigation was supported in part by a National Science Foundation Grant, NSF-GK2521 entitled, "Some New Developments in Work Sampling," Paper was handled by Work Measurement/Methods/Ergonomics/Health Services Department.

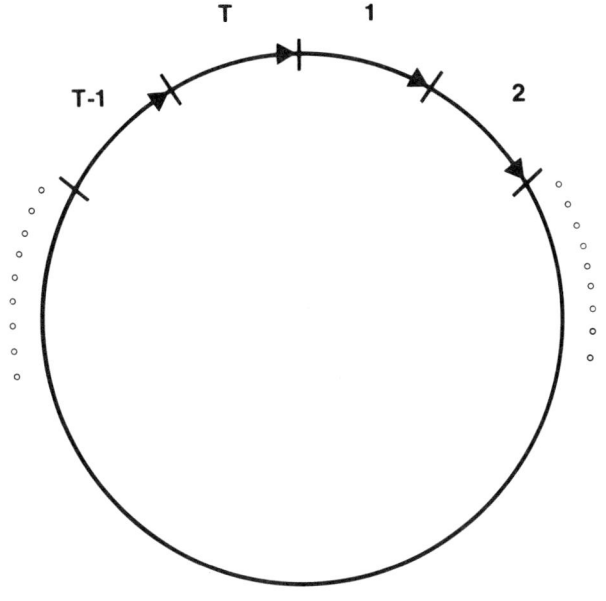

Fig. 1. Loop arrangement of time points in the interval (1,T).

stochastic processes (e.g., see Parzen [10]) which was also used in a related sampling study by Shapiro and Silverman [11]. This theorem considers random samples $(X_i; i = 1, 2, \ldots, N+1)$ taken from an exponential distribution with a mean (interarrival or observation time) of T/N, and the resulting sample times are then taken as $t_1 = x_1$, $t_2 = x_1 + x_2$, etc. If these times are taken as acceptable only if $t_N \leq T < t_{N+1}$, then the joint density function of the resulting sample times is identical to the density function of the order statistics obtained from a SRS drawn from a uniform distribution on the continuous interval $(0,T)$.

The above theorem assumes no minimum spacing ($M = 0$), however, it does suggest an alternative RRS procedure[1] for $M > 0$. In this procedure, sample times are obtained by adding realizations of independent, exponentially distributed random variables to the minimum spacing, M. The expected value of each of the random portions (X_j), is taken as $E(X_j) = [(T/N) - M]$, so that the expected value of the time between each of the samples that are generated remains equal to T/N. To facilitate exposition of this procedure, all variables will be assumed to be integer-valued.

The Exponential Sampling Procedure for RRS

(1) Generate $u_1, u_2, \ldots, u_{N+1}$ according to
$u_1 = 1$, and
$u_j = u_{j-1} + M + x_j; j = 2, 3, \ldots, N+1$

where x_j has an exponential distribution with mean $[(T/N) - M]$;

(2) Repeat step (1) until the following test is passed
$$u_N + (M-1) \leq T < u_{N+1}.$$

(3) Generate a uniform random integer on the interval $[1, T]$, call it R, and compute the shifted times defined as

$$u_j^* = u_j + R; j = 1, 2, \ldots, N, \text{ so that } R + 1 \leq u_j^* \leq T - M + 1 + R.$$

(4) Let $u_j^{**} = [(u_j^* \text{ modulo } T) + 1]$, so that $1 \leq u_j^{**} \leq T$.

(5) Order the u_j^{**} values to obtain the random sample times, $1 \leq t_1 < t_2 < \ldots < t_N \leq T$, with minimum spacing M.

The set of u_j's as generated in step (1) must by construction have the specified minimum spacing, M, and the desired average spacing (T/N). The test, $u_N + (M-1) \leq T$ in step (2), together with assignment of $u_1 = 1$, insures that $T - t_N + 1 \geq M$. Also the requirement that $u_{N+1} > T$ in step (2) follows directly from the theorem cited above. Then, to establish a steady state of the underlying Poisson process, step (3) in effect randomizes the origin by "spinning the loop," which was arbitrarily fixed by assigning $u_1 \equiv 1$. The last two steps reduce the u_j^{**}'s to integer values in the range $[1,T]$, and orders them to give sample times with the desired properties.

This procedure suffers from the fact that the first two steps may be repeated many times before proceeding to step (3), if M is close to T/N. This can be remedied by using the conditional distribution of the exponential generator. It can be shown[2] that the cumulative conditional distribution function of the waiting time, t_j, to the jth event in a Poisson process, given that there are exactly N events in the time T, is given by

$$F_{[t_j | N(T)=N]}(x) = P[t_j \leq x | N(T) = N] = \sum_{\alpha=j}^{N} b(\alpha),$$

where $b(\alpha)$ is the binomial probability of α events in a sample of size N with probability (x/T). For $j = 1$, this function can be expressed in closed form as:

$$F_{[t_1 | N(T)=N]}(x) = 1 - (1-x/T)^N.$$

Letting ν denote a random number, uniformly distributed on the interval $(0,1)$, the above can be inverted and used to generate the first time as:

$$t_1 = T - T\exp[(\ln \nu)/N].$$

---

[1] An illuminating queueing model analogous to RRS can be described. Consider a single channel service facility with no queue allowed, constant service time equal to M, and random customer arrivals with rate equal to $N/(T-NM)$. The arrival times of customers *actually served* by the service channel are then the RRS observation times. This model also holds for the limiting conditions defined by SRS and SyRS.

[2] This was pointed out to the authors by Mr. Sanji Arisawa, at North Carolina U.

Subsequent times can be generated from the distribution function

$$F_{[t_j = t_{j-1} + x \mid N(T - t_{j-1}) = N - j + 1]}(x) = 1 - [1 - x/(T - t_{j-1})]^{N-j+1}.$$

This function can be inverted to give

$$x = (T - t_{j-1}) \{1 - \exp[(\ln \nu)/(N-j+1)]\}$$

and since $t_j = t_{j-1} + x$, we obtain

$$t_j = T - (T - t_{j-1}) \{\exp[(\ln \nu)/(N-j+1)]\}; \quad j = 2, 3, \ldots, N. \tag{1}$$

Thus, to generate a random sample of exactly N observations in the range $[1, T]$, with minimum spacing M, the following procedure should then be used, in which the variables are treated as continuous rather than integers as was assumed above solely for ease of exposition

Modified Exponential Sampling Procedure for RRS

(1) Generate $u_1, u_2, \ldots, u_N$ according to
$u_1 \doteq 0$, and $u_j = E_j + M(j-1); j = 2, 3, \ldots, N$
where $E_j = (T - NM) - (T - NM - E_{j-1})\{\exp[(\ln \nu)/(N-j+1)]\}$,
and $E_1 = 0$.

(2) Generate a uniform random number on the interval $(0, T)$, call it R, and compute the shifted times defined as $u_j^* = u_j + R; j = 1, 2, \ldots, N$, so that $R \leq u_j^* \leq T - M + R$.

(3) Let $u_j^{**} = [(u_j^* \text{ modulo } T) + 1]$ truncated, so that $1 \leq u_j^{**} \leq T$.

(4) Order the u_j^{**} values (integers) to obtain the random sample times, $1 \leq t_1 < t_2 < \ldots < t_N \leq T$, with minimum spacing M.

For this procedure, the limiting cases for M equal to 0 and T/N are worthy of note. First, for the case $M = 0$, the above expression for E_j reduces to that given in Eq. (1) for SRS, and for the case $M = T/N$, all of the E_j's are identically zero, as required by SyRS.

The use of this modified procedure is illustrated in Table 1 for the case where a sample of size $N = 4$ is to be drawn with a minimum spacing of $M = 10$ time units, from a study period of $T = 100$ time units. The entries in the ν_j column are random numbers on the interval $(0,1)$, expressed to only two decimals for ease of exposition. The values in the u_j^* column were computed from $u_j^* = u_j + R$, where $R = 76.58$ is the uniform random number generated on the interval $(0, T=100)$ to "spin the loop." This value of R was deliberately chosen to illustrate how the minimum sample time $t_1 = 1$, arises in this procedure.

Table 1: Illustration of the modified exponential sampling procedure for a sample of size $N=4$, from a study period of size $T=100$, with a minimum space of $M=10$.

j	ν_j	$e^{(\ln \nu_j)/(5-j)}$	E_j	u_j	u_j^*	u_j^{**}	t_j
1	--	--	0	0	76.58=R	77	1
2	.45	.77	13.80	23.80	100.38	1	33
3	.27	.52	35.98	55.98	132.56	33	47
4	.83	.83	40.06	70.06	146.64	47	77

Error Variance for RRS

If a SRS of size N with replacement is used to estimate π for a finite population of 0's and 1's denoted by X_t, $t = 1, 2, \ldots, T$, then regardless of the pattern of 0-1 episodes, the unbiased estimator $\hat{\pi}$, and its variance are given by Eq. (2), where N_s is the number of times the process is observed to be in the 1 state.

$$\hat{\pi} = N_s/N, \text{ and } \sigma_{\hat{\pi}}^2 = \pi(1-\pi)/N. \tag{2}$$

For SyRS, it can be shown that the variance of $\hat{\pi}$ is given by Eq. (3), where $\rho(\alpha M)$ is the auto-correlation function for samples with a spacing of (αM) time units, σ^2 is the variance of a stationary 0-1 process X_t, and $M = T/N$ is the fixed time interval between consecutive (systematic) sample points. Cochran [2] has shown that for populations with a convex, non-increasing, non-negative auto-correlation function,

$$\sigma_{\hat{\pi} \text{SyRS}}^2 = \frac{\sigma^2}{N}\left[1 + \frac{2}{N}\sum_{\alpha=1}^{N}(N-\alpha)\rho(\alpha M)\right] \tag{3}$$

the error variance for SyRS is equal to or less than the error variance for SRS or StRS. Stemmler [12] obtained the same result without requiring non-negativity of the autocorrelation function. Hannan [5] obtained general forms for the variance of the means of SyRS and SRS using standard spectral theory. Hannan did not consider any particular population types but merely stated some general conditions under which one form of sampling or the other is likely to be superior. A similar but not quite equivalent comparison between SRS and SyRS may be inferred from Grenander and Rosenblatt [4]. Kume [8] has studied the error variance for SyRS of various periodic processes by expanding X_t in a Fourier series to approximate ρ in Eq. (3). It has been shown by Kahn [7] that it is not possible to evaluate ρ in this manner for RRS. In a related study on the estimation of the two parameters in an alternating (0,1) Poisson process, it was concluded by Brown, et al. [2] that mathematical analysis appears intractable, and use was made of Monte Carlo studies. For these reasons, the same approach was used in this study to estimate the error variance of RRS relative to SRS.

Table 2: Sampling efficiency for restricted random sampling.*

Experiment parameters	Experiment number 1	2	3	4	5	6	7	8
π	.248	.24	.253	.24	.25	.24	.10	.10
T	1200	1200	2880	2400	4800	4800	1200	2400
$SI=T/N$	25	25	60	50	100	100	25	50
$E(1)$; $E(0)$	62;188	12;38	22.65	12;38	14;42	6;19	25;225	10;90
SI/PP	.10	.50	.69	1.0	4.0	4.0	.10	.50
No. of Runs N_R	200	300	50	200	200	200	200	200
Exponential Populations								
Minimum spacing 0 (SRS)	1.15	0.95	0.95	1.0	1.15	0.95	1.0	1.0
20	1.45	1.1	1.25	1.2	1.45	0.95	1.25	1.05
40	2.5	1.8	1.4	1.25	1.5	1.0	2.1	1.35
60	6.15	2.05	2.15	1.8	1.3	1.05	2.6	1.2
80	7.55	2.05	1.75	1.2	2.4	0.95	3.7	1.25
90	10.0	2.45	1.85	1.45	1.15	1.05	5.05	1.6
100 (SyRS)	10.5	2.1	1.65	1.45	1.2	1.1	4.1	1.4
StRS	5.85	1.9	1.65	1.4	1.2	1.1	3.3	1.4
Normal populations								
Minimum spacing 0 (SRS)	1.0	1.05	0.95	0.9	1.5	1.05	0.95	0.95
20	1.4	1.2	1.4	1.2	0.9	0.95	1.35	0.95
40	2.4	1.4	0.85	1.1	0.85	0.85	2.05	1.25
60	4.45	1.5	1.1	1.1	0.95	1.15	2.8	1.0
80	8.0	1.15	1.65	0.6	0.95	0.9	4.55	0.9
90	7.0	0.9	2.05	0.4	0.95	0.95	6.4	0.95
100 (SyRS)	12.4	0.9	1.75	0.25	1.3	0.65	6.4	0.75
StRS	5.95	1.35	1.15	1.0	1.0	1.0	3.0	1.15
Periodic populations								
0 (SRS)	1.05	0.9	0.8	0.9	1.15	0.8	1.2	0.9
20	1.35	1.25	1.3	1.05	1.0	0.9	1.4	1.15
40	2.5	1.6	1.2	1.45	0.85	0.9	2.4	0.95
60	5.05	1.5	0.85	1.2	1.0	0.95	3.1	1.0
80	6.5	0.95	2.8	0.5	1.0	1.4	7.25	0.8
90	5.55	0.4	2.0	0.15	1.65	1.0	8.35	0.4
100 (SyRS)	1.45	0.06	9.1	0.02	2.4	0.02	∞	0.04
StRS	6.75	1.5	1.0	1.0	1.0	1.0	∞	1.1

* Data rounded to nearest .05 value, except in cases where the data was less than 0.1.

Simulation Model

To represent a wide range of process behavior, three basic population types were chosen for the random durations of the alternating zero and one states: exponential, normal, and periodic distributions. The elapsed time between changes from one state to another are strictly determined in the periodic case, "random" in the exponential case, and somewhere between these extremes in the case of the normal distribution. For the normal distribution, the standard deviation was chosen to be 15 percent of the mean, as compared to 0 and 100 percent for the periodic and exponential distributions, respectively. This value of the standard deviation, as well as other population characteristics of the simulation experiment, shown in Table 2, were chosen to be representative of conditions likely to be encountered in work sampling.

Each of the eight columns in Table 2 constitutes an experiment described by the parameters π, SI/PP, $E(1)$ and $E(0)$, defined as follows:

$E(1)$ = mean duration of the process state 1 episodes

$E(0)$ = mean duration of the process state 0 episdoes

$PP = E(1) + E(0) = $ Period of the Population or Process being simulated

$\pi = E(1)/PP = $ fraction of time the process is in state 1

$T = $ duration of the study (minutes)

$N = $ number of sample observations to be taken in the study ($N=48$ will be used in all experiments in this simulation exercise).

$SI = T/N = T/48 = $ Sampling Interval, or the average time between sample observations.

$N_R = $ number of Runs that were conducted and averaged to obtain the simulation results in Table 2, for each of the 8 experiments.

Each of the (N_R) runs consisted of the generation of random realizations of the process, X_t, $t = 1, 2, \ldots, T$, using exponential, normal, and periodic distributions for the durations of the alternating zero and one episodes. Each of these random realizations of the process were then observed at $N = 48$ RRS times chosen with minimum spacing (M) from 0 to 100 percent of T/N.

For each experiment, the RRS sample variance given by Eq. (4) was computed for each of the 3 population types and minimum spacing values.

$$V(\hat{\pi}_{ij})_{RRS} = \frac{1}{N_R} \sum_{i=1}^{N_R} (\hat{\pi}_{ij} - \pi_i)^2 \quad (4)$$

where $\pi_i = $ actual fraction of time the ith process realization is in state one, $i = 1, 2, \ldots, N_R$, for each of the 3 population types

$\hat{\pi}_{ij} = $ the sample estimate of π_i using the jth level of the minimum spacing, $j = 1, 2, \ldots, 7$.

A theoretical error variance for SRS was also computed once for each of the 3 population types, using Eq. (5). This computation was based on the average value of π_i for the N_R runs constituting an experiment, which is denoted by $\bar{\pi}$.

$$V(\hat{\pi})_{SRS} = \bar{\pi}(1-\bar{\pi})/N \quad (5)$$

The results of the simulation study are given in Table 2 in terms of relative efficiency defined by the ratio of the variances, given by Eq. (5)/Eq. (4). This ratio can be interpreted as an effective increase (or decrease) in the sample size using RRS compared with SRS. For example, a ratio of 2.0 means that N observations using RRS are equivalent to $2N$ observations using SRS. The simulation program was validated by testing for lack of bias in the estimates, and by testing the results in Table 2 which should theoretically be equal to unity for SRS, i.e., $M = 0$.

Subsequent to the above computer runs, interest was generated in the use of StRS, as treated in Part I of this series [9]. The relative efficiency ratios given in Table 2 in the rows marked StRS, were obtained by a separate simulation of each of these 8 experiments, using 100 runs each, applying Eqs. (6) and (7) below to estimate variances for StRS and SRS. The values given in Table 2 are the ratios of Eq. (7) to Eq. (6).

$$V(\hat{\pi})_{StRS} = \frac{1}{100} \sum_{i=1}^{100} \left[\frac{1}{48} \sum_{j=1}^{48} \pi_{ij}(1-\pi_{ij})/48 \right] \quad (6)$$

$$V(\hat{\pi})_{SRS} = \frac{1}{100} \sum_{i=1}^{100} \bar{\pi}_i(1-\bar{\pi}_i)/48 \quad (7)$$

where $\pi_{ij} = $ value of the parameter obtained for the jth stratum of the ith simulation run, and

$$\bar{\pi}_i = \sum_{j=1}^{48} \pi_{ij}/48 \text{ for } i = 1, 2, \ldots, 100.$$

Simulation Results

The simulation results for the exponential population follow Cochran's theorem in that the efficiency of SyRS ($M=100\%$) is always equal to or greater than that for SRS ($M=0\%$) or for StRS. The theorem does not apply to RRS, but in the simulation results the efficiency generally increases as SyRS is approached on the minimum spacing scale. This effect is suggested by the theorem and seems reasonable to expect. In general, for the larger values of M, the efficiency decreases as the ratio SI/PP decreases. This behavior can be explained on the basis that the autocorrelation function also decreases as SI/PP decreases. The efficiency also decreases as π decreases; this can be explained on the basis that the variance among the strata π_{ij} values also decreases as π decreases. What is important to note here is that a *minimum spacing of 60 to 90 percent has about the same efficiency as* SyRS *and* StRS.

The populations generated using the normal distribution have a damped sinusoidal autocorrelation function, and hence do not follow Cochran's theorem. In this case the effect of M on efficiency depends very markedly on the the value of SI/PP. First, for odd values of SI/PP, i.e., 0.69 and 1.786, variation in M has little or no effect. For experiments 1 and 7 where $SI/PP = 0.1$, the efficiency increases sharply as M increases, due to the sinusoidal nature of the autocorrelation function, as well as the fact that the variance among the strata π_{ij} values is large in these experiments. For $SI/PP = 0.5$, 1.0 and 4.0, however, the efficiency generally decreases as M increases, due to the coincidence of the sampling interval (T/N) and the process period (PP)

Peculiar as it may seem, the latter relationship is not monotonic. For $SI/PP = 0.5$ and to a lesser extent for 1.0, the efficiency rises first and then falls off as M increases so that intermediate values of M are better than 0 or 100 percent. This effect was found to be statistically significant on the basis of a significant quadratic effect for the factor M in a least squares regression analysis of these data. Again, it is important to note that a *minimum spacing of 60 to 90 percent does an excellent job of capturing the advantages of SyRS and StRS while tending to avoid the disadvantages of SyRS, when applied to a process having normally distributed 0-1 episodes.*

Although it is unlikely that an exactly periodic population would be encountered in practice, almost-periodic populations do exist. (Usually in these cases, the process period is known before the study is carried out, and should therefore be avoided in selecting a sampling interval.) The results in Table 2 indicate a drastic loss in efficiency as M increases for $SI/PP = 0.5$, 1.0 and 4.0. This loss, of course, is caused by sample points occurring in phase with the oscillatory behavior of the population, which results in a large error variance. For the very low value of $SI/PP = 0.1$, or for the odd values of 0.69 and 1.786, the efficiency improves as M increases. The most important observation here is again that for *minimum spacing of 60 to 90 percent, the serious loss in efficiency for* $SI/PP = 0.5$, *1.0 and 4.0 is avoided, while the advantage is retained for 0.1, 0.69 and 1.786.*

Conclusions

The Modified Exponential Sampling Procedure given above is an efficient method of generating RRS observation times, with minimum spacing over the entire range from 0 to 100% to T/N. The results of the simulation study indicate that RRS does, indeed, combine most of the advantages of both SyRS and StRS, and at the same time it tends to avoid the disadvantages of SyRS. In particular, the simulation study of the population types chosen in this study as being representative of work sampling applications, indicates that an excellent compromise is achieved by choosing a minimum spacing in the range from 60 to 90 percent of T/N. In particular, the error variance associated with StRS is comparable to RRS with a minimum spacing of about 70 percent, and both methods are almost as efficient as SyRS. RRS adds a very useful form of sampling to the array of methods discussed in Part I [9] of this series of papers on work sampling methodology.

References

[1] Brown, M., H. Solomon, and M. A. Stephens, "Estimation of Parameters of Zero-One Processes by Interval Sampling," *Operations Research*, **25** 3, 493-505 (May-June 1977).

[2] Cochran, William G., "Relative Accuracy of Systematic and Stratified Random Samples for a Certain Class of Populations," *Annals of Math Stat.*, **17**, 164-77 (1946).

[3] Cochran, William G., *Sampling Techniques*, John Wiley and Sons, Inc., New York, Chapter 8 (1964).

[4] Grenander, Ulf and Murray Rosenblatt, *Statistical Analysis of Stationary Time Series*, John Wiley and Sons, Inc., New York, 57-59 (1957).

[5] Hannan, E. J., "Systematic Sampling," *Biometrika*, **49**, 281-283 (1962).

[6] Hines, William W. and Joseph J. Moder, "Recent Advances in Systematic Activity Sampling," *The Journal of Industrial Engineering*, **16** 5, 295-302 (Sept-Oct, 1965).

[7] Kahn, Henry D., "Restricted Random Activity Sampling," MS Thesis, Dept. of Industrial Engineering and Systems Analysis, U. of Miami (1969).

[8] Kume, Hitoski, "On the Spectral Analysis of 0-1 Process," *Technology Reports of the Seikei University*, No. 3, Shinjukuku, Tokyo (1965).

[9] Moder, J. J., "Selection of Work Sampling Observation Times: Part I Stratified Sampling," *AIIE Transactions*, this issue.

[10] Parzen, Emanuel, *Stochastic Processes*, Holden-Day, Inc., San Francisco, 139-141 (1962).

[11] Shapiro, Harold S. and Richard A. Silverman, "Alias-Free Sampling of Random Noise, *Soc. Indust. Appl. Math.*, **8** 2, 225-248 (1960).

[12] Stemmler, Ronald E., "Some Extensions in the Theoretical Structure of Sampling from Divariate Two-Valued Stochastic Processes," PhD Thesis, Georgia Institute of Technology (1971).

Dr. Joseph J. Moder is Professor of Management Science and Industrial Engineering at the University of Miami. He received his BS degree from Washington University, his MS and PhD degrees in Chemical Engineering from Northwestern University, did Post Doctoral work in Statistics and Operations Research at Iowa State University and Stanford University, and was a Visiting Professor of Engineering Production at the University of Birmingham. Professor Moder is a member of AIIE, ORSA, TIMS, AIDS, ASA and PMI,

Mini-Micro Computers

Gary E. Whitehouse, P.E. — Column Editor

Preparatory Computer Assistance for a Work Sampling Study

By David M. Rhyne, P.E.
Auburn University
and Douglas K. Freeman

Work sampling is increasingly being used as an effective work measurement system in both manufacturing and service settings. As more jobs are created in industry groups like government, personal and professional service businesses, health care, transportation and banking, the applicability of work sampling will continue to grow. Work measurement for a range of uses, from productivity measurement to a quasi-quality control program, will necessitate some form of work sampling.

The advantages of work sampling over traditional time study analysis procedures provide additional justification for its acceptance and application. This work measurement system enables a single observer to gather essential data on several workers who may be performing a variety of tasks in different locations within a facility. And, the entire process is accomplished with a minimum expenditure of time and other resources.

The power of the work sampling technique lies in the "theory of probability;" that is, a few observations taken at random from a large group (population) tend to represent the characteristics of the group. If the number of observations is large enough, the characteristics observed will differ only slightly from the true characteristics of the group (population).

The key feature of the observed sample is its randomness, which is maximized by taking the number of observations required to ensure a stated level of statistical confidence in the results of the study. Starting the sampling trips from random points minimizes the effect of sampling bias.

The purpose of this study is to provide the work measurement practitioner with a program that will generate the essential randomness of a work sampling program. It is assumed that the following components have been defined:
☐ The problem to be studied.
☐ The objectives of the study.
☐ The categories of activity that will be observed.
☐ The desired statistical confidence of the study and the allowable level of accuracy for the observed categories.

The remaining items of the preparatory phase of a work sampling study include:
● Determining the appropriate sample size of random observations. From this calculation, the number of sampling days needed can be found.
● Determining the random times at which observations will be made. In a setting where the observer remains essentially stationary, at the appropriate clock time he glances over the area and makes a mental snapshot of the workers or equipment studied. In a setting where the observer travels through an area or several areas, these random times correspond to the clock times when the sampling trips begin.
● Determining the random starting points. If the observer travels through a sizable geographical area, it is essential that each sampling trip start from a defined point that is randomly selected from among several alternative points.

BASIC can be used

Although it is possible to obtain the above requirements manually, a BASIC program presented in this study can generate these items. Indeed, numerous firms in industry, as well as consulting engineering and computer software companies, have already developed similar programs. Articles in previous issues of *Industrial Engineering* have addressed the topic of using computers to generate input for work sampling. However, as an update on the topic and to help meet the need of users without computer program assistance, this study is provided.

The supporting program for this study features a menu selection with the following output:
☐ The required number of random work sampling observations. The user provides the input for the study's statistical level of confidence, the level of accuracy for the principal observed control category and the estimated percentage of occurrence of the observed control category obtained in a pilot study.
☐ An additional menu that prepares observation records or forms for the work sampling study. Inputs such as the number of required samples, the length of the study (in days), the observation workday, the number of work categories to be sampled (up to 25) and parameters on the form construction are required for the observation

record. The number of persons or groups being observed is also provided along with whether the personal break times will be included in the sampling day. The observation record will contain the random times at which observations will be made as well as the random starting point from which the sampling trip will begin.

☐ The listing of a matrix of random observation times. For each hour of the sampling day, a number of random sample times is provided. This listing of random times would be extremely helpful to a sampler for a study where little or no sampling travel would be required. As the geographical area to be covered increases, a decreasing number of the random times are applicable for sampling trip times.

☐ The provision for an analysis of the work sampling results. Although this evaluative analysis is not part of the preparatory phase of the study it has been provided for the user. This computer-assisted analysis takes the tabulated totals of observations by category and computes the percentages of the total and the control limits for the observation-count of each category at the given level of statistical confidence.

The principal feature of a work sampling study that measures the work of personnel or equipment is its randomness. After the initial definitions, which were cited earlier, have been made and a pilot study has been performed to track the count of at least one category, an adequate sample size can be determined.

Randomness from this point forward is critical. Sensitivity to randomness in terms of the sample size, observation times, trip starting points and the accurate recording of observations will enhance the reliability and utility of the study's results. The program that follows provides the user with options for obtaining these critical items.

For further reading:

Erwin, Walter W. *Work Sampling Procedure*. Industrial Engineering Services, 1984.

Lindenmeyer, Carl R. and Melinda M. Sykes. "Computer-Aided Work Sampling: Description and Application," in *Proceedings, 1982 Annual Conference of the Institute of Industrial Engineers*, 1982.

Whitehouse, G. E. and D. A. Washburn. "Work Sampling Observation Generator," *Industrial Engineering*, Vol. 13, No. 3 (1981), pp. 16-18.

The authors wish to thank Steve Stuckwisch, Department of Mathematical Sciences, Auburn University, for his consultation on the algorithm for a probability curve.

Appendix:

```
5 COLOR 14,1,1:CLS
7 KEY(1) ON:ON KEY(1) GOSUB 9
8 GOTO 10
9 NP=1-NP:RETURN
10 RANDOMIZE VAL(RIGHT$(TIME$,2))
20 DIM SAM(25),Y(50),Z(50),MEN(50),DAT(50),ACT$(25),
   BRKS(20,4),HOLI(20,4)
25 DIM TEMP(3),DATE(3),DAY$(7),UL(3),LL(3),MON(12),
   TIM(20),DA(3)
30 N=44:GOSUB 8000:Z$=CHR$(12)+Z$:GOSUB 8150:LLINE=4:
   LC=1:GOSUB 8300
40 FOR I=1 TO 5:FOR J=1 TO 10:
   Z$=RIGHT$(STR$(INT(RND*10^4 MOD 999)+1001),3)+" "
45 GOSUB 8150:NEXT J:GOSUB 8300:NEXT I
50 Z$="Choose one of the above numbers ":GOSUB 8150:
   GOSUB 8200:A=DIGIT
60 IF A<0 OR A>999 THEN N=3:GOSUB 8000:
   Z$=Z$+"1 to 999":LC=0:GOSUB 8150:GOTO 50
65 IF A=0 THEN 40
70 RANDOMIZE A
500 DEF FNA(X)=TEMP(2)+TEMP(1)*(1+(59*(1-TEMP(4))))
600 N=49:GOSUB 8000:DATE(1)=1:DATE(2)=1:DATE(3)=86
650 TITLE$="FreeEnterprizes-PCAWS Ver 1.4"+
    "    -Example Work Study-"
700 FOR I=1 TO 50:READ DAT(I):NEXT I:FOR I=1 TO
    DAT(14):READ ACT$(I):NEXT I
760 DATES$="1/1/86":DAY$(0)="Wednesday"
780 FOR I=1 TO DAT(14):READ SAM(I):NEXT:
    FOR I=1 TO 7:READ DAY$(I):NEXT I
810 FOR I=1 TO 12:READ MON(I):NEXT
1000 L=45:MENU=4:MEN(0)=26:MEN(1)=21:MEN(2)=42:
     MEN(3)=23:MEN(4)=24
1101 TWO=18:GOSUB 7000:TWO=0
1105 IF DIGIT=5 THEN PRINT "system":BEEP:SYSTEM
1110 N=DIGIT+20:IF DIGIT=2 THEN N=42
1120 GOSUB 8000:Z$="     You have chosen   "+Z$:
     LC=0:GOSUB 8150
1130 L=DIGIT+20:IF DIGIT=2 THEN L=42
1150 ON DIGIT GOSUB 2000,3000,4000,5000:GOTO 1000
2000 REM        Determine Sample Size
2010 MENU=3:MEN(0)=27:MEN(2)=19:MEN(1)=18:MEN(3)=17:
     GOSUB 7000
2020 GOSUB 5500:SAMM=0
2030 FOR SA=1 TO DAT(14):SAM=SAM(SA)/100
2040 SAM=INT(Z*Z*SAM*(1-SAM)/(DAT(18)+.000001)^2+
     .499999)
2050 IF SAM>SAMM THEN SAMM=SAM
2080 NEXT SA:DAT(4)=SAMM
2090 N=38:GOSUB 8000:Z$=Z$+STR$(DAT(4)):GOSUB 8150:
     GOSUB 8300:RETURN
3000 REM        FORM PREPARATION
3010 MENU=15:MEN(1)=4:MEN(2)=7:MEN(3)=8:MEN(4)=10:
     MEN(5)=5:MEN(6)=6
3020 MEN(7)=33:MEN(8)=31:MEN(9)=16:MEN(10)=15:
     MEN(11)=13:MEN(12)=14
3030 MEN(13)=11:MEN(14)=12:MEN(15)=32:MEN(0)=27:
     GOSUB 7000:GOSUB 3100
3035 LP=LPW:LPW=0:SAMPLES=0:PAGE=0:DATE0$=DATES$
3040 IF SAMPLES>=DAT(4) THEN GOSUB 8300:LP=0:
     GOSUB 8300:RETURN
3041 PAGE=PAGE+1:IF LP=0 THEN GOSUB 8300:GOTO 3045
3042 LP=0:N=49:GOSUB 8000:GOSUB 8150:GOSUB 8200:LP=1
3044 IF A$<>"" THEN TITLE$=A$
3045 GOSUB 3600:Z$=TITLE$:LC=0:GOSUB 8150
3050 FOR I=6+DAT(14) TO DAT(15)-1 STEP 2+DAT(14):
     GOSUB 8300
3052 IF DAT(33)>1 THEN Z$="TRIP" ELSE Z$="VIEW"
3053 Z$=Z$+"        TIME!!":II0=DAT(13)*2-2*(4-
     DAT(13))*(DAT(13)<4)
3055 FOR J=20+II0 TO DAT(16) STEP 1+II0:
     SAMPLES=SAMPLES+1
3056 GOSUB 3200:IF ER=-5 THEN ER=0:GOTO 3052
3057 Z$=Z$+CHR$(65+INT(RND*DAT(33)))+" "+A$
3058 Z$=Z$+SPACE$(-2*(DAT(13)-4)*((DAT(13)-4)>0))+"!!"
3059 NEXT J:Z$=Z$+"COLUMN!!":LC=0:GOSUB 8150
3060 IF I>6+DAT(14) THEN 3066 ELSE C$="OBS. UNIT!"
3063 FOR J=20+II0 TO DAT(16) STEP 1+II0:
     FOR K=1 TO DAT(13):C$=C$+"!"+CHR$(64+K)
3064 NEXT K:C$=C$+SPACE$(-2*(4-DAT(13))*(DAT(13)<4)) +
     "!":NEXT J
3066 Z$=C$+"! SUMS !!":LC=0:GOSUB 8150:
     FOR J=1 TO DAT(14):Z$=ACT$(J)+"!":
3068 FOR K=20+II0 TO DAT(16) STEP 1+II0:
     FOR L=1 TO DAT(13):Z$=Z$+"! "
3070 NEXT L:Z$=Z$+SPACE$(-2*(4-DAT(13))*(DAT(13)<4)) +
     "!":NEXT K
3072 Z$=Z$+"!     !!":LC=0:GOSUB 8150:NEXT J:NEXT I
3074 Z$=CHR$(12):GOSUB 8150:GOSUB 3040
3100 REM     INITIALIZE FOR FORM SET UP
3105 FOR II=1 TO 3:DA(II)=DATE(II):NEXT
3110 TMIN=DAT(6)*60-TBRKS
3120 TMAX=TMIN/DAT(4)*DAT(7):
     IF DAT(8)>TMAX THEN DAT(8)=TMAX
3130 REAL=DAT(5)*60+TMAX/2
3140 LEAR=(DAT(5)+DAT(6))*60
3190 RETURN
3200 REM     SET A$ TO TIME
3210 TEMP=REAL+(TMAX-DAT(8))*(RND-.5):GOSUB 3300:A$=Y$
3220 MAX=TMAX
3270 TEMP=0:REAL=REAL+MAX:
     IF ABS(REAL-LEAR)<10 THEN GOSUB 3500
```

```
3280 RETURN
3300 REM           CONVERT TEMP TO Y$
3310 T=INT(TEMP/60):IF T>12 THEN T=T-12
3320 T=T*100+(INT(TEMP) MOD 60)+90000!
3330 Y$=STR$(T):Y$=MID$(Y$,3,2)+":"+RIGHT$(Y$,2)
3390 RETURN
3500 REM           DATE MAINTAINANCE
3510 DA(2)=DA(2)+1
3515 IF DA(2)<=MON(DA(1)) THEN 3520
3517 DA(2)=1:DA(1)=DA(1)+1:
     IF DA(1)>12 THEN DA(1)=1:DA(3)=DA(3)+1
3520 IF NOT(DA(2)>29 AND DA(3)/4=INT(DA(3)/4)) THEN 3550
3525 DA(2)=1:DA(1)=DA(1)+1:
     IF DA(1)>12 THEN DA(1)=1:DA(3)=DA(3)+1
3550 DATE0$="":FOR II=1 TO 3:TEMP(II)=DA(II)
3555 DATE0$=DATE0$+RIGHT$(STR$(TEMP(II)),1-
     (TEMP(II)>9))+"/":NEXT
3557 DATE0$=LEFT$(DATE0$,LEN(DATE0$)-1):
     GOSUB 7300:DAY$(0)=DAY$(TEMP)
3560 GOSUB  3110:Z$="New Day":LC=0:GOSUB 8150:
     GOSUB 8300:I=I+2
3570 ER=-5
3590 RETURN
3600 REM           FORM HEADER
3610 IF DAT(12)<>0 THEN 3620
3615 Z$=SPACE$((LEN(STR$(PAGE))+DAT(16))/2-
     2)+"Page"+STR$(PAGE)
3617 Z$=Z$+SPACE$((DAT(16)-LEN(STR$(PAGE)))/2+2):
     GOTO 3640
3620 II=LEN(DAY$(0))+LEN(STR$(PAGE))+LEN(DATE$)
3630 Z$=DAY$(0)+SPACE$((DAT(16)-II)/2-
     4)+"Page"+STR$(PAGE)
3635 Z$=Z$+SPACE$((DAT(16)-II)/2)+DATE0$
3640 LC=0:GOSUB 8150:GOSUB 8300
3650 RETURN
4000 REM           Accuracy Analysis
4010 MENU=4:MEN(0)=27:MEN(1)=4:MEN(2)=20:MEN(3)=17:
     MEN(4)=32:GOSUB 7000
4020 GOSUB 5500:LP=LP0:LP0=0
4025 IF  DAT(4)=0 THEN MENU=1:MEN(1)=4:
     GOSUB 7000:GOTO 4025
4027 GOSUB 8300:N=48:GOSUB 8000:GOSUB 8150:GOSUB 8300
4030 FOR SA=1 TO DAT(14):SAM=SAM(SA)/100
4040 RANGE=INT(Z*SQR(SAM-SAM^2)/DAT(4))*100)/100
4050 Z$=ACT$(SA)+SPACE$(6)+CHR$(9)+
     STR$(INT(SAM*DAT(14)*100)/100)
4055 Z$=Z$+CHR$(9)+STR$(INT(SAM*10000)/100)+CHR$(9)
4060 Z$=Z$+STR$(INT((SAM-RANGE)*10000)/100)+CHR$(9)+"    "
4070 Z$=Z$+STR$(INT((SAM+RANGE)*10000)/100):LC=0:GOSUB 8150
4080 NEXT SA:LP=0:RETURN
5000 REM           R T G
5010 MENU=5:MEN(0)=27:MEN(1)=9:MEN(2)=16:
MEN(3)=6:MEN(4)=5:MEN(5)=32:GOSUB 7000
5020 DAYS=DAT(9):LENGTH=INT((DAT(16)-
     9)/6):HOU=DAT(6):STA=DAT(5)-1:LP=LP0:LP0=0
5050 FOR D=1 TO DAYS:LC=0:
     Z$=CHR$(10)+"Day"+STR$(D)+CHR$(10):GOSUB 8150:LC=1
5060 FOR H=STA TO STA+HOU-1
5065 Z$=RIGHT$(STR$((H MOD 12)+101),2)+":"+"00 ¦ ":
     GOSUB 8150
5070 FOR T=1 TO LENGTH:
     TIM(T)=INT(RND*60)+INT(RND*60)*100:NEXT T
5100 FOR I=1 TO LENGTH:FOR J=I+1 TO LENGTH
5120 IF TIM(I)>TIM(J) THEN A=TIM(I):TIM(I)=TIM(J):TIM(J)=A
5140 NEXT J:Z$=MID$(STR$(TIM(I)+10000),3,2)+":"
5145 Z$=Z$+RIGHT$(STR$(TIM(I)+10000),2)+"  ":
     LC=1:GOSUB 8150
5150 NEXT I:LC=0:Z$="¦":GOSUB     8150:LC=1:
     NEXT H,D:LP=0:RETURN
5500 REM           NEWTUNS METHOD ON PROBABLITY CURVE
5510 X=0: W=0: ER=1E-37: P=(DAT(17)+1)/2
5520 Y=1/(.231645*ABS(X)+1):Z=1/SQR(2*3.14159*EXP(X*X))
5530 Q=Y^4*1.33027-Y^3*1.82126+Y^2*1.78148-
     Y*.356538+.319382
5540 Q=Q * Y : Z: IF   X < 0   THEN Q=1 - Q
5550 IF   ABS(Q + P - 1) <= ER   THEN 5590
5560 X=X + (Q + P - 1) / Z
5570 IF   W > 5   THEN ER=ER + .01: W=0 ELSE W=W + 1
5580 GOTO 5520
5590 Z=X: RETURN
7000 REM           MENU
7005 IF MENU<2 THEN DIGIT=1:LC=1:GOTO 7100
7006 MENU=MENU+1:MEN(MENU)=25+TWO
7007 GOSUB 8300:N=MEN(0):GOSUB 8000:A$=Z$:N=L:GOSUB 8000
7008 Z$=CHR$(9)+Z$+A$:GOSUB 8150
7009 IF NP=1 THEN 7080
7010 GOSUB 8300:FOR M=1 TO MENU
7020 N=MEN(M):GOSUB 8010:LC=M MOD 2
7025 IF Y(N)=0 THEN Z$=STR$(M)+". "+Z$+CHR$(9):GOTO 7040
7030 Z$=STR$(M)+". "+Z$+": ":IF Y(N)<=0 THEN 7037
7035 Z$=Z$+STR$(DAT(N))+CHR$(9):GOTO 7040
7037 N=N:GOSUB 7200:Z$=A$+Z$+CHR$(9)
7040 IF M<10 THEN Z$=" "+Z$
7050 GOSUB 8150:NEXT M
7070 GOSUB 8300
7080 N=1:GOSUB 8000:LC=1:GOSUB 8150:GOSUB 8200
7090 IF DIGIT<1 OR DIGIT>MENU THEN N=2:GOSUB 8010:
     GOSUB 8150:GOTO 7080
7093 IF MEN(DIGIT)=32 THEN LP0=1
7095 IF Y(MEN(DIGIT))=0 THEN RETURN
7097 IF Y(MEN(DIGIT))<0 THEN GOSUB 7500:GOTO 7099
7098 GOSUB 7100
7099 IF LLINE>20 THEN LLINE=0:GOTO 7007 ELSE 7010
7100 CHOICE=DIGIT: INDEX=MEN(CHOICE)
7110 N=MEN(CHOICE):GOSUB 8010:
     Z$="  Enter a value for "+Z$+": "
7120 GOSUB 8150:GOSUB 8200
7130 IF NOT(DIGIT<INT(Y(INDEX)/100) OR DIGIT>Z(INDEX))
     THEN 7140
7135 N=3:GOSUB 8010:LC=0:
     Z$=Z$+STR$(INT(Y(INDEX)/100))+" to"
7137 Z$=Z$+STR$(Z(INDEX))+".":GOSUB 8150:GOTO 7110
7140 DAT(MEN(CHOICE))=DIGIT:RETURN
7200 REM          SPECIAL MENU CASES Y(N)<0, CREATE STRING
7210 ON ABS(Y(N)) GOSUB 7230,7240,7240,7240,7250:RETURN
7230 Z$=STR$(DAT(14)):RETURN
7240 Z$=STR$(DAT(N)):RETURN
7250 Z$=DATES$:RETURN
7300 REM          DAY OF WEEK FROM DATE
7310 TEMP=365*(1900+TEMP(3))+TEMP(2)+31*TEMP(1)-
     31:YEAR=TEMP(3)+1900
7320 IF TEMP(1)>=3 THEN TEMP=TEMP-INT(.4*TEMP(1)+2.3)
     ELSE YEAR=YEAR-1
7330 TEMP=TEMP+INT(YEAR/4)-15:TEMP=
     INT(-TEMP/7)*7+TEMP+7:RETURN
7340 TEMP=INT(-TEMP/7)*7+TEMP+7:RETURN
7500 REM          Sample Percentage
7510 ON ABS(Y(MEN(DIGIT))) GOSUB 7520,7700,7600,
     7800,7860:RETURN
7520 REM        VALUES FOR PERCENTAGE OF SAMPLES
7525 IF DAT(14)<1 THEN GOSUB 7800
7527 GOSUB 8300:N=37+ONE:GOSUB 8000:LC=0:GOSUB 8150:
     N=47:GOSUB 8000
7528 LC=0:GOSUB 8150:GOSUB 8300:SM=0
7530 FOR SA=1 TO DAT(14)
7540 N=34+ONE:GOSUB 8000:Z$=STR$(SA)+". "+ACT$(SA)+Z$
7545 Z$=Z$+"<"+STR$(SAM(SA))+" >":GOSUB 8150
7550 GOSUB 8200:
     IF NOT(DIGIT<0 OR DIGIT>100) THEN 7560
7555 N=3:GOSUB   8000:LC=0:Z$=Z$+"1 to   100":
     GOSUB 8150:GOTO 7540
7560 IF LEFT$(A$,3)="ERR" THEN GOSUB 7600:GOTO 7540
7570 IF LEFT$(A$,3)="DON" THEN RETURN
7575 IF NOT(SM+DIGIT>100) THEN SM=SM+DIGIT:GOTO 7580
7577 N=41:GOSUB 8000:Z$=Z$+STR$(100-SM)+" >":LC=0:
     GOSUB 8150:GOTO 7540
7580 SAM(SA)=DIGIT
7590 NEXT SA:RETURN
7600 REM          ENTER TIMES FOR BREAKS, LENGTHS
7610 GOSUB 8300:DA=0:N=11:GOSUB 8000:C$=Z$:N=28
7615 GOSUB 8000:Z$=CHR$(9)+C$+Z$:LC=0:GOSUB 8150:
     N=47:GOSUB 8000
7617 LC=0:GOSUB 8150:GOSUB 8300:N=46:GOSUB 8000:
     C$=Z$:TBRKS=0
7620 DA=DA+1:DE=INT(DA/2)
7621 IF DA/2<>DE THEN B$="Start:":ER=0
     ELSE B$="Length:":ER=-1
7622 IF DA/2<>DE THEN   Z$=STR$(DA/2+.5)+".       "+C$+B$
     ELSE Z$="          "+C$+B$
7625 GOSUB 8150:GOSUB 8200:IF A$="DONE"
     THEN DAT(11)=INT(DA/2):RETURN
7628 Z$=":":GOSUB 7900
7629 IF ER=1 THEN ER=0:GOTO 7621
7630 FOR I=1 TO 3:BRKS(DA,I)=TEMP(I):NEXT I
7680 BRKS(DA,3)=FNA(I)
7690 IF DA/2=DE THEN TBRKS=TBRKS+BRKS(DA,3)
7695 GOTO 7620
7700 REM          ENTER DATES FOR HOLIDAYS
7710 GOSUB 8300:DA=0:N=29:GOSUB   8000:C$=Z$:N=28:
     GOSUB 8000:Z$=CHR$(9)+C$+Z$
7715 GOSUB   8150:N=47:GOSUB 8000:LC=0:GOSUB 8150:
     GOSUB 8300
7720 DA=DA+1:DE=INT(DA/2):IF DA/2<>DE THEN B$=" From:"
     ELSE B$=" To:"
7722 IF DA/2<>DE THEN   Z$=STR$(DA/2+.5)+".       "+C$+B$
     ELSE Z$=CHR$(9)+C$+B$
7725 GOSUB 8150:GOSUB 8200:IF A$="DONE"
     THEN DAT(12)=INT(DA/2):RETURN
7726 Z$="/":ER=0:GOSUB 7900
7727 IF NOT((TEMP(3)<DATE(3))  OR (TEMP(3)=DATE(3)  AND
     TEMP(1)<DATE(1))  OR  (TEMP(3)=DATE(3)  AND
     TEMP(1)=DATE(1) AND TEMP(2)<DATE(2))) THEN 7729
7728 ER=1:Z$=CHR$(9)+"A date after "+DATES$:
     LC=0:GOSUB 8150
7729 IF ER=1 THEN ER=0:GOTO 7722
7730 FOR I=1 TO 3:HOLI(DA,I)=TEMP(I):NEXT I:GOTO 7720
7800 REM          ENTER ACTIVITIES
7810 ACT=0:GOSUB 8300:N=40:GOSUB 8000:LC=0:
     GOSUB 8150:N=47:GOSUB 8000
7815 LC=0:GOSUB 8150:GOSUB 8300
7820 ACT=ACT+1:Z$="Activity"+STR$(ACT)+": "+ACT$(ACT):
     GOSUB 8150:GOSUB 8200
7840 IF A$="" THEN 7820 ELSE IF A$="DONE"
     THEN DAT(14)=ACT-1:RETURN
7850 ACT$(ACT)=LEFT$(A$+SPACE$(9),9):Z$=ACT$(ACT):
     GOSUB 8150:GOSUB 8300
7855 GOTO 7820
7860 REM          STARTING DATE
7870 ER=0:N=30:GOSUB 8000:Z$=Z$+DATES$+"]:":GOSUB 8150:
     GOSUB 8200
7875 Z$="/":GOSUB 7900:IF ER=1 THEN 7870
7880 DATES$="":FOR I=1 TO 3:DATE(I)=TEMP(I)
7885 DATES$=DATES$+RIGHT$(STR$(DATE(I)),1-(DATE(I)>9))+"/":
     NEXT I
7890 DATES$=LEFT$(DATES$,LEN(DATES$)-1):GOSUB 7300:
     DAY$(0)=DAY$(TEMP)
7895 RETURN
7900 REM  MENU UTILITY: IN:Z$="/",":"
     AND A$="STRING"; OUT:TEMP(1-3)
7910 TEMP(4)=1:W=1:FOR TEM=1 TO 3:TEMP(TEM)=0
7920 IF W=1 THEN TEMP(TEM)=VAL(A$):W=0
7930 FOR TE=1 TO LEN(A$)
7940 IF NOT(MID$(A$,TE,1)=Z$) THEN 7950
```

```
7945 A$=RIGHT$(A$,LEN(A$)-TE):TEMP(4)=TEMP(4)+1:
     W=1:GOTO 7960
7950 NEXT TE
7960 NEXT TEM
7965 IF Z$="/" AND TEMP(4)<3 THEN ER=1:RETURN
7966 IF ER>-1 AND Z$=":" AND TEMP(4)<2 THEN ER=1:RETURN
7967 IF ER=-1 AND TEMP(4)>2 THEN ER=1:RETURN
7970 IF Z$="/" THEN 7971 ELSE 7972
7971 LL(1)=1:UL(1)=12:LL(2)=1:UL(2)=31:LL(3)=1:
     UL(3)=99:GOTO 7975
7972 LL(1)=1:UL(1)=12:LL(2)=0:UL(2)=59:LL(3)=0:UL(3)=59
7975 IF ER=-1 AND TEMP(4)=1 THEN LL(1)=0:UL(1)=59
7980 FOR TEM=1 TO 3
7981 IF NOT(TEMP(TEM)<LL(TEM) OR  TEMP(TEM)>UL(TEM))
     THEN 7984
7982 N=3:GOSUB 8000:Z$=Z$+STR$(LL(TEM))+" to"+
     STR$(UL(TEM))
7983 LC=0:GOSUB 8150:ER=1
7984 IF ER=-1 AND TEMP(4)=1 THEN TEM=5
7985 NEXT TEM: IF TEMP(4)<>1 THEN TEMP(4)=0
7987 IF ER>-1 AND Z$=":" THEN IF TEMP(1)<DAT(5)
     THEN TEMP(1)=TEMP(1)+12
7988 IF ER=-1 THEN ER=0
7990 RETURN
8000 REM         UTILITIES
8010 REM         GET
8020 RESTORE:FOR D=1 TO N:READ Z$,Y(D),Z(D):
     NEXT:RETURN
8150 REM         PRINT
8151 IF NP=1 THEN RETURN: '**********************************
8152 FOR LPP=1 TO LEN(Z$)
8155 IF MID$(Z$,LPP,1)<>CHR$(10) THEN 8157
8156 IF LP=1 THEN LPLINE=LPLINE+1 ELSE LLINE=LLINE+1
8157 NEXT LPP
8160 IF LP+LC=2 THEN LPRINT Z$;:RETURN
8170 IF LP+LC=0 THEN PRINT Z$:LC=1:
     LLINE=LLINE+1:RETURN
8180 IF LP=0 THEN PRINT Z$;:LC=1 ELSE LPRINT Z$:
     LPLINE=LPLINE+1
8190 RETURN
8200 REM         INPUT
8210 INPUT  Z$:DIGIT=VAL(Z$):LLINE=LLINE+1:A$="":
     FOR INPU=1 TO LEN(Z$)
8230 A=ASC(MID$(Z$,INPU,1)):IF A>96 THEN A=A AND NOT(32)
8240 A$=A$+CHR$(A):NEXT:RETURN
8300 LC=0:Z$="":GOSUB 8150:RETURN
9000 REM         BANK OF DATA
9010 DATA "     Enter the number for your   choice
     : ",1,1  :'01
9020 DATA "           This number is not listed above.
     ",1,1    :'02
9030 DATA "        A number from ",1,1
9040 DATA "Number of Samples ",1,10000
9050 DATA "Start of Day ",1,24
9060 DATA "Hours in the Day ",1,24
9070 DATA "Study Length(Days)",1,2000
9080 DATA "Least between time",1,1000
9090 DATA "Length in Days ",1,90
9100 DATA "Days in a week ",1,7
9110 DATA "Break times,length",-3,12
9120 DATA "Day-Date on Forms ",0.001,1
9130 DATA "Positions ",1,25
9140 DATA "Activities ",-4,50
9150 DATA "Lines per page ",10,999
9160 DATA "Columns per line ",20,999
9170 DATA "Desired Confidence    ",0.001,1
9180 DATA "Study Accuracy        ",0.001,100
9190 DATA "Study Percentages    ",-1,100
9200 DATA "Sample Percentage     ",-1,100
9220 DATA "Determine Sample Size",0,0
9230 DATA "Print out Information",0,0
9240 DATA "Accuracy Analysis     ",0,0
9250 DATA "Random time generator",0,0
9260 DATA "Table is now complete",0,0
9270 DATA "All   Available  - Table  Level   One",0,0
9280 DATA "  - Table Level Two",0,0
9290 DATA "  - Table Level Three",0,0
9300 DATA "Dates for Holidays",0,0
9310 DATA "The Studies Starting Date [",0,0
9320 DATA "Starting Date ",-5,1
9330 DATA "Output data to printer",0,0
9340 DATA "Starting points ",1,25
9350 DATA " has a Sample Percentage of : ",1,1
9360 DATA "Enter the Number for the percentage in ",1,1
9370 DATA " has an Accuracy of :",1,1
9380 DATA " Samples for Activities -
     Table Level Three",1,1
9390 DATA " Minimum Number of Samples should be",1,1
9400 DATA " Accuracy for Activities  -
     Table Level Three",1,1
9410 DATA " Activities - Table Level Three",1,1
9420 DATA " Total percentage can not exceed  100
     <remaining",1,1
9430 DATA "Prepare WS Forms",0,0
9440 DATA "Return to DOS ",0,0
9450 DATA "PCAWS - Preparatory Computer Assistance for
     Work Sample Study FreeEnterprizes 2/85 Ver 1.4
     Written by D.K. Freeman",0,0
9460 DATA "",0,0
9470 DATA "Break Time ",0,0
9480 DATA "Type DONE when done.",0,0
9490 DATA "Activity       No of Obs.    %      LCL%
     2(SIGMA)  UCL%",0,0
9500 DATA "Prepare Printer,  Enter a TITLE and strike
     return",5,6   :'49
9600 DATA 0,0,0,821,9,   9,10,0,5,5,    0,1,5,6,65
9610 DATA 00,.75,.03,0,0  5,5,0,0,0,    0,0,0,0,0
9620 DATA 2,3,4,5,0       0,0,0,0,0,    0,0,0,0,0,
     0,0,0,0,0, 0,0
9640 DATA "Basic Job","Planning ","Delay     "
9641 DATA "Travel    ","Other    ","Personal   "
9650 DATA 14,21,18,26,10,11
9660 DATA "Sunday","Monday","Tuesday","Wednesday"
9665 DATA "Thursday","Friday","Saturday"
9670 DATA 31,28,31,30, 31,30,31,31, 30,31,30,31
```

David M. Rhyne, P.E., is an assistant professor in the Department of Management, Auburn University. He holds a BSIE and an MS in engineering administration from the University of Tennessee and a PhD in engineering management from Clemson University. Rhyne is a senior member of IIE.

Douglas K. Freeman received a BSEE from Auburn University in 1985. He has experience as a design engineer in telephony. A member of IEEE, Freeman's computer experience and interests include hardware design, robotics, EM radiation and exhaustive model programming.

WORK/ACTIVITY SAMPLING-CONTEMPORARY DESIGN ANALYSIS METHODOLOGY AND APPLICATIONS: PART II - WORK SAMPLING CALCULATIONS REVISITED

Elinor S. Pape
Industrial Engineering Department
University of Texas at Arlington
Arlington, Texas 76019

ABSTRACT

A procedure for computing work sampling interval estimates of proportions and sample sizes is presented that incorporates some realistic assumptions about the work situation and about the manner in which data is collected. This procedure uses traditional work sampling calculations as a starting point and progresses to alternate formulas that allow for correlations between members of a work crew and for random variations from day to day (sampling period to sampling period).

INTRODUCTION

Consider a_i to be a single random observation of a single worker. Then a_i takes the value 1 with probability p and takes the value 0 with probability $1 - p$; that is a Bernoulli trial. The value 1 indicates that the worker is performing the activity of interest.

If N observation times are selected at random and N a_i's are recorded their sum $X = \Sigma a$ has a binomial distribution with parameters N and p. Then $E(X/N) = p$ and $V(X/N) = p(1-p)/N$ so it can be stated with (approximately) 95% confidence that

$$\hat{p} - 1.96\sqrt{\frac{\hat{p}(1-\hat{p})}{N}} < p < \hat{p} + 1.96\sqrt{\frac{\hat{p}(1-\hat{p})}{N}} \quad (1)$$

where $\hat{p} = X/N$, and hence to produce an interval $\hat{p} \pm \Delta$ will require

$$N = \frac{\hat{p}(1-\hat{p})(1.96)^2}{\Delta^2} \text{ observations.}$$

This is the classic work sampling theory that has been popularized and heavily utilized by Brisley [1] and many others.

This paper questions (a) whether p is really the parameter for which an estimate is desired, and (b) whether X is, in practice, really binomially distributed and hence whether $\hat{p}(1-\hat{p})/N$ is a good variance estimator.

Interval Estimate of What?

Interval (1) is a confidence interval, or interval estimate, or the proportion of the single worker's time occupied by the activity of interest during the time period of the study. This is rarely ever the proportion for which an estimate is desired. More often the practitioner recognizes that the proportion of time occupied by the activity of interest is subject to random variation over the period of the study and also varies from worker to worker engaging in the activity. The sampled proportions are viewed as realizations of a random variable and a confidence interval is generally desired for the mean of that random variable. Recognition of the proportions as a random variable requires modeling of the variation and many different models can be considered.

Different models result from different assumptions about the process being studied and the differences may well cause the practitioner to feel insecure. However the classic model is also based on assumptions and the assumption that p is constant from time period to time period is difficult to defend. There are some cases [3] when the use of interval (1) may be justified despite random variations in p due to high variance of alternate variance estimators.

Distribution of X

A Single Worker - When only a single worker or machine is being studied and the true proportion is subject to random variation, the number of times the worker is observed performing the activity, X_j, might well be assumed to have a binomial distribution with parameters n_j and p_j. However $\Sigma X_j = X$ would not have a binomial distribution. The expectation of X would be $\Sigma n_j p_j$ and its variance would be $\Sigma n_j p_j (1 - p_j)$. Then if the p_j are assumed to be independent realization of a random variable with mean π and variance σ^2 the expectation of X becomes $N\pi$, where $N = \Sigma n_j$, and the variance of X becomes $N\pi(1-\pi) + (\Sigma n_j^2 - N)\sigma^2$. The binomial variance estimator used in interval (1) estimates only the first term of this expression and is, hence, too small if σ^2 is non-zero.

A Crew of Workers - When several, say K, workers are sampled on the same observation round with the idea of pooling those observations into a single estimate, the K observations cannot legimately be considered independent. There are cases [6] in which use of interval (1) is still justifiable but those cases are difficult to identify in practice. If the true proportion for the kth worker is p_k and if the kth and k'th workers are not independent but have a correlation coefficient of $\gamma_{kk'}$, the distribution of X becomes more complex. Let X_i be the number of crew members observed performing the activity of interest on the ith round. $E(X_i) = \Sigma p_k$ which can be defined equal to Kp but the variance of X_i is a complex function of p_k's and $\gamma_{kk'}$'s. The distribution of X_i is binomial only if the p_k's are equal and the γ's are 0.

If p = .2 and K = 5 the binomial distribution of X_i would look like Figure 1. It's mean value is 1. Four other possible distributions of X_i each with a mean value of 1 are shown in Figures 2-5. Figure 2 shows the 5 workers idle in unison - on every observation round we would find either 0 idle or 5 idle. In Figure 3 at most 3 workers are idle at one time, that is 2 must always be working. Figures 4 and 5 represent realistic distributions in cases where the 5 workers might easily, though erroneously, be considered independent. In Figure 4 the variance of X_i is greater than the binomial variance; a pattern typical of mild positive correlations between workers, in Figure 5 the variance is less than the binomial variance, a pattern typical of negative correlations between workers and/or of greatly different p values for different workers.

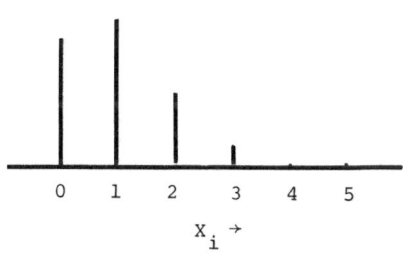

Figure 1. Binomial

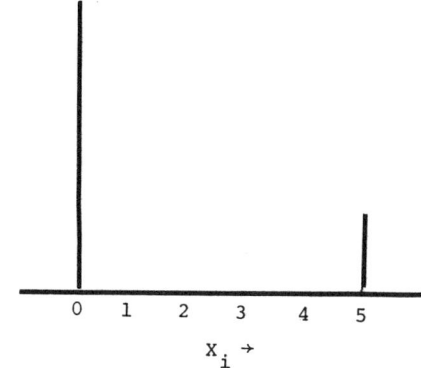

Figure 2. Totally Correlated

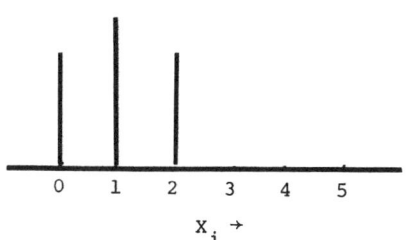

Figure 3. $X_i \leq 2$

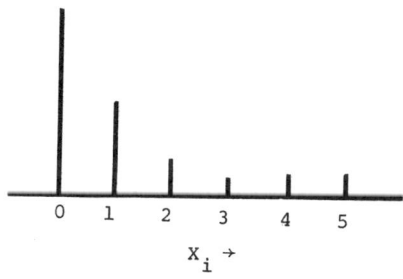

Figure 4. Positively Correlated

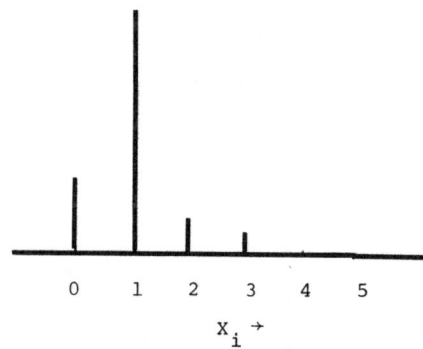

Figure 5. Negatively Correlated

ESTIMATOR	EXPECTED VALUE OF ESTIMATOR
$V_1 = \dfrac{\hat{p}(1-\hat{p})}{N} = \dfrac{s^2(a_{ijk})}{N}$	$\dfrac{\Pi(1-\Pi)}{N-1} - \dfrac{IK\sigma_p^2}{N(N-1)} - \dfrac{\sigma_v^2}{N(N-1)}$
$V_2 = \dfrac{\Sigma\Sigma(\hat{p}_{ij}-\hat{p})^2}{IJ(IJ-1)} = \dfrac{s^2(X_{ij})}{KN}$	$\dfrac{\sigma_v^2}{N} + \dfrac{(J-1)IK\sigma_p^2}{(IJ-1)N}$
$V_3 = \dfrac{\Sigma(\hat{p}_j-\hat{p})^2}{J(J-1)} = \dfrac{s^2(X_j)}{IKN}$	$\dfrac{\sigma_v^2}{N} + \dfrac{IK\sigma_p^2}{N}$

where $s^2(X_{ij}) = \dfrac{\sum\sum_{ij}^{JJ} X_{ij}^2 - (\sum\sum_{ij}^{JJ} X_{ij})^2/IJ}{IJ-1} = \dfrac{\Sigma\Sigma(X_{ij}-\bar{X})^2}{IJ-1}$

and $s^2(X_j) = \dfrac{\sum_j^{J}(\sum_i^{I} X_{ij})^2 - (\sum_j^{J}(\sum_i^{I} X_{ij}))^2/J}{J-1} = \dfrac{\Sigma(X_j-\bar{X})^2}{J-1}$

Table 1. Candidate Estimators

A SIMPLE ALTERNATE MODEL

In order to model correlated work crews in the face of randomly varying proportions the basic modeling unit is considered to be X_{ij}, the number of members of a crew of size K who are performing the activity of interest during the ith observation round of the jth day (random period) of the study. To simplify the model K, the crew size; I, the number of observations in a day; and J, the number of days of the study are all considered to be constant. Later in this paper computational adjustments for unequal crew size are given.

The conditional expectation of X_{ij} is Kp_j, and the expectation of p_j is π while its variance is σ_p^2. The conditional variance of X_{ij} is Kv_j and the expectation of v_j is σ_v^2. Then the overall expectation of X_{ij} is $K\pi$ and the overall variance of X_{ij} is $K\sigma_v^2 + K^2\sigma_p^2$. Note that this model does not assume any binomial distribution. Different restrictive assumptions about X_{ij}, based on binomials, can be incorporated into this model by stating v_j in terms of p_j; e.g. if independent with equal p_j's, $v_j = p_j(1-p_j)$; if equicorrelated $v_j = (1 - \gamma + k\gamma)p_j(1-p_j)$. If v_j is a function of p_j, σ_v^2 will be a function of σ_p^2. But in all cases Var $X/N = \sigma_v^2/N + IK\sigma_p^2/N$. This is the variance of $\hat{p}$ for which an estimate is desired. Three candidate estimators are proposed, the binomial estimator, V_1; the sample variance of the X_{ij}, V_2; and the sample variance of the $X_j = \sum_i X_{ij}$, V_3. These estimators and their expected values are given in Table 1.

The expected values of V_2 and V_3 will be equal if $\sigma_p^2 = 0$, that is, if the proportion being estimated does not change from day to day, i.e. random period to random period. The expected value of V_1 would equal the other two if $\sigma_p^2 = 0$ and if all members of a crew had identical p proportions and if all observations of crew members were completely independent. If a single worker is sampled K = 1 and $V_1 = V_2$ because σ_v^2 would equal $\pi(1-\pi) - \sigma_p^2$.

The effects of various conditions, or modeling assumptions, can be found by studying Table 1 or very quickly from Table 2 in which the effects are summarized.

CONDITION	V_1	V_2	V_3
Difference in Crew Members	Large	✓	✓
Positive Correlations	Small	✓	✓
Negative Correlations	Large	✓	✓
Random Effects	Small	Small	✓

Table 2. Effects on Estimators

Looking only at the expected values V_3 would seem to be the preferred estimator. But V_3 may be a very high-variance variance estimator if J is small. If data only exists for four or five days V_3 would be an unreliable estimator even though it would be unbiased. The fact that the equation defining V_3 averages out well over hundreds of uses is small comfort to the practioner who is looking for a single result. Since an S^2 with a small number of degrees of freedom is a poor estimator of σ^2 the formulas for sample size based on V_2 and V_3 given in Table 3 are of little value early in a study.

ESTIMATOR	J, # DAYS OF STUDY (I rounds per day) (K members in crew)
V_1	$Qp(1-p)/IK = J_1$
V_2	$Q[\sigma^2(X_{ij})]/IK^2 = J_2$
V_3	$Q[\sigma^2(X_j)]/I^2K^2 = J_3$

$Q = (Z_{\alpha/2}/\Delta)^2$

$S^2(X_{ij})$ is an estimate of $\sigma^2(X_{ij})$

$S^2(X_j)$ is an estimate of $\sigma^2(X_j)$

Table 3. Length of Study to Achieve $\hat{p} \perp \Delta$ with confidence $1 - \alpha$

A SENSIBLE APPROACH TO THE USE OF THIS MODEL

The immediate objective of a work sampling study is a confidence interval for a proportion of the sort

$$\hat{p} - C_{\alpha/2}\sqrt{V} < \pi < \hat{p} + C_{\alpha/2}\sqrt{V}, \quad (2)$$

where V is V_1, V_2, or V_3 and $C_{\alpha/2}$ is a $1 - \alpha/2$ percentage point of a normal, or t, distribution. $C_{\alpha/2}$ is generally taken as 1.96 when a 95% confidence interval is desired but some practitioners prefer to use simply 2.

Start with a rough estimate of p and use it to find a preliminary estimate of the study size by making J_1 in Table 3 as large as feasible bearing in mind that it may need to be increased from the first estimate to achieve the desired accuracy. The crew size, K, will be determined by the situation and then I, the number of observation rounds per day will be $Qp(1-p)/J_1K$.

After several days of the study, when IJ > 30, calculate V_2 and V_1. If $V_2 > V_1$ the study may need to be extended to achieve the desired accuracy, if $V_2 < V_1$ it may be shortened. Estimate J_2 from Table 3 using $S^2(X_{ij})$ to estimate $\sigma^2(X_{ij})$. The value J_2 is the new length of study in days. This can be repeated whenever an update is desired. If V_2 consistently exceeds V_1 the increase may be due to σ_p^2 or to positive correlations among crew members. If it is largely due to σ_p^2 the statistic V_2 will still tend to under estimate Var $\hat{p}$. V_3 should be calculated after about 10 days, but 10 is still so small that the quantity $Z_{\alpha/2}$, a percentile point from a normal table, in Table 1 should be replaced with $t_{\alpha/2}$, an equivalent percentile point from a t distribution with $J - 1$ degrees of freedom. If and when 30 days of data exist the $Z_{\alpha/2}$ value is appropriate.

When $V_2 < V_1$ crew effects are possibly making σ_V^2 smaller than an expected binomial variance. It is still possible that σ_p^2 is large enough to make V_3 significantly larger than V_2, but not likely.

Computer programs (and calculator program) can be written to calculate J_1, J_2 and J_3, and interval (2) using V_1, V_2 and V_3; when J < 2, V_1 and J_1 should be used; when 2 < J < 10, V_2 and J_2 should be used; when $10 \leq J \leq 30$ max (V_2, V_3) and max (J_2, J_3) should be used; and when 30 < J, V_3 and J_3 should be used

COMPUTATIONAL ADJUSTMENTS

Adjustment for Unequal Sample Sizes

When the crew size and number of observation rounds per day are not constant the crew size on the ith observation round of the jth day will be represented as K_{ij} and the number of rounds on the jth day as I_j.

Then $V_2 = [\Sigma\Sigma X_{ij}^2/K_{ij} - X^2/N]/N(\Sigma I_j - 1)$

and $V_3 = [\Sigma X_j^2/N_j - X^2/N]/N(J-1)$

where $N_j = \sum_i K_{ij}$ and $N = \sum_j N_j$.

The expected values of V_1, V_2 and V_3 are also subject to adjustment but the summary in Table 2 is valid.

Adjustment for Stratification

If stratification of work sampling observations, i.e. systematic sampling, can be achieved without incurring a bias in $\hat{p}$ (without worker manipulation) the effects are all beneficial. The variance of $\hat{p}$ is reduced and the expectation of V_3 is reduced accordingly. Unfortunately the expectation of V_1 and V_2 are actually increased somewhat. The nature and extent of these changes are complex and are not modeled in this paper.

One form of stratification that causes V_3 to become unbiased, too large, is when the J random effects are accompanied by repeatable fixed effects, e.g. day of the week effects. The Monday, Tuesday,... Friday effects inflate V_3 but cancel out in the expression for the variance of $\hat{p}$. Think of the days as replicated determinations of several different treatments and the solution to this problem is clear from analysis of variance theory.

Calculate ΣX_{ij} and ΣK_{ij} for each of the L days of the week, call them $X_{(\ell)}$ and $N_{(\ell)}$. Find Adj = $\Sigma X^2_{(\ell)}/N_{(\ell)} - X^2/N$ and adjust V_3 to V'_3 by the equation

$$V'_3 = [(J-1)V_3 - Adj/N]/(J-L)$$

Then V'_3 will be an unbiased estimate of Var $\hat{p}$.

When stratification occurs within the random period, as hours or shifts of a day, no such correction is indicated. Only when the X_j's can be divided into several groups, each group associated with a different fixed effect, should the variance between groups be eliminated.

SUMMARY

For years work sampling has been an effective management tool. The model that has been used is simple and generally gives good results, not because the model really fits the sampling methods generally used, but because of compensating errors. The variance estimator traditionally proposed, $\hat{p}(1-\hat{p})/N$ is frequently at the same time too small and too large. It may often be close to unbiased, but there is no guaranty.

In slide rule days the use of any model other than the traditional one would not be justified. The alternate models require more arithmetic than the improvement was considered to be worth twenty years ago. However more and more programs are being written every day to summarize work sampling data on a daily basis. Work sampling clipboards can now electronically store data for later computer consumption. For a computer one variance calculation is as easy as another so the 1980's are the time to bring work sampling theory up to date.

In this paper three variance estimators are proposed each related to a necessary-sample-size formula. It is suggested here that one start with the first estimator (the traditional one), change to the second after 2 days, and change to the third sometime between the 10th and 30th day of the study.

REFERENCES

[1] Brisley, C. L., "Work Sampling," *Industrial Engineering Handbook*, editor H. B. Maynard, McGraw-Hill, 1956.

[2] Kumar, K. R. and Pape, E. S., "Selecting a Test Level for Random Effects," *Communications in Statistics*, Vol. 7, No. 7, Fall 1978.

[3] Kumar, K. R. and Pape, E. S., "Determination of Level for a Test of Variation in a Bernoulli Process," submitted for publication in *Technometrics*, 1979.

[4] Moder, J. J. and Halladay, W. J., "Work Sampling Applied to Long Cycle Operations Performed by a Variable Labor Force," *Journal of Industrial Engineering*, 1956, Vol. VII, No. 4.

[5] Moder, J. J., "Activity Sampling with Applications to Time Standard Estimation," *Journal of Industrial Engineering*, 1967, Vol. 18, No. 1.

[6] Pape, E. S., "Correlated Binomials: Work Sampling of Crews," submitted to *AIIE Transactions*, 1979.

BIOGRAPHICAL SKETCH

Elinor S. Pape is an Associate Professor of Industrial Engineering at the University of Texas at Arlington where she has been employed since 1972. She holds a B.S. degree in I.E. and M.S. and Ph.D. degrees in statistics all from Southern Methodist University. She is a senior member of AIIE in which she currently serves as Vice President of the Dallas Chapter, and she is an active member of the American Statistical Association and the Society of Women Engineers.

V. Predetermined Time Systems

Predetermined time systems (PTSs) are used to estimate the amount of time required to complete tasks in production, service industries, clerical areas, and many other fields. Defining the time required to complete a task helps the analyst measure the performance of individuals, estimate the costs of various tasks, and choose the best method for minimizing required time.

The Industrial Engineering Terminology Standard Z94.12 defines predetermined time system as "an organized body of information, procedures, techniques, and motion times employed in the study and evaluation of manual work elements. The system is expressed in terms of the motions used, their general and specific nature, the conditions under which they occur, and their previously determined performance times."

PTSs have countless applications; choosing the system that is best for a particular application is important. The PTSs covered in this chapter include packages that integrate PTSs with CAD/CAM software, utilizing the available information to perform several types of analyses.

ADVANTAGES AND LIMITATIONS

Although predetermined time standards have existed for more than forty years, many firms still use the stopwatch time study technique; each of the two methods has its place.

PTSs have the following advantages for both direct and indirect labor activities: 1) developing methods and cycle times in advance of actual production; 2) improving existing methods; 3) developing time formulas for standard data; 4) eliminating the need for performance rating; 5) establishing consistent work standards; 6) providing a detailed record for operator training; 7) analyzing the ability of people; and 8) comparing the time for alternative methods. Computer-aided systems offer improved accuracy as well as greater speed in determining and applying standards. Such systems are particularly appropriate where short-cycle, repetitive operations are involved, or where high productivity losses or excess labor complaints occur.

PTSs also have several limitations. First, using MTM as an example, many professionals argue that the individual elements are not independent of one another, and that a particular motion will influence the preceding and following motions. In such a case, the total time calculated for a given operation will be inaccurate. Independence of motions can be extremely difficult to determine in systems that are broken down into very detailed motions. In less detailed systems, individual motions are combined into a more general movement, and the independence of movements becomes somewhat easier to determine. Second, PTSs are not universally applicable. Level I systems are of great value in short-cycle, repetitive jobs, while higher-level systems are not particularly suited to such jobs. Finally, PTSs require the analyst's judgment in applying the data and interpreting the rules associated with the specific PTS.

SUMMARY OF ARTICLES

The lead article, "Comparison of Predetermined Time Systems" by Chester L. Brisley, provides a thorough introduction to the topic. Brisley begins by providing an historical background, emphasizing the contributions of Frederick W. Taylor, Henry L. Gantt, Frank B. Gilbreth, A.B. Segur, and the team of Ralph Presgrave and Gerald B. Bailey. He then explicates a number of major PTSs. These include Motion Time Study (MTS), developed by the General Electric Company; Joseph Quick's Work-Factor System; the Maynard Operational Sequence Technique (MOST), developed by Kjell Zandin for H.B. Maynard and Company; Modular Arrangement of Predetermined Time Standards (MODAPTS), an Australian-created system; Master Standard Data (MSD), developed by Serge A.

Birn Company; and Douglas M. Towne's Cue system. Finally, the article covers the International Methods-Time Measurement family of data systems: MTM-1, MTM-2, MTM-3, MTM-GPD, M4, MTM-V, MTM-M, and MTM-C.

In "Today's International MTM Systems—Decision Criteria for Their Use," Karl Eady presents an updated overview of the systems comprising the MTM family. After showing how each family member fits into a classification scheme, he gives the characteristics of each system, emphasizing accuracy levels, speed of application, and degree of methods description provided. He concludes by offering highly useful guidelines for selecting the appropriate system for a specifc work situation.

"A High Level Predetermined Time Standard System and Short-Cycle Tasks" by Abu Masud, Don Malzahn, and Scott Singleton examines the accuracy of high level synthetic time standard systems for establishing performance estimates for short-cycle tasks. These systems are widely accepted by the industrial community for tasks with cycle times greater than one minute, primarily because of the systems' speed and simplicity of application. Given the desirability of obtaining accurate performance estimates for short-cycle tasks, the authors tested the MODAPTS system on tasks of less than 0.5 minutes. The study showed that MODAPTS can be successfully applied to short-cycle tasks with certain exceptions.

"Robot Time and Motion System Provides Means of Evaluating Alternate Robot Work Methods" by Shimon Y. Nof and Hannan Lechtman presents a methodology for robot work analysis and performance measurement. Analogous to MTM, RTM (robot time and motion) is based on standard elements of fundamental work motions. The RTM system attempts to model the work of robots by utilizing knowledge of their mechanical design, control, and work patterns. The authors note that the system is user friendly, with the capacity of systematically specifying a work method for a given robot; applying computer aids to evaluate a specified method and compare it to alternative methods; and repeating methods evaluation for alternative robot models. The bulk of the article is devoted to an exhaustive analysis of the system's three major components: RTM elements, robot performance models, and an RTM analyzer.

COMPARISON OF PREDETERMINED TIME SYSTEMS (PTS)

Chester L. Brisley, Ph.D.

Department of Engineering and Applied Science

University of Wisconsin-Extension

Milwaukee, Wisconsin

With respect to work measurement, a considerable amount of effort has been focused in trying to determine how much time is required for a person to perform a given job. Today we are striving to analyze some of the predetermined time systems.

In preparation for this presentation, we have visited and had discussions with many authorities on the subject of work measurement and predetermined time systems. These pioneers have invested their time over a period of years in motion-time relationships. As we proceed we shall give credit for the contribution of these people as we discuss the various systems.

There are certain principles and rules with regard to predetermined time systems that we would like to discuss which have been debated over many years by some of the outstanding pioneers and current researchers in the subject of time measurement.

Work and Rest Increases Productivity

A principle emphasized by Taylor was that of overcoming fatigue by working and resting short periods. This was demonstrated at the Bethlehem Steel Company in which a Pennsylvania Dutchman, named Schmidt, proved that he had endurance by virtue of the fact that he would run a mile to work in the morning, work twelve hours, and run back home in the evening after the day's work was completed. During the evenings he also was working on a new home that he was building. Prior to going to work in the morning, he would work in his garden. Schmidt was selected from a group of 75 men. He was earning $1.15 per day loading pig-iron onto railroad cars. Taylor asked Schmidt, "How would you like to earn $1.85 per day?" Schmidt answered, "That will be a long time before I can possibly earn that much." Wage increases at that time were on the order of five cents per day and increases didn't come frequently. Taylor had the privilege, as a consultant, to raise wages when he increased productivity. He told Schmidt that he would pay him $1.85 per day to load pig-iron exactly as he told him to do. He worked when he was told to work and rested when he was told to rest, and subsequently loaded 47 half-tons a day instead of the 12 half-tons he had previously loaded. With this work and rest pattern he seemed no more fatigued than at the previous rate.

Written Instructions of Methods to be Employed

Henry L. Gantt also made contributions. It was obvious that the influence of Taylor's work upon Gantt's thinking was very great. He observed the need of scientific investigation in the study of managerial problems; the careful and detailed written instruction of the workers on methods of performance; the manner of working conditions and the various tools and implements to be used; and the planning of work. All these are restated as essentials to reach the end of high wages and low labor costs per unit of production.

Training in Correct Methods

Frank B. Gilbreth, like Gantt, placed a great deal of emphasis upon the personnel factor in production in much of what he considered as fundamental in the training program. He emphasized that correct methods of work should be taught from the first day even though the completed job was badly done. He maintained that it was bad practice to teach an apprentice to do perfect work because in his efforts to get the final job correct, he would get into bad habits of laying brick with too many unnecessary motions. Instead of following such a procedure, he taught the apprentice to lay bricks with speed and with the least number of motions from the first day, even if it was necessary to have a brick layer go over his work as fast as he laid the brick to make his work right.

Measurement of Motions

By placing properly ruled paper in the motion path as motion pictures were taken, it became possible to determine the distance through which each motion was made. This technique constituted a very great refinement in Motion Analysis and went a great deal further than anything that had been developed previously.

Gilbreth also used a cyclograph consisting of small electric bulbs which were fastened to the fingers of the operator. These bulbs flashed at regular intervals and by taking pictures with a stereoscopic camera, the movements of the hands and fingers were recorded on the plate of the camera in three dimensions. Since the flashes occurred at known time intervals, it was possible to determine the time involved and the distance covered by the movement between each dot appearing in the photograph. Gilbreth made the first attempt at measuring motions.

Mr. Segur started working on his Motion-Time-Analysis in 1924 by analyzing micromotion films taken of expert operators during World War I. These films were originally taken with the view of discovering a means of training blind and other handicapped workers to perform useful industrial tasks after the war. The films were made of workers who were the best available in their industry. At the time these analyses were made, Gilbreth's motion classification was available as an aid. In fact, Segur worked with Gilbreth on this project of training the blind and handicapped soldiers. [1]

Segur placed great emphasis on the fact that a person should be an expert by having a great amount of practice in performing an operation. He established what he called the Segur Law: "Within reasonable limits, the time required by experts to perform a fundamental movement is a constant." An expert was a person with sufficient practice opportunity which may vary from 500,000 to 2,000,000 cycles. Segur defined reasonable limits to be within $\pm$ 10%.

Motion-Time-Analysis employed the Gilbreth's therbligs along with detailed finger, hand and body motions.

Emphasis on Eliminating Lost Motion When Training Operators

He believed unswervingly that motions should be very detailed so that minute lost motions could be eliminated. He was of the opinion that MTM and Work Factor were too gross. Therefore, his emphasis was the necessity of a great amount of training of the operator and the eliminating of all losses that might occur in the motion pattern. He would have the operator perform the operation properly according to the methods he stipulated, eliminating the various losses in the motion pattern as shown in the current method.

Emphasis on Beginning and End-Points of Motions

On July 27, I met with Ralph Presgrave in the Royal York Hotel in Toronto. Ralph Presgrave and Gerald B. Bailey were connected with J. D. Woods & Gordon, Ltd., a management consulting firm in Toronto.

These two men developed Basic Motion Timestudy, BMT. A unit of motion, called Basic Motion, relates to standard time data. Variable factors such as distance, precise care, weight handled, etc., are specified numerically, according to the requirements of the type of motion being used. The significant emphasis is "the end point of a Basic Motion--a full stop--is clear-cut and therefore easy to recognize."

A book, Basic Motion Timestudy, published by McGraw-Hill Book Company, 1958, describes the theory of motion identification and timing with useful methods of applying the resulting time data to manual activities.

Product Design, Operator Training, Workplace Layout

The Motion Time Survey (MTS) is a proprietary PTS at General Electric Company. Mike Womeldorph reviewed the details of MTS with me.

At General Electric, they recognize that economic efficient operations and quality are dependent upon:

1. Product Design

2. Operator Training

3. Workplace Layout

A training program of four weeks is set up in Schenectady to instruct trainers. Of this four weeks approximately four days are spent in lecture and the balance of the time spent in analyzing projects brought from the GE Divisions by those being trained. Those being trained are encouraged to instruct personnel within their divisions in MTS including the members of the union. If they can train members of the union to question industrial engineers to ask: "Did you allow for this move or that move?"--rather than saying: "You didn't allow enough time."--they feel they have accomplished their purpose.

The time of MTS is measured in minutes per 100. For example, a Contact Get, similar to a Pick-up Grasp (G1A) in MTM, is 0.18 minutes per hundred Contact Gets. This is a simplification over 0.0018 minutes per Contact Get. A G1A in MTM would have a time value of 0.0012 minutes. While this standard for MTS is about 50 per cent looser than MTM, Mike Womeldorph found that the average as demonstrated recently while involved with the U.S. Government MIL Standard 1567 showed a difference of MTS being slightly tighter than MTM.

Learning

GE has established that a new operator would require 100,000 trials to meet the published standard values. For example, in the traditional Pegboard Demonstration (with 30 holes) employed in Work Simplification courses, 100,000 pegs would be placed in 3333.33 boards to represent the required learning.

At GE, an experience leveling percentage may be added or subtracted from the standard dependent upon the degree of learning opportunity on the operation. The number of trials in a six-month period determines the experience level selected.

In job shop operations GE STANDARD DATA is used. This standard data is derived from MTS.

WORK-FACTOR SYSTEMS

In Westboro, Massachusetts, on August 10, I met with Joseph H. Quick at his beautiful country home. The name Joseph H. Quick goes with the founder of the Work-Factor PTS. He is currently semi-retired and Chairman Emeritus of Science Management Corporation, a management consulting firm and the parent organization of the WOFAC Company that specializes in the Family of Work Factor Systems.

Research work started in 1934 on this PTS. [2] It has been estimated that thirty-five million man-hours of factory and office work were measured in the application of the original data in 1938.

Work Factors

The Work-Factor System has as its objective the analyzing and cataloging of "work-factors" as they affect the time to make manual motions. A "Work-Factor Unit" is equal to 0.0001 Work-Factor Time Minute.

Mental Processes

A. B. Segur had some fascination with the time for mental processes but it was Joseph Quick who devoted considerable research to this area. Chapter 13 comprising 64 pages in the book, Work Factor Time Standards, presents much of this research. Later, the Mento-Factor System was added to the Work-Factor Family of Systems. Data for Mento-Factor were obtained in laboratories, visual inspection, printing and engineering departments.

The other systems in Work-Factor follow:

	WORK-FACTOR SYSTEM	COMPETITIVE WITH:
1.	Detailed Work-Factor System	MTM-1 (but more detailed)
2.	Mento-Factor System	
3.	Ready Work-Factor System	MTM-2 (but more detailed)
4.	WOCOM System	4M
5.	Brief Work-Factor System	M-3
6.	Work-Factor Standard Data System	

FIGURE 1

Each year the Science Management Corporation has seminars in various geographic sections of the United States to discuss applications of these systems by various users. Emerson Boepple has presented some data regarding the comparison of predetermined time systems at some of these meetings. In his talk, "What's New in Predetermined Time Standards," he presented a comparison of some Level 2 and 3 Systems. With his permission, I am reproducing a table presenting the system, number of elements, number of time values, the name of the time unit used and the minimum decimal minute employed in the system.

LEVEL	SYSTEM	ELEMENTS	NO TIME VALUES	TIME UNIT MINUTES
1	MTA DETAILED	18	1000	0.00001
	WORK-FACTOR	8	735	0.0001
	MENTO-FACTOR	14	730	0.0001
	MTM-1	8	265	0.0006
2	READY			
	WORK-FACTOR	9	125	0.001
	MTM-2	11	39	0.0006
	MODAPTS	13	21	0.00215
3	BRIEF			
	WORK-FACTOR	5	32	0.005
	MTM-3	4	10	0.0006

FIGURE 2

UNIVATION SYSTEMS

In 1964, Mr. Willard Kern organized Management Science, Inc. to develop and market computer software for Industrial Engineering applications utilizing his new concepts. Mr. P. D. Dumbauld, a Professional Engineer with many years experience in time study data development and the application of predetermined time systems immediately joined MSI. Upon joining MSI, Mr. Dumbauld was thoroughly trained in the development of computer programs and systems.

The totally integrated systems offered by MSI are named "UnivAtion Systems" and are made up of the following modules:

SYSTEM	PURPOSE	DATE COMPLETED
UnivEl	Time and Data Based Generator - Translator	1965
UniForm	Develops multi-variant models	1966
UnicComp	Simplified Algebraic Language	1967
UniPlan	Provides for micro-metric assy. planning and balancing	1970
MultiComp	Provides mass automatic updating	1973
VariComp	Provides automatic preparation of input data	1973
RTG (Routing)	Network Mfg. Planning data base	1973
PAR (Performance Audit & Review)	Labor performance and control	1975
UniCost	Automatic preparation and maintenance of operational Standard Costs	1976

FIGURE 3

Video Tape Recording

MSI adopted a policy in late 1966 which dictated that before installing the new systems in any new company or new operations, a thorough detailed analysis through the use of video taped studies as documentation was necessary.

UNIVEL VERSUS MTM-1 AT A. O. SMITH CORPORATION

A User's Conference is scheduled annually to discuss progress in the UNIVATION SYSTEMS. Gile Tojek, at A. O. Smith Corporate Manufacturing and Engineering Staff, presented a comparison of UnivEl and MTM-1 at one of these meetings. We quote a portion of this study.

Selection of Plant

To begin with, we chose a validation site that we thought would be most representative of the bulk of work done at A. O. Smith. Local and corporate management agreed to the selection of one of our metal fabricating plants. This plant is a measured day work shop where the major activities are blanking and forming of parts, limited assembly of these parts, and large amount of welding.

The team selected to conduct the study displayed our attempt to avoid bias towards either system. It included the supervisor of Industrial Engineering at the plant and three of his industrial engineers. One of these engineers in particular was designated to do all of the MTM-1 and UnivEl studies. The manager of Industrial Engineering from our main plant along with two of us from the Corporate Staff were there to observe and critically analyze the work being done.

We tried to eliminate deviation in the two techniques by having the same I.E. analyze the job during the same period by UnivEl and MTM-1.

Long Moves

Because of the very long parts, we had numerous moves of 60 to 70 inches with a great deal of body assist. The result in these situations was some deviation between the two systems. We have accepted the UnivEl times in these cases as being correct. Our reasoning is several fold. First, the very long moves with large internal body assist are not really typical. Second, where these moves do exist, it is strictly judgment as to how much body assist there really is.

Weight

The final difference we found between UnivEl and MTM-1 is there by design. It is also one that we felt was deserving of some concern. It is a difference in the consideration of weight when two hands are holding or moving an object. A comparison of typical UnivEl and MTM-1 elements plotted against weight shows that they follow each other closely.

Indeed, there is no difficulty in one-hand handling of an object. The rules, however, differ when the second hand is brought in. UnivEl says use the total object weight. MTM-1 says, if the object is held in two hands, use the effective weight held by each hand. If the amount differs, use the larger of the two. So, MTM-1 divides the weight in two when two hands are used. Since these two plots follow so closely, it is obvious that following different rules will yield different results.

Since we have always followed the MTM-1 rule and we wish to match its concept of normal, we have chosen to continue following that rule rather than the UnivEl rule. This is not to say one is right and the other is wrong. Management Science put up several arguments as to why UnivEl was correct.

What is really important to shop application is that the maximum study deviation was only 4.2% and overall, the average was within 0.2% of properly applied MTM-1.

MAYNARD OPERATION SEQUENCE TECHNIQUE (MOST)

Information regarding this proprietary technique was presented to me by Kjell Zandin, Division Vice President of H. B. Maynard and Company, Inc. This technique was developed by the Swedish Division of H. B. Maynard and Company, Inc., under Zandin's direction.

MOST System is:

1. A comprehensive combination of a manually applied work measurement technique - MOST Work Measurement Systems.

2. Systematic application procedures for this technique - MOST Computer Systems.

3. A computerized version of the total system - MOST Computer Systems.

The most common sequence represents the light manual work of handling objects. Three types of sequence models have been established.

1. General Move

2. Controlled Move

3. Tool Use

The first fixed activity sequence - "General Move" - is defined as moving all objects from one location to another freely through the air.

This can account for as much as 35% of the work of a machine operator and even more for an assembly worker. This activity is represented by the following sequence of letters or sub-activities:

A B G A B P A

A - Action Distance (mainly horizontal hand or body motions)
B - Bend (mainly vertical body motions)
G - Grasp
P - Position

The variations for each subactivity are indicated by an index figure, for instance:

A_6 B_6 G_1 A_1 B_0 P_3 A_0

A_6 - Walk 3-4 steps

B_6 - Bend and arise

G_1 - Grasp one object with one hand

A_1 - Move within reaching distance

B_0 - No bend

P_3 - Position and adjust object

A_0 - No return move

All values on this and other index tables are derived from MTM-2. . . . The time value for the sequence is obtained by adding the index numbers and multiplying the sum by ten.

6+6+1+1+0+3+0 = 17 (17 x 10 = 170 TMU)

Other Activity Sequence Models

"Controlled Move" is a sequence which is applicable when the object retains contact with another object during the move i.e. a lever, crank or push button.

"Tool Use' covers not only conventional hand tools like wrenches, screw drivers, gauges, writing tools, etc., but also fingers and mental processes.

Three (3) higher level special sequence models have been developed for the use of material handling equipment.

1. Jib cranes

2. Bridge cranes

3. Wheeled trucks (fork lifts, stackers, hand trucks, etc.)

The difference between the levels is the multipliers. Identical index numbers are applied on all levels.

-- Basic sequence models (basic MOST) = multiplier 10

-- Bridge cranes and wheeled trucks = multiplier 100

-- Job preparation, etc. = multiplier 1,000

Precision of MTM-1, MTM-2 and MOST--Dr. Brinckloe

William D. Brinckloe, Ph.D., has written a paper, "How Precise are the Methods-Time Measurement Systems?" In this paper he evaluates the system error determined statistically based on the frequency of the different motions involved in the various systems: MTM-1, MTM-2 and MOST.

He points out that the balancing time of MTM-1 and MTM-2 are critically dependent on the distribution of motion frequencies. Balancing time is defined as the sum of the theoretical individual motions, at whatever their relative accuracy, it would take to add up to a total with five percent relative accuracy, at 95 percent confidence. The balancing time of two predetermined motion time systems thus gives a relative measure of accuracy for comparing one with the other.

Number of elements to produce balancing time

$$n = \left(\frac{r}{.05}\right)^2$$

r = the relative error of a typical element in whatever PTS we are considering.

It is concluded from his analysis that MTM-1 is about twice as precise as MTM-2, which in turn is about twice as precise as MOST. Rounding the balance times as measures of system precision, we would get:

MTM-1	800 TMU
MTM-2	1600 TMU
MOST	3200 TMU

Modular Arrangement of Predetermined Time Standards (MODAPTS)

Mr. David Moran, Manager, Industrial Engineering Group of Peat, Marwick, Mitchell & Co., gave me guidance on this part of the presentation. Much of the background was written in a paper, "MODAPTS--A New Work Measurement System," prepared by R. J. Shaw, New York Office of Peat, Marwick, Mitchell.

MODAPTS is a registered trademark. MODAPTS has been registered under the terms of the Universal Copyright Convention and also under Australian law on behalf of the Australian Association for Predetermined Time Standards & Research (AAPTSR).

The basic unit in MODAPTS is a simple finger movement with the finger doing nothing special.

All other activities are expressed in terms of this finger movement or module. . . . There are only eight different values - 0, 1, 2, 3, 4, 5, 17 and 30 mods.

The basic mod value is .129 seconds.

The eight mod values are applied to 21 types of activities derived from movements of fingers, limbs, body, and eyes. . . . (See Figure 4)

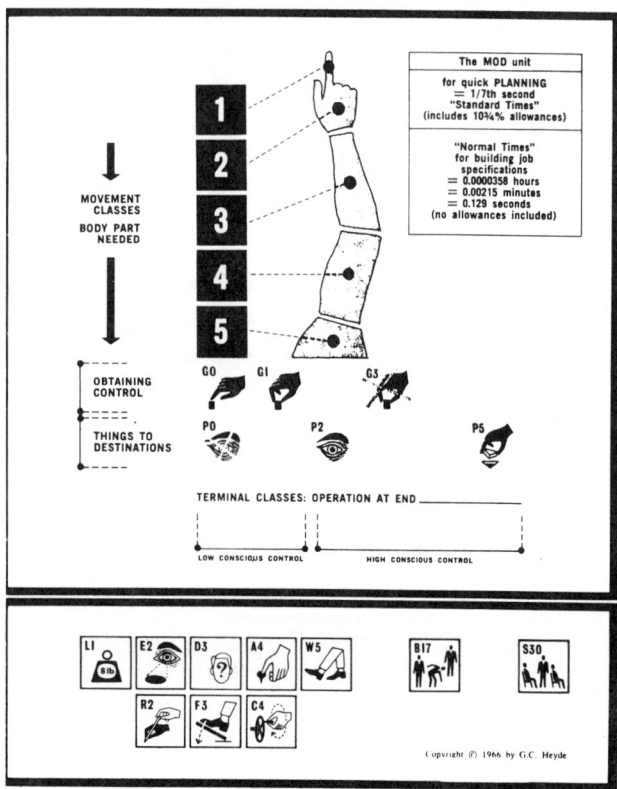

FIGURE 4

In MODAPTS the first identification is the class of movement, and second tag or label is that which happens at the end of the movement--or the "terminal activity."

There are two categories of terminal activities--Obtaining Control and Things to Destination.

Each category has three different mod values which are selected based upon the type of terminal activity.

Movement classes and terminal activities are paired so that the analyst has an immediate coding of the action and the time value without having to refer to a table or chart.

Master Standard Data (MSD)

Serge A. Birn Co., Inc., developed MSD in the late 1950's. Harold Nance, President of Serge Birn Company, furnished me the details of this technique. It was developed to set standard MTM-based data on manually controlled operations where production was less than 100,000 units per year or a few thousand units per week. Between production runs, the operator would lose most of the skill he or she would develop. Statistically, a very high percentage of work in industry falls within this limited practice category. MSD was developed by studying statistically all motions; and, consequently, because many motions studied occur rarely, they can be ignored.

MSD comprises the most common motions B, C, D reaches; all the Grasps with the exception of the G1C and the non-symmetrical positioning; A, B, and C Moves; P1 and P2 Positions, T..S Turns; both Releases and Apply Pressure. Likewise, simultaneous motions are unlikely with the lack of practice opportunity, except those that can be made regardless of practice; therefore, a simplified Simultaneous Motion Chart was constructed. Also, six simplified tables combined with decision charts established the tables: OBTAIN - O with degree of control; PLACE - P with Other Hand, General, Loose and Close positioning; ROTATE; USE; FINGER SHIFT; AND BODY MOTIONS.

MSD became one of the first higher level horizontal data systems. Much of the work done by Serge A. Birn Co. contributed toward development of MTM-GPD; and MSD is also similar to MTM-2.

CUE

CUE is the name of the work measurement system developed by Dr. Douglas M. Towne, Vice President, General Analysis, Inc., Rolling Hills Estates, California. He has programmed MTM data into the Texas Instruments TI-59 handheld programmable calculator.

Towne has spent a considerable amount of time programming MTM, Work Sampling and Machining Data on program cards. These programs are proprietary and offered to business and industry for a consulting fee. The TI-59 has available five buttons to handle ten user labels. $\boxed{A}$ to $\boxed{E}$ and by pressing a 2nd key, these same buttons will handle programs $\boxed{A'}$ to $\boxed{E'}$. Therefore, higher level MTM data is directly inputted into the calculator either in observing an operation or simulating an operation while estimating the time.

A 12" Get would be inputted by pressing buttons 1 & 2, then A, or 12A, a 6" Move would be 6B, an Apply Pressure (AP-1) 7E.

Besides the accuracy evaluations of General Analysis, Inc., three national companies have now made independent comparisons of CUE to their MTM-1 studies. A summary of the results (combined) is shown in Figure 5.

CYCLE TIME (MIN.)	ABSOLUTE AVG. DIFFERENCE, %
≤.02	7.7
.021 - .04	4.0
.041 - .06	2.6
.061 - .08	2.5
.081 - .10	2.3
>.10	2.3

FIGURE 5

METHODS-TIME MEASUREMENT
(MTM-1, MTM-2, MTM-3, MTM-GPD
4M, MTM-V, MTM-M, and MTM-C)

At each MTM conference, we emphasize the International MTM family of work measurement data systems.

MTM-1   Dr. Harold B. Maynard (those of us who knew him well called him "Mike"), Gustave J. Stegemerten, and John L. Schwab developed this predetermined time standards system. It is the first level detailed MTM system.

MTM-2   MTM-2 was developed by the International MTM Directorate. It is based on MTM-1 and comprises the second level of the MTM Family with 39 time values. MTM-2 has a speed of analysis twice that of MTM-1 but a somewhat lower precision in time prediction.

MTM-3   MTM-3 was also developed from MTM-1 by the International MTM Directorate. It has 10 time values. MTM-3 is the third level of the MTM Family of systems and has a speed of analysis which is seven times that of MTM-1. It can be used in situations where a less detailed methods description is required and reduced precision can be tolerated. RADAC: RApid Data And Computer-Aided Calculations of Standards uses MTM-3 for long cycle, short run work. Robert Perry from Tektronix, Inc., developed RADAC.

MTM-GPD   MTM-GPD is the first general purpose data system developed from MTM-1. It is universal in part and functional in part. The universal portion of the system contains 14 elements and 149 time values.

4M   (Micro-Matic Methods and Measurement) Computerized Work Measurement System--developed in Westinghouse and now widely applied, this system has been adopted and marketed by the MTM Association. The computer-aided analysis process retains MTM-1 accuracy and description while providing a faster speed of analysis, reduced human error, and other advantages, including calculations of various indices which measure the effectiveness of the methods employed.

MTM-V   (Work Measurement of Machine Shop)--This work measurement standard data based on MTM was developed for machine-tool users. MTM-V contains time values for handling and adjusting work pieces of any weight and size, including machine tool setup attaching crane hooks, and other mechanical handling equipment. Process-controlled activities are not included. Only manual elements are analyzed.

MTM-M   MTM-M is specifically designed for use where assembly is performed under magnification (micro assembly). It has proved highly advantageous not only for time determination but for method improvement in this type of work.

MTM-C   This Clerical Data System continues to be developed in order to make available a full clerical work measurement system at three levels of job description, precision, and speed of analysis.

<u>MTM Family of Systems</u>

Karl Eady presented us with some classifications for the MTM Family Tree into the following systems:

1. Generic - Intended to be used and understood by all--Not restricted or particular in application MTM-1, MTM-2, MTM-3.

2. Functional - Adapted to a particular type of activity.

    Clerical Use, Tool Use - Micro assembly
    MTM-C, MTM-V, MTM-M, MTM-GPD

3. Specific Systems

> Developed for a particular industry or organization, such as Banking - Construction. XYZ Company
>
> Standard Data Systems - Not generally available outside a particular company.

Eady showed the MTM Family in relation to the basic MTM-1 system. The higher level systems have fewer elements but some accuracy is sacrificed.

At the 1978 MTM Spring Conference in Los Angeles, William C. Arnwine and William F. Fielder, Jr. presented papers regarding work measurement standards accuracy. Some of the excerpts from these two presentations follow:

Bill Fielder presents key points in comparing MTM-1, MTM-2, MTM-3 and MTM-V.

> 1. Accuracy varies with the size of the standards produced from the application of each of these systems.* For example, at any particular length of standard such as one minute (1600 TMU).

MTM-SYSTEM	%
MTM-1	±5
MTM-2	±10
MTM-3	±15
MTM-V	±30

FIGURE 6

> 2. All four systems have an accuracy of ±5% referred to as "balance time" at some particular level of non-repetitive cycle time.

MTM-SYSTEM	MINUTES	TMU
MTM-1	1	1,600
MTM-2	4	6,667
MTM-3	10	16,000
MTM-V	36	60,000

FIGURE 7

> Thus, standards accuracy varies with the particular MTM system used and with the total non-repetitive cycle time produced by these systems. MTM-1 is the most accurate, and MTM-V is the least accurate currently in use.

*This "Law of Averages" is applicable to all PTS, not just the MTM Family of Systems.

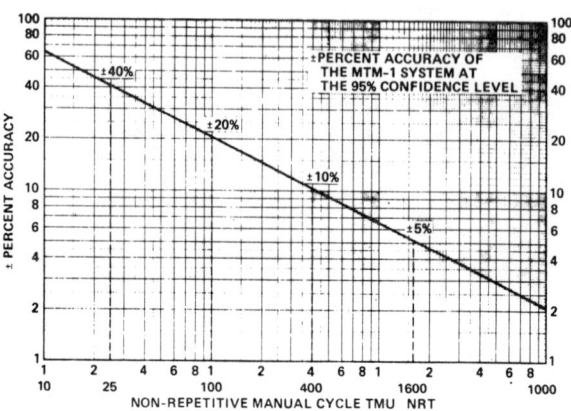

FIGURE 8

WHERE DO WE GO FROM HERE

We believe that the only way we can continue relentlessly to refine our systems is to wrestle with questions about these PTS. To find the answer to some of our questions, we immediately think of performing an experiment. We realize that the validity of the results depends upon our sample size and a careful design of the experiment. We are striving to offset future criticism of our results that the experiment was ill-designed or badly executed.

The project upon which we expect to embark follows:

PTS Olympics

We will have a PTS Olympics at several geographic locations in which we have the analysts of the various systems in one room, operating at peak performance for a period of time, observing loop motion picture films. The objective being to have all analysts making similar type studies at the same time period rather than having an opportunity later to review "Rules of Thumb" and "massage" the analysis.

Statistical Accuracy

We shall attempt to have a sufficient number of observations of each PTS to obtain a statistical accuracy, precision, speed of application and confidence as it relates to each PTS and "Standard Data Systems and Their Construction" originally published by Finnish MTM Association. English version by Karl Eady, and as discussed by Bill Arnwine and Bill Fielder at the MTM Association Spring Conference, Los Angeles, California, April 20, 1978. These papers are referred to in this talk. We shall also make a frame count of the film taken at 24 frames per second to compare the various elements of Reach, Move, etc., to the various PTS analysis.

REFERENCES

[1] A. B. Segur, "Motion-Time-Analysis," *Industrial Engineering Handbook*, Second Edition, H.B. Maynard, Editor-in-Chief, (McGraw-Hill Book Co.), Cahpter 5, pp. 107-124.

[2] Joseph H. Quick, James H. Duncan, and James A. Malcolm, Jr., *Work-Factor Time Standards*, (McGraw-Hill Book Co., 1962), pp. 48-49.

BIOGRAPHICAL SKETCH

Dr. Chester L. Brisley is a Professional Engineer and a graduate in Industrial Engineering with a Ph.D. and M.S. from Wayne State University in Detroit. He graduated from General Motors Institute, Flint, Michigan, in 1939 and received his B.S. degree in Industrial Engineering in 1946 at Youngstown State College, Youngstown, Ohio.

Dr. Brisley has served on teams sponsored by the State Department, giving technical assistance in West Berlin, Japan, Indonesia, Korea, and India.

Currently, Dr. Brisley is Professor and Associate Chairman, Department of Engineering, University of Wisconsin-Extension. Formerly, he was director of industrial engineering for a number of industries, as well as being involved with two consulting firms.

He is affiliated with the following professional and engineering societies: AIIE (American Institute of Industrial Engineers), Awarded Fellow Membership at 1968 Convention, Tampa, Florida, and Past President, Detroit and Milwaukee Chapters; WSPE (Wisconsin Society of Professional Engineers), President, 1978-79.

TODAY'S INTERNATIONAL MTM SYSTEMS
--DECISION CRITERIA FOR THEIR USE

Karl Eady
MTM Association for Standards and Research

Here is an updated overview of the systems comprising the MTM family covering the additions to the family in recent years. System classification is discussed with specific information on how each family member fits into the classification scheme. Characteristics of each system are described with emphasis on accuracy levels, speed of application and degree of methods description provided. Finally, some guide lines are suggested for making the selection of the appropriate system to measure a specific work situation.

Until 1960 the term "Methods-Time Measurement", or MTM, identified a single system for analyzing and measuring work through the use of incremental bits of activity to which predetermined time values had been assigned. The system proved to be quite versatile and universal, readily applicable to just about any kind of work activity existing in the industrial world.

This very versatility, moreover, made it an ideal base for the development of additional systems for use in more specialized categories of activity at savings in time with greater ease of application. As a consequence, in 1977, many MTM-based systems or techniques exist, each of which was developed for a specific purpose and use. Development of these new systems has not been restricted to this country. Many of the new systems were developed in Europe. MTM has truly become international.

The term MTM now identifies a concept rather than a single system. The original system is now known as MTM-1 and other systems have both number and letter suffixes: This paper lists the international MTM family of systems currently in use together with a brief history of the development of each. More importantly, the paper discusses the criteria which should be examined when decisions must be made as to which system or systems are appropriate for use in any given work environment.

Before we can discuss intelligently the parameters and criteria for system use, we should sort out our systems in a more-or-less methodical fashion

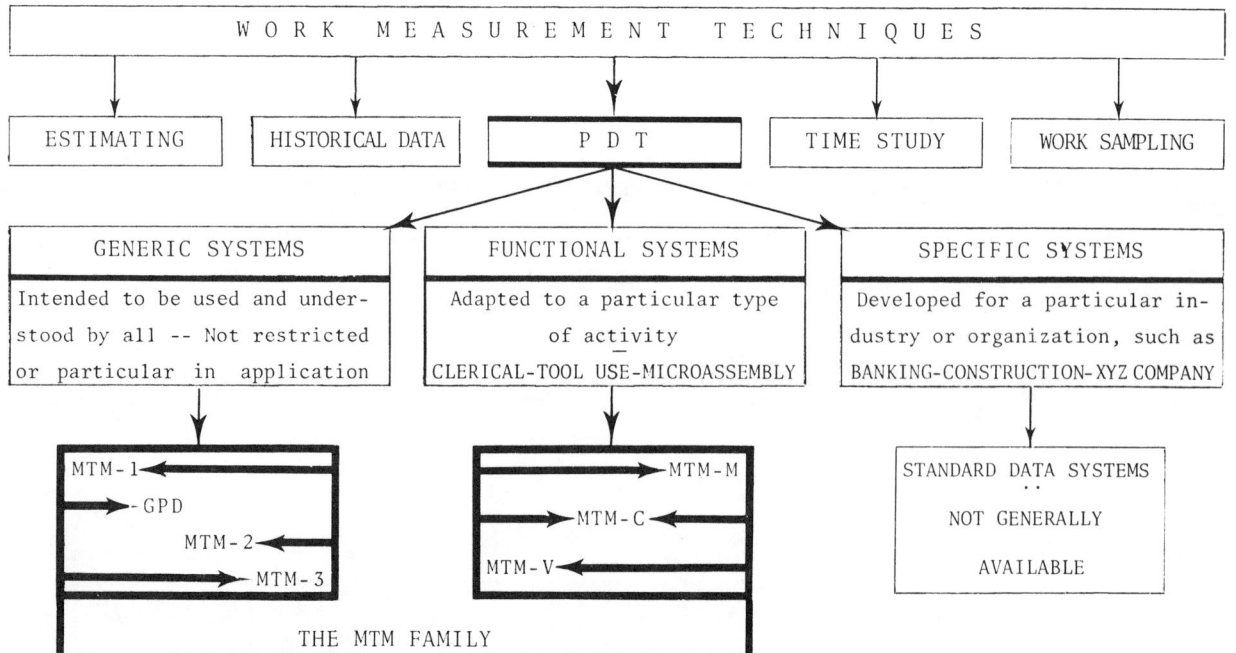

Figure 1. The Family Tree

and put them in their proper perspective. But before we can do this sorting job we need to take a look at work measurement techniques in general and predetermined time techniques in particular.

The Family Tree

Although Figure 1 doesn't look like a tree, it does have a trunk with branches and twigs. With a little study we can see from this chart just where the MTM family fits into the picture. We will also get an idea of how big this family really is.

When we investigate the broad spectrum of work measurement techniques we find that there are many. Figure 1 lists just a few with which most of us are quite familiar. The field is by no means exhausted, however, since we do not include all of the techniques for measuring machine cycles or process times which certainly have an impact on manual work measurement. The five techniques shown here, however, will suffice for our purpose - tracing the genealogy of the MTM family.

System Classification

We may proceed a step further, now, and take a look at Predetermined Time Systems in particular. Some sort of classification scheme is necessary before we can do a thorough job of this, however, and for the purposes of this presentation I have classified PDT systems in two ways - first, by application universality and then by element complexity. How general the system is in application, I believe, is the more significant method of classification, so let's take a closer look.

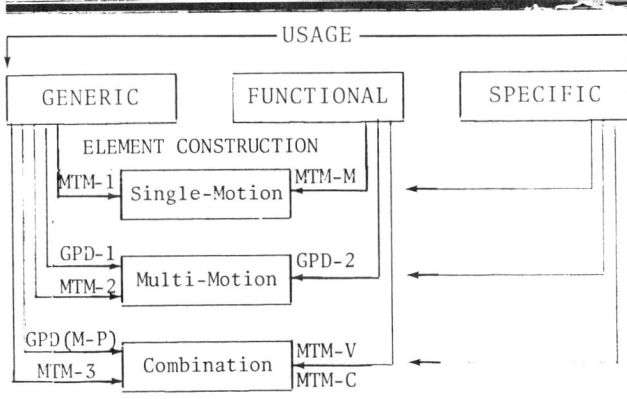

Figure 2. System Classification

Usage

Here we note that PDT systems may be classified as generic, functional, or specific. By a generic system we mean one which is intended to be used and understood by all users of work measurement and which is not restricted or particular in application. A functional system is one which is adapted to a particular type of activity, such as clerical, tool use, or micro assembly. Finally, a specific system is one which is developed for a particular industry or organization, such as banking, construction, electronics, or simply XYZ Company.

Specific systems are often referred to as standard data systems and are not generally available since they are for the most part proprietary. We may also observe that these three classes of PDT systems are not mutually independent, Figure 3. By this we mean that functional systems may be constructed from either generic or other functional systems or from a mixture of the two.

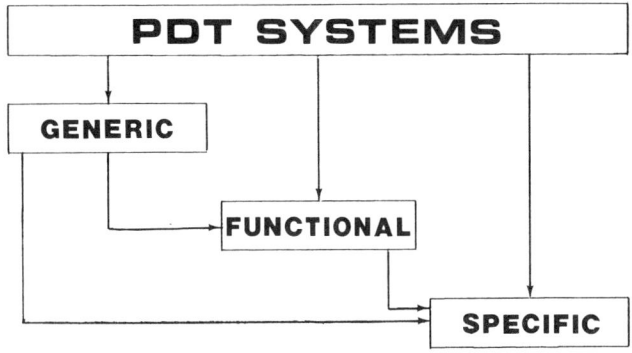

Figure 3. Interdependent Systems

Generic Systems. Now let's see where the members of the MTM family fit into these classes. The generic systems are the ones with which we are most familiar. They include MTM-1, some parts of GPD, MTM-2, and MTM-3. The term MTM, without the addition of a number or a letter, denotes the MTM philosophy or approach, including all those Association-recognized techniques which are derived from MTM-1, emphasize methods description, and use the same time notation - the TMU.

The names of the elements of a generic system are quite general to the degree that in themselves they do not reveal the nature of the activity being analyzed. Typical elements in such a system would be called REACH, MOVE, PUT, HANDLE, etc.

Functional Systems. The functional systems are not quite as universal in application as the generic systems. They are meant to be used in some areas of activity but would be difficult to use properly in other areas. The family members in this class are MTM-V, MTM-M, and MTM-C.

In general, functional MTM-based systems are identified by letter, whereas generic systems carry a number. The sole exception is GPD which has both generic and functional elements.

MTM-V requires a note of explanation. In order to keep our notations straight we should explain that the "V" in MTM-V is not the Roman numeral 5, but rather it is the letter Vee which stands for the

Swedish word for Machine tools - "verktygmaskiner". To summarize, as far as the MTM family is concerned we can generally recognize generic systems by the numeric suffix and functional systems by the letter suffix.

Elements of the functional system, MTM-V are of two types - simple and complex. The simple elements are manual elements not requiring the use of a tool. They are called INSPECT, ROTATE, OPERATE and HANDLE. Complex elements involving tool usage are called FASTEN/LOOSEN, MEASURE, GAUGE, PROCESS, MARK, and COUPLE.

One of the primary uses of MTM-V is the development of specific systems involving time formulas and tables of standard data.

Just a word about MTM-M to illustrate its position as a functional system. One of the typical patterns bears the notation IIET-G4. The letters stand for "Inside-to-Inside Empty Tweezer - Grasp". The motion pattern may be described as a moving of a pair of tweezers within a magnified field with the purpose of grasping an object with the tweezers and picking it up. The digit "4" refers to the distance moved and the precision required in picking up the object.

MTM-C is going through its gestation period. It is being developed as a functional system for measuring all degrees of clerical work from desk-top activities to the operation of today's office equipment - typewriters, calculators, key punches, reproduction machines. It is expected that MTM-C will be available toward the end of this year.

4M DATA. Many of you are no doubt familiar with another MTM family member - 4M DATA. 4M DATA is the acronym for Micro Matic Methods and Measurement. It is not a separate, unique system of work measurement, but is merely a computer-aided means of applying MTM-1.

It is important that 4M DATA be brought into the picture because it will solve many problems for us when we consider system characteristics in an effort to make a decision as to choice of system to use.

Summarizing our discussion on system classification thus far can best be done by first looking at the MTM family of systems as an important part of PDT systems and then taking another look at our family tree where we get it all together. (Figure 1).

Element Construction

From a technical viewpoint we may also classify PDT systems by the level of element complexity. No standards currently exist for precisely defining these levels so we have put together our own concepts which took form as we began using these systems in practical situations. No single system will have all of its elements defined at the same level of comprehensiveness, but most of them will be. Classification by element complexity is of great help in understanding the relative occurrences of the systems in the family and in determining situations in which each level can be used efficiently.

Referring to Figure 2, let's take a closer look at how the family members fit into the level of element complexity. The single-motion level is the basic level. Each element consists, for the most part, of a single motion in terms of MTM-1. In addition to MTM-1, a majority of the MTM-M elements can be thought of as single-motion.

Perhaps the most frequently occurring single-motion element is MOVE. A study made by the Swedish MTM Association during the development of MTM-2 analyzed over 22,000 motions performed in the United States, Sweden and the United Kingdom. The analysis showed that MOVE accounted for just about a third of all motions performed. In MTM-1 the element MOVE contains four variables which allows one to describe over 3,500 different MOVES.

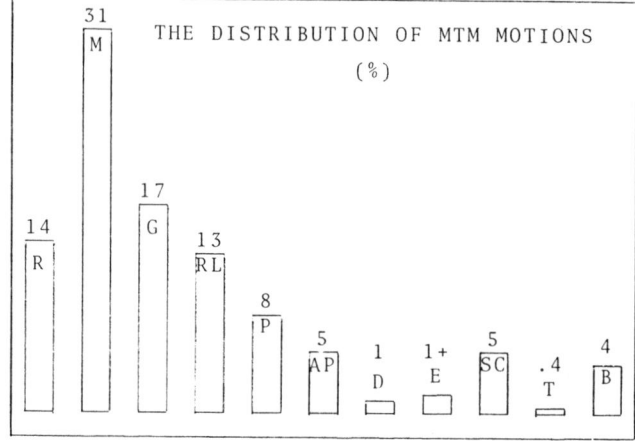

Figure 4. Distribution of Motions

To completely describe a MOVE, one must be able to specify a definite distance, weight of the object being moved, place at which the object will be released and conditions which prevail prior and subsequent to the MOVE. The MTM-1 system contains 23 separate single-motion elements. Next to MOVE, the REACH element contains over 600 possible values depending on the proper combination of distance, destination and acceleration.

Single-Motion Elements. A table of the MTM-1 system reveals a total of 23 elements and almost 5,000 time values with which to describe the kind of manual work which is a part of any day-to-day activity. It is obvious that the decision-making process is quite involved and much time is required to analyze an operation. Experience has indicated that, with MTM-1, a one-minute operation (exclusive of non-manual time) could well require 6 hours to analyze. Consequently, we may conclude that MTM-1 is an expensive system to use, but with very precise results - more precision, probably, than is required in many instances.

Multi-Motion Elements. Somewhere along the analysis trail it became evident that the wheel was continuously being re-invented. By this I mean that certain motion patterns were consistently being repeated. Why, asked the analyst couldn't these motion patterns be identified and used as a block when they occurred, rather than rebuilding each time with single-motion elements? Experimentation and implementation assured the analyst that this indeed could be done. So the first multi-motion systems came into being.

The Swedish MTM Association has constructed a unique diagram which looks suspiciously like it might be the digestive system of an automated cow. However, it does serve to illustrate the kinds and frequencies of recurring motion patterns. The width of the various pathways have a direct relationship to the frequency of occurrence. This flow diagram was found to be highly useful in the construction of higher level systems of data.

Multi-motion elements are those which consist of two or three basic motions. Family systems at this level are generic GPD, MTM-2, and functional GPD.

The construction of a multi-motion element from single-motion elements is illustrated by the elements REACH, GRASP, and RELEASE forming the element GET, and the elements MOVE and POSITION forming the element PUT (PLACE). These two multi-motion patterns account for the majority of time values in both generic GPD and MTM-2. In MTM-2 they account for 82 percent of the 39 time values.

The relationship between MTM-1 and MTM-2 is shown in Figure 5. The 23 elements of MTM-1 are now included in the 9 elements of MTM-2 - and the 4,988 time values have been reduced to 39.

Combination Elements. The next higher level in element construction is achieved in the same fashion. We can combine two or more multi-motion elements into a more comprehensive pattern which we call a combination element. Note that we have several members of our family composed of combination elements. Two of these are generic systems and one is functional. The most recent of our generic systems is MTM-3. It is a complete system which, with two restrictions, can be used by itself to analyze a work situation.

The build-up of a combination element is illustrated by a GET and a PUT combined to form a HANDLE. If we look at the complete process, we can readily trace the combination HANDLE element into its basic single-motion components: REACH-GRASP-MOVE-POSITION-RELEASE.

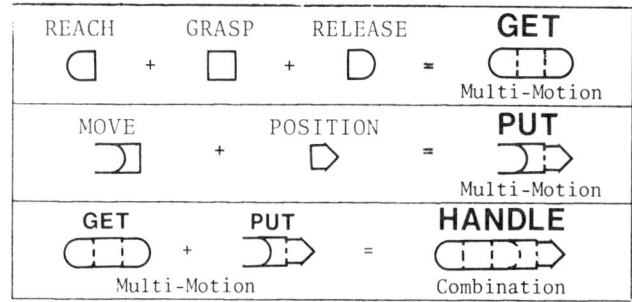

Figure 6. Element Build-up

Now let's tie MTM-2 and MTM-3 together. We have just observed that HANDLE, an element of MTM-3 is composed of a REACH, GRASP, MOVE, POSITION, and RELEASE. Consequently, in terms of MTM-2, it is a combination of GET and PUT. In order to do a complete analysis job with MTM-3, we must also have the MOVE-POSITION pattern to function by itself. (Figure 7.)

The Building Block Principle

It is characteristic of the MTM family of systems that they are completely compatible with each other. By this we mean that the systems may be used interchangeably in a work analysis. As a matter of fact, when using MTM-3, motion patterns having a frequency of 10 or more, such as in pounding a nail, must be analyzed with MTM-1 or MTM-2. And MTM-2 or MTM-3 elements can often be required when using MTM-V.

The Building Block concept can be seen in action in a series of slides which show the motions involved in removing a pen from its holder, making a mark with it, and laying it aside. In terms of MTM-1 we are demonstrating REACH, GRASP, MOVE, POSITION, MOVE, and RELEASE.

The same series of slides can also illustrate a GET and two PUTS.

Finally, the slides can illustrate a HANDLE and a

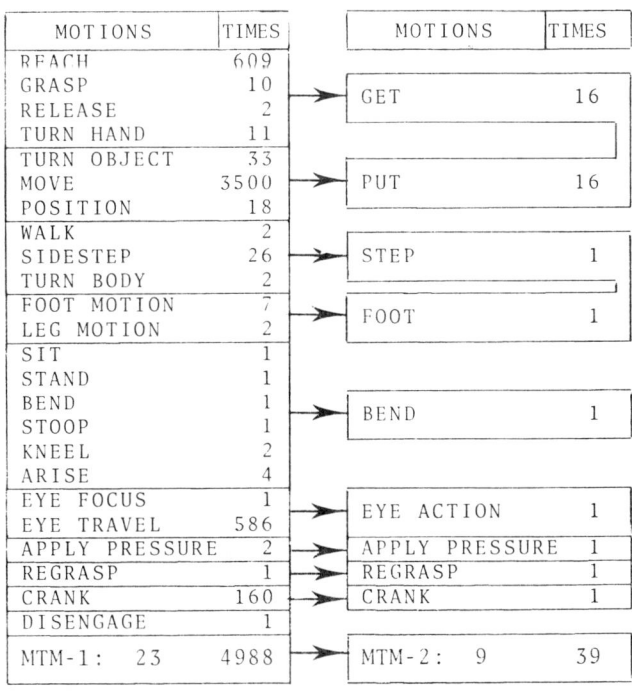

Figure 5. The MTM-1 and MTM-2 Relationship

TRANSPORT.

The three analyses are shown in Figure 8. Obviously, we are using idealized motion patterns. In practice, using actual work patterns, time values may differ more widely depending upon the length of the job being analyzed.

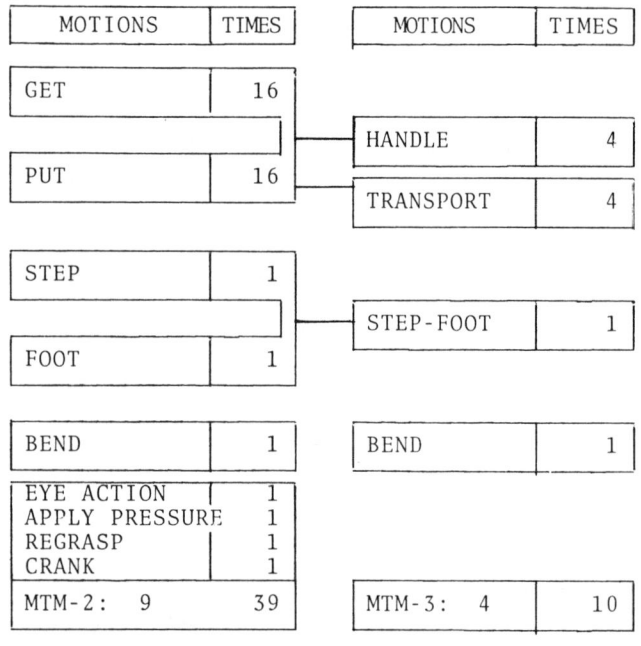

Figure 7. MTM-2 and MTM-3 Relationship

MTM-1		MTM-2		MTM-3	
REACH	.62	GET	.83		
GRASP	.07				
MOVE	.37			HANDLE	1.22
POSITION	.20	PUT	.54		
MOVE	.44			TRANSPORT	.58
RELEASE	.07	PUT	.40		
1.77		**1.77**		**1.80**	

Figure 8. Using Building Blocks

System Selection

When MTM-1 was the only technique available in the "family", system choice was a matter of determining whether MTM would do the job satisfactorily or whether some other means of work measurement should be used. The choice was likely to be between stopwatch study and predetermined time systems, although other techniques, such as work sampling and estimating were also considered.

This choice must still be made today. But the job doesn't end there, if a Predetermined Time System is chosen as the way to go. Most Predetermined Time Systems today exist on several levels. As we have been made aware thus far, MTM is no exception. If the choice was MTM, then a further decision must be made. Shall we use MTM-1, MTM-2, MTM-3, MTM-V, MTM-M, MTM-C or 4M? In some cases, the decision is simple. If, for example, we want the optimum MTM system for measuring sub-miniature assembly work done in a microscopic field, we need look no farther than MTM-M. On the other hand, we generally are faced with a more complex decision process. Many factors must be considered in making the decision. The final result may very well be that, under a given set of circumstances, several systems can be chosen to meet the varying demands of the measurement arena.

Demands on the Measurement System

Perhaps the first step in choosing a work measurement system is to establish the use or uses to which the system will be put. These uses may vary anywhere from cost estimating to developing work standards for an incentive payment plan. Other uses could very well be developing manning requirements, scheduling production, improving methods or preplanning new product manufacturing.

Each of these uses may place different demands on the various systems in terms of their operating characteristics.

Let us first examine these characteristics for each of the MTM systems and then develop some examples which will demonstrate how a particular system will meet the demands required of it.

Decision Criteria

Several system characteristics must be critically examined if we are to make an intelligent and economic choice of measurement system. Among them are total system accuracy, speed of application, level of method description and training costs. Additional parameters include a complaint probability factor and a productivity loss factor. These two parameters are presented and developed in an article "More Evaluation Parameters for MTM Systems," by Bayha, Hancock and Langolf, appearing in the International MTM Journal, Volume 2, No.1.

Accuracy

System Accuracy. After a good deal of conjecture and experimentation, Dr. Walton Hancock, of the University of Michigan, devised a means for determining the precision of the MTM-1 system. The results of his work were published in the MTM Journal, Volume XV, No. 3. In his work he compared the MTM-1 system element by element with a theoretical ideal system in which there would be an infinite number of values to choose from in measuring work. Simply stated, his conclusion was that in 95 cycles out of 100 the MTM-1 system would come within + 5 percent of the "true" value of a work cycle of 375 TMU (0.225 minutes). At one minute cycles the accuracy would be + 2.5 percent.

In 1972 the Swedish MTM Association published a monograph which paralleled the work of Hancock. The determination of the accuracy of the MTM-1 system in this case, however, was based on the frequency distribution of some 22,000 motion patterns which had been used in the development of MTM-2 and MTM-3. The results of this study indicated that the accuracy at 95-percent confidence limits was + 5 percent at cycles of 450 to 930 TMU (0.27-0.558 minutes). Since MTM-1 is the base from which all other family systems were developed, it was now possible to make comparisons as to the accuracy of each system.

System accuracy (also called system error) is predominantly a function of the average length of a motion or element of the system. The average length of an MTM-1 motion was found to be 8.7 TMU (0.005 minutes) by Ulf Åberg in a study of 7,174 motions having a total time value of 62,445 TMU. A later study of 22,000 motions by the Swedish MTM Association yielded an average motion length for MTM-1 of 8.1 TMU.

Application Accuracy. The accuracy of a work measurement system is not dependent solely upon the statistical accuracy of the system itself. Application accuracy is also a factor.

Application error is the variation in analysis times by different analysts utilizing the same MTM system to analyze a given work situation. It can be caused by a number of factors, such as incorrect data card usage, incorrect motion distance, improper assignment of the degree of difficulty and missed motions.

Balance Time. When system and application error are combined, the total accuracy is expressed in a unit called "Balance Time - the non-repetitive cycle time in TMU at which variations up to + 5 percent may be expected 95 percent of the time. Stated another way, it is the cycle time at which 95 out of 100 analyses of a given job would fall within + 5 percent of the "true" value for that job.

This accuracy can be graphically shown on an accuracy chart such as Figure 9. You will note from this chart that the Balance Time for MTM-1 is at approximately 1,600 TMU. It must be emphasized at this time that these accuracy figures can be

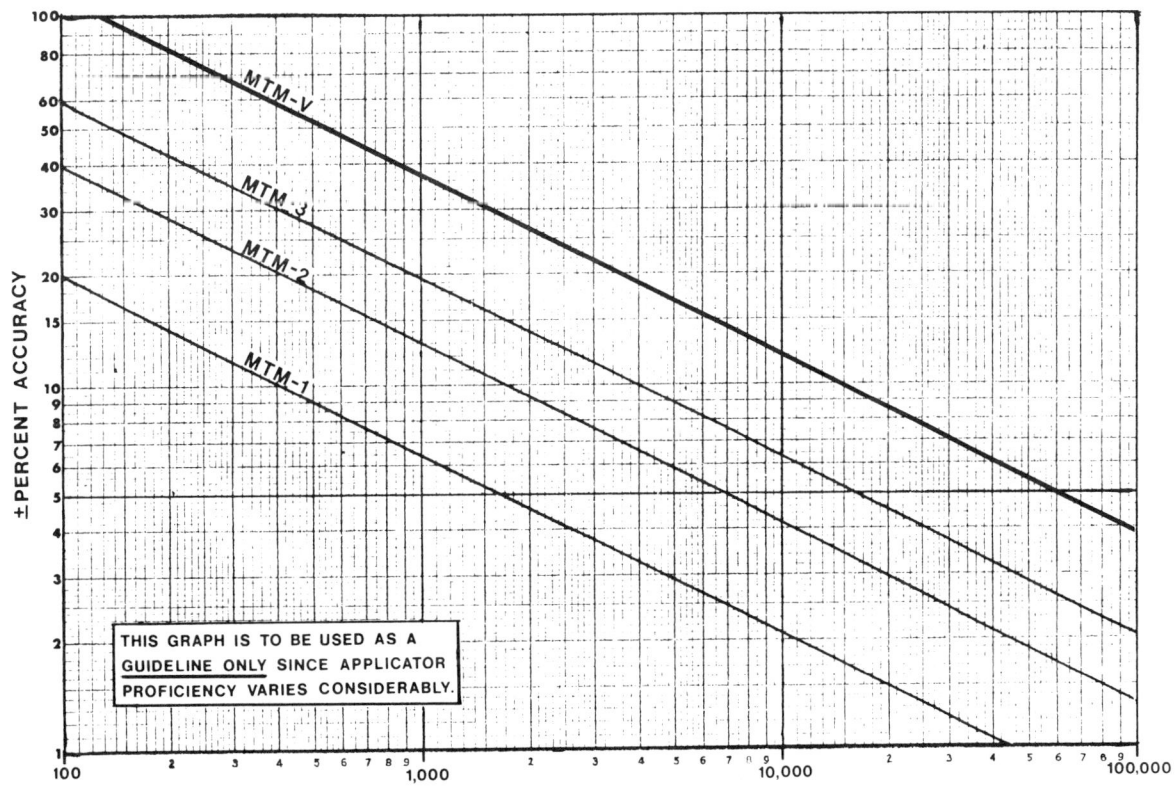

Figure 9. MTM System Accuracy

used only as a guideline since application proficiency varies considerably.

The accuracy chart is constructed to show the total system variation at any given non-repetitive cycle time. For example, the MTM-1 accuracy for 1,000 TMU (36 seconds) is + 6.4 percent. As the work cycle increases in length the total variation decreases at a logarithmic rate. At 10,000 TMU (6 minutes) the accuracy is slightly above the 2 percent level. It must be kept in mind that these accuracy figures are given for a confidence level of 95 percent.

Absolute Accuracies. Accuracies for most of the other family members of the MTM system have been established. These include MTM-2, MTM-3 and MTM-V. The accuracy for MTM-M is being evaluated at the present time, and has not yet been added to the chart. MTM-C, of course, is not yet complete. Figure 9 is a chart of the total system accuracies for MTM-1, MTM-2, MTM-3 and MTM-V.

The chart shows "absolute" accuracies - comparison with an ideal system which has no system or applicator deviation. (Such a system, of course, does not exist). When calculating the absolute accuracy of any MTM-based system, the system and applicator deviation of MTM-1 is included.

The Balance Time for MTM-1 system error alone was found to vary between 375 and 930 TMU dependent upon the motion frequency distribution that was used in its computation. When application error is added to this figure the Balance Time is found to be 1,600 TMU, or approximately 1 minute. These Balance Times may be read directly from the accuracy chart, Figure 9, along the horizontal 5-percent line. They vary from 1,600 TMU for MTM-1 to 60,000 TMU for MTM-V.

When looking at the accuracy of the various systems for a non-repetitive cycle time of 1,000 TMU (36 seconds) the results show, for MTM-1 an accuracy of + 6.4 percent; for MTM-2, + 13 percent; for MTM-3, + 19 percent; and for MTM-V, + 37 percent. These accuracies may be read directly along the vertical 1,000-TMU line.

Relative Accuracy. Before we leave the subject of system error I would like to make one further point. Since we are dealing here with MTM Systems rather than Predetermined Time Systems generally, it is very helpful to be able to express accuracy for the various high-level systems in relation to MTM-1, the base to which all MTM Systems can be traced. If we express this relative error in terms of Balance Time we can make the following statements: (Figure 10)

MTM-2 will yield times within + 5 percent of MTM-1 values at non-repetitive cycles of approximately one minute at 95-percent confidence levels.

MTM-3 will yield times within + 5 percent of MTM-1 values at non-repetitive cycles of approximately four minutes at 95-percent confidence levels.

MTM-V will yield times within + 5 percent of MTM-1 values at non-repetitive cycles of approximately 24 minutes.

At cycles of 1,000 TMU (36 seconds) the relative accuracies are + 6.6 percent for MTM-2, + 12.6 percent for MTM-3 and + 30.6 percent for MTM-V, all at 95-percent confidence limits.

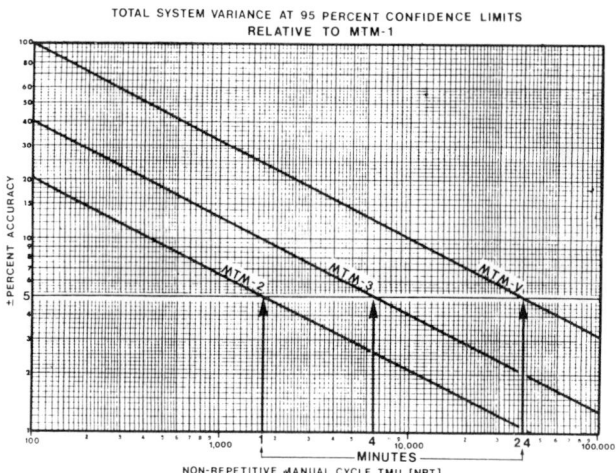

Figure 10. System Accuracies Relative to MTM-1

Speed of Application

Application speed is directly related to the size of the individual motions or elements making up the measurement system. It was previously mentioned that MTM-1 has an average motion length of 8.1 to 8.7 TMU. No similar weighted figures are available for the other family systems, but the average length of the 39 time values on the MTM-2 Data Card is 14 TMU, while the MTM-3 Card has an average element length of 29 TMU. The average length of an MTM-V element is 260 TMU.

Experiments have suggested that the following figures may be used as guidelines when comparing the application speeds of the MTM family systems:

MTM-1 requires 350 times the cycle time for analysis
MTM-2 requires 150 times the cycle time for analysis
MTM-3 requires 50 times the cycle time for analysis
MTM-V requires 15 times the cycle time for analysis

Relatively speaking, MTM-2 is approximately twice as fast to apply as MTM-1, MTM-3 is seven times as fast and MTM-V is 23 times faster than MTM-1. 4M DATA is three times faster than manually applying MTM-1 - fifteen to twenty times faster when using an abbreviated input format.

Level of Method Description

The third decision criterion which is inherently a characteristic of the measurement system is the level of methods description. This criterion, like accuracy and speed of application, is directly related to element or motion length. Because of this fact, MTM-1 has the highest level of method description while MTM-3 and MTM-V have the least.

To illustrate this characteristic let us describe a simple event in each of the systems. Our earlier example of making a mark with a pen will do nicely. The following chart shows the number of conventions required to describe the event in each system.

System	Conventions
MTM-V	1
MTM-3	2
MTM-2	3
MTM-1	6

Although the conventions, even in the case of the single one in MTM-V, do an adequate job of describing the operation, it is obvious that opportunities for method improvement are not as apparent as they are in MTM-1 or MTM-2 descriptions.

Training Costs

Because of the possibility of turnover in measurement analysts the cost of training can be an important factor in system selection. These training costs can usually be determined from the length of the training courses for the various systems. The minimum requirements as they now are constituted are summarized as follows:

System	Minimum Course Hours	Prerequisite
MTM-1	105	None
MTM-2	72	None
MTM-2	40	MTM-1
MTM-3	24	MTM-2
MTM-V	40	MTM-1 or -2
MTM-M	40	MTM-1 or -2
4M DATA	32	None
GPD	80	MTM-1

Complaint Probability

According to Bayha, Hancock and Langolf (Reference Number 3) a complaint probability factor can be derived for each system. This factor is a function of system accuracy and performance rate requirements. The authors have defined the complaint threshhold as that performance level over 100 percent at which complaints start. The higher the complaint threshhold, the lower the expected number of complaints will be.

Figure 11 is a curve which indicates the complaint probability under conditions when 110 percent performance is expected. If, for example, MTM-1 were used to set standards on operations having an average cycle time of 1,000 TMU (36 seconds), the expected accuracy would be ± 6.4 percent. Entering the complaint probability chart at this point would yield a complaint probability factor of 0.07 complaints per standard. If MTM-2 had been used, the expected accuracy would be ± 13 percent and the complaint probability would be 0.13 complaints per standard. In other words, the more accurate the measurement system, the fewer the number of complaints which can be expected.

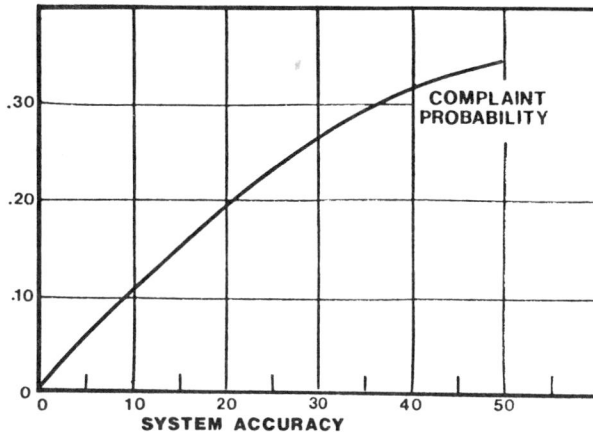

Figure 11. Complaint Probability

To illustrate how this complaint probability factor may be used in system selection, let us make the following assumptions:

1. The measurement system is MTM-2.
2. The average job length is 1,200 non-repetitive TMU.
3. 75 new jobs are processed each month.
4. The cost of settling complaints averages $400 each for industrial engineering and industrial relations work.
5. The required performance rate is 110 percent.

The expected accuracy for MTM-2 at 1,200 TMU is ± 12 percent. Entering the complaint probability chart for 110-percent threshhold at 12 percent yields a complaint probability of 0.12 complaints per standard. The monthly cost of handling complaints, then, would be 0.12 x 75 x $400 = $3,600 per month.

If MTM-1 were used as the measurement system, the expected accuracy would be ± 6 percent and the complaint probability factor would be 0.07 complaints per standard. The cost of handling complaints would then be 0.07 x 75 x $400 = $2,100 per month, a reduction of $1,500 per month.

Productivity Loss Factor

MTM work measurement systems are all based on the performance level for average skilled operators generally referred to as the 100-percent level. According to the normal curve of distribution, only 5 percent of the average industrial population cannot attain this level while the average capability over the working day is probably about 120 percent. In other words, it may be said half of the operators can exceed the average level by 20 percent or more and 5 percent can exceed the MTM standard by 40 percent.

These productivity expectations, however, are seldom realized in the real world because of, among other factors, the complaint probability factor previously discussed. If complaints are corrected by lowering the standards involved, no compensating raises of possibly "loose" rates can be expected since complaints against "loose" rates are practically unheard of. Therefore, the average of all standards suffers lowering to some degree and accordingly a loss of productivity in comparison to the original mix of standards is likely to occur.

Figure 12 is a curve showing the expected rate of productivity as a function of the accuracy of the measurement system. In the previous example, when MTM-2 is used the productivity rate is 99 percent. This is found by entering the chart at a system accuracy of 12 percent and reading off the productivity at 99 percent.

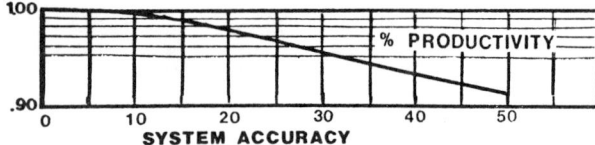

Figure 12. Productivity Loss Factor

When MTM-1 is used as the measurement system the expected accuracy is $\pm$ 6 percent and the rate of productivity would be 99.9 percent.

In order to calculate the cost for loss of productivity in the foregoing example, some additional assumptions are required:

6. The labor rate, with fringes, is $6.00 per hour.
7. The productive labor force is 150 employees.
8. The average hours per month is 170.

The probability loss factor when MTM-2 is used is
$1 - \frac{100}{99} = -.010101$

The monthly labor cost is 150 x 6.00 x 170 = $153,000.

The cost of lost productivity is $153,000 x .010101 = $1,545 per month.

With MTM-1 the probability loss factor is
$1 - \frac{100}{99} = -.001001$, and the cost of lost productivity is $153,000 x .001001 = $153 per month.

Defining Time Periods and Accuracy Requirements

We have, by no means, exhausted the material which could be examined with regard to decision criteria, but at least we have some appreciation of the magnitude of the task. Before we can explore a means of tying all these criteria together, we need to explore the uses to which measurement systems are put and the accuracies that these uses may demand from a given system.

When the measurement system is used for several purposes, the most exacting use should determine the accuracy requirements for the system. Sometimes it is useful, and possibly more economical, to use several systems, each tailored to different uses. In this way the expense of using an overly precise system is minimized.

Production Planning

The loading period is generally used as the total figure which must be known when choosing the proper measurement system for this purpose. The loading period generally ranges from daily to bi-weekly, or from 8 to 80 hours. The desired accuracy ranges from $\pm$ 5 to $\pm$ 20 percent.

Incentive Payment

The wage payment interval often is used as the total time and generally ranges from 40 to 160 hours. The demand on accuracy may fall between $\pm$ 5 and $\pm$ 15 percent, but is usually set at $\pm$ 5 percent. Because of labor contract considerations it may be advisable to use the average job length as the total time interval.

Line Balancing

The time-per-station statistic of line production is generally chosen as the total time (T). This makes T equal to the length of the operation. The required accuracy may fall between $\pm$ 5 and $\pm$ 10 percent.

Methods Improvement

The average length of an operation is chosen as the total time (T). The required accuracy will fall between $\pm$ 5 and $\pm$ 10 percent. When greater variations are permitted the system finally selected may not have the level of methods description necessary for making meaningful methods improvements.

Conclusion

The task of choosing the right measurement system for the right job has always been with us, but it has become more complex in recent years. We now must choose not only a branch from the tree, but also a twig from the branch. And new twigs are popping up and growing all the time.

Our intent today has been to present some ideas which might make the task, if not less complex, at least more rational and economically sound to accomplish. In order to make a proper selection it is necessary to understand the nature and inner workings of the systems available to us. We hope that our discussion concerning the classification of PDT systems and their characteristics has been helpful in this respect.

We have also presented some of the current thinking with respect to the selection of the appropriate MTM measurement technique for a given set of circumstances. We hope we have been successful, at least to a degree, in accomplishing our objectives.

A HIGH LEVEL PREDETERMINED TIME STANDARD SYSTEM
AND SHORT-CYCLE TASKS

Abu Masud, Ph.D. and Don Malzahn, Ph.D.
Department of Industrial Engineering
Wichita State University, Wichita, Kansas 67208

Scott Singleton, Boeing Military Airplane Co.
Wichita, Kansas 67210

INTRODUCTION

High level synthetic time standard systems have reached a level of wide acceptance by the industrial community. Generally, these systems are considered appropriate only for tasks with cycle times greater than 1 minute [1, 2, 4, 8, 13, 18]. The popularity of the high level systems rests primarily on their speed and simplicity of application. There exist many situations in which a highly accurate estimate of performance for short cycle tasks may be desired. This research examines the accuracy of a high level synthetic time standard system for establishing performance estimates for short cycle tasks.

The impetus for this particular research stems form our work in rehabilitation engineering at the Wichita State University Rehabilitation Engineering Center. Part of the work done there involves the functional vocational evaluation of individuals with severe neuromuscular impairments, and the design of appropriate task modifications required for placement in mainstream industry [2, 5, 10, 11]. One of our objectives is to develop individual specific synthetic time standards systems so that the economic viability of various modification strategies can be evaluated before placement on the job [17]. The work presented here involves only data from able-bodied individuals. In part it examines the feasibility of high level individual specific synthetic time standard systems for tasks with cycle times less than 0.5 minutes.

THE MODAPTS SYSTEM

MODAPTS is an acronym for " MODular Arranged Predetermined Time Standards" [6]. MODAPTS is an Australian developed time system that is based on the premise that larger body sections take longer to move than smaller sections. For example, in this system it takes twice as long to move a hand as it does to move a finger. It takes three times as long to move the forearm as it does a finger, and it takes four times as long to move the whole arm outward. From this simple framework, MODAPTS has built an entire system of predetermined macro time standards. The focus of this research is to examine the simple conceptual scheme of MODAPTS applied to short-cycle tasks.

There were a number of reasons for choosing MODAPTS as the system to be studied. First, it is simple to apply. Second, Modapts lends itself more readily to future applications once a short-cycle synthetic time standard system is developed. This is due to the relatively few number of motions needed by MODAPTS to describe an activity. Third, MODAPTS has already been used in Australia at Centre Industries to develop time-standards for workers with physical handicaps [7].

MODAPTS has a wide range of applications including some specially developed for specific activities. However, the tasks used for this project required only a fraction of the MODAPTS capabilities for their documentation (very few synthetic applications require the full range of any reputable standard system). The MODAPTS activities applied were all of the common Move (M), Get (G), and Put (P) elements, the Decide (D) element, and the Juggle (J) element. The times associated with each movement are multiples of the MODAPTS unit of time measurement, the "MOD", which is equivalent to 0.129 seconds for short-cycled tasks.

The notation for MODAPTS documentation deserves some explanation. The following two sample documentations will serve as examples.

Task	MODAPTS Code	Total MODs
Example 1:	D3, 2(M2P2)	= 11.0 MODS
Example 2:	M3P0,J2,D3,0.5(M2P2+M3P2)	= 12.5 MODS

The M2, M3, P0, P2, D3, and J2, are abbreviations for the MODAPTS activities. The complete list of MODAPTS activities used in this study is given below.

M2 = Move 2	G1 = Get 1	P2 = Put 2
M3 = Move 3	G3 = Get 3	P5 = Put 5
M4 = Move 4		
M5 = Move 5	J2 = Juggle 2	D3 = Decide 3

In the documentations, the activities are separated by commas, and the MOD values of the activities (represented by the numerals in the activity expressions) are totaled. These totals are then directly converted into expected times for the activity by the MOD conversion factor (0.129

seconds/MOD). For Example 1, the time is equal to 1.419 seconds (= 11.0 MODS X 0.129 seconds/MOD); and for example 2, it is 1.6125 seconds.

THE AVAILABLE MOTIONS INVENTORY (AMI)

The AMI [10, 14, 17] was well suited for the testing and validation of short-cycle time standards because the AMI is basically a battery of short-cycle tasks. Some of the added advantages of the AMI were that it was a well documented, reliable (test re-test adjusted R-squares greater than .85) ability evaluation system and significant amounts of data were accessible for analysis purposes. It also involved a wide range of commonly required motions.

THE AVAILABLE MOTIONS INVENTORY (AMI)

The AMI [10, 14, 17] was well suited for the testing and validation of short-cycle time standards because the AMI is basically a battery of short-cycle tasks. Some of the added advantages of the AMI were that it was a well documented, reliable (test re-test adjusted R-squares greater than .85) ability evaluation system and significant amounts of data were accessible for analysis purposes. It also involved a wide range of commonly required motions.

The AMI battery of tests is divided into 6 basic categories: switches, settings, rates, strength, assembly, and reaction. The strength, reaction, and rate tests were not be used. The strength measures were not used because the standard time systems do not differentiate in weights of less than 6 pounds, and the forces in AMI tests are far below that level. The reaction times are not used because they are not estimated directly by the MODAPTS system. Finally, the rates are not applicable to this study because subjects are instructed in this section of the test to perform at their maximum speed on the same devices used in the settings section of the evaluation. Consequently, the only tests of the AMI used for research involve the switch, settings, and assembly sections. The total number of individual tests retained was 41.

The AMI tests individuals as to their capacity in not only different activities but also the same activity in different orientations to the body. The AMI has 3 basic height positions lower-horizontal position, lower-vertical position, and upper-vertical position. In addition, each vertical position has 3 possible horizontal positions: the front-center position, the left side-panel positions, and the right side-panel positions. All AMI tests are performed independently by each hand. The order of testing is first the dominant hand and then the subordinate hand.

All of the test results are scored in units of activations per minute as the standard performance measurement. The reasoning for this evaluation procedure is that a zero indicates that the person is not able to perform a specific task. This has implications for persons with disabilities but is not a factor for able-bodied subjects. If the results were listed in time units, then a person who could not perform a task would take an infinite amount of time.

THE MODELING PROCEDURE

Several models were hypothesized and then evaluated by regression analysis. The results of the regression were then examined to validate or nullify each model and its underlying hypothesis. The models were tested by two forms of significance tests. The first was the Student's t-distribution test which tests the significance of each of the individual terms estimated for the regression equations. If more than one term was independently found insignificant then a joint test of significance was conducted for these terms (see [12] for details of the joint test). The second was the F-distribution test which yields a measure of the significance of each of the regression equations as a whole. Only the significant terms were retained to develop a new reduced equation. Then, the new equation's predicting power was examined. It was the predicting power that this research project attempted to optimize. The R-squared is the pure indicator of predicting power, but the adjusted R-squared, which is the R-squared adjusted for the number of variables in the model, was used as the indicator in this research. The reason was that the adjusted R-squared incorporates a measurement of efficiency (in terms of variables used) in predicting power of the regression models. A final predicting model was considered a good predictor if it had an adjusted R-squared of at least 0.85.

The data for these experiments was collected as part of the standardization process for a physical performance evaluation system, the Available Motions Inventory (AMI). Data from 50 able-bodied individuals was collected, 25 males and 25 females. The age ranged from 19 to 55. Standards were developed for a sub-set of the AMI tasks with the MODAPTS synthetic time standard system and observed performances compared.

ANALYSES AND RESULTS

Notation

The following notational scheme will be used to explain the models, analyses and results:

   Y = dependent variable
     = AMI test results in activation/minute

   X = independent variable
     = MODAPTS developed activations/minute

Z1 = indicator variable
 = 1.0 for setting tests
 = 0.0 for switch and assembly tests

Z2 = indicator variable
 = 1.0 for assembly tests
 = 0.0 for switch and setting tests

Note that the base-line model in the above scheme was the model with only switch tests when both Z1 and Z2 equal 0.0. Also, MODAPTS developed values are normally in seconds; however, for this analysis, we have transformed them to the equivalent activations/minute units to have consistency with AMI measurement units.

Consistency Across AMI Tasks

One of the desirable features of a standard time system is that its elemental time values, such as P2 and G1, etc., be valid across different types of tasks. To test this for MODAPTS, we hypothesized the following general linear model:

$$Y = a + b*X + Z1*(c + d*X) + Z2*(e + f*X) \quad (1)$$

where a, b, c, d, e, and f are regression coefficients. The regression results indicated that Equation (1) was significant at = 5%. Except for the coefficient b, all other coefficients were not significant individually or collectively. A new regression run was made with only the b coefficient and the result was the following equation:

$$Y = 0.96783\ X \quad (s=0.01296\ and\ d.f.=40) \quad (2)$$

This model passed all the usual validity checks and has an adjusted R-square of 0.9927. Equation (2) indicates that MODAPTS times are excellent predictors of actual times on AMI tests but they prescribe less time than observed from test subjects. One explanation for this may lie in the fact that MODAPTS, like other standard time systems, is designed for long-cycle activities and does not directly include learning effects.

Test for Handedness and Model Generality

The data used for deriving the parameter estimate of equation (2) came from 50 able-bodied persons using their dominant hand for performing the AMI tasks. However, a question that arose at this point in the analysis was "is equation (2) applicable as a general model for all able-bodied persons?" and, if yes, then "is the equation valid for both the dominant and the subordinate hands?" To answer these, data was collected from six new subjects using their dominant and the subordinate hands separately. By combining these two data, both hand data was developed for each of the 6 subjects. Next, predicted activations/minute were generated by using equation (2) and used in the following three regressions run with each subject data separately and with all six subjects' combined into one group (a total of 21 regression runs).

$$YD = a * W \quad (3)$$
$$YS = a * W \quad (4)$$
$$YB = a * W \quad (5)$$

where: W = Equation (2) predicted act./minute
YD = AMI test act./minute for dominant hand
YS = AMI test act./min. for subordinate
YB = AMI test act./minute for both hands

The results were tabulated in Table 1. If equation (2) was applicable across subjects and across hands, then all estimates of parameter "a" should be statistically equal to 1.0, i.e. the 95% confidence intervals should include 1.0. From Table 1, note that when data for both hands was combined, Equation (2) was not appropriate for 5 out of 6 subjects. Observe that, for the dominant hand, performance Equation (2) was slightly better since Equation (2) was derived from the dominant hand data of the 50 original subjects. When the data

Subject	Hand	Estimate of "a"	95% Conf. Interval
5029	D	1.04325	0.98931 to 1.09720
	S	1.03125	0.98357 to 1.07893
	B	1.03724	1.00200 to 1.07248
5044	D	1.11898	1.06459 to 1.17336
	S	1.04853	1.00042 to 1.09665
	B	1.08369	1.04733 to 1.12005
5054	D	0.96048	0.88834 to 1.03262
	S	0.90594	0.84531 to 0.96658
	B	0.93316	0.88667 to 0.97965
5056	D	1.01520	0.95469 to 1.07572
	S	0.84377	0.81005 to 0.87749
	B	0.92933	0.89052 to 0.96814
5065	D	1.00960	0.95183 to 1.06737
	S	0.91632	0.86369 to 0.96894
	B	0.96287	0.92328 to 1.00246
8883	D	0.83146	0.77457 to 0.88835
	S	0.76792	0.72221 to 0.81363
	B	0.79965	0.76327 to 0.83602
All six combined	D	0.99647	0.97045 to 1.02238
	S	0.91896	0.89612 to 0.94180
	B	0.95764	0.94009 to 0.97520

Table 1. Prediction Accuracy of Equation (2).

for all 6 subjects was combined, the handedness effect was even more pronounced. In this situation, Equation (2) was an acceptable predictor of only the dominant hand performance of the subjects as a group. Thus, two significant conclusions can be made from the results in Table 1: (a) there was a significant handedness effect and (b) Equation (2) can not be used directly for a specific individual without making an individual specific correction.

These two conclusions are supported by the results shown in Table 2. Table 2 results were obtained in a manner very similar to those for Table 1 with the difference being in the way in which predicted values (variable W) were determined. For Table 2, the predicted values were obtained from Equation (2) by estimating the parameter "a" value individually for each of the subjects (using their dominant hand data only). Note the pronounced handedness effect on MODAPTS' effectiveness in predicting actual performance for the subordinate hand. The differences between MODAPTS and actual as indicated in Table 1 results are due to both the handedness and the subject specific effects; the results in Table 2, however, indicates only the handedness effect because the individual differences have been removed through the estimation of Equation (2) parameters for each individual subject.

Subject	Hand	Estimate of "a"	95% Conf. Interval
5029	S	0.98850	0.94280 to 1.03420
	B	0.99424	0.96046 to 1.02802
5044	S	0.93705	0.89405 to 0.98004
	B	0.96847	0.93597 to 1.00096
5054	S	0.94322	0.88009 to 1.00635
	B	0.97156	0.92316 to 1.01996
5056	S	0.83113	0.79792 to 0.86434
	B	0.91541	0.87718 to 0.95364
5065	S	0.90760	0.85548 to 0.95973
	B	0.95372	0.91450 to 0.99293
8883	S	0.92358	0.86861 to 0.97856
	B	0.96174	0.91799 to 1.00549

Table 2. Prediction Accuracy of Equation (2) when Adjusted for Individual Subjects

SUMMARY AND CONCLUDING REMARKS

In this paper we have reported the results of the analyses performed to answer the question "Can a high level Predetermined Time Standards system, such as MODAPTS, be applied to short-cycle tasks?" The answer is yes but with some adjustments. It has been demonstrated that the handedness effect and individual differences are pronounced when short-cycle jobs are analyzed with MODAPTS. We suspect that similar conclusions can be made for other systems as well. From the point-of-view of the applicability to physically handicapped persons, this may not be that much of a setback since they need to be evaluated as to their remaining capability before placement on a job. If that evaluation is made on AMI then the needed adjustments are fairly straight forward.

BIBLIOGRAPHY

1. Aberg, Ulf. "Sequence of Motions in Manual Workshop Production." Journal of Methods Time Measurement 9 (November-December 1963): 1-12.

2. Bridges, Clark D. Job Placement of the Physically Handicapped. New York: McGraw-Hill Book Company, pre 1946.

3. Buffa, Elwood S., and Lyman, John. "The Additivity of the Times for Human Motor Response Elements in a Simulated Industrial Assembly Task." Journal of Applied Psychology 42 (1958):379-83.

4. Crossan, Richard M., and Nance, Harold W. Master Standard Data. New York: McGraw-Hill Book Company, 1972.

5. Hardy, Richard E. and Cull, John G. Vocational Evaluation for Rehabilitation Services. Springfield: Charles C. Thomas--Publisher, 1973.

6. Heyde, G. C. (Chris). MODAPTS Plus. Sydney, Australia: Heyde Dynamics Pty. Ltd., 1983.

7. Hume, Bruce C. "The MODAPTS approach to Vocational Evaluation." In Vocational Evaluation for Rehabilitation Services, pp. 167-240. Edited by Richard E. Hardy and John G. Cull. Springfield: Charles C. Thomas--Publisher, 1973.

8. Karger, Delmar W., and Walton, M. Hancock. Advanced Work Measurement. New York: Industrial Press Inc., 1982.

9. Lane, Robert E., and Stelter, Leonard. "Selection and Implementation of a Predetermined Time System." Assembly Engineering, October 1972, pp. 26-27.

10. Malzahn, Don. "Functional Evaluation for Task Modification Using the Available Motions Inventory." in Functional Assessment in Rehabilitation Halpren, A. S. and Fuhrer, M. J. ed.s, Paul H. Brookes Publishing Co., 1984, pp. 131-144.

11. Mink, J. A. "MTM and The Disabled." The Journal of Methods Time Measurement 2 (1975):23-30.

12. Pindyck, Robert S. and Rubinfeld, Daniel L. Econometric Models and Economic Forecasts. 2nd ed. New York: McGraw-Hill, Inc., 1981.

13. Quick, Joseph H.; Duncan, James H.; and Malcolm, James A., Jr. Work-Factor Time Standards: Measurement of Manual and Mental Work. Haddonfield, New Jersey: The Work-Factor Company, Inc., 1962.

14. Rahimi, Mansour and Malzahn, Don. "Task Design and Modification Based on Physical Ability Measurement". *Human Factors*. Vol 26, No. 6, December, 1984.

15. Sanfleber, H. "An Investigation into some Aspects of the Accuracy of Predetermined Motion Time Systems." *The International Journal of Production Research* 6 (1967):25-45

16. Schmidtke, Heinz, and Stier, Fritz. "An Experimental Evaluation of the Validity of Predetermined Elemental Time Systems." *The Journal of Industrial Engineering* (Dortmund, Germany) 12 (May-June 1961):182-204.

17. Singleton, Scott K. "Evaluation of MODAPTS as a Short-Cycle Standard Time System." M.S. thesis, Wichita State University, June 1984.

18. Zandin, Kjell B. *MOST Work Measurement Systems*. New York: Marcel Dekker, Inc., 1980.

BIOGRAPHICAL SKETCH

Dr. Abu Masud, Assistant Professor of Industrial Engineering at Wichita State University, is active in research in the areas of multiple objective decision making, application of operations research techniques to industrial problems, and rehabilitation engineering. He is co-author of the textbook "Multiple Objective Decision Making - Methods and Applications" published by Springer-Verlag. He is a senior member of IIE and a member of ORSA, TIMS, ASEM, and IEEE Computer Society.

Dr. Don Malzahn, Associate Professor of Industrial Engineering at Wichita State University, has been active in research regarding task modification for persons with physical impairments for over 10 years. He is a senior member of IIE and a member of HFS, ACRM, and RESNA.

Scott Singleton has received both his B.S.I.E. and M.S. in Engineering Management Science from Wichita State University. He is currently employed at the Boeing Military Airplane Company and is involved with research concerning simulation based training for skilled manufacturing personnel. He is a member of IIE.

* The research reported here was supported in part by the the National Institute of Handicapped Research through grant #G008300069.

Robot Time And Motion System Provides Means Of Evaluating Alternate Robot Work Methods

By Shimon Y. Nof
Purdue University
and Hannan Lechtman
International Harvester Inc. Science and Technology Laboratories

We now have the ability to design not only the work system, its components and its environment, but also the structure and capabilities of the (robot) operator. The flexibility, variety and work potential of robots present substantial new challenges to industrial engineers.

The use of industrial robots raises two overlapping design issues: Out of the variety of robot models and configurations, which particular one is most suitable for a given job; and what is the best method of carrying out the tasks assigned to the selected robot. In order to reach an effective solution, those two questions must be answered in parallel.

Nof, Knight and Salvendy (see "For further reading") describe a skills analysis approach to identifying the skills and abilities a robot operator will need to best perform the tasks of a given job.

Another traditional IE approach to work system design is performance measurement. The purpose of this article is to describe a method for robot work analysis and performance measurement. The method, called RTM (robot time and motion), was first presented in 1978 and since then has been undergoing development at Purdue University. Analogous to MTM (methods time measurement), which has long been in use for human work analysis, RTM is based on standard elements of fundamental work motions. However, since the RTM method is developed for robots—i.e., machines—an attempt is being made to model their work by utilizing knowledge about their mechanical design and work patterns.

After describing the RTM method, the concept behind it and its current capabilities, we will illustrate how it is applied in actual, practical performance evaluation.

RTM methodology

In practice, a robot task method has to be planned and analyzed before the precise details of a particular method are programmed. A number of alternative methods have to be examined. Programming the complete details of several alternatives in a robot control language merely to evaluate and compare them is a tedious, lengthy and uneconomical approach.

Furthermore, in order to evaluate several robot models and select the most suitable one, it is preferable to be able to evaluate them first, and not acquire all of them for experimentation.

The RTM methodology provides a high level, user friendly technique with the following capabilities:
☐ Systematically specifying a work method for a given robot in a simple, straightforward manner.
☐ Applying computer aids to evaluate a specified method by time-to-perform, number of steps, positioning tolerances and other requirements so that alternative work methods can be compared.
☐ Repeating methods evaluation for alternative robot models.

The RTM system is comprised of three major components: RTM elements, robot performance models and an RTM analyzer. The system has been implemented and experimented with using two main robot types, the Stanford Arm equipped

T^3 performing insertion task: Picking up base part and moving it to center of table.

Robot inserts Peg No. 1 in center hole of base. Hole diameter is 25 mm, and peg diameter is 20 mm at point of insertion. With T^3's position accuracy of ±1.27 mm, sensory guidance is not needed for this insertion.

Robot picks up Peg No. 2 from fixture for insertion in back of Peg No. 1. Center hole of Peg No. 1 is 25 mm, and Peg No. 2 diameter is 20 mm. Following this movement, the T^3 returns to its original position.

with touch and force sensing and Cincinnati Milacron's T³.

The Stanford Arm is a small, all electric prismatic robot that can carry a payload of not more than three kilograms. The T³ is a large hydraulic articulated robot, equipped with a pneumatic double gripper, that can carry a payload of up to 50 kg.

Extensive laboratory studies have been carried out to test and validate the RTM methodology. The main measure of performance used has been the method time. At present, additional robot types are being modeled and other performance measures added to extend the scope of the original RTM system.

The ten general work elements defined in the RTM method are shown in Table 1. They are divided into four major groups:

☐ *RTM Group 1:* Motion elements, representing arm movements and

Table 1: RTM Symbols and Elements

Element No.	Symbol		Definition of Element	Element Parameters
1	Rn		*n-segment reach:* Move unloaded manipulator along a path comprised of n segments	Displacement (linear or angular) and velocity or Path geometry and velocity
2	Mn		*n-segment move:* Move object along path comprised of n segments	
3	ORn		*n-segment orientation:* Move manipulator mainly to reorient	
4	SEi		*stop on position error*	Error bound
4.1		SE1	Bring the manipulator to rest immediately without waiting to null out joints errors	
4.2		SE2	Bring the manipulator to rest within a specified position error tolerance	
5	SFi		*Stop on force or moment*	Force, torque and touch values
5.1		SF1	Stop the manipulator when force conditions are met	
5.2		SF2	Stop the manipulator when torque conditions are met	
5.3		SF3	Stop the manipulator when either torque or force conditions are met	
5.4		SF4	Stop the manipulator when touch conditions are met	
6	VI		Vision operation	Time function
7	GRi		*Grasp an object*	
7.1		GR1	Simple grasp of object by closing fingers	
7.2		GR2	Grasp object while centering hand over it	Distance to close/open fingers
7.3		GR3	Grasp object by closing one finger at a time	
8	RE		Release object by opening fingers	
9	T		Process time delay when the robot is part of the process	Time function
10	D		Time delay when the robot is waiting for a process completion.	Time function

reorientation with or without a payload.

☐ *RTM Group 2:* Sensing elements, which represent sensory actions by robots that are equipped with position, force or vision inputs.

☐ *RTM Group 3:* Gripper or tool elements that depend on the type of robot end-effector (i.e., the robot working "hand" or fingers).

☐ *RTM Group 4:* Delay elements that add specific time durations according to processing and waiting requirements.

Each element is specified with certain parameters, such as distance moved, angles of rotation or positioning error bounds. For each type of robot the details and parameters may vary. Robots without sensory feedback capabilities will, of course, have no sensing elements.

Robot performance models

Within the framework of the RTM system a number of modeling approaches are possible for each individual element. Four modeling approaches are identified and briefly explained here.

Element tables

The simplest modeling approach, which follows the original MTM approach for human work methods, applies a set of tables with estimates for each element according to particular parameters. Tables are developed based on laboratory experiments with the robot type for which data are prepared.

For instance, Table 2 shows times for the element STOP-ON-ERROR as defined by positioning error bounds of 1.0, 0.2 and 0.05 cm.

Table 3 contains time data for elements REACH (R1) or MOVE (M1) by the T^3. (Note that for the T^3, unlike the Stanford Arm, velocity is an external parameter; also, T^3 REACH and MOVE elements are identical, since carried weight does not affect performance time.)

The advantage of this approach is its simplicity. Once tables are developed for a variety of robot types, a user can retrieve the time estimates for a task method evaluation either manually or by using a computer aid such as the RTM system. With regard to the predictive accuracy of such tables, in our experiments the deviations of predicted time from actual measured time for a variety of realistic tasks were within about ±5%.

Regression equations

This modeling approach is considered useful when observation values are random or when a complicated functional relationship has to be approximated. Both conditions apply in the case of robot work. With regard to value randomness, robots, as machines, are highly consistent in their operation. However, experiments have shown that robot performance varies with dynamic changes in electric power or hydraulic pressure.

Regression equations developed for the RTM system have been found to yield a predictive accuracy of about ±5%, similar to that of the tables approach.

Motion control

While the previous two modeling approaches have traditionally been applied in time and motion studies,

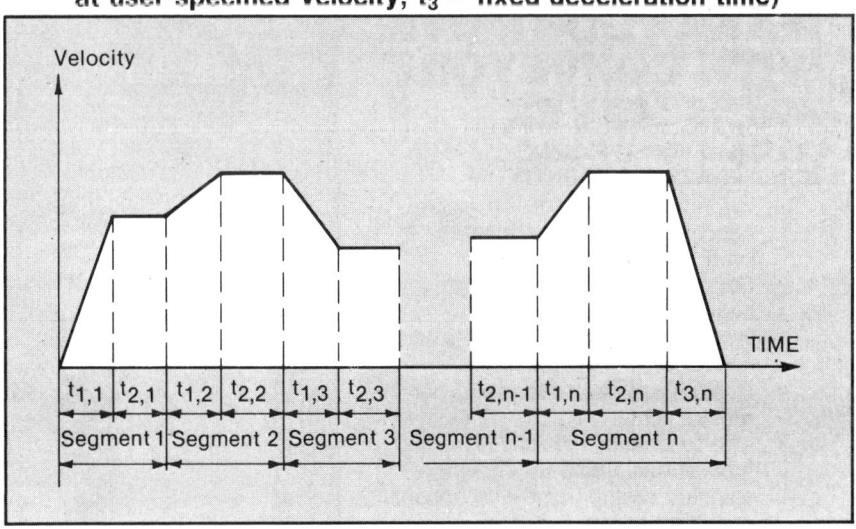

Figure 1: Velocity Pattern in General, Multisegment T^3 Motions (t_1 = fixed acceleration time; t_2 = motion at user-specified velocity; t_3 = fixed deceleration time)

Table 2: RTM Table for STOP-ON-ERROR (SE) by the Stanford Arm

Error Bound (cm)	Time (sec)	Description
1.0	0.1	Bring the arm to rest to within the position error tolerance specified
0.2	0.3	
0.05	0.8	

the following two can be used when knowledge about the engineering design of a robot is available (either from robot designers or by analysis and experimentation).

Models of motion control are mainly useful for the group of RTM motion elements, RTM Group 1. These models can also be applied for a certain gripper and finger motions for elements in RTM Group 3.

A typical motion control is applied according to predesigned patterns of the robot velocity. One such pattern combines three main sequences of motion velocity: acceleration; motion at user-specified velocity; deceleration. In the case of the T^3, the times allotted for acceleration and deceleration are fixed and identical, about ¼ second. Using kinematic formulas, the total motion time is found according to

$$T(\text{total motion time}) = \frac{S}{V} + \frac{1}{4} \text{ for } S > \frac{V}{4}$$

where S is the total distance moved and V is the user-specified velocity.

It is assumed in this case that the robot starts from a stationary position (at V = 0) and moves to a stop.

In certain short motions, the robot does not have sufficient time to reach the user-specified velocity. In the ideal case, this happens wherever

$$S \leqslant \frac{V}{4}$$

However, in practice, several other factors have to be considered, such as some overhead for computation, minimum time to prevent shocks and jerks, etc. As a result, the model for the T^3 motion time can be summarized as follows, as we found in our laboratory experiments:

T(total motion time for elements R1, M1 in seconds) =

□ For $S > \frac{V}{2.857}$:

$$T = \frac{S}{V} + 0.365$$

□ For $S \leqslant \frac{V}{2.857}$:

T = 0.413 for S < 6.25 cm
T = 0.610 for 6.25 ≤ S < 17.50 cm
T = 0.802 for 17.50 ≤ S < 43.75 cm

where S is the total distance moved in centimeters and V is the user-specified velocity in cm/sec.

For multi-segment motions (see Fig. 1) in which the robot moves from point to point without stopping, more complicated analysis leads to the general equation for the T^3:

T(total motion time for elements Rn, Mn, ORn in seconds) =

$$\frac{n}{8} + 0.24 - \frac{1}{8} \sum_{i=2}^{n} \frac{V_{i-1}}{V_i} + \sum_{i=1}^{n} \frac{S_i}{V_i}$$

where n is the number of segments along the path.

Other factors that are considered during the computation in the RTM are limitations on motions that entail excessive rotation and combined

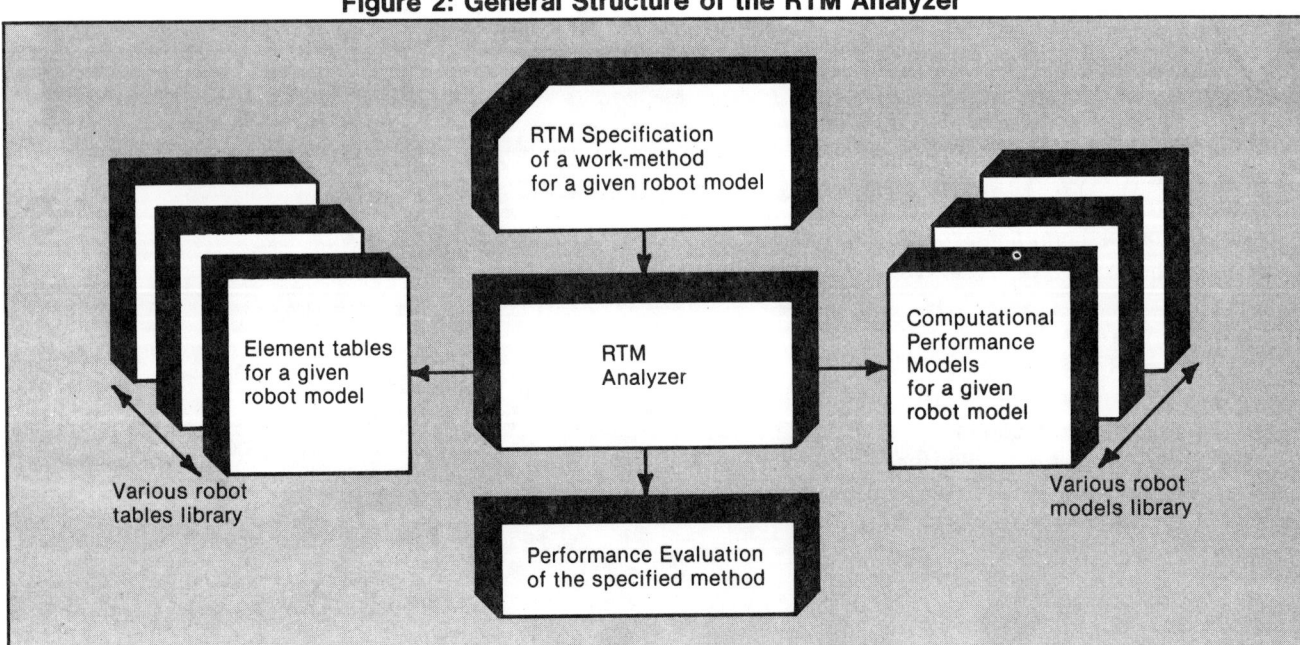

Figure 2: General Structure of the RTM Analyzer

rotation and linear motions. The predictive accuracy of this modeling approach in our experiments has been in the range of −2% to +3%.

Path geometry

Robot motions can also be modeled by specifying the motion path geometry. Based on robot joint and link velocities, the motion time can be computed.

In essence, the distance, linear or rotational, that each joint or link has to move relative to its preceding link in order to reach a specified position for the end-effector can be figured out. The time it takes to complete the motion will depend on the longest individual link or joint motion time.

Path geometry computation programs have been implemented in the RTM system for the Stanford Arm and the T^3. Their predictive accuracy, as found in our experiments, is between −2% and +12%. It should be noted that the detailed input data required for the specification of each motion path are usually not available during advanced planning stages.

The RTM analyzer

The RTM analyzer has been developed to provide a means of systematically specifying robot work methods with direct computation of performance measures. The general structure of the analyzer is shown in Figure 2.

The input to the RTM analyzer includes control cards (e.g., task and sub-task titles, type of robot and type of model to apply) and operation statements. The statements specify particular parameters for RTM work elements, one element per statement.

A control structure in the RTM analyzer provides further capabilities in terms of work method specification. It includes the following:

☐ REPEAT blocks, which permit specification of repeated groups of operations. For instance, when a robot has to insert a peg in a hole in five consecutive parts using an identical set of motions, then push a control button once, the five insertions can be specified once in a REPEAT block.

☐ PARALLEL blocks, which specify sub-tasks that are performed in parallel, either by multiple arms or by a robot and its sensory equipment. The longest performance time will be the one considered for the overall performance. Certain parallel operations are not allowed because they are too complicated or simply infeasible.

☐ "GO TO" transfer statements, which define branching between and within sub-tasks based on combinations of simulated status signals. This capability provides for a simulation of probabilistic work situations. With it, the practical work of a feedback-controlled robot or a decision-making robot can be studied.

Applying RTM

To illustrate the application of the RTM method, an insertion task performed by the T^3 is described. The task was studied by the RTM analyzer and then actually measured for comparison using a counter timer in the laboratory.

The objective of the insertion task performed by the T^3 was to insert two pegs, one at a time, into a base part. One peg had to be inserted first, then the smaller peg inserted into the back of the first one. Figure 3 shows the layout of the robot operation.

The work method to be analyzed includes 26 steps. Initially, the two pegs are placed in a fixture and the base part is placed on the work table. The robot starts by grasping and moving the base to the table center, then picks up peg number 1 and inserts it, picks up peg number 2 and inserts it and returns to its initial position.

Table 4 contains the RTM input data for this task. Note that the distances shown in Table 3 are approximate, since a user typically would not have precise details at this evaluation stage. Two delay elements are specified in the input (operations 9 and 18) to simulate the robot waiting for an external signal—e.g., opening of a safety gate.

Results of the RTM analysis are shown in Table 5. For each RTM element the calculated time (in msec) is given next to the time estimated from element tables. In the last column the actual measured time is listed. The standard RTM output includes a column for distance movements and angles rotations; however, in this illustration all movement times were dominated by the distance movements, and the rotation

Table 3: RTM Times for REACH (R1) or MOVE (M1) by T^3

Distance to Move (cm)	Time (sec)				
	At 5.0 cm/sec	At 12.5 cm/sec	At 25.0 cm/sec	At 50.0 cm/sec	At 100.0 cm/sec
1	0.4	0.4	0.4	0.4	0.4
30	6.4	2.8	1.6	1.0	0.8
100	21.3	8.7	4.5	2.4	1.4

Table 4: RTM Input Data for Insertion Task by the T³

Statement	Comments
T³	Robot type
Insertion Task	Task title
*Conditions	
....	Input of condition signals, if any were used
*	
[REPEAT 1 To 26 SEVEN TIMES]	Repetition command could be stated here
1 R1 5, 7.5	Reach 7.5 cm at 5 cm/sec to start position above base
2 GR1	Grasp base
3 M1 5, 5.0	Raise base
4 M1 25, 48.0	Move to above assembly position
5 M2 25, 12.5, 5, 5.0	Bring base down in two segments movement
6 RE	Release it
7 R1 25, 23.0	Rise
8 R1 25, 55.0	Reach fixture
9 D 2	Wait 2 sec (for continuation)
[IF SIGNAL. NE. VALUE) GO TO END]	Conditional branching could be used
10 R1 5, 15.0	Bring fingers above peg 1
11 GR1	Grasp peg 1
12 M1 5, 15.0	Raise peg 1
13 M1 25, 55.0	Move it to above base
14 M2 25, 11.0, 5, 9.0	Insert peg in two steps
15 RE	Release peg 1
16 R1 25, 22.0	Rise
17 R1 25, 50.0	Reach above peg 2
18 D 2	Wait 2 sec
19 R1 5, 12.5	Bring fingers above peg 2
20 GR1	Grasp peg 2
21 M1 5, 12.5	Raise it
22 M1 25, 50.0	Move it above base
23 M2 25, 10.0, 5, 5.0	Insert it in two steps into peg 1
24 RE	Release peg 2
25 R1 25, 17.5	Rise
26 R1 25, 60.0	Reach start point

angles have been omitted.

The distance data given in Table 5 are the actual displacements moved by the robot end-effector in performing the task. These data were used for time calculations in the motion control models of the T³. The data calculated from element tables, shown in column 6 of Table 5, were determined according to the estimate specifications provided in the input as shown in Table 4.

The time estimates shown in Table 5 for the individual elements are often close, but sometimes highly inaccurate. However, from the summary results in the table there is clearly no difference between the calculated time estimates and the actual time for the complete task.

The difference between the table estimates and the actual time of the complete tasks is +5.6%. The good overall predictive ability of the RTM as demonstrated here may result from the fact that the majority of the estimates used are very close.

Once a task method is analyzed, a user can modify certain parameters of specified elements, change the sequence of operations, etc., to establish the best method for use with a given robot model.

For instance, in the example above the double gripper could be applied to grasp both pegs at once, then move and insert them together. The next planning stage will be to compare the performance of the T³ with that of alternative robot models.

Work in progress

The RTM method has been developed for industrial robot work analysis and implemented in the RTM system. Since robot work design involves the concurrent selection of the best robot model and the best work method for it, the system must permit the performance analysis of a variety of robot models.

Currently, the RTM system is based on a set of general work elements; performance models of several kinds, which were implemented and tested for two robot models; and the RTM analyzer, which interprets RTM input data and directly performs calculations of time estimates.

Further work is in progress to develop performance models for other robot types and to extend the development to simultaneous work by several robots. Additionally, the RTM analyzer will be used to provide input estimates for a design optimization method for assembly robot stations. We expect the RTM method to be applied in industry in the near future.

Table 5: RTM Analysis Results for Insertion Task by T^3

Operation Number	RTM Element	Distance (cm)	Angle (degrees)	Velocity $V\left[\dfrac{cm}{sec}\right]$	Calculated Time (msec)	Time from Tables (msec)	Measured Time (msec)
1	R1	8.25	—	5	2000	1854	2003
2	GR1				10	10	36
3	M1	5.10	—	5	1385	1365	1390
4	M1	48.18	—	25	2292	2169	2299
5	M2	13.18 5.75	— 	25 } 5 }	1543	2173	1526
6	RE				10	10	36
7	R1	23.5	—	25	1305	1268	1293
8	R1	55.75	—	25	2594	2564	2590
9	D				2000	2000	2000
10	R1	15.00	—	5	3360	3372	3368
11	GR1				10	10	37
12	M1	15.00	—	5	3360	3372	3367
13	M1	55.75	—	25	2595	2564	2590
14	M2	10.75 9.25	— —	25 } 5 }	2120	3041	2099
15	RE				10	10	37
16	R1	21.75	—	25	1233	1268	1243
17	R1	50.25	—	25	2373	2169	2371
18	D				2000	2000	2000
19	R1	13.50	—	5	3075	3372	3076
20	GR1				10	10	37
21	M1	13.50	—	5	3075	3372	3075
22	M1	50.25	—	25	2372	2366	2370
23	M2	10.75 5.00	— —	25 } 5 }	1290	2173	1271
24	RE				10	10	37
25	R1	17.50	—	25	1067	1122	1073
26	R1	59.25	—	25	2731	2761	2724
		Total (msec):			43830	46405	43948
		Difference from measured:			0%	5.6%	

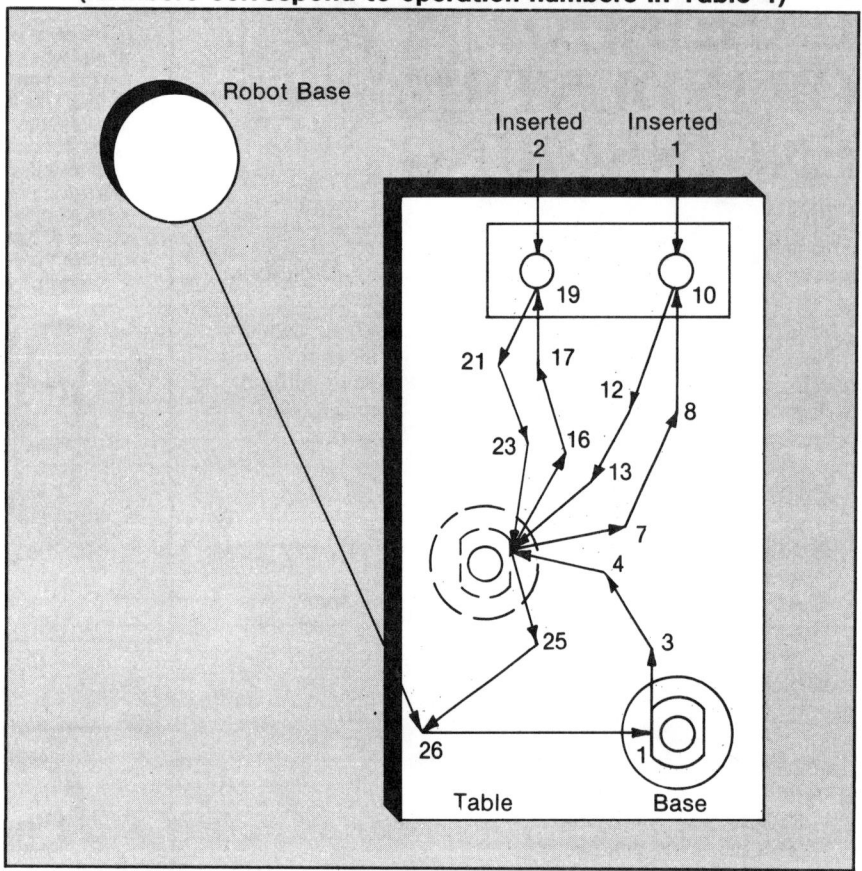

Figure 3: Layout of Insertion Task
(Numbers correspond to operation numbers in Table 4)

For further reading:

Lechtman, H., "Robot Performance Models Based on the RTM Method," unpublished MS Thesis, School of Industrial Engineering, Purdue University, West Lafayette, IN, May 1981.

Lechtman, H., and Nof, S.Y., "A User's Guide to the RTM Analyzer" and "Robot Performance Models Based on the RTM Method," Technical Reports, Research Program on Advanced Industrial Robot Control, School of Industrial Engineering, Purdue University, West Lafayette, IN, 1981.

Nof, S.Y.; Knight, J.L.; and Salvendy, G., "Effective Utilization of Industrial Robots—A Job and Skills Analysis Approach," *AIIE Transactions*, Vol. 12, 1980, pp. 216-225.

Nof, S.Y., and Paul, R.L., "A Method for Advanced Planning of Assembly by Robots," *Proceedings of SME Autofact-West*, Anaheim, CA, November 1980, pp. 425-435.

Paul, R.L, and Nof, S.Y., "Human and Robot Task Performance," *Computer Vision and Sensor-Based Robots*, G.G. Dodd and R. Lothar (Eds.), Plenum Press, New York, 1979.

Paul, R.L., and Nof, S.Y., "Work Methods Measurement—A Comparison Between Robot and Human Task Performance," *International Journal of Production Research*, Vol. 17, No. 3, 1979, pp. 277-303.

This article is based in part on research supported by the National Science Foundation under Grant APR78-27404, "Advanced Industrial Robot Control Systems." **IE**

Shimon Y. Nof is an associate professor of industrial engineering at Purdue University. His teaching and research areas include production system design and control, industrial information systems and robotics. Recently, he returned from a research leave at Tel Aviv University's Department of Industrial Engineering. Nof is a senior member of IIE.

Hannan Lechtman is a robotic research associate at International Harvester's Science and Technology Laboratories. He is involved in the evaluation and development of languages, sensor systems and control strategies for assembly robot application in the automation technology department. He also serves as a consultant for robot applications in the manufacturing operation. Lechtman received a BSc in industrial engineering and management from the Technion, Israel, Institute of Technology and an MSIE from Purdue University. He is a member of IIE and the RIA of SME.

VI. Standard Data and Mathematical Applications

Standard data has been defined in MIL-STD-1567 as "a compilation of all the elements that are used for performing a given class of work with normal elemental time values for each element. The data are used as a basis for determining time standards on work similar to that from which the data were determined without making actual time studies." Daniel O. Clark, in "Standard Data and Its Maintenance Today," defines standard data as "a synthesis of operations of elements of work, analyzed and arranged for ready application to a range of equipment, parts, or operations." Another definition by Clark adds clarity: "Standard data is the average time for a single element or a group of elements that represents a specific unit and method of work with a specified range of accuracy, constructed so that it can be classified and coded for application and retrieval as required." As mentioned in Chapter I, standard data is not a fundamental work measurement technique, because it relies on one or more of those techniques for its establishment. Most commonly, standard data is developed from time study. Standard data also includes the development of normal times through the use of mathematical equations, tabulated elemental times, alignment charts, and other techniques that determine time measurement by applying previous experience. The above definitions would also imply that predetermined time systems are a form of standard data. While technically true, most practitioners consider predetermined systems to constitute a fundamental work measurement technique by the nature of their specialized development.

The principal advantages of standard data systems are that they afford a fast and consistent methodology to estimate normal time values, and that they can be tailored to meet the needs of a particular organization. A major disadvantage is that standard data systems have a high development cost. Over time, however, a properly developed system will be cost-effective for companies that perform a substantial number of repetitive operations or suboperations.

DEVELOPING THE STANDARD DATA SYSTEM

Obtaining reliable standard data requires a consistent development methodology. The following steps constitute accepted practice: 1) study existing conditions; 2) standardize methods and work place; 3) determine and code all work elements; 4) determine element times; 5) test for validity; and 6) audit and maintain the system.

A preliminary study of existing conditions should precede any development of standard data. One purpose of the study is to determine whether the methods, facilities, and total work environment are standardized, or if corrective action is required before beginning to collect data. Standard times will apply only for work place conditions under which they were developed. Standard data should be established for types of operations independent of the product. If possible, the same methods should be used for various products; at the least, a minimum number of different methods should be employed.

Work elements should be determined not only in terms of the operation for which the time is to be set, but in relation to the family or group of operations. For example, in standard data development using time study, there are many situations where an element in an individual operation could be ended at more than one point. Any of the end points would probably be satisfactory for an individual time study, but if the study is to be used for standard data purposes, the element end point should be exactly the same for all the operations in which the element appears. By recognizing similar elements in all the operations of the group before any time studies are made, the finalization of the standard data is easier and the results more accurate. Before defining an element as similar to others, the analyst should make certain that all the elements utilize the same method and similar equipment.

As standard data is developed, it should be indexed and filed. In this activity, it is necessary to separate constant elements from variable elements. A constant element is one for which the allowed normal time will remain approximately the same for any task within a specific range. A variable element is one for which the allowed normal time will vary for a given task. Constant elements are generally stored under machine or process, for ease of reference. Variable data can be stored or may be expressed in terms of a mathematical equation. Where standard data are subdivided so as to cover a given class of operation, it may be possible to combine constants with variables, thus allowing for quick reference that will express the total allowed time to perform a given operation. Setup elements also can frequently be combined. Typically, however, standard data are not combined but are left in their basic elemental form, thus allowing greater flexibility in developing future time standards.

Finally, most standard data systems require ongoing maintenance to ensure that standards are based on defined methods and work place configurations.

TYPICAL APPLICATIONS

Standard data application begins with the analyst subdividing the entire motion sequence into elements (manual and machine) that are provided for by the system. Then the engineer or analyst uses the standard data system to obtain normal times for each element. Manual and machine times can be summed to provide the total average cycle time of the task being studied. The analyst usually simplifies the application process by using work sheets or computer routines. These provide documentation that can save time in finding errors and in modifying the process, since only the changed elements will need to be corrected. Computerization also enables persons with limited experience and training to apply the data successfully.

SUMMARY OF ARTICLES

"Standard Data and Its Application Today" by Daniel O. Clark presents an overview of the subject. The article stresses the importance of conducting a preliminary study to determine the commonality and coding of activities, and their constants, variables, and frequencies. The study also assists in maintaining accuracy requirements, establishing the number of work data development levels, setting application formats, and creating computerized storing and retrieval of the data. Arguing that the long-range value of standard data depends on the effectiveness of the maintenance program, the author discusses computer programs that provide maintenance capabilities such as "mass change" to update files, standards, operations, and parts list.

In "Standard Data Coding Techniques," William H. Bostion states that the key to any standard data system, manual or computerized, is a uniform coding system structured to provide quick identification of standard data units. Such a system allows for correct filing and easy retrieval of analyzed elements, thus simplifying application. The author provides guidelines to be followed during the coding system development in order to ensure successful standard data. He also provides examples of five standard data coding technique options: sequential numeric filing, coded numeric, simple mnemonic, financial/functional, and ten digit mnemonic.

"Standard Data Development and Application," also by Bostion, evaluates a work measurement program's compliance with MIL-STD-1567. The author addresses five basic concerns: 1) the accuracy and confidence validation of the existing standard data base; 2) documentation for supportability and traceability; 3) identification and support of personal, fatigue, and delay allowances; 4) the development techniques currently employed, as well as available systems technology; 5) the cost associated with implementation.

"Meeting the Requirements of MIL-STD-1567 and Lockheed-Georgia Computerized Standard Data Development System" by Lawrence Aft and Thomas Merritt discusses the use of multiple regression-based standard data. The authors present an extensive case example to illustrate the use of the system and discuss the advantages and limitations of the procedures used.

"An Easier Way to Measure Variable Standards" by Michael D. Stevens presents a mathematical approach to establishing work standards in a multivariate-type environment by identifying specific variables and constants within the work cycle. Stevens gives an example that mathematically determines standards for a simple manufacturing operation. The equation describing the operation can be linear or nonlinear, depending on the operation being modeled. Mathematically determined standards can be applied to several labor-setting-standard and time-study-computation tasks. The author notes that this procedure is particularly useful when the time to perform a specific operation is a function of a large number of variables.

In "Low Cost Measurement of Indirect Labor Productivity," Blair H. Schlender and Harvey H. Smerilson discuss a methodology that combines conventional work measurement techniques with mathematical techniques to establish production standards. They present three case histories that utilize regression equations for the standard times. Typically, the equation can be used to measure performance or to make estimates. In order for the regression analysis to produce a good equation, the authors caution, the data received must be accurate and, if used as a standard, must be adjusted for productivity and pace.

"A Comparison of Alternative Time Slotting Methods for Indirect Time Standards" by E. Emory Enscore, Jr., Kenneth Knott, and Benjamin W. Niebel discusses the use of different statistical distributions and cluster analysis as a basis for calculating time slot sizes. The authors explain the most common approaches of assigning benchmark jobs to slots utilizing different statistical distributions: uniform, normal, gamma, and log-normal. This article also presents a methodology to utilize cluster analysis to establish the assignment of benchmark jobs into specific slots.

"An Analytical Approach to Designing and Testing Time Slotting Systems" by Knott, Enscore, and Jeya Chandra presents a method whereby time slotting scales can be designed and their characteristics analyzed according to a priori requirements. To provide a practical framework for time slotting, the authors review the concepts of comparative estimating and identify the types of error which may result from setting time standards by this technique. They then state the basic generalized equations which describe these errors and outline the equations' construction. The article provides these equations, as they relate to a log-normal distribution, and a computer program to make the associated calculations. Finally, an actual time-slotting scale is designed and analyzed to illustrate the methodology.

STANDARD DATA AND ITS MAINTENANCE TODAY

Daniel O. Clark
Western Regional Director
MTM Association for Standards and Research
Newport Beach, California

ABSTRACT

Standard Data is an important consideration in work measurement today. The Industrial Engineering Manager is being asked to present total requirements and costs for long range work measurement programs utilizing standard data elements as their base. With the entrance of advanced computer hardware and structural programming concepts, standard data can be developed, filed and maintained for immediate access to all related company requirements.

We should understand the importance of conducting a preliminary study. With information derived in a detailed preliminary study, the commonality and coding of activities, their constants, variables and frequencies will be unveiled. It will also assist in accuracy requirements, the number of work data development levels, application formats, and computer buildup and retrieval concepts for the data.

Computer programs can provide maintenance capabilities such as "mass change" to update files, standards, operations, parts lists, etc., that allows instant access to updated data. The long range value of standard data depends upon a directly related maintenance program.

INTRODUCTION

The proper development of Standard Data, its effective application and long term maintenance, is the ultimate goal in work measurement today. Companies that developed standard data systems 20 to 30 years ago have been without a formal maintenance program. They now are finding that the use of those systems in today's changing industrial world is a major problem. The old data does not conform to the present methods, procedures, tools and products; therefore, it is of little value. The original investment in the data system can easily require duplicating in bringing it up to date, or the selection of a new work measurement system can cost the same or more today.

A chart depicting the cost and time involvement for standard development is shown in Figure 1.

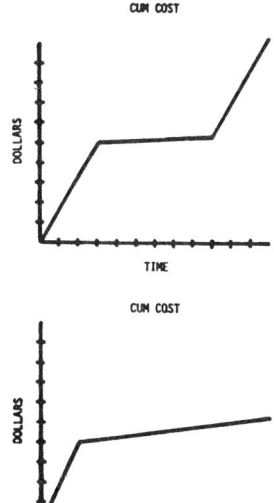

Figure 1.

Data from five companies, consisting of small to large work measurement groups showing costs of development for different time periods display the same common slopes. The bottom line results of this chart shows a 2 to 1 cost factor for those companies who do not carry out an engineered standard maintenance program.

To embark on a planned program to establish a successful standard data system in a company, the industrial engineer must consider some serious questions. Why work measurement? What are the needs of a work measurement program? Does everyone understand standard data? What accuracy is required for the established data elements? What system and approach should be used for the development? Will the system be worked manually or via a computer? How is the data to be applied? What coding system will be utilized? Are the funds available for a long range maintenance of the system?

The aforementioned questions are just some of the points to consider in planning a standard data project. Many of these questions will be addressed in this paper.

STANDARD DATA

Definitions

There are many definitions of standard data that are acceptable. The Military Standard on Work Measurement (USAF 1567) was released in June of 1975 with a definition as follows:

> Standard Time Data: A compilation of all the elements that are used for performing a given class of work with normal elemental time values for each element. The data are used as a basis for determining time standards on work similar to that from which the data were determined without making actual time studies.

One of the MTM Association's application training course manuals has a definition for a standard data system as follows:

> A synthesis of operations or elements of work, analyzed and arranged for ready application to a range of equipment, parts, or operations.

A definition that encompasses the material that is presented here would be:

> Standard Data is the average time for single element or a group of elements that represents a specific unit and method of work with a specified range of accuracy, constructed so that it can be classified and coded for application and retrieval as required.

Standard Data therefore, eludes to "commonality of work". This phrase should always be remembered during development.

Advantages and Disadvantages

The advantages outweigh by far the disadvantages of a standard data program. Some advantages of Standard data are listed as follows:

- Reduced cost of standard's development.
- Greater consistency and accuracy.
- Increased coverage of low volume work.
- Establishment of standards in advance of production.
- Ease of maintenance

Disadvantages of standard data can be listed as follows:

- Initial cost of development.
- Effects of averages and frequencies.

Computerization

Standard data elements relate directly to both the structured programming and the free-format style when working with computer input. In the 4M data computer program, the standard data element file is the key file of the system. Elements are then developed using the 4M input codes as well as bringing them from the outside as process time or user elements. The elements are stored in the element master file and used directly in the development of operations, part numbers or studies. This is shown in Figure 2.

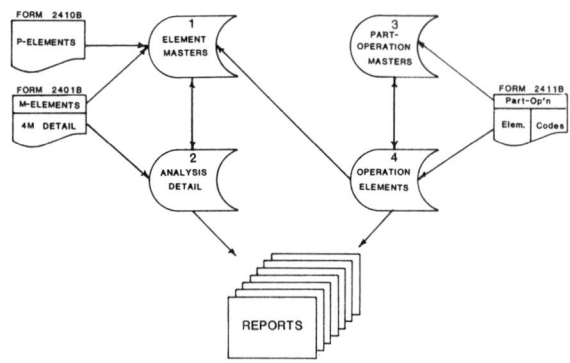

4M DATA-MOD II INPUT/OUTPUT FLOW

Figure 2

In the free-formated ADAM Program that works with a desk-top micro-computer, the element structure shown in Figure 3 presents the build-up possiblity into multi-levels of data for operations.

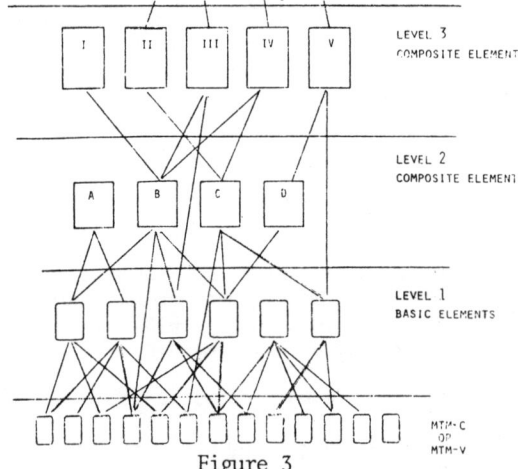

Figure 3

The maintenance updating features of computers can revise an element due to product or material design requirements. This is done via the "mass change capability" function and makes revisions to all common elements of related operations. Utilizing a computer for maintenance updating, a good standard data program receives a benefit of having updated data for a company's daily use.

DATA CONSTRUCTION STEPS

The construction of a standard data system proceeds in a logical sequence of events. They are listed below:

- Preliminary Study
- Accuracy Requirements
- Measurement Technique(s)
- Coding System
- Data Collection
- Application Techniques
- Validation
- Maintenance

The basic information and the foundation are derived in the first four steps above. The importance of these steps is paramount in that it establishes the total system for the future, and the maintenance requirements. The premise upon which standard data elements can be used to describe and prescribe times for actual elements is that the deviation between the standard data time and the standard time derivable from an individual analysis will not exceed the accuracy limits set for the system.

Preliminary Study

The Preliminary Study consists of becoming familiar with the job area, collecting production information on jobs and establishing factors having an affect upon them. From this information the coverage and the end purpose of the standard data can be defined. Here is where an investigation is made to define the systems accuracy, conditions for data application and methods description requirements. This information becomes the basis for determining the procedure which must be followed to develop the standard data system. It must not be based solely on the data pertaining to present situations but also should consider the plans and forecast of the future when available.

Previously, the subject of frequencies was listed as a disadvantage of a standard data system. The preliminary study effort will expose frequencies involved with the method of operation, number of parts, and the variable considerations of different production orders. Frequencies can be involved within an element of time as well as in the build-up development of operations. The overall understanding of the effect of frequencies provides data for the many decisions to be made concerning standard data construction.

During the preliminary study, the analyst must be keenly aware of the length of time for an element or operation of work to be completed. A main concern is the common elements of work that are seen throughout a given work area. The study should be continued in this way until information is available to apply the 80/20 concept. This occurs when 80% of the total activity is found in 20% of the elements or operations to be studied. An example of this concept is shown in Figure 4.

80/20 CONCEPT			
		% OF TOTALS	
WIRE PREP TASKS		No. of TASKS	STD HRS
1. CUT WIRES		4	16
2. PACKAGE WIRE KITS		4	15
3. PIG-TAIL PREP.		4	14
4. CRIMP CONTACT		4	18
5. TIN WIRE ENDS		4	12
6. STAMP ENDS		4	12
	SUB TOTAL	24%	87%
7 - 26 OTHERS		76%	13%
	TOTAL	100%	100%

Figure 4

The knowledge derived in the preliminary study will indicate the largest size of work element that will be most economical to handle and yet meet the demands on the system which have been set for it. If the accuracy demands are high it may be necessary to use a very detailed technique such as MTM-1. However, a high level less accurate measurement technique could be used in certain work and its corresponding components would be of larger size. This concept is true with standard data elements relative to the accuracy requirements and the work classification level.

Work can be divided into eight or more levels, each having increasingly larger work components in the level beneath it. An example of work classification is as follows:

Level	Level Name	Level Description
9	Work Type	Maintenance
8	Equipment	Vehicle Maintenance
7	Process	Service Tires
6	Operation	Change Tire
5	Sub-Operation	Remove Tire from Wheel
4	Element	Remove Lugs
3	Sub-Element	Remove one Lug
2	Motion Sequence	Loosen Lug
1	Basic Motion	Move Wrench to Lug

Classifying work into levels enables the analyst to choose the proper level for data collection consistent with the accuracy demands.

The Preliminary STudy is of the utmost importance in the development of data in a new area. The time given the analyst to investigate and understand the product, processes, methods and work place involvement will enhance a standard data project.

Accuracy

A standard data system is a collection of data elements whose contents and development times are known. The elements are put together to form a standard time. The premise upon which standard time elements can be used to prescribe times for actual elements is that the deviation between the standard data time and the standard time derivable from an individual analysis will not exceed the accuracy limits ($\pm$ %) set for the system. Figure 5 illustrates this concept.

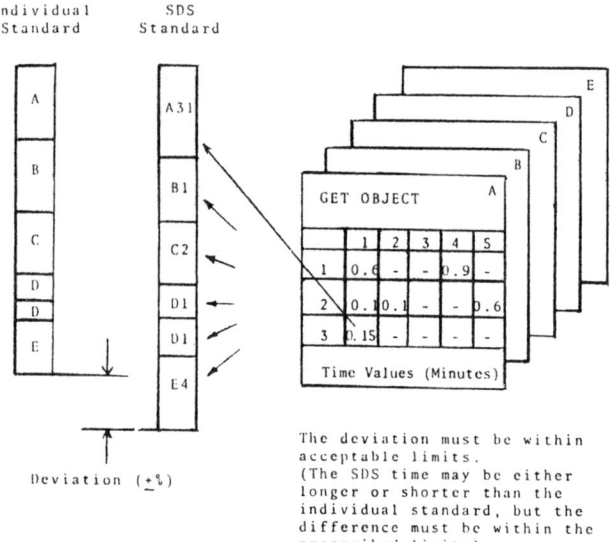

Figure 5

We have previously discussed the 80/20 concept with reference to the common element for standard data use and its approximate length in time. The concern is with the elements to be established at this base level. This will determine the system to use or the accuracy of the elements to be built into the standard data system standard as shown in the above Figure 5.

Let us first discuss accuracy in terms of Total System Accuracy.

System accuracy (also called system error) is predominantly a function of the average length of the data element of the system. But the accuracy of a work measurement system is not dependent solely upon this statistical accuracy of the system. Application accuracy is also factor. Application accuracy is the variation in time derived by different analysts in the same system to analyze a given work situation. It can be caused by a number of factors, such as incorrect application of data times, improper assignment of the particular value and missed elements.

Absolute accuracy is the comparison with an ideal system that has no system for applicator deviation (such a system, of course, does not exist). An accuracy chart is constructed to show the total system variation at given non-repetitive cycle time.

For example, the MTM-1 accuracy for 1,000 TMU (36 seconds) is plus or minus 6.4%. As the cycle of work increases in length the total variation decreases at a logarithmic rate. At 10,000 TMU (6 min.) the accuracy is slightly above the 2% level. It must be kept in mind that these accuracy figures are given for a competence level of 95 per cent. These examples can be seen on the chart displayed in Figure 6.

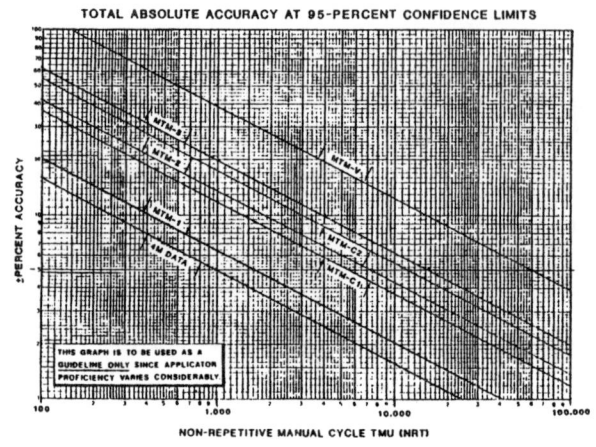

Figure 6

Today, formulae are available for the derivation of given elements of time to determine the element deviation. It is expressed as a function of the total deviation times the square root of the total time to be measured divided by the element times and divided by the number of its identical repetitions.

Measurement Technique(s)

The selection of a measurement technique relates directly to the element, the operation, the work levels established and the decision on accuracy required. The speed of application of the particular system is taken into consideration in terms of engineering time and costs. The level of methods description is always a concern when full traceability and maintenance of the system is planned. These considerations are fully evaluated to determine the proper work measurement system to be selected at the corresponding work level it will be placed relative to the total operation.

Coding

A coding system is necessary to establish a disiplined system for quick identification of standard data elements in order to provide for correct filing and easy retrieval of analyzed elements for multiple use in various areas of application.

The coding system must be structured to provide the following:

The broadest possible field of application

with the flexibility necessary for summarization at various levels.

. Ease of understanding and application while providing adequate information to identify a well defined standard data unit.

. A consistent format to insure filing and retrieval requirements.

. Ease of access to standard data maintenance programs with utilization of "where-used" and "mass-change" capabilities.

. A common "language" of communications among engineers and analysts for data construction and application in different functions and/or plans of a company.

The primary objective in coding is to follow a logical and consistent coding system in order that retrieval from the data base will be made easier. Each standard data element should be coded with respect to its specific content definition.

The coding of standard data is usually achieved via an alpha-numeric approach. Three to ten places are commonly used which gives a larger number of codes when utilizing the A to Z and 1 to 0 combinations available. Examples of coding systems are shown below.

Example 1.

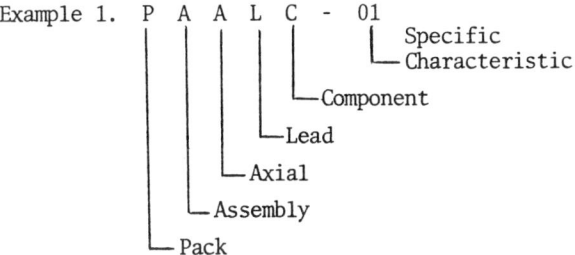

Example 2. P MA R ACSH 01

Ten is considered a maximum code with reference to the number of places for alpha-numeric coding. The code represents:

P - PC Board Assembly
MA - Mechanical Assembly
R - Run Time
A - Assemble
CS - Captive Screw
H - Hardware
01 - Element 1

Although this is a long coding system it does allow maximum classifications of work activity and it is found that these are mnemonic in nature for application.

Utilizing a computer, many benefits are derived. The retrieval system is instantaneous and the use of the files are available for all qualified users.

Data Collection

Once the appropriate measurement technique has been chosen, element data collection can begin. The information derived from the preliminary study becomes exceedingly meaningful and assists in the design fo the data collection event.

As data are collected it will be necessary to define the variables of the element and those factors which may change in value. Not only must the variables be defined, the upper and lower limits must also be established. The differences between elements and their time values can be analyzed to decide whether one time value can be retained to establish all work situations or if the element variables must be further subdivided to provide additional elements.

When time values have been established for each individual data element, it may be possible to combine elements that have the same variables into larger measurement blocks. Again, the importance of frequencies and their consideration toward long term maintenance is very important in the development of standard data elements.

Application Techniques

The standard data system with time information is generally put into the form of a standard application or calculation sheet which records the steps required for calculating the standard. However, other forms such as data tables, curves, charts and formulas can also be used. Regardless of the form, simple and precise direction for applying the standard data system must be developed and made available to all users. It is also important that all back-up material is properly filed for ready reference and retrieval in the total standard data system. An example of a standard calculation sheet is shown in Figure 7 deriving a resultant staffing requirement.

STANDARD CALCULATION SHEET	OFFICE SERVICE DEPARTMENT		
TASK	SDS (Hours/Unit)	VOLUME per day	STANDARD Hours/day
1. Incoming Mail Handling			
1.1 Open, sort and stamp	0.35 hr/100	700	2.45
1.2 Sort and stamp personal mail	0.12 hr/100	200	.24
1.3 Logging	1.80 hr/100	50	.90
2. Inter-office Mail Handling			
2.1 Sort Inter-office Mail	0.80 hr/100	400	3.20
2.2 Deliver and Collect	0.40 hr/round	6	2.40
3. Duplicating Service			
3.1 Duplicating	0.13 hr/100	1200	1.56
3.2 Sorting and Stapling	0.08 hr/100	700	0.56
4. Outside Messenger Service	0.25 hr/trip	20	5.00
5. Outgoing Mail Handling			
5.1 Mail parcels	0.14 hr/parcel	5	.70
5.2 Mail letters	0.15 hr/100	300	.45
6. Supervisory and Other Daily Tasks Independent of Volume	11.20 hr/day	1	11.20
Add standard hours for all tasks	TOTAL NORMAL HOURS:		28.66
Add hours needed for allowances	ALLOWANCE (15%):		4.30
Hours required at 100% performance	TOTAL STANDARD HOURS:		32.96
Divide by level of performance	PERFORMANCE LEVEL:		88%
Actual hours to be staffed	TOTAL ACTUAL HOURS:		37.45
Divide by length of work day	DAILY HOURS PER PERSON:		7.5
Staff required for given volumes	STAFFING REQUIREMENT:		5

Figure 7

An application instruction sheet for a manufacturing process is displayed in Figure 8.

Figure 8

The usual practice is to make up a standard data manual with all of the necessary information for engineered standards application. This manual is a working tool for the applicator.

Today's computers can easily be used to develop and apply standards. They are very effective for standard data element application. The elements are stored in a file and retrieved as needed for the further development of operations and working standards. The computer makes all calculations, stores formulas for quick application use, applys allowances, and gives final times per part, hour, unit, etc.

Total computer application of standards occurs when all the required data is stored in the computer in such a way that maintenance of the data can be kept current. The computer will display all of the data elements per request via the coding system established. The elements can be changed individually or by the mass change capability of computer programs. The end result of good computerization programming is to provide updated information on a daily basis to all people in the company as needed.

Validation

The standard data system must be tested and validated in actual practice and any biases which have been revealed need to be corrected. The validation also serves a final test of the accuracy of the data applied to actual time or individual developed standards. Validation is essential to disclose the effects of methods changes and incorrect construction during the standard data development period.

Maintenance

The standard data system should be audited periodically to see if the data is still reliable. Methods can change as well as parts and, of course, the frequencies. This should be scheduled on a fixed or random basis and selected from jobs based on the previously discussed 80/20 principle. The data being applied can be compared with current methods to reveal the immediate changes. Data elements should be retrieved and changed accordingly and relocated in the total filing system.

If the standard data has been constructed with full traceability, the elements can be updated to conform with any change reported or detected within the total standards system's parameters.

Good traceability and good filing procedures go hand and hand in a manual system. The importance of a complete coding system has a direct relationship to the filing of data and its subsequent retrieval today. Computerization approaches improve the speed of all maintenance features.

System maintenance is critical to a developed standard data program. With no active maintenance, deterioration sets in, the original investment suffers, and the question is heard "Why work measurement?".

CONCLUSION

We have presented some of the current thinking with respect to construction and maintenance of standard data. An outstanding program includes an effective preliminary study producing the smallest number of required data elements that are responsive to accuracy demands. Coding concepts that allow ease of filing and retrieval for standards application and maintenance are also an essential ingredient.

We have also recommended the utilization of state-of-the-art computers as an aid in the total process. This overall concept of standard data provides management an effective tool and reduces long range work measurement costs.

REFERENCES

1. Eady, Karl, Standard Data Systems and Their Construction, MTM Association for Standards and Research, 1977.

2. Clark, Daniel O. "Standard Data Development with MTM," Proceedings of the 1980 MTM Fall Conference

3. Eady, Karl, "Today's International MTM Systems-Decision Criteria for Their Use," Proceedings of the AIIE 1977 Spring Annual Conference.

4. (USAF) MIL-STD-1567, 30 June 1975.

BIOGRAPHY

Daniel O. Clark is Western Regional Director of the MTM Association for Standards and Research. He is an International MTM Instructor and past member of The Board of Trustees of the International MTM Directorate. Mr. Clark has served AIIE as V.P. of Region XII, Director of the Aerospace Division, Region XII Chairman of WM&ME Division and President of El Camino Real Chapter. He is the recipient of the "Phil Carrol Achievement Award" and "Outstanding Service Award" both given by AIIE Work Measurement and Methods Engineering Division. Mr. Clark is a contributing author of the soon to be published, Handbook of Industrial Engineering. He is a graduate of Drexel University and a registered Professional Engineer in the state of California.

STANDARD DATA CODING TECHNIQUES

William H. Bostion
Senior Industrial Engineer
Westinghouse Electric Corporation
Manufacturing Systems & Technology Center
Columbia, Md.

ABSTRACT:

The key to any standard data system, manual or computerized, is a uniform coding system structured to provide quick identification of standard data units in order to provide for correct filing and easy retrieval of analyzed elements for multiple utilization in various areas of application.

Before the actual development of a standard data system is begun, one must define the conditions under which the system is to be used and the purposes for which the standard data is applicable.

In order for any standard data system to be a practical and economic means of defining work content the elemental data should be structured and coded to provide for maximum utilization. The coding technique selected must not be based on conditions pertaining to the present situations, but must also consider the plans and forecasts of the future.

The theory of any standard data coding system, manual or computerized, is to develop a uniform structure to provide for correct filing and easy retrieval of analyzed elements during the standard data generation phase with the overall objective for simplistic utilization at the application level.

The coding system structure will vary depending on the degree of diverse operations performed within a given facility, organization, or company. However it should be structured to provide the following:

o   The broadest possible field of application required with the flexibility necessary for summarization at various levels.

o   Ease of understanding and application while providing adequate information to identify a well defined standard data unit.

o   A consistent format to insure filing and retrieval requirements for, economical utilization of common data elements.

o   Utilization with other work measurement techniques.

o   Simplification of standard data maintenance with utilization of "where used" and "mass change" capabilities.

o   A common language of communications, among industrial engineers, for data construction and application is different functions and/or plants of a company.

Several rules should be followed when developing a coding system to insure the short and long term requirements. The following rules should be included in your initial evaluation.

o   The primary objective in coding is to follow a logical and consistent coding system in order that retrieval from the data base will be made easier. This, in turn, reduces the time needed for work measurement analysis and increases consistency of final time standards.

o   Always try to use one of the established codes because generation of more codes than are absolutely necessary tends to make retrieval more difficult. Establishment of additional codes should be controlled by a designated individual, i.e., a data coordinator. He/she should determine whether to use an existing code or to add another category to the code list.

o   Each unit of standard data should be coded taking into account <u>only</u> the specific content of that element.

Additional rules may be necessary to perform the intended accomplishments and/or requirements of your specific operations.

The coding technique options available are almost unlimited and normally developed to respond to individual company needs or in the case of computerized programs restricted or limited to the field size allocated in the software programming. Both techniques have caused considerable problems when attempting to convert a work measurement system from one technique to another. Alternatives available range from basically a "no-code" technique to a coding system structured to provide the required details for specific organizations (financial or responsibility), activities and/or functions in numeric, alpha, alpha/numeric formating.

A recent study of ten different coding systems (see Exhibit #1) revealed that:

- The code field size ranged from 5 positions to 23 positions and

- that 5 systems were numerical with the other 5 being mnemonic.

    A major reason for this mix was the dates that the individual systems were developed, 4 of the 5 mnemonic systems were developed for use with computers within the last several years.

Evaluation of these demonstrated the need for flexibility dependent on the individual requirements and scope of application in addition to the predominant influence of the work measurement technique being employed. However each system was structured to provide specific identification of the standard data element/operation content.

The following examples will demonstrate five standard data coding technique options:

- Sequential Numeric Filing - Is where each data unit is numbered in sequence as it is added to a file.

- Coded Numeric Code - A specific meaning is assigned to each digit in the code.

- Simple Mnenomic Code - Letters of the code serve as an aid to memory. A good example of this is certain state license plates where the first two letters of the county name are the first two characters of the license number.

- Financial/Functional Code - This code is typical of those used by organizations which tie their work measurement data into their cost accounting and reporting systems.

- Ten Digit Mnenomic Code - The code is intended to provide certain smarts associated with the data element (s) for quick identification and broadfield of application.

We will now take a more detailed look at the five coding system.

SEQUENTIAL NUMERIC FILING is essentially a no code system. Technically, it is sequential numeric filing. Under this system each data unit is assigned a locater number and filed in numerical order. The locater numbers are issued as data units. A file will normally contain a locater number, a description of the data unit, supporting data, and the unit time value.

Example 1
Sequential Numeric Filing

Locator Number	Description	Value
00001	Assemble Component Board	0.0075
00002	Solder Resistor Leads to Board	0.0095
01074	Prepare Leads for Solder	0.0018

The advantages of this system are:

1. Broad field of application
2. Easy to understand and apply
3. Consistent format

The disadvantages are:

1. Difficult to summarize
2. Limited compatibility with other systems
3. Easy to duplicate descriptions
4. Difficult to make mass changes

The second example is a coded numeric code. A coded numeric code is a simple refinement of straight sequential numbering systems. Blocks of numbers are set aside for specific units, activities or departments. Example 2 shows how the code is used for level one MTM-C. The code consists of six digits divided into three fields. The first of these fields is composed of three digits which denote the purpose of the data unit. The first digit of this field designates the work category involved. Examples of these categories in MTM-C are get/place, open/close, fasten/unfasten, organize/file and read/write. The second digit in this field denotes the action involved such as: GET, OPEN, LOG, LOCATE, ERASE, and PREPARE. The third digit in the field shows any distances involved or special conditions. Examples are: 1 - SMALL, SHORT, LOOSE; 2 - MEDIUM, APPROXIMATE; 3 - LARGE, TIGHT; 4 - ADDITIONAL. The second field is composed of two digits and names the object involved in the action. The first digit in this field names the exact object involved and might be objects such as: SHEET, STACK, PAD, STAPLE, WORD, LETTER. The second digit in the field qualifies the object by size or any other property which might be appropriate such as: SMALL, MEDIUM, LARGE, ADJUSTABLE, and NON-ADJUSTABLE. The third and final field is composed of a single digit which designates any aid or control which may be involved. These may be: the other hand, letter opener, longhand, print, manual, electric and stool.

Example 2
Coded Numeric Code

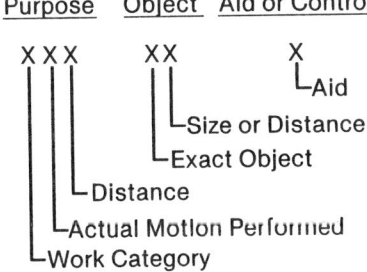

The advantages of this code are:

1. Relatively short code
2. Broad field of application
3. Consistent format
4. Can use common language
5. Easy to computerize

The disadvantages are:

1. Applicator must have an index to the code.
2. Limited size of data file
3. Only provides for element description.

SIMPLE MNENOMIC CODE

The third coding system is a simple mnemonic code. Mnemonic is an adjective meaning to aid the memory. For this particular code, the characters used in the code are the first letter of a word or a combination of letters of words describing the action or object. Our third example consists of seven characters divided into two fields. The first field is composed of five characters and is a description of the activity taking place. This description is composed of two parts with three characters describing the object of the action.

Example 3
Simple Mnemonic Code

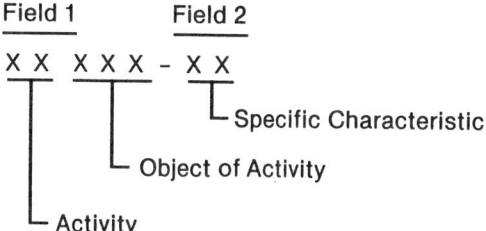

PA - Pack Assembly
LR - Lawn Raking
CF - Card Filing

Examples of objects are:

ALC - Axial Lead Component
TML - Terminal
GDN - Garden Rake
SIG - Signature

Combined these activities and object codes would make first fields such as:

PAALC - Pack Assembly Axial Lead Component
PATML - Pack Assembly Terminal
LRGDN - Lawn Raking Garden Rake
CFSIG - Card Filing Signature Card

The second field consists of two digits following a hyphen which designate special characteristics such as size, class, polarity, or any other identifier.

The advantages of this code are:

1. Ease of understanding within its field of use
2. Economical use of common data
3. Compatible with other systems
4. Easy to maintain.

Disadvantages are:

1. Limited field of application
2. Limited flexibility
3. Possible to duplicate codes
4. Not common language
5. Applicator must memorize code elements

Example 4 is a Financial/Functional code. This code is typical of those used by organizations which tie their work measurement data into there cost accounting and reporting systems. This tie in requires that the financial data be retrievable with the work measurement data. This type of code is generally made up of a prefix which provides the financial data and a suffix which provides the work measurement data description. This particular code is composed of nine characters. It has both alpha characters and numeric digits. The prefix which provides the financial data is composed of five characters as follows:

    One alpha character - Division or product line
    One numeric digit - Activity location
    One numeric digit - Cost Center
    One numeric digit - Budget Center
    One alpha character - Process

The suffix which provides the functional data description data description is composed of four characters as follows:

    One alpha character - Machine family
    One numeric digit - Machine size/capacity
    Two numeric digits - Machine number

Example 4
Financial/Functional Code

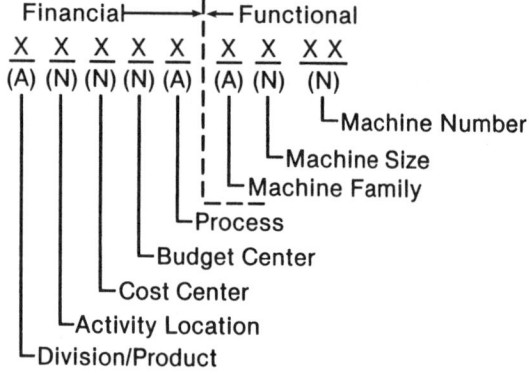

The advantages of this code are:

1. Broad field of application
2. Flexibility
3. Simple to maintain
4. Uses some common language

The disadvantages of this code are:

1. Limited use of common data
2. Not compatible with other systems
3. Need cross indices to various parts of the code

Example 5 is a 10 digit mnemonic code which is divided into three basic fields. The prefix field establishes a data set with major and minor categories. The study field will identify the function and activity of the data element. The element field is a variable.

    Note: The alternatives listed in the example are not all inclusive of the options available for utilization with this technique.

Standard Data Coding

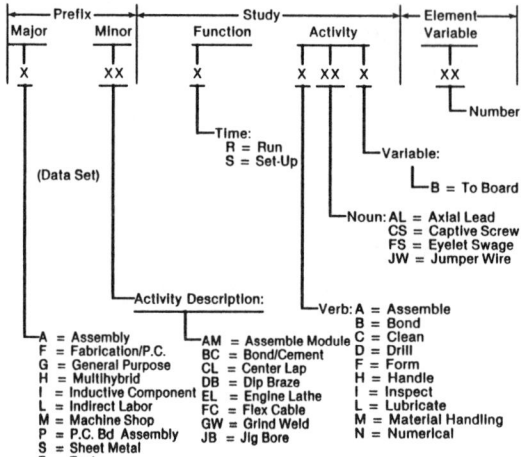

Standard Data Coding - Example

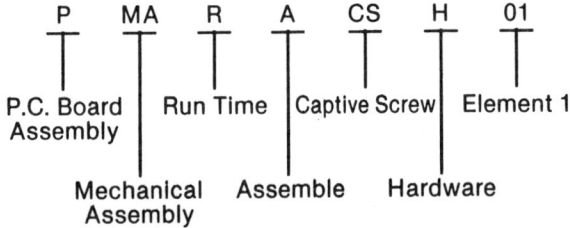

The advantages of this code are:

1. Broad field of application
2. Flexibility
3. Consistent format
4. Compatibility with other codes
5. Ease of maintenance
6. Ease of mass change

The disadvantages of this code are:

1. Length of code
2. Need of indices to field codes
3. Does not always use common language

While these five examples differ considerably in make up and complexity, there are certain element that all have in common. These are:

1. An activity description
2. An object designator
3. Some qualifier of variable

Standard data codes may be any of the forms discussed or a combination of several forms. Slight variations of these should cover the majority of coding needs.

SUMMARY

To say that the standard data coding technique is perhaps one of the most important aspects of a work measurement system would be an under rstatement of its potential. The success of any standard data system, whether manual or computerized, is very dependent on the coding structure. While no one technique will meet everyones needs, the options can be tailored to fit your specific requirements.

REFERENCES

1. MTM Association, Standard Data Systems and Their Constuction.

2. MTM Association, Standard Data Development Technology Advancement Seminar Materials.

BIOGRAPHICAL SKETCH

William H. Bostion, Senior Industrial Engineer, Westinghouse Electric Corporation, Manufacturing Systems and Technology Center, Columbia, Maryland. Mr. Bostion is currently involved in the cost/benefits analysis of technology modernization programs. Prior to this he was responsible for the work measurement activities, new planning teams, and tactical industrial engineering. He is a Senior Member of I.I.E., Past President of the Baltimore chapter, and Past Director of the I.I.E. Aerospace Division. Mr. Bostion is a General Membership Representative to the Methods/Time Measurement (MTM) Association and currently serves as Chairman of the Subcommittee on Coding Techniques and is part of a research and development project on standard data development Technology.

Exhibit #1

MTM Research and Development Project #3 - Standard Data Development Technology

Source of Coding System	Number of Positions	System Type	Fields					Compatible With 4 M - Mod II
			#1	#2	#3	#4	#5	
MSTC-Maynard Standard Time System	10	Numeric	Activity (3N)	Object Group (2N)	Object (2N)	Occurrence Freq.Grp. (1N)	Running Number (2N)	No
Northern Telecom	5	Mnemonic	Study Activity (5A)	Element Number (2N)	N/A	N/A	N/A	Yes
MTM Association Training Course	6	Numeric	Purpose (3N)	Object (2N)	Aid/Control (1NO)	N/A	N/A	Yes
Department of Defense	23	Mnemonic	Data Source (3A)	Occupation Code (3N)	Quality & Function (3A)	Element Source (7 A/N)	Data Element (7 A/N)	No
Northrop	5	Mnemonic	Work Center /Activity (2A)	Specific Category (3N)	N/A	N/A	N/A	Yes
H.B. Maynard Example #1	8	Numeric	General Area (2N)	Detailed Area (2N)	Specific Item (2N)	Client Company (2N)	N/A	Yes
H.B. Maynard Example #2	6	Numeric	General Area (2N)	Detailed Area (2N)	Specific Item (2N)	N/A	N/A	Yes
Litton Lamp/ Yes AMTD	8	Numeric	General Area (2N)	Detailed Category (2N)	Specific Item (2N)	Table Symbol (2N)	N/A	N/A
ESD-Element Standard Data	7	Mnemonic	Element Category (3A)	General Category (2A)	Operation (2 A/N)	N/A	N/A	Yes
Westinghouse D&ESC Balto.	10	Mnemonic	Prefix Major/ Minor (3A)	Study Function/ Activity (5A)	Variable Element Number (2N)	N/A	N/A	Yes

STANDARD DATA DEVELOPMENT AND APPLICATION

William H. Bostion
Supervisor, Industrial Engineering
WESTINGHOUSE DEFENSE AND ELECTRONIC SYSTEMS CENTER

Westinghouse Electric Company, Defense and Electronic Systems Center, Baltimore, Md. began evaluation of their work measurement program's compliance with Mil-Std-1567 (an Air Force specification on work measurement) in 1978. The basic concerns included: (1) the accuracy and confidence validation of the existing standard data base; (2) documentation for supportability and traceability; (3) identification and support of personal, fatigue, and delay allowances; (4) the development techniques currently employed, as well as available systems technology; and (5) the cost associated with implementation.

There was total management agreement that the D&ESC needed a work measurement system that not only complied with Mil-Std-1567, but exceeded certain requirements to provide quantitative information for (1) developing budgets, plans, and schedules; (2) detailing manpower and facility requirements; (3) measurement of costs and productivity; and (4) providing a consistent baseline for cost estimating.

The existing work measurement group consisted of two individuals with basic responsibilities for I.E. System Programs, i.e., cost/expense reduction program, productivity reporting, suggestion system coordination, etc, in addition to the administration of the work measurement program. The actual development of standard data was the responsibility of the floor Industrial Engineer. This data was forwarded to the work measurement group for inclusion in the standard data system, after review for content, accuracy, and format.

The existing standard data consisted of 128 sets of data covering PC board fabrication, PC board assembly, assembly and wiring, machine shop, sheet metal, inductive components, mechanical assembly, and multichip hybrid package operations. Only eight of the data sets were structured in a worksheet format, while the rest contained selection elements and tables, which necessitated that a data applicator be well versed in the process plan and the standard data structure.

Review of the current system revealed the need for a formal procedure to audit the work measurement program on a regular schedule.

In addition, we found that the current technique for accumulating direct labor costs was structured to identify only three major categories.

Based on the work measurement systems review, a proposed work measurement plan and schedule was developed with recommendations for implementation and incorporation.

The work measurement recommendations included:

- o Establishment of a centralized work measurement group
- o Review of currently available programs and technology
- o Development of a formal audit procedure
- o Conducting I.E. training programs (monthly)
- o Compiling direct labor distribution consistent with standard data structure
- o Increased interface with factory systems (computerized programs) and cost estimating.

The work measurement plan/schedule was structured to reflect concurrent activities, key points for standard data, work measurement interface and training program. The plan/schedule included:

a. <u>Standard Data</u>

- o Audit procedure - development, validation, and training
- o Coverage reports - semiannual analysis
- o New programs and technology
- o Worksheet development - of existing data

- o Development technique review - 4M, UNIVEL, and MOST
- o Computerized application - of standard data
- o Development staffing - centralized group
- o Development technique selection
- o Development technique training - work measurement
- o Test strategy analysis - required by Mil-Std-1567
- o Upgrade existing data - detailed analysis
- o Computerized existing data - for application enhancement

b. <u>Work Measurement Interface</u>
- o Methods improvement - floor Industrial Engineers
- o Estimating and budgeting
- o Manpower planning and scheduling
- o Realization factor(s) - development of P,F&D allowances
- o Reporting controls and methods
- o Performance procedures - revised display
- o Data supportability - variance analysis
- o System audits and reports
- o Work measurement committee meetings
- o Factory systems - computerized programs (MIS)

c. <u>Development Technique and Review</u>
 (1) Systems analyzed - 4M, MOST and UNIVEL
 (2) Key Evaluation Points
 - o System Cost
   - Software
   - Hardware
   - Training
   - Application
   - Support
   - Validity Testing
 - o Savings
   - Development Technique
   - Development Time
   - Methods Improvement
   - Update Capabilities - Maintenance
   - Traceability
   - Audit
 - o Reports - Output
   - Productivity
   - Coverage
   - Variance Analysis
 - o Precision
   - Accuracy Limits
   - Confidence Levels
   - Effective Controls
   - Application
 - o Utilization - Interface
   - Cost Estimating
   - Routing Development
   - Budget Development
   - Scheduling
   - Productivity Measurement
 - o Industrial Engineer Acceptance
   - Aerospace Users

d. <u>Work Measurement Staffing</u>

A review of the existing standard data was conducted to determine:

- o The amount of supportability available, by standard data set.
- o The estimated time to analyze, using the appropriate system, each standard data set.
- o Required data not currently available in the existing standard data base.

A separate study was conducted to determine the distribution of direct labor hours to develop a priority ranking for analysis of standard data sets. Utilizing these studies we were able to identify:

- o The needed staffing for a centralized work measurement group consistent with the overall program requirements
- o Specific areas of responsibility
- o Required areas of expertise

e. <u>Status</u>

We currently have five individuals assigned to the work measurement group for development of standard data. One of the individuals also serves as the data coordinator, responsible for the review and filing of standards to the data base. The I.E. systems programs were consolidated into a separate group with additional responsibilities.

f. Development Technique Selection and Training
   (1) Development technique review - we consulted with:
       o Aerospace Companies
       o Commercial Companies
   (2) Including visits to:
       o Aerospace Companies
       o Commercial Companies
   (3) Attendance at Conferences and Meetings:
       o MTM Association
       o 4M Users
       o MOST Users
       o A.I.I.E. - Aerospace Division
       o A.I.I.E. - National
   (4) Status
       o The review was completed June 1979

g. Development Technique Selection

We selected the 4M system in July 1979 based on the selection criteria mentioned earlier; however, we conducted a clinical study of the MOST system and elected to lease the basic and machining modules in March 1980.

h. Development Technique Training

We chose to train 10 persons in the 4M system using A.T. Kearney Co., mainly because of their experience in similar operations. Training was completed in October 1979 and A.T. Kearney was retained for a 12 week implementation phase. Only six persons participated in the MOST system training, conducted by H.B. Maynard, in our facilities. We also retained H.B. Maynard during the implementation phase, completed in March 1980.

i. System Status
   o We currently have 4M MOD I (Batch) on IBM 3031
   o We were the Beta Test Site for 4M MOD II (on-line/interactive) which was completed on December 15, 1980 in IMS on our IBM 3031.
   o We currently have the MOST program (including the Machining Module) on our IBM 3031.

j. Concept

Conceptually, we addressed the work measurement program in several phases:
   (1) Standard data development:
       o structure existing data (manual) in worksheet format
       o Load existing data into MOST computer program
       o 4M MOD I analysis of PC Board Assembly, wire preparation and PC Board fabrication
       o Converting MOD I data to MOD II data
   (2) Standard data application:
       o Manual worksheets
       o Computer application feasibility
       o Convert tables to formula
       o Ease/consistency of application
       o Traceability/documentation
       o Maximum utilization/coverage

While accuracy and confidence levels of standard data are a prime concern in the technique(s) employed in the development phase, the major concern is in the discipline associated with the application of the developed standards. When discussing standard data, there should be a definite separation of development and application.

Coding

The key to any standard data system, manual or computerized, is a uniform coding system structured to provide quick identification of standard data units, in order to provide for correct filing and easy retrieval of analyzed elements for multiple utilization in various areas of application. The coding system structure will vary depending on the degree of diverse operations performed within a given facility, organization, or company. However, it should be structured to provide the following:

o The broadest possible field of application required with, the flexibility necessary for summarization at various levels;

o Ease of understanding and application while providing adequate information to identify a well defined standard data unit;

o A consistent format to ensure filing and retrieval requirements for economical utilization of common data elements;

o Utilization with other work measurement techniques;

o Simplification of standard data maintenance with utilization of "where used" and "mass change capabilities;"

o A common language of communications, among industrial engineers, for data construction and application in different functions and/or plants of a company.

Several rules should be followed when developing a coding system to ensure the short and long term requirements. The following rules should be included in your initial evaluation:

(1) The primary objective in coding is to follow a logical and consistent coding system in order that retrieval from the data base will be made easier. This consistency in turn reduces the time needed for work measurement analysis and increases consistency of final time standards.

(2) Always try to use one of the established codes because generation of more codes than are absolutely necessary tends to make retrieval more difficult. Establishment of additional codes should be controlled by a designated individual, i.e., a data coordinator. This coordinator should determine whether to use an existing code or to add another category to the code list.

(3) Each unit of standard data should be coded taking into account only the specific content of that element.

Additional rules may be necessary to perform the intended accomplishments and/or requirements of your specific operations.

The coding technique options available are almost unlimited and normally developed to respond to individual company needs, or in the case of computerized programs, restricted or limited to the field size allocated in the software programming. Both techniques have caused considerable problems when attempting to convert a work measurement system from one technique to another.

Alternatives available range from basically a "no-code" technique, to a coding system structured to provide the required details for specific organizations (financial or responsibility), activities and/or functions in numeric, alpha, or alpha/numeric formatting.

A recent study of 10 different coding systems revealed that:

(1) the code field size ranged from 5 positions to 23 positions, and

(2) That 5 systems were numerical with the other 5 being mnemonic. A major reason for this mix was the dates that the individual systems were developed, 4 of the 5 mnemonic systems were developed for use with computers within the last several years.

Evaluation of these systems demonstrated the need for flexibility dependent on the individual requirements and scope of application, in addition to the predominant influence of the work measurement technique being employed. However each system was structured to provide specific identification of the standard data element/operation content.

STANDARD DATA DEVELOPMENT

The standard data being developed utilizes the most cost effective technique of evaluation, based on cycle times and frequency of occurrence, with:

o Short-term methods improvement being implemented prior to the element/operation analysis

o Long-term methods improvements being documented for incorporation later (with a feedback system and the analysis being conducted on the current techniques).

We began 4M MOD I analysis of our Printed Wiring Assembly elements and/or operations in November 1979. In order to develop a consistent technique for standard data construction, we assigned the entire Work Measurement Group to this task with a central data coordinator.

Currently we have individuals assigned to specific areas of functional responsibility to retain scheduled objectives for standard data development.

The 4M MOD I analyzed data is being restructured into our 4M MOD II program with additional elements being added because of new equipment and processes.

STANDARD DATA STRUCTURE

The standard data is structured and coded to ensure the maximum utilization of analyzed elements in compiling the task or operation times to reduce analysis time.

Variable allowances, i.e., Personal, Fatigue, and Delay, are not included in the standard base times; however, they will be applied in the application process dependent on the applicable conditions.

o The Westinghouse policy, in the measured daywork system, was to include a 5% personal, fatigue, and delay allowance in the standard data time(s), in addition to any other applicable allowances which varied between operations. This caused concern to many of the operating functions because of the conditions associated with special areas.

o Utilizing DoD data, we conducted a P.F.&D. analysis of 37 operating areas and concluded that we needed 22 different allowances. We were able to reduce the need to 5 allowances by consolidating functions based on averaging and percent of utilization.

STANDARD DATA APPLICATION

The first objective was to enhance the current manual system by constructing the existing standard data into a worksheet format. This provided a traceable/documented system for routing times based on the variables, i.e., frequency, process data, and allowances, associated. A project was implemented to resolve the structure of the standard data sets, as a result of the audit conducted earlier. Of the 128 data sets, we were able to restructure an additional 110 data sets into worksheet format. The decision was made to retain the associated tables and to follow-up with applicable formula techniques when we upgraded the specific standard data.

A manual file system was implemented to provide traceability at the individual part number level at the time of standards application, with provisions for filing of any changes to the original routing, with documentation as to the cause for revision.

The second phase was to load the existing standard data into the MOST program utilizing a lease arrangement with H.B. Maynard in Pittsburg, Pa. The intent was to provide the computer capability for a disciplined approach with the where-used and mass change features.

Phase three included establishment of:
o A coding technique to identify the system utilized to develop the standard data;
o An application system data base residence to interface with our short- and long-term objectives;
o A documented application technique that would provide traceability of applied standard data with applicable frequencies, process times, and allowance variables.

We will be utilizing parallel techniques for the application process with our WICAPP (Westinghouse Computer-Aided Process Planning) program, establishing the allowed times, based on the standard data in some areas, with the balance being established with our Computerized Work Measurement Program.

Our long term objective is to have a totally integrated, on-line interactive Work Measurement, Process Planning, and Factory Systems program.

Standard Data "Applicator" Manuals will be compiled by area(s) of application to simplify utilization of the applicable standard data. The manuals will contain, as a minimum the:
o Organization of manual
o Technical documentation
o Workplace data
o Quality and safety requirement
o Associated controls
o Exhibits
o Definitions
o Application procedure
o Standard data masters
o Detailed data for set-up and run elements

SUMMARY

Determine what, if any, specifications and/or manufacturing process procedures may be applicable so that you can better evaluate the method being employed.

Establish your standard data coding and structure based on the degree of diverse operations performed within your organization.

Regardless of the technique utilized to develop standard data, be sure to review the program and objectives with the area management, industrial engineer, and union, if applicable.

Discuss the basic ground rules and explain the techniques in terms that clearly identify exactly what will occur in analyzing an element/operation, how the standard data will be applied at a later date, and identify any do's or don'ts for the area.

BIOGRAPHICAL SKETCH

William H. Bostion, Supervisor, Industrial Engineering, Westinghouse Defense and Electronic Systems Center, Baltimore, Maryland. Mr. Bostion is currently responsible for the Work Measurement and I.E. Systems activities, which includes the computerized predetermined time systems, performance reporting, and coordination of the Industrial Engineering Training Program. He is a senior member of A.I.I.E., past President of the Baltimore Chapter and currently serves as Director Elect for the Aerospace Division. Prior to joining Westinghouse, he was senior engineer for National Industries for the Handicapped, Washington, D.C., assisting sheltered workshops in costing techniques, overhead burden development, and development of manufacturing capabilities. His background includes industrial engineering, cost estimating and cost control with Bendix Communications Division, and Cutler Hammer. Some of his accomplishments include publication, by the Department of Labor, of Procedures on Cost Estimating and Overhead Burden Development in Sheltered Workshops and completion of a position paper on Development of (P.F.&D.) Personal, Fatigue, and Delay Allowances.

Meeting the Requirements of MIL-STD 1567
and
Lockheed-Georgia Computerized Standard Data Development System

Lawrence Aft, Professor, Southern Technical Institute
Thomas Merritt, Supervisor of Computer-Aided Manufacturing, Lockheed-Georgia Company

The purpose of Military Standard 1567 is, "to assist in achieving increased discipline in contractors' work measurement programs with the objective of improved productivity and efficiency in contractor industrial operations." (1) Specifically, the standard requires the setting of time standards, using recognized industrial engineering procedures, that are accurate within plus or minus 25 percent or plus or minus 10 percent with at least a 90 percent confidence level.

Lockheed-Georgia Company has elected to comply with Mil-Std 1567 through the use of a multiple regression based standard data analysis and has had significant success meeting the accuracy and statistical confidence requirements of Mil-Std 1567. Standard data systems organize standard times from a number of related jobs into a data base from which the standard times for similar jobs may be constructed or synthesized. This paper will illustrate the use of the system and discuss the advantages and limitations of the procedures used.

The purpose of Military Standard 1567 is, "to assist in achieving increased discipline in contractors' work measurement programs with the objective of improved productivity and efficiency in contractor industrial operations." (1) Specifically, the standard requires the setting of time standards, using recognized industrial engineering procedures, that are accurate within plus or minus 25 percent or plus or minus 10 percent with at least a 90 percent confidence level.

Lockheed-Georgia Company is a manufacturer of a low volume product with over 11,000 different component parts, each requiring, on the average, several machining operations, as well as numerous assembly operations. The government has required conformance to a tailored version of MIL-STD 1567 which Lockheed accomplishes through the use of the recognized industrial engineering procedure of standard data, specifically multiple regression based standard data analysis.

Standard data systems organize standard times for a number of related jobs into a data base from which the standard times for similar jobs may be constructed or synthesized. (2)

> Once a time standard has been developed for a given job, the job obviously need not be re-analyzed for its standard time on each occasion it is performed; and it is logical to keep a systemized record of the time for that job. In this way, the need to direct attention to the fine details of that job on repeated occasions can be avoided; effectively, the analyst is freed to study other jobs and thus enlarge the coverage and scope of his efforts. (3-687)

Specifically, the ANSI Standard, Z-94 for Industrial Engineering Terminology defines standard data as,

> a structured collection of normal time values for work elements codified in tabular or graphic form. The data is used as a basis for determining time standards on work similar to that from which the data was collected without making additional studies. (5)

Reasons for standard data development, in addition to compliance with Mil-Std 1567 include the productivity of industrial engineers responsible for setting time standards. Standard data can reduce the time and labor required to set standards. Standards set using standard data are also more consistent. "They are developed from multiple time studies, performed by different analysts, on many different occasions. Random errors...are minimized." (1-215)

Standard data permits the analyst to synthesize standards for jobs that are not yet in operation. Time standards can be set before production begins on a new job. (4) Much greater attention is paid to methods study when standard data is used. "The development of good standard data requires the work measurement engineer to devote much greater attention to the underlying causes and factors affecting the times for various elements than is the case when individual rates are set." (3-688)

Although there are numerous advantages to the development and use of standard data, it does represent a significant investment. Any organization which desires to use standard data should, as a first step, make sure that the developmental costs can be justified. Once this is done, the standard data system can be developed.

Standard data systems are based on the concept that the time required to perform an operation can be predicted from certain variables that describe the process. If these variables are designated x, then algebraically,

$$t = f(x_1, x_2, x_3, \ldots x_n)$$

Depending upon the nature of the functional relationship, the appropriate equation can be derived through regression analysis. It is generally suggested that the equation developed have the following characteristics. (Although these may seem to be unusual constraints, they are practical guidelines to make the acceptance of the standards so developed easier to "sell.")

The relationship developed should be linear and fit the general multiple linear regression format

$$t = b_0 + b_1 x_1 + b_2 x_2 + \ldots + b_n x_n)$$

where the x's are the variables, b_0 is the intercept, and the remaining values of b are the coefficients. If an obvious second order relationship exists, such as time being a function of length times width, or area, then area should be specified as one of the variables, returning the relationship to a linear basis.

The values of the coefficients should be positive. There is a common but incorrect interpretation among many uninformed people that a negative coefficient means that less time is required as a variable increases. It is much easier to "sell" equations in which the coefficients are positive.

When a standard data relationship is developed in this manner it often involves considerable expertise. In order for an acceptable relationship, at least a relationship that will meet the requirements of Mil-Std 1567, to be developed, it is necessary for the analyst to be able to identify variables that influence time in a linear fashion. Sometimes, these variables remain linear over selected ranges of potential values. For example, when certain parts that must be machined exceed the capacity of one machine tool, they might be processed on a larger tool or special purpose tool that has a much shorter or longer run time. Thus, it is imperative that strict limits regarding the use of formulas be specified.

Due to the difficulty of this analysis, the use of standard data for setting time standards has often been limited. Lockheed-Georgia Company has developed a computer based standard data system that has removed much of the drudgery associated with the multiple regression procedure. The features of this computer system are described in the following text.

The Lockheed-Georgia Standard Data Development System allows user entry into one of three modes during each execution. The three modes are:

1. Regression/Nomograph/Mil-Std 1567 Test/Axis Graphing
2. Mil-Std 1567 Test only
3. Nomograph only

Once the mode selection is made, a menu of functions is offered to the user. The menu functions are explained in the following text. After completion of a menu function, the program always returns the user to the menu for selection of another function or termination of the program. Most of the functions contain sub-menus to offer the user additional options within the function selected.

A. DATA ENTRY FUNCTION:

This function allows the interactive construction of regression or Mil-Std 1567 test files. The computer prompts the user for variable type data entry if Regression/Nomograph/Mil-Std Test/Axis Graphing mode is chosen. If the 'Mil-Std Test Only' mode is chosen the computer prompts for observation and predicted time values. If the 'Nomograph Only' mode is chosen the computer prompts the user to enter a linear equation.

The data entry routine also allows for data entry from a computer file previously created and stored interactively for later use.

Regression and Mil-Std files can be readily expanded using features of the data entry function. In addition, the data entry function allows for entry of variable names, user name, and other text to be displayed on output.

B. DATA EDITING FUNCTION:

This function allows the user to change any data entered in the data entry routine. This function operates in all three modes.

C. OBSERVATION SEGREGATION FUNCTION:

This function allows the user to 'turn on' or 'turn off' a range of observations. Those turned off will not be considered in a subsequent regression or Mil-Std 1567 test. This capability allows the user to make several different equations or tests from a single data entry file. Also, 'what if' situations are easily handled by selectively 'turning on' or 'turning off' certain ranges of observations. This function operates in only the first two modes and the status of the observation 'switches' is stored with the data so execution at a later time can pick up where it was left at termination.

D. MULTIPLE REGRESSION FUNCTION:

The regression function allows the user to manually select any combination of variables for a regression analysis or to have the computer exhaustively regress every possible combination of the user entered variables.

The sum of the squares method is used and a linear equation and other statistics about the regression are returned for each variable combination.

A Mil-Std 1567 test is automatically performed on all regressions.

E. NOMOGRAPH FUNCTION:

The nomograph function allows the user to create 'quick read' charts composed of one, two, or three independent variables. The user can control such parameters as variable minimum and variable maximum values, variable output type (integer or real), spread sensitivity (i.e., 10%, 20%, ... 100% increasing increments of variable values), position of each variable in chart, and method of obtaining first variable value (user supplied or automatic calculation).

The nomograph is created using the equation generated in the regression function in Mode #1 or the equation supplied by the user in Mode #3.

In Mode #1, it is possible to perform the Mil-Std 1567 test on a nomograph. When this is done, the chart is automatically applied in the same manner as a person would obtain values from the chart.

F. MIL-STD 1567 TEST:

The Mil-Std 1567 function is accessible from either Mode #1 or Mode #2. This test compares the observed values to the predicted values and determines the confidence level at up to 15 different user supplied accuracy limits. The 'Student's T Statistic' is used to express the confidence level.

G. AXIS GRAPHING FUNCTION:

This function allows the user to graphically display the results of a regression so as to determine such things as break points where actually more than one linear equation exists in a data set and how the points are clustered along the equation line.

The equation line, a + % accuracy line, and a - % accuracy line are displayed with the individual points scattered around the lines.

This function is accessible only from Mode #1.

SAMPLE DEVELOPMENT SESSION:

The following is an example of how an engineer would use the above described system to develop a standard.

Once the engineer has taken a representative sampling of time studies and summarized them into normal hours, he goes to a computer terminal and executes the development system program. Using the data entry routine the engineer enters all his time study data and checks his input for errors. If any errors are found, he goes to the editing function and corrects them. The next step will be to run a regression. The engineer will probably choose to automatically run all possible combinations of variables in regressions and from the results choose the most promising regression for further analysis. The engineer will now iterate around the regression, editing, observation segregation, and axis graphing functions in search of the most precise equation meeting Mil-Std 1567 to define his data set. Once this equation has been found he can now make a nomograph. Within the nomograph function, the engineer can try different combinations of parameters in an effort to create the most reasonable chart meeting Mil-Std 1567 to display his data set. At this point the engineer can choose to store his data and terminate.

In the event that no reasonable equations can be found, the engineer can store his data and terminate. At a later date the stored data can be read into the system and new data from further studies added to it. Then the program may be executed again in an attempt to develop a satisfactory standard data formula and nomograph.

References

1. Aft, PRODUCTIVITY, MEASUREMENT AND IMPROVEMENT, Reston, Reston, Virginia, 1983.

2. Abruzzi, "Developing Standard Data for Predictive Purposes," JOURNAL OF INDUSTRIAL ENGINEERING, November, 1952

3. Karger, Delmar W., and Bayha, Franklin H., ENGINEERED WORK MEASUREMENT, Industrial Press, New York, 1977.

4. Mundel, MOTION AND TIMESTUDY, Prentice Hall, Englewood Cliffs, NJ, 1978.

5. ANSI Standard Z-94, "Industrial Engineering Terminology," American National Standards Institute, 1972.

An easier way to measure variable standards

A fresh method for measuring multivariable labor standards is developed using a mathematical approach.

MICHAEL D. STEVENS
A.T. Kearney, Inc.
Cleveland, OH

If mathematical expressions can be used in predicting multivariable business conditions, sales forecasts, and an assortment of economic conditions, why can't they be applied as an aid in determining multivariable labor standards? This is the premise that led to an easier method for measuring labor standards in industry.

Traditional methods for establishing work time standards involve the use of stop watches, predetermined time systems, work sampling studies, historic standard rates, and a miscellaneous assortment of quasi-engineered standards. These methods usually reflect the time values based upon a given set of variables at the moment of observation or calculation (predetermined standards). They do little to compensate for outside variables and consequently have to be recalculated for each subsequent similar type of work activity. An example of this is the analysis of every product configuration or weight allied to a foundry operation.

The advent of standard data tables have eliminated some otherwise needed observations by graphing the existing data in order to forecast the impact of one or two variables on the work activity. When there are more than two independent variables on a single dependent variable, this method becomes very cumbersome and expensive to implement as well as maintain. The use of formulas have been historically limited and normally used in conjunction with or in designing of standard data tables, e.g., machine shop data.

These traditional methods of approach have been sufficient in the establishment and prediction of work standards based upon only two variables, but an estimated 60% of all manufacturing industries have products with characteristics containing three or more variables. This complication requires that a tremendous amount of industrial engineering time be spent in establishing and maintaining work measurement data. It does not include the poor plant efficiency levels experienced by those nonstandard plants with low production requirements and high product variability characteristics (job shops).

Mathematical approach

The mathematical approach to establishing work standards in a multivariate type environment consists of identifying specific variables and constants within the work cycle. A matrix is a good tool for identifying the range of product variation. It also indicates the total amount of labor standards that can be set by the model. Figure 1 illustrates a major motor manufacturer's matrix for an operation containing over 156,000 variable combinations. This identification stage is the most critical part of the system because the accuracy of the model will depend upon how good these variable and constants were documented.

After identifying the variable and constant elements, the next step is to establish time values. Constant time elements (no time variation) can be combined into a single total constant group. Variable time elements should be studied at both extremes and in the middle in order to determine if there is a linear or nonlinear

Michael D. Stevens is manager of manufacturing services, A.T. Kearney Inc., Cleveland, OH. He has managed industrial engineering departments in several manufacturing plants in addition to owning a company. Stevens has degrees in industrial engineering, industrial technology, and systems and finance from North Central Technical College, and Ohio State and Ohio Universities. He is a member of AIIE, AMA and SME.

Figure 1. Preparation of motor coils for winding in which there are 153,600 product variables.

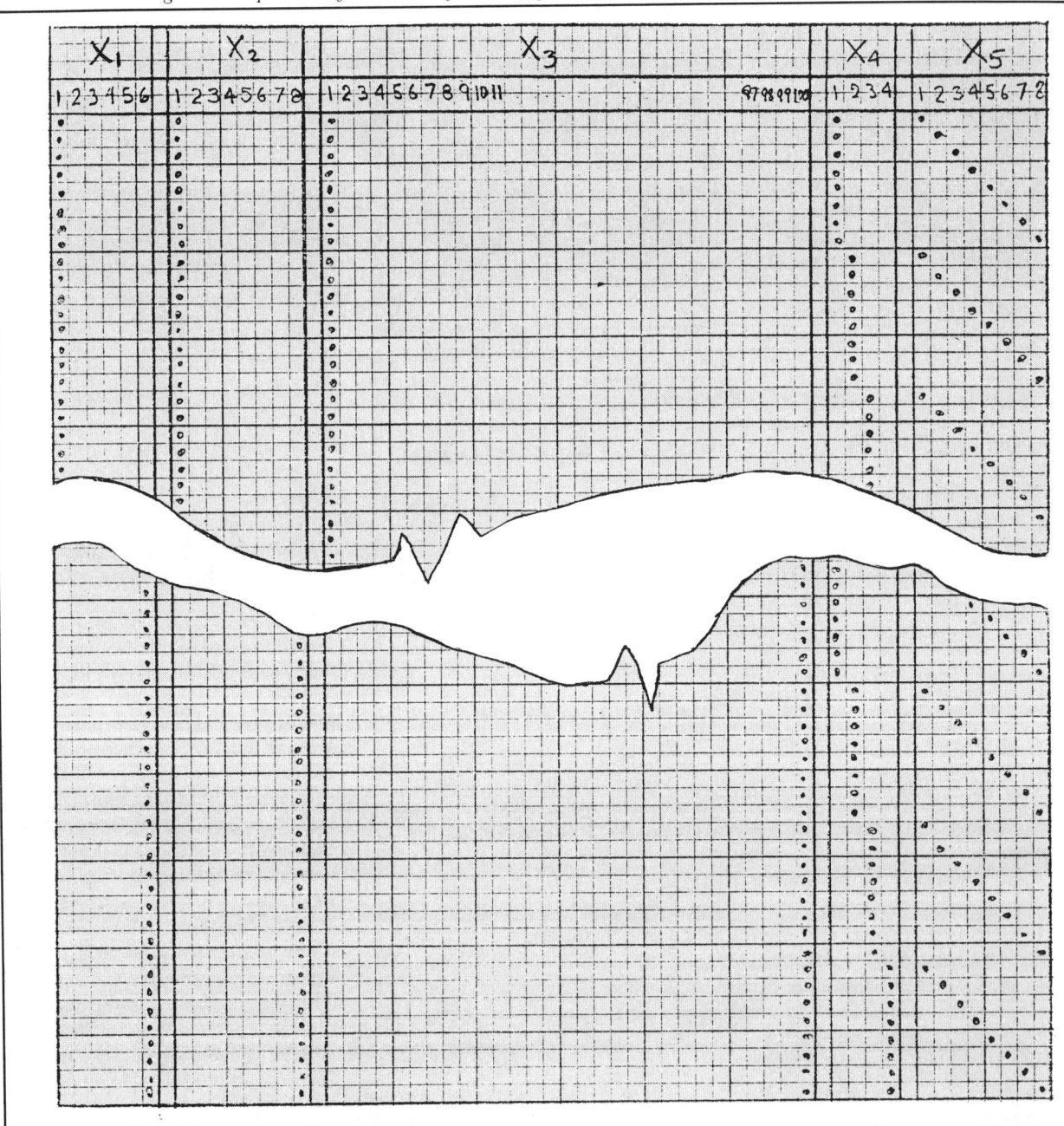

X_1 = Coils per group—(6 variations)
X_2 = Number of wires—(8 variations)
X_3 = Number of turns—(100 variations)
X_4 = Number of coils—(4 variations)
X_5 = Number of groups—(8 variations)
X_6 = Constant—(No variation)

NOTE: There are 153,600 product variables within this operation.

relationship. If it is linear, then a constant time coefficient can be established. For example, if we consider the operation *wind coils* from the sample operation, we can see that one variable affecting the time is the number of coils to be wound (X_2). This element has a constant time coefficient (a_2) for winding each coil. Thus a simple linear relationship is noted ($a_2 X_2$) for one part of the mathematical model. In the unusual case of a nonlinear relationship, a quadratic equation will accurately describe the needed time values.

The accumulation of all elements and their corresponding coefficient values can now be arranged in an algebraic model as in the sample and tested. Tests can be performed by comparing an actual time study with the model's computational results upon a given set of variables. The difference should be no more than 3%. Excessive deviances can usually be attributed to incorrect variable times, missing variables, or inconsistently rated time studies. Once the tests have proved the model as being valid, you may transfer it to a minicomputer which allows quicker computations and is readily accessible.

Advantages of MDS

Mathematically determined standards (MDS) can be applied to any situation of setting labor standards and time study computations. The advantages of the system become evident when the amount of time to perform any particular operation is a function of a large amount of variables or when the range of variation of any individual variable is relatively large. In other words, any attempt to establish standard times using traditional time study techniques will require an excessive amount of industrial engineering time due to the fact that many observations will be needed for every different combination of variables and for every particular variable value.

Other advantages of this system include: accurate production and cost estimating of new products; fast implementation, thereby immediately increasing plant efficiency levels; 100% labor standard coverage; reduction or elimination of books, records, time study sheets, etc., associated with traditional techniques; easy computerization and adaptation to existing standard application systems; establishment of a more economical technique, thus allocating valuable industrial engineering time towards other activities; and consistent pace rate determination.

The mathematically determined standards method is not a theory for establishing labor standards, but rather a useful tool that has been successfully employed by several leading manufacturing companies. In addition to typical direct labor operations, it can also be used in measuring indirect work activities such as those in maintenance departments, warehousing operations and hospital services. However, it should be noted that, although MDS is a powerful statistical tool, it should only be used with proper knowledge and in conjunction with a thorough understanding of the problem. **IE**

Mathematically determined standards applied to coil winding.

Sample operation—A manufacturer of motors has an operation consisting of manually winding copper wires into a coil, taping the groups of coils and wrapping insulation around them (9100 variable combinations).

Sample Model—The elements and coefficients in this operation are:

a_1 = Time to tape one group (linear relationship)
x_1 = Number of groups to be taped (1-13 range)
a_2 = Time to wind one coil (linear relationship)
x_2 = Number of coils to be wound (1-35 windings)
a_3 = Time to insulate one wire (linear relationship)
x_3 = Number of wires to be insulated (1-20 wires)
a_4 = Time to turn machine on/off and fill out production specifications
v_1 = Unknown standard

Arranging the above elements into a workable model we have:

$$v_1 = a_1 x_1 + a_2 x_2 + a_3 x_3 + a_4$$

Sample Data (Example Calculation for a_1)—a_1 (tape) has a range of 0.25 (tape one group) to 3.25 (tape 13 groups) standard minutes. Total time range is 3.00 minutes (3.25 − 0.25). This three-minute range divided by the total number of deviations, less one (12) is the standard time to tape one group.

$$a_1 = \frac{3 \text{ standard minute range}}{12 \text{ deviations}} = 0.25 \text{ standard minute deviation}$$

Utilizing similar procedures, the following coefficients are:

a_2 = 0.018 standard minute/deviation
a_3 = 0.092 standard minute/deviation
a_4 = 1.385 standard minutes/deviation

Sample Computation—Model XYZ has the following electrical specifications (from bill of materials):

9 groups = X_1 = 9 assigned value
29 coils = X_2 = 29 assigned value
6 wires = X_3 = 6 assigned value

Substituting values from Model XYZ into the following formula gives the standard minutes per piece:

$V_1 = a_1 X_1 + a_2 X_2 + a_3 X_3 + a_4$
$V_1 = (0.25)(9) + (0.018)(29) + (0.092)(6) + 1.385$
$V_1 = 4.71$ standard minutes/Model XYZ

Note: The coefficient relationships in this sample model have been linear in nature. Curved data relationships can be accurately calculated with a quadratic formula.

LOW COST MEASUREMENT OF INDIRECT LABOR PRODUCTIVITY

Blair H. Schlender
Harvey H. Smerilson
Martin Marietta Aerospace
Orlando, Florida

Abstract

The measurement of productivity and setting of standards on Indirect Labor has always been an area of concern for management. At the Orlando Division of Martin Marietta Aerospace, the measurement of indirect labor, including professional employees has been developed using a low cost technique that can be applied to any group of individuals who work together on the same tasks to produce a common output.

The procedure used was a combination of basic Industrial Engineering techniques used in a new, unique manner. This technique uses task analysis, work sampling and regression analysis to arrive at a standard time for a given task. The standard can then be applied to measure performance, or used for estimating.

Introduction

Two major functions of Management are planning and control. In order to be able to plan and control fully, all labor costs - including indirect labor costs - must be considered. At Martin Marietta Aerospace, Orlando Division, a low-cost method of estimating and measuring indirect labor was developed to assist Management in its planning and control functions. This paper will outline the approach used, conclusions reached and three case studies that will illustrate different possible outcomes.

Types of Measurement Considered

Prior to the start of our study, several known measurement systems were reviewed. Since indirect labor, such as engineers, does not produce a smooth flow pattern of motions resulting in a finished product, stop watch study as well as MTM was ruled extremely difficult to apply. Engineers <u>do</u> produce a pattern of motions resulting in a finished product, but it is not always a smooth flow. The key words here are "smooth flow." Any engineer will have a certain amount of thinking time which could not easily be measured by stopwatch due to its variance. This time could be averaged and thought of as process time. The problem then becomes: How much of an engineer's work is process time and how much is physical motion? It was decided that a new fundamental approach was necessary in the beginning.

Task Analysis

One basic Industrial Engineering tool called upon to help determine the proper measurement system to use was task analysis. There are many techniques available for performing a task analysis. These include review of job descriptions, observing the worker over a period of time, observations by category of work at random times, and a task analysis sheet prepared by the worker himself. A sample of a task analysis sheet, used by Facility Design Engineers in our study, is shown in Figure 1. The engineer filled in the required information such as job number, activity code and task code, then blocked in the correct amount of time. These sheets became step one in our data collection process and provided a good understanding of the job.

Work Sampling

Task analysis provided an overview of the job being performed. However, it was important to have an accurate measure of productive time, which task analysis did not provide. One good method of providing this measurement was work sampling. To provide a coordination with the task analysis being conducted, we chose to perform a work sample by category. Activity descriptions were chosen to parallel those on the task analysis sheet. To achieve a 95 percent confidence level, the alignment chart in the H. B. Maynard Industrial Engineering Handbook (Third Edition, Section 3-52, Figure 4-3) was used to determine the number of observations required. Pace was determined at the same time work sampling was being accomplished. An experienced observer was able to do both at the same time. All observations were made at random times. Besides providing productive versus nonproductive time, our work sample studies provided a good estimate of the amount of time spent on the major tasks, since they were done by category over a long span of time.

Data Analysis

After enough work samples were conducted to obtain a 95 percent confidence level, the study was analyzed along with the task analysis sheets. It was at this point we were able to determine what the measurable output was. This, of course, varied for each group being studied. Significant points

MARTIN MARIETTA

DESIGN ENGINEERING JOB CONTENT ANALYSIS

Form D-1169 Apr74

EMPLOYEE NAME ______________________ DATE ______________

ITEM	ACTIVITY CODE	0742	0800	0900	1000	1100	1200	1230	1300	1400	1500	1600	1612	JOB OR PED NUMBER
01														
02														
03														
04														
05														
06														
07														
08														
09														
10														
11														
12														
13														
14														
15														
16														
17														
18														
19														
20														

TASK CODES

ACTIVITY CODES		ENGINEERING FUNCTIONS		DRAFTING FUNCTIONS		SUPPORT FUNCTIONS	
01	NEW DESIGN	AA	INVESTIGATE	BA	SPREAD LEAD	CA	LIAISON
02	MOD OF EXISTING DESIGN	AB	SPECIFICATIONS	BB	____________	CB	VERIFY CONTRACT DOC
03	FIELD CHANGES	AC	FINAL APPROVALS			CC	FILLOUT FORMS
04	CONSULATIONS	AD	V. A.'s, ETC.			CD	RUN PRINTS
05	GENERAL	AE	____________			CE	____________

Record all time to nearest 0.1 hours (6 minutes)

Figure 1.

to watch for were actual time spent on assignments outside the job classification, under-employing and avoidable-delay times. An example of time outside the job classification would be time reported and seen in a general, plant-wide safety meeting. An engineer doing routine drafting could be considered under-employed. An engineer waiting twenty minutes for transportation when another means of transportation was available was experiencing avoidable-delay time. With a percent of available time known, and measurable output known, our next step was to collect a count of the measurable output and relate this to the total available time. Two methods of data analysis were investigated for use. These were linear programming and multiple linear regression.

Multiple Linear Regression

Multiple linear regression is a method of solving n unknowns with m equations. If we consider our different tasks as the n unknowns and our weeks of data as our m equations, we can solve for an average time to accomplish a given task. We have had a great deal of success using the "Stepwise Regression - Version of May 2, 1966, Health Sciences Computing Facility, UCLA." In general terms, the counts of each task are added together and equated to the total time used in completing the tasks. For example, if an engineer has six different tasks which he performs, and works a 40-hour week, this would be the foundation of our data equations. If, in a given week, he had only 37 hours available to work and performed (4) task 1, (7) task 2, (12) task 3, (2) task 4, (30) task 5 and (1) task 6 - we would form the following equation:

$$4\times\text{task }1 + 7\times\text{task }2 + 12\times\text{task }3 + 2\times\text{task }4 + 30\times\text{task }5 + 1\times\text{task }6 = 37$$

It would be necessary to run several weeks to obtain an average time. As long as the nature of the work remains constant, the more weeks obtained, the better the results. Usually, between 13 and 26 weeks data is enough to provide statistically sound results. The output of the regression program will provide a coefficient of each task as well as a constant. If the constant is zero, or very close to zero, the coefficient will be the time it takes to accomplish the task. Of course this is only true if the data is shown to be statistically sound with a high multiple R, a low error of estimate and good results from an F Ratio or Student T test. Figure 2 shows a sample regression output from our Stepwise Regression computer analysis.

BMDO2R - STEPWISE REGRESSION -
HEALTH SCIENCES COMPUTING FACILITY, UCLA

VARIABLE		MEAN	STANDARD DEVIATION
FOLD	1	82.19231	84.78227
250,S	2	93.46153	96.53535
SHIPP	3	14.03846	17.39252
POST	4	143.53845	157.83388
RM,S	5	159.15384	164.83813
SHTYP	6	22.26923	33.91788
UPS	7	28.61537	31.44714
1149S	8	15.96154	17.49394
PLATE	9	22.80768	39.76421
LABPO	10	2941.34595	3514.84937
STAMP	11	12.53846	15.45962
HOURS	12	126.66147	129.58243

MULTIPLE R   0.9977
STD. ERROR OF EST.   10.5063

VARIABLE		COEFFICIENT
(CONSTANT		0.0
FOLD	1	0.06175
SHIPP	3	0.56158
POST	4	0.15860
RM,S	5	0.41878
SHTYP	6	0.24289
1149S	8	0.57960
PLATE	9	0.10245
LABPO	10	0.00242

Figure 2.

Conclusions

When a statistically sound regression equation is formed, the system is operational, easy to maintain and low in cost. The equation can be used to measure performance or to make estimates. This facilitates planning and control by Management. The data being collected and measured for performance can also be used to build a new data base. Although the task analysis sheets and work sampling are no longer required, either can be resumed if there is a situation that requires investigation, such as a sharp change in performance.

Pitfalls

Up to now, we have explained the basic system and how easy it is to use. There are a couple of pitfalls to watch for. In order for the regression analysis to produce a good equation for use, the data received must be accurate. Experience on the part of the analyst is needed here to monitor the system. Another point to remember is that the data collected, if used as a standard, must be adjusted for productivity and pace to be meaningful. Otherwise, the performance is based on past history only. The regression equation, when developed, is multiplied by the percent productivity from the work sample study. This product is then, in turn, multiplied by the determined pace. To this product we add any allowance. The resulting sum is then compared to the available hours for a performance rating. A complete example of this will be shown in our case history examples. When defining tasks for the regression analysis, it is best to select numbers of the same order of magnitude to avoid a large constant. A negative appearing in the regression formula could be used for performance, but it would not be valid for estimating. A negative number can usually be combined and eliminated by use of a related task. These pitfalls should be considered warnings only and not shortcomings of the system. The advantages of this system far outweigh the disadvantages, as can be seen by our following case histories.

Case History I

An easy to understand example was the development of standards for application to Design Engineers within the Facility Engineering Department. The first step was a task analysis study by use of the form shown in Figure 1.

Following our outlined procedure, a work sample was conducted and a percent productivity and pace determined. It was determined that our finished product was a job consisting of a drawing, a writeup, material order lists, specifications, or any combination of these. To have a meaningful standard, it was necessary to measure the effort that went into the finished product with proper weighing of the output.

After reviewing the work sample study and task analysis sheets, it became clear that the design group was really three smaller groups of engineers. The three groups were the electrical, structural and mechanical sections, all reporting to the same supervisor. It was decided to gather the data by subgroup. This was done to provide information about the individual groups.

The next step was to develop an equation to describe the work. The tasks were determined to be sheets of written instructions for specifications, completed forms and completed drawings. Here, a standard for each task - known as an independent variable - had to be determined and all others equated by some ratio factor to the standard. It had to be determined how many words and lines per page would be the basis of the standard for measurement. Also, size and complexity of the drawings had to be equated to each other. This was particularly true when many different sizes of drawings were used and when standard drawings were used and revised to fit a job. Since the smallest drawing used was an 8-1/2 by 11-inch sketch, this was chosen as a base measure with all others becoming a multiple of the base. For instruction sheets and forms, one completed form and sheet were chosen as the base, which then allowed multiple and fractional use. Since size and degree of difficulty varied on forms, each was equated to the other.

For the regression of Design Engineers, X_1 was chosen to represent one standard size writeup sheet, consisting of 36 lines of handwriting. X_2 was chosen to represent one completed material list sheet, consisting of 19 items. As previously mentioned, x_3 was one completed 8-1/2 x 11-inch sketch. The developed formula for the regression for a job of 54 lines of handwritten instructions, 52 items on a material list and one 17 x 22-inch drawing would be:

$$1.5x_1 + 2.7x_2 + 4x_3 = \text{hours to complete the job}$$

This formula was derived by dividing each variable output by the chosen base count. The hours to complete the job were the total job time as reported by the engineer on the job con-ent analysis sheet.

Our output yielded some constant, k. We also had coefficients for x_1, x_2 and x_3. Our formula for measuring performance was:

$$k + C_1 \Sigma x_1 + C_2 \Sigma x_2 + C_3 \Sigma x_3 = \text{(Raw Standard)}$$
(Productivity) (Pace)/Adjusted Hours

Each group had its own formula. To complete the standard for the entire section, the developed standards were combined as were the total available hours. The adjusted hours were the hours available to perform the job. Any allowances were multiplied in the formula after "Pace."

In our study of Design Engineers, it was observed through work sampling that the engineers seemed to be spending a lot of time on the boards, drafting. This was an example of under-employment. To verify the percent of time shown by work sampling, a review of the job content analysis was made. This also revealed that a great deal of time was spent for drafting. Since one of our dependent variables in the regression analysis dealt directly with finished drawings, it was a rather simple task to further verify the drafting time mathematically. The next step was a meeting with engineering supervision, wherein the problem of using highly skilled engineers for lesser skilled drafting tasks was discussed.

Case History II

Another example worth noting was that of our Shipping Analyst. Shipping Analysts at Martin Marietta, Orlando, are semi-professional employees. Th-ir job is to prepare the paperwork and expedite all shipments to be made. Due to the nature of their work and the individuals involved, a task analysis sheet was not used in this study. Instead, review of their job descriptions along with continuous observations over a period of time was used. Next, a work sample study by category was conducted. Since this was a relatively small group, the work sample study along with continuous observations served to provide enough information to determine the measurable output.

The work of the Analyst all related back to the number of folders of instructions, Form DD250's, Form DD1149's, shipping postings, return material request forms, shipping request forms, UPS items, Government-Furnished Equipment items and typing chores. The point worth noting here was the fact that the forms could not be equated to one another and had to be entered into our regression equation as separate variables. This was due to the differences in the work involved prior to reaching a finished product. Our time for processing a Form DD149, for example, was double the time for processing a return material request form. Each week, a count of completed forms and other measurable items was made and compared to a standard obtained by regression analysis. Manpower in this area can be planned based on the known shipping load. The study of the Shipping Analyst enabled us to make some method changes which helped reduce this department by 50 percent. Improved performance was also a factor in helping reduce the manpower requirements.

Case History III

The third case history is one which is still in the development process. It has proven a particularly frustrating area to measure and has required a different approach from the two previously related studies. The department for which measurement is being developed is the Manufacturing Engineering Department. The department is divided into six operating sections: Manufacturing Test Engineering, Test Equipment Fabrication, Electrical and Electronic Production Engineering, E & E Development and Planning, Structural Tool Design and Planning, and Advanced Manufacturing Technology.

Several years ago a performance measurement system was attempted, based upon multiple linear regression. The performance reporting was short-lived. If a single reason were selected for its demise, it would be incorrect selection and evaluation of tasks and, therefore, the regression values developed.

In the meantime, a related activity was undertaken by the Industrial Engineering Department. This activity was the development of an estimating manual for the Manufacturing Engineering Department.

The development of this manual was an Industrial Engineering activity, but it was performed with active Manufacturing Engineering participation. The manual was divided into sections for each of the areas of operation within the department. The listed activities were broken down into segments/values that were similar in size or smaller than the attempted regression standard approach. For each identified activity a value per occurrence or a ratio relating this activity to another readily recognizable activity was provided. Both the activity list and activity time value were coordinated with the supervision from each Manufacturing Engineering group. When a negotiated agreement on values was reached, the activity list and values were published in the Estimating Manual.

The current effort to begin a measurement system for Manufacturing Engineering was undertaken with the concept of providing data with double usage, i.e., data for validating the Estimating

Manual and data to establish performance measurement. However, Management was unwilling to wait for the design of a data collection system and the collection of sufficient data before measuring the performance of the Manufacturing Engineering Department. It was, therefore, decided to institute an interim informal measurement program based on the values collected weekly from the area being studied, and to measure the output against the values from the Estimating Manual as earned value. While this would not provide an objective performance measurement value, it would provide a week-to-week indicator or index to display the trend of the measured group's activity.

Since the Estimating Manual existed, the task analysis effort was a review of the Estimating Manual elements for applicability to the measurement program. If, through observation and interview, the elements appeared usable, the task analysis was completed. If the elements did not appe usable or the particular activity was not yet covered by the manual, proper measurement elements were generated by observation and interview.

This effort has worked for those Manufacturing Engineering activities to which it has thus far been applied. We have accomplished the development of performance measurement with the use of minimum manpower by use of pre-existing and related information. In turn, we realized a double benefit by validating the Estimating Manual values while estimating performance measurement.

A COMPARISON OF ALTERNATIVE TIME SLOTTING METHODS FOR INDIRECT TIME STANDARDS

E. Emory Enscore, Jr., Kenneth Knott, Benjamin W. Niebel
Department of Industrial and Management Systems Engineering
The Pennsylvania State University
University Park, PA 16802

ABSTRACT

The use of different statistical distributions and cluster analysis as a basis of calculating time slot sizes, for comparative estimating purposes, is discussed and compared.

INTRODUCTION

The problem of determining the numerical standard that will be representative of each group of operations contained in each time slot has been studied by a number of researchers. It is a problem that is becoming more important because of the increasing amount of indirect labor in all areas of business and industry [6]. In order to establish time standards for this type of labor, techniques such as Universal Indirect Labor Standards (UILS) are used. For such a technique, a range of time for the indirect jobs to be covered in the system must first be determined. For example, all jobs taking 30 hours or less might be covered. Jobs taking more than 30 hours would be handled differently. Next the range of times must be divided into n slots. The size of n has been studied [3] and will be assumed to equal 20 throughout this paper. The last information needed for the UILS system is a representative sample of jobs for which detailed standards have been developed. These jobs are generally referred to as benchmark jobs. The number of benchmark jobs required is a guess but will generally number in the hundreds [4]. Developing these standards is costly and time consuming but is necessary. The accuracy of these standards will ultimately dictate the accuracy of the UILS system.

These benchmark standards are next arranged in increasing order based on time and assigned to the 20 slots. It is this assignment that is the subject of this paper.

SLOTTING BASED ON STATISTICAL DISTRIBUTIONS

The most common approaches of assigning the benchmark jobs to the 20 slots involves the use of different statistical distributions. The use of four different distributions, uniform, normal, gamma, and log-normal, are discussed below.

(1) Uniform Distribution [4] -- The traditional distribution used to assign benchmark jobs to slots in UILS systems has been the uniform. The reason for this is that the assignment is very simple. If we have n slots and k benchmark jobs arranged in ascending order, then the first slot (n=1) has the first k/n (assumed to be integer) benchmark jobs assigned to it and the slot time is just the average of the first k/n benchmark standards. The second slot (n=2) would have the next k/n benchmark jobs, ..., and the nth slot would have the last k/n benchmark jobs.

(2) Normal Distribution -- Ferri and Niebel [3] introduced the use of the normal distribution for assigning benchmark jobs to slots. They demonstrated that the use of the normal distribution was superior to the uniform distribution. However, the use of the normal requires more knowledge of statistics and the use of normal tables which may be considered a disadvantage. Since the normal assumes that our time axis would range between $-\infty$ and $+\infty$, we must truncate the distribution. If $\pm 3\sigma$ were selected as the end points, this would account for 99.87% of the area under the normal curve. The range of each interval, r, would be calculated as

$$r = \frac{(+3.0) - (-3.0)}{n}$$

The number of benchmark jobs assigned to slot 1 (and slot n because the normal is a symmetric distribution) would be

$$NJ(1) = NJ(20) = \frac{P(-3.0 \leq Z \leq -3.0+r)(k)}{.9987}$$

$$= \frac{P(3.0-r \leq Z \leq 3.0)(k)}{.9987}$$

Similar calculations would be made for the other n-2 slots. The advantage of using the normal distribution is that it assumes that the number of jobs in each slot would not be the same. Very often in indirect work, there are many jobs which would take less than one hour to do and very few that would require more than eight hours. This is the reason for the normal outperforming the uniform.

(3) Gamma Distribution -- An actual plot of the benchmark job standards for many types of indirect labor yields a curve which is positively skewed, not symmetric. For this reason Enscore and Niebel [2] demonstrated the use of the gamma distribution for assigning benchmark jobs to slots. The results obtained were superior to both the normal and uniform distribution results. However, a major disadvantage of using the gamma is the high level of statistical knowledge required. Also, the probabilities associated with slot intervals <u>cannot</u> be obtained from tables but must be approximated using an integral approximation procedure which requires the use of a computer.

(4) Log-Normal Distribution -- Like the gamma distribution the log-normal distribution is positively skewed. However, it has the advantage in that probabilities associated with slot intervals <u>can</u> be obtained from tables. The use of the log-normal distribution still requires some statistical sophistication. A brief development of the log-normal distribution is given below to show how the necessary probabilities are obtained [5].

The probability density function of the log-normal distribution is given by:

$$f(x) = \begin{cases} \frac{1}{\sqrt{2\pi}\,\beta} x^{-1} e^{-(\ln x - \alpha)^2/2\beta^2} & \text{for } x, \beta > 0 \\ 0 & \text{elsewhere} \end{cases}$$

Thus the probability that a log-normal random variable, x, lies between a and b would be:

$$P(a \leq x \leq b) = \int_a^b f(x)\,dx \quad (A)$$

If we let $y = \ln x$ and thus get $dy = \frac{dx}{x}$, and substitute this into equation (A), we get

$$P(a \leq x \leq b) = \int_{\ln a}^{\ln b} \frac{1}{\sqrt{\pi}\,\beta} e^{-(y-\alpha)^2/2\beta^2}\,dy$$

which is the normal distribution with $\mu = \alpha$ and $\sigma = \beta$. If we let $F(z)$ be the distribution function for the standard normal distribution, we can find the probability as:

$$P(a \leq x \leq b) = F(\frac{\ln b - \alpha}{\beta}) - F(\frac{\ln a - \alpha}{\beta})$$

The mean and variance of the log-normal distribution are:

$$\mu = e^{\alpha + \beta^2/2}$$
$$\sigma^2 = e^{2\alpha + \beta^2}(e^{\beta^2} - 1)$$

Thus we can find estimates for α, $\hat{\alpha}$, and β, $\hat{\beta}$, by finding the mean and variance of our benchmark jobs and using the equations:

$$\hat{\beta} = \sqrt{\ln\left(\frac{\sigma^2}{e^{2\ln\mu}} + 1\right)}$$

$$\hat{\alpha} = \ln\mu - \hat{\beta}^2/2$$

Since the log-normal's range is zero to infinity, we must truncate, say at time equal to t. Our range (width) of each of n slots, r, would be

$$r = (t-0)/n$$

The first NJ(1) benchmark jobs would be assigned to slot 1 where

$$NJ(1) = \frac{P(0 < x \leq 0+r)}{P(0 < x \leq t)}(k)$$

$$= \left[\frac{F(\frac{\ln(r)-\hat{\alpha}}{\hat{\beta}}) - F(\frac{\ln(0^+)-\hat{\alpha}}{\hat{\beta}})}{F(\frac{\ln(t)-\hat{\alpha}}{\hat{\beta}})}\right](k)$$

and in general for slot i we have:

$$NJ(i) = \frac{P((i-1)r \leq x \leq (i)r)}{P(0 < x \leq t)}(k)$$

$$= \left[\frac{F(\frac{\ln(ir)-\hat{\alpha}}{\hat{\beta}}) - F(\frac{\ln[(i-1)r]-\hat{\alpha}}{\hat{\beta}})}{F(\frac{\ln(t)-\hat{\alpha}}{\hat{\beta}})}\right](k)$$

ANALYSIS OF STATISTICAL DISTRIBUTIONS

In both the Ferri-Niebel paper [3] and Enscore-Niebel paper [2], 270 benchmark maintenance job standards (taken from the <u>Engineering Performance Standards Machine Shop and Machine Repair Handbook for Public Works Maintenance</u>) were used to demonstrate the accuracy of the slotting distributions. After assigning each of the 270 benchmark jobs to one of 20 slots by the uniform, normal or gamma distribution, a twenty-five week simulation was done to determine the accuracy of the distribution dependent UILS system. For each of the twenty-five weeks, jobs were randomly selected until the sum of the actual standard times exceeded or equaled 40 hours. Next the UILS system standard for each job selected was determined and the weekly sum calculated. (Remember that the time standard for a slot is just the average of the benchmark job standards assigned by the statistical distribution to that slot.) It is assumed that the individual doing the slotting of the random job is perfect in his task. The measure of accuracy is absolute percent error calculated for each week as

$$\left|\frac{\text{Actual Standard Time} - \text{UILS System Time}}{\text{Actual Standard Time}}\right| \times 100\%$$

The results of this simulation with the number of weeks equal to 1000 is given in Table 1 for the uniform, normal and gamma distributions.

Table 1 Simulation Results

Slotting Technique	Average Absolute Percent Error	Standard Deviation of Abs. Percent Error
Uniform Distribution	8.406	7.651
Normal Distribution	1.900	1.525
Gamma Distribution	2.571	2.202
Log-Normal Distribution	3.180	2.593
Cluster Analysis	1.144	0.862

In order to do a similar analysis of the log-normal distribution the following steps are necessary:

Step 1 -- Calculate the mean and standard deviation of the benchmark job standards.

$$\mu = 2.08 \quad \sigma = 2.91$$

This yields

$$\hat{\beta} = 1.0415 \quad \hat{\alpha} = 0.1901$$

Step 2 -- Truncate the log-normal distribution at t=25 hours. (The same as was done for the gamma distribution.) The slot width, r, is calculated as

$$r = (25-0)/20 = 1.25 \text{ hours}$$

Step 3 -- Calculate the probabilities and number of benchmark jobs associated with each of the twenty slots. For slot 3 the calculations would be:

$$P(2.50 \leq x \leq 3.75) = \frac{F(\frac{\ln(3.75)-0.1901}{1.0415}) - F(\frac{\ln(2.50)-0.1901}{1.0415})}{F(\frac{\ln(25.0)-0.1901}{1.0415})}$$

$$= \frac{F(1.087) - F(0.697)}{F(2.908)} = \frac{0.1044}{0.9982} = 0.1045$$

$$NJ(3) = (0.1045)(270) = 28.22$$

Similar calculations are done for each of the 20 slots and the results are given in Table 2.

Step 4 -- Perform a chi-square goodness of fit test with the null hypothesis that the log-normal distribution with $\alpha=0.1901$ and $\beta=1.0415$ provides a good fit of the data. The last fifteen slots are combined for the test since the expected frequency is less than five for each of them. Our calculated value for chi-square is

$$\chi^2 = 6.47$$

and $\chi^2_{.05,3} = 7.815$. (The degrees of freedom is calculated as 6-3 where we lose a degree of freedom for using n=270, $\mu=2.08$ and $\sigma=2.91$.) Thus we fail to reject our null hypothesis at the 0.05 level of significance and conclude that the log-normal distribution with $\alpha=0.1901$ and $\beta=1.0415$ provides a good fit.

Step 5 -- Use the expected frequencies of Table 2 for the assignment of benchmark jobs to slots and then do the 1000 week simulation as previously described. The results of the simulation are given in Table 1 for ease of comparison with the other three distributions.

INTRODUCTION OF CLUSTER ANALYSIS

Cluster analysis is a statistical concept used to compare data points and sort or categorize (slot) like elements into groups or clusters (time slot). The techniques of cluster analysis have been used successfully in such areas as biology, psychology, medicine, economics and marketing research.

Once the data has been collected on which cluster analysis is to be applied, the user must define which measure of similarity or likeness to be used. Examples of these measures include squared euclidean distance, product moment correlation, error sum of squares, variance and dispersion. The cluster analysis procedure depends on whether the number of clusters is unknown or specified. If unknown, the general classification of cluster analysis procedures to be used is called hierarchical. If specified, the classification is called nonhierarchical. Basically, the hierarchical procedures start by looking at n clusters with one data point each and work down to one cluster with n data points and decides based on the similarity measure which clustering is best. The best known of these procedures is Ward's Method using an error sum of squares similarity measure. The nonhierarchical procedures start with an initial assignment of the n data points to k clusters and then move the data points from cluster to cluster in order to optimize the chosen similarity measure. Throughout the process the number of clusters remains fixed at k. These procedures are sometimes referred to as relocation procedures because of the relocation of data points among the k clusters. A popular relocation procedure is k-mean with an error sum of squares similarity measure.

Table 2 Probabilities, Actual and Expected Frequencies of Benchmark
Jobs Based on Log-Normal Distribution

Slot No.	Slot Range (Hours)	Probability	Actual Frequency of Benchmark Jobs in Slot	Expected Frequency of Benchmark Jobs in a Slot
1	0.00 - 1.25	0.5137	150	138.69
2	1.25 - 2.50	0.2448	50	66.09
3	2.50 - 3.75	0.1045	29	28.22
4	3.75 - 5.00	0.0523	13	14.12
5	5.00 - 6.25	0.0291	11	7.86
6	6.25 - 7.50	0.0175	6	4.73
7	7.50 - 8.75	0.0112	2	3.02
8	8.75 - 10.00	0.0074	2	2.00
9	10.00 - 11.25	0.0051	2	1.38
10	11.25 - 12.50	0.0038	0	1.03
11	12.50 - 13.75	0.0026	0	0.70
12	13.75 - 15.00	0.0020	2	0.54
13	15.00 - 16.25	0.0014	1	0.38
14	16.25 - 17.50	0.0013	0	0.35
15	17.50 - 18.75	0.0008	1	0.22
16	18.75 - 20.00	0.0007	0	0.19
17	20.00 - 21.25	0.0006	0	0.16
18	21.25 - 22.50	0.0005	0	0.14
19	22.50 - 23.75	0.0004	0	0.10
20	23.75 - 25.00	0.0003	1	0.08

SLOTTING BASED ON CLUSTER ANALYSIS

It was decided that the relocation procedure k-mean would be used to assign benchmark jobs to our UILS slots. This procedure was selected because it yields tight clusters with the property that each cluster center represents the constituent cases at a high level of similarity. The measure of similarity used was error sum of squares (within group sum of square distances). The steps of the procedure are as follows:

Step 1 -- Assign the n benchmark jobs to k clusters (slots). This can be a random assignment if desired.

Step 2 -- Calculate the total sum of the squared distances from each job to the centroid of its parent cluster for the initial assignment.

Step 3 -- An iterative process is done with each benchmark job being removed from its parent cluster and placed in the k-1 other clusters. The job is relocated in the cluster for which the squared distance is the smallest. This is done for each of the n jobs in each iteration. The iterations stop when there is no relocation during an iteration. The mathematical statement for relocating a benchmark job x can be given as:

The error sum of squares is reduced by switching x from its parent cluster P to cluster Q if

$$\frac{(n_P - 1)}{n_P} DSQ(P-x, x)$$

exceeds

$$\frac{n_Q}{(n_Q + 1)} DSQ(Q, x)$$

where

n_P = number of jobs in cluster P

n_Q = number of jobs in cluster Q

DSQ(P-x,x) = the squared distance of x to the centroid of cluster P and x <u>not</u> used in the calculation of the centroid.

DSQ(Q,x) = the squared distance of x to the centroid of cluster Q.

Step 4 -- Repeat steps 1-3 for a number of initial assignments of the jobs to the slots.

Step 5 -- Select the assignment which produces the minimum error sum of squares.

In order to perform the relocation cluster analysis procedure, a computer program entitled

CLUSTAN [7] was used. The program was developed at the Universities of St. Andrews and London. The data used was the 270 benchmark jobs previously used for slotting with statistical distributions. The results of the clustering are given in Table 3. The clustering results were then tested using the simulation program and the results are included in Table 1.

Table 3 Assignment of Benchmark Jobs to Slots via Cluster Analysis

Cluster No.	Initial Assignment Job Numbers*	Final Assignment Job Numbers*
1	1 - 14	1 - 21
2	15 - 28	22 - 41
3	29 - 42	42 - 81
4	43 - 56	82 - 114
5	57 - 70	115 - 146
6	71 - 84	147 - 171
7	85 - 98	172 - 193
8	99 - 112	194 - 218
9	113 - 126	219 - 234
10	127 - 140	235 - 239
11	141 - 153	240 - 248
12	154 - 166	249 - 254
13	167 - 179	255 - 261
14	180 - 192	262
15	193 - 205	263
16	206 - 218	264 - 265
17	219 - 231	266 - 267
18	232 - 244	268
19	245 - 257	269
20	258 - 270	270

*Jobs are arranged in ascending order based on time.

ANALYSIS OF RESULTS

Examination of Table 1 reveals some interesting results. It is clear that the uniform distribution is outperformed by the other four slotting techniques. However, the normal distribution outperforms the gamma distribution, which is contrary to the results obtained by Enscore and Niebel [2]. There are two reasons that may account for this turn around:

(1) The sample size used for this paper was 1000 weeks instead of the twenty-five weeks used for their results.

(2) The normal distribution was truncated at $\pm 3.0\sigma$ for this paper versus the $\pm 2.7\sigma$ used for their results. This adds to the accuracy of the normal distribution, especially in the tails.

The normal distribution appears to be the superior distribution but cluster analysis appears to be the best slotting technique.

Assuming that the absolute percent errors are normally distributed for each of the five slotting techniques, we can use the F-distribution to test the equality of variances, two techniques at a time. The test is as follows:

Step 1 -- Select the null hypothesis to be $\sigma_i^2 = \sigma_j^2$ (i=1,...,5 and j=1,...,5, i≠j) and the alternate hypothesis to be that one variance is larger than the other.

Step 2 -- Using a significance level of $\alpha=0.05$, the region of rejection of the null hypothesis is

$$F > F_{0.05}(999;999) = 1.11$$

Step 3 -- Calculate the F with the larger variance as the numerator and the smaller variance as the denominator. Using the standard deviations of Table 1, F is calculated for the 10 pairs of the 5 techniques. The F values are given in Table 4.

Table 4 F Ratios for Testing of Variances

Technique	Uniform	Normal	Gamma	Log-Normal	Cluster Analysis
Uniform		25.17	12.07	8.71	78.78
Normal			2.09	2.89	3.13
Gamma				1.39	6.53
Log-Normal					9.05
Cluster Analysis					

Step 4 -- For each pair, we reject the null hypothesis and accept that the larger variance of the pair is greater than the smaller variance.

CONCLUSIONS

Based on these tests we can make the following claims:

(1) The slotting techniques can be ranked best to worst as cluster analysis, normal, gamma, log-normal, and uniform.

(2) The use of the uniform distribution cannot be justified based on its ease of use. The error associated with its use is too large.

(3) The normal, gamma and log-normal distributions all yield low average absolute errors and small variances.

(4) The ease of computing the probabilities associated with the log-normal should justify its use over the gamma when a positively skewed distribution is desired.

(5) Cluster analysis is the best of the five slotting techniques but its use is complicated by the fact that it is a relatively new technique which is unknown to many Industrial Engineers.

The results of this paper are based on one set of maintenance data. It would be desireable that future research concentrate on using other data sets from different types of indirect work. The authors feel that the ranking of the statistical distributions may change but that cluster analysis will consistently stand as the best slotting technique.

REFERENCES

[1] Anderberg, M. R., Cluster Analysis for Application, Academic Press, New York, 1973.

[2] Enscore, E. E. and B. W. Niebel, "Reliable Indirect Labor Standards Achieved Through Computerized Computation of Mean Slot Values of the Gamma Distribution," presented at the Third National Conference on Computers and Industrial Engineering, Orlando, Florida, October 22-24, 1980.

[3] Ferri, F. and B. W. Niebel, "Universal Maintenance Standard System," Industrial Engineering, Vol. 8, No. 2, February, 1976, pages 26-29.

[4] Maynard, H. B., Industrial Engineering Handbook, Second Edition, McGraw-Hill, Inc., New York, 1963.

[5] Miller, I. and J. E. Freund, Probability and Statistics for Engineers, Second Edition, Prentice-Hall, Inc., Englewood Cliffs, New Jersey, 1977.

[6] Niebel, B. W., Motion and Time Study, Fifth Edition, Richard D. Irwin, Inc., Homewood, Illinois, 1972.

[7] Wishard, D., CLUSTAN IC Users Manual, Third Edition, Program Library Unit, Edinburg University, Edinburg, Scotland, 1978.

BIOGRAPHICAL SKETCH

Dr. E. Emory Enscore, Jr., is Associate Professor of Industrial Engineering at The Pennsylvania State University. He received his BSIE from North Carolina State University and MSIE and Ph.D. from Penn State. His research interests are in production and inventory control, computerized plant layout, simulation and scheduling. In addition to being a Senior member of IIE, Dr. Enscore is a member of ASME, ASEE, ORSA and Sigma Xi and is a registered Professional Engineer in Pennsylvania.

Professor Knott is in the Department of Industrial and Management Systems Engineering, The Pennsylvania State University, and is responsible for the Continuing Education Program in Industrial Engineering. He has an M.S. in Industrial Engineering from The Pennsylvania State University, and a Diploma for Graduate Studies in Engineering Production from the University of Birmingham, England. His current research activities are in work measurement optimization, productivity analysis, and methods time measurement. He is a Senior Member of IIE, and belongs to the Institution of Mechanical Engineers, the Institution of Production Engineers, the Methods Time Measurement Association and the Federation of Productivity Services.

B. W. Niebel is Professor Emeritus of Industrial Engineering at The Pennsylvania State University. He is active in doing consulting work for major national and international firms and in writing several new books. Professor Niebel is a Fellow of IIE and has served as its Vice-President of Education and Student Affairs. He is also a member of Sigma Tau, Tau Beta Pi, Alpha Pi Mu, Sigma Xi, ASEE, SAM, and SME.

AN ANALYTICAL APPROACH TO
DESIGNING AND TESTING TIME
SLOTTING SYSTEMS

by

Kenneth Knott, Ph.D., P.E.
Associate Professor

E. Emory Enscore, Ph.D., P.E.
Associate Professor

and

Jeya Chandra, Ph.D.
Assistant Professor

Department of Industrial Engineering
The Pennsylvania State University

ABSTRACT

An analytical approach to designing time-slotting scales is described. A numerical example to illustrate this method is presented together with the advantages resulting from this approach over previous approaches.

INTRODUCTION

Comparative Estimating is a work measurement technique in which a single time value is used to represent the time standards of all tasks whose standard times lie, or are judged to lie, in a defined time range. This time range is known as a time-slot and the single, representative time is known as the slot-time. Together, these time-slots and slot-times are referred to as a time-slotting scale.

To design a time-slotting scale the disposition of each time-slot and the value of its associated slot-time must be calculated. Several methods making such calculations [2, 3, 4, 5, 6, 7] have been described in detail elsewhere and some of these methods have been subjected to investigation using computer simulation. Unfortunately, this computer simulation hid the fact that some of these approaches resulted in serious distortion and, in fact, incorrect results.

The purpose of this paper is to present an analytical basis to the theory of slotting, so that time-slotting scales can be designed and their characteristics analyzed according to a priori requirements. So as to provide a practical framework for time-slotting, the concepts of Comparative Estimating are reviewed and the types of error which may result from setting time standards by this technique are identified. The basic, generalized equations which describe these errors are stated and their construction outlined. These equations, as they relate to a log-normal distribution, and a computer program to make the associated calculations are given. Finally, an actual time-slotting scale is designed and analyzed to illustrate the method.

TIME-SLOTTING AND COMPARATIVE ESTIMATING

Time-slotting and Comparative Estimating are more widely used in maintenance management than perhaps elsewhere. A text-book by Lewis [11] gives comprehensive treatment of the use of Comparative Estimating but not of the design and analysis of the time-slotting scales.

Several authors [2, 3, 4] suggest that the Time-slotting scale should be designed upon a data-base of sample work-task times taken from the work area being studied. The methods proposed by these authors led to allocating these work-tasks incorrectly and cannot, therefore, be recommended. Furthermore, the behavior of the resulting scales were tested by simulation and gave performance measures which were not entirely clear.

The data-base is important however in as much as a number of work tasks are selected from it, which are then used as benchmarks. These benchmarks are selected by the familiarity which exists in the organization with their work content. These benchmarks are arranged systematically in columns (time-slots) according to their standard times. They can, in this way, be used for estimating time standards of new, unmeasured work tasks. These new work-tasks can be allocated into the appropriate column and the slot-time used as the time standard.

ELEMENTS OF THE SLOTTING SCALE

The work-task times in the work area being applied are assumed to follow a continuous distribution. Onto this distribution the time-slotting scale must be constructed. The elements of this scale are illustrated in Figure 1. No particular distribution is assumed at this time, so that a generalized form is being considered.

If the time-slotting scale contains n time-slots, the disposition of time-slot i can be defined by lower and upper time boundaries, t_{i-1} and t_i, with slot-time equal to T_i for $i = 1, 2, ..., n$. Further, the lowest time considered by the system has the restriction $t_0 \geq 0$. Similarly, the highest work task time considered by the system would be t_n. The value of t_n does not have to be equal to the highest value of the task times in the data base of sample work tasks. Some truncation is desirable and should be reflected in calculations. Finally, the probability density function of the work task time, t.

Errors in Time-Slotting Scales

This paper recognizes three types of error

System Error (Accuracy) - the result of using improper assumptions and/or design procedures.

Random Error (Precision) - the result of using a single slot-time in the place of the standard times of the work tasks in the time-slot.

Allocation Error - the result of the estimator allocating unmeasured work tasks into the wrong time-slot.

In this paper only Systematic and Random Errors are considered. These are illustrated in Figure 2, where it will be seen that the accuracy of the system is inversely proportional to systematic error and the precision is inversely proportional to the random error.

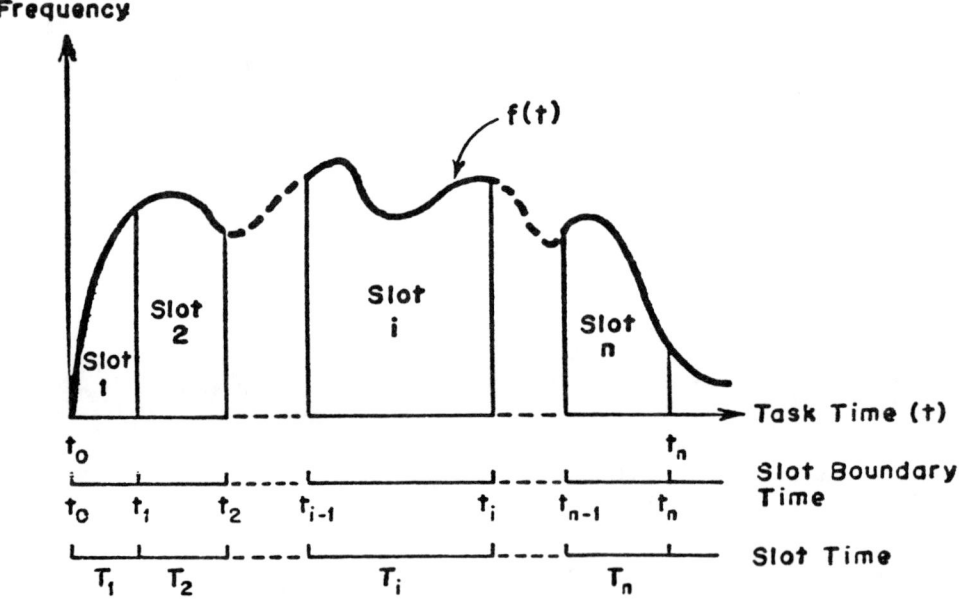

Figure 1: Elements of a Time-Slotting Scale

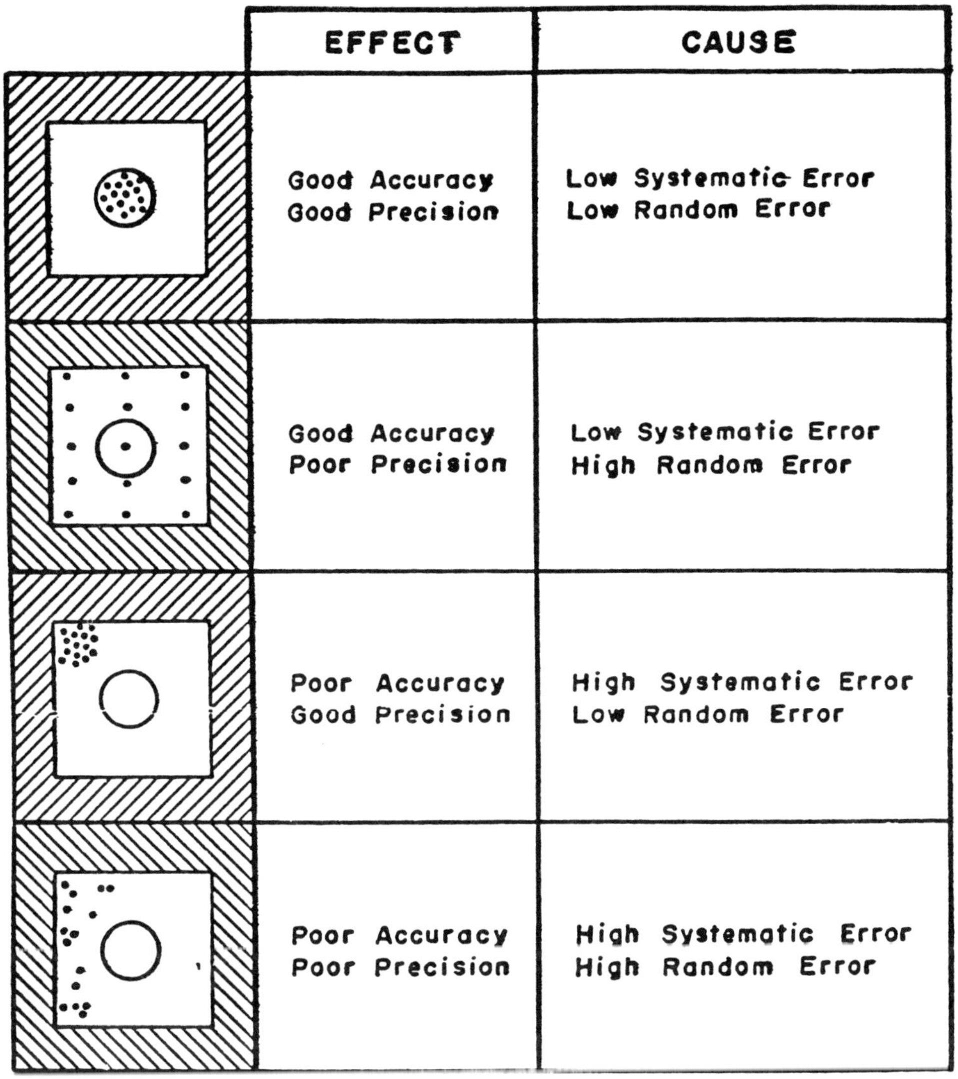

EFFECT	CAUSE
Good Accuracy Good Precision	Low Systematic Error Low Random Error
Good Accuracy Poor Precision	Low Systematic Error High Random Error
Poor Accuracy Good Precision	High Systematic Error Low Random Error
Poor Accuracy Poor Precision	High Systematic Error High Random Error

Figure 2: Effect of Errors on Precision and Accuracy of a Time-Slotting Scale

The Systematic Error of a Single Time-Slot

A systematic error occurs if the standard time of a work task is t and if $t_{i-1} \leq t \leq t_i$ the work task is allocated to time-slot i. The value used for T_i should be selected so that there is no long term bias in the time prediction of the work tasks in time-slot i.

The only time in which the use of the mean of t_{i-1} and t_i is acceptable is when the distribution is uniform. Elsewhere, the distribution of times in the time-slot must be considered in the calculations. Take, as an example, the case shown in Figure 3 where the distributions form the plane figure ABCD. The value of T_i is the distance of the centroid of ABCD from the axis at which t=0.

A systematic error will occur in time-slot i when a slot-time other than T_i, as determined above, is used. Let this incorrect slot-time be $\hat{T}_i$ where $T_i \neq \hat{T}_i$. In which case there will be a systematic error in time-slot i, identified as ε_i where

$$\varepsilon_i = f\ (T_i - \hat{T}_i) \qquad (1)$$

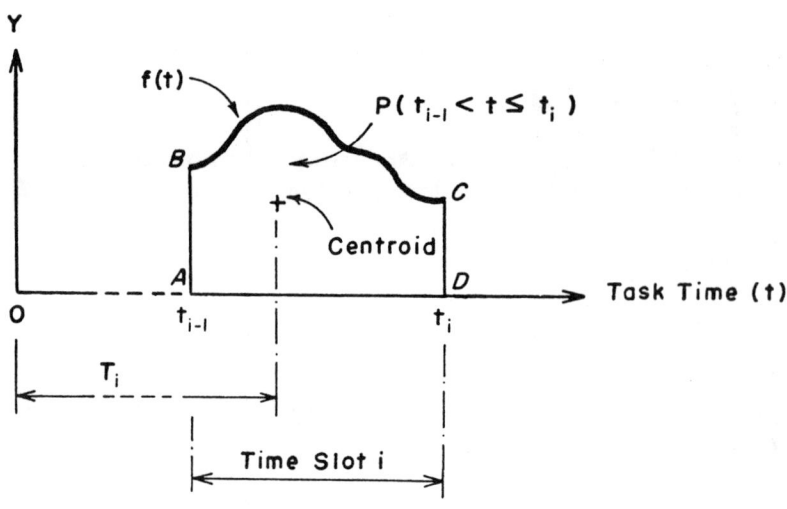

Figure 3: Determination of the Slot-Time, T_i, Using the Centroid

System Error or Accuracy

The system error is the result of the sum of the systematic error of each time-slots. As a matter of convenience, it can be expressed as the percentage of that time which is calculated for a single slot for the range $t_n - t_o$. If this "single slot-time" is given the symbol T' and the system error e' then

$$e' = \frac{\sum_{i=1}^{n} \varepsilon_i}{T'} \times 100 \qquad (2)$$

The measure of System Error, or accuracy, given in Equation (2) represents a percentage bias in the long term average of jobs in the work area.

Random Errors and System Precision

One measure which has been used by several investigators [1, 5, 7] to express the random errors in work measurement is the coefficient of variation. This appears to be particularly useful in analyzing time-slotting.

It will be seen from Figure 4 that the variance which exists in time-slot i will be dependent upon the slot width and the value used for the slot-time. If a value of $\hat{T}_i$ rather than T_i is used then a different variance will result. Similarly, if the width of the time-slot is increased then the variance will also increase.

Where $\hat{T}_i$ is used, the form of the equation will be the same as when T_i is used for the slot-time. In view of this T_i will be used in subsequent equations. It must be recognized that $\hat{T}_i$ could be substituted.

The standard deviation for the system, σ', is the weighted sum of the standard deviation of individual time-slots, σ_i. The system precision, p' can then be stated as the coefficient of variation of the system, expressed as a percentage. Thus

$$p' = \frac{100 \ \sigma'}{T'} \qquad (3)$$

A NUMERICAL EXAMPLE

Work-time distribution have been found to have a positive skew. Analysis on the generalized equations was carried assuming that these distributions were Log Normal [9, 10]. This allows calculations to be made from standard normal tables. The generalized form of the equations and the specific equation for the log normal distribution are given in Table 1.

The histogram of the data base of work task times used in the following calculations is shown in Figure 5. The estimates of the mean and standard deviations of this distribution are 2.08 hours and 2.91 hours respectively from this estimates of the parameters of the log normal distribution are calculated as

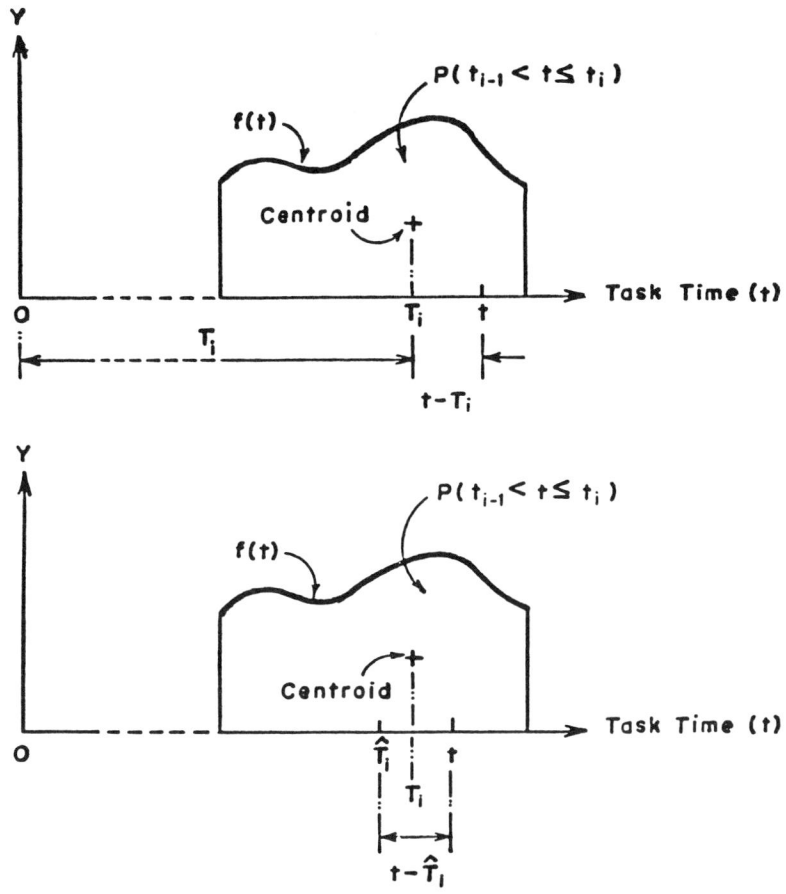

Figure 4: Determination of the Precision of a Single Time-Slot, i

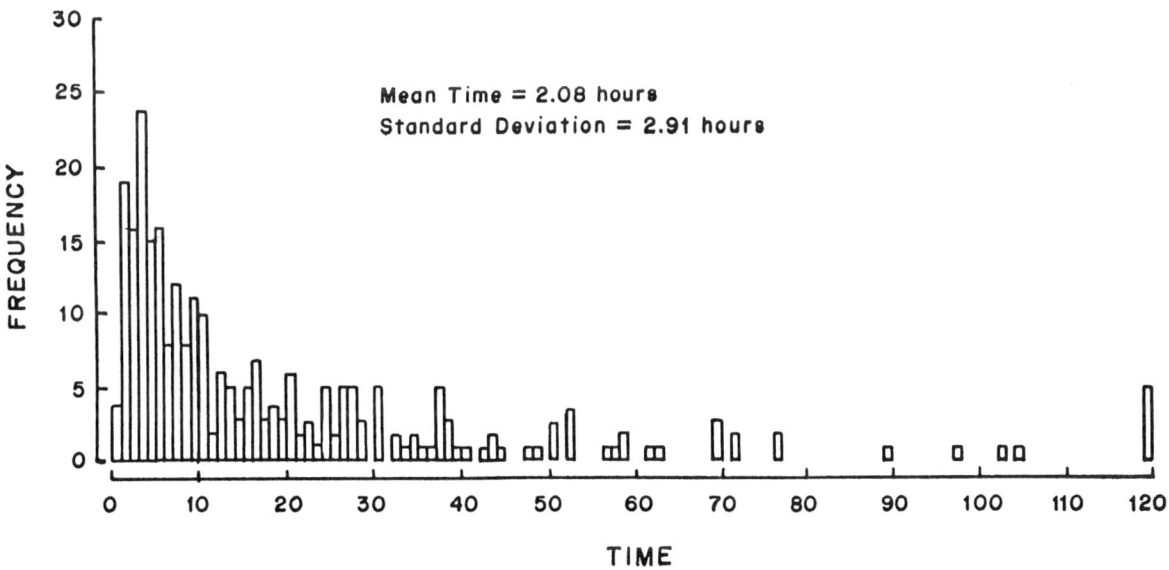

Figure 5. Histogram of Data Base (10)

Table 1: Summary of Equations Used in Designing
and Examining Time Slotting Scales

Eq. No.	Equation
I	$A_i = F\left[\dfrac{\ln t_i - \alpha}{\beta}\right] - F\left[\dfrac{\ln t_{i-1} - \alpha}{\beta}\right]$
II	$t_i = e^{\beta F[1A_i]}$ (when $t_o = 0$)
III	$T_i = \dfrac{e^{(\alpha+\frac{\beta^2}{2})}\left\{F\left[\dfrac{\ln t_i - (\alpha+\frac{\beta^2}{2})}{\beta}\right] - F\left[\dfrac{\ln t_{i-1} - (\alpha+\frac{\beta^2}{2})}{\beta}\right]\right\}}{F\left[\dfrac{\ln t_i - \alpha}{\beta}\right] - F\left[\dfrac{\ln t_{i-1} - \alpha}{\beta}\right]}$
IV	$\epsilon' = 100 * \dfrac{D}{H} \sum_{i=1}^{n} G_i$
V	$D = F\left[\dfrac{\ln t_n - \alpha}{\beta}\right] - F\left[\dfrac{\ln t_o - \alpha}{\beta}\right]$

Table 1. (Cont.)

Eq. No.	Equation
VI	$H = e^{(\alpha+\frac{\beta^2}{2})}\left\{F\left[\dfrac{\ln t_n - (\alpha+\beta^2)}{\beta}\right] - F\left[\dfrac{\ln t_o - (\alpha+\beta^2)}{\beta}\right]\right\}$
VII	$\sum_{i=1}^{n} G_i = e^{(\alpha+\frac{\beta^2}{2})}\left\{F\left[\dfrac{\ln t_i - (\alpha+\beta^2)}{\beta}\right] - F\left[\dfrac{\ln t_{i-1} - (\alpha+\beta^2)}{\beta}\right]\right\} - \hat{T}_i\left\{F\left[\dfrac{\ln t_i - \alpha}{\beta}\right] - F\left[\dfrac{\ln t_{i-1} - \alpha}{\beta}\right]\right\}$
VIII	$\rho' = \sum_{i=1}^{n}\left[e^{2(\alpha+\beta^2)}*\left\{F\left[\dfrac{\ln t_i - (\alpha+2\beta^2)}{\beta}\right] - F\left[\dfrac{\ln t_{i-1} - (\alpha+2\beta^2)}{\beta}\right]\right\}*\left\{F\left[\dfrac{\ln t_i - \alpha}{\beta}\right] - F\left[\dfrac{\ln t_{i-1} - \alpha}{\beta}\right]\right\}\right.$ $\left. - e^{2(\alpha+\beta^2)}\left\{F\left[\dfrac{\ln t_i - (\alpha+\beta^2)}{\beta}\right] - F\left[\dfrac{\ln t_{i-1} - (\alpha+\beta^2)}{\beta}\right]\right\}^2\right]^{\frac{1}{2}} * \left[\dfrac{F\left[\dfrac{\ln t_n - \alpha}{\beta}\right] - F\left[\dfrac{\ln t_o - \alpha}{\beta}\right]}{e^{(\alpha+\frac{\beta^2}{2})}\left[F\left[\dfrac{\ln t_n - (\alpha+\beta^2)}{\beta}\right] - F\left[\dfrac{\ln t_o - (\alpha+\beta^2)}{\beta}\right]\right]}\right] * 100$

$$\hat{\beta} = \sqrt{\ln \left(\frac{\sigma^2}{e^{2\ln\mu}} + 1 \right)}$$

$$= \sqrt{\ln \left(\frac{2.91^2}{e^{2\ln 2.08}} + 1 \right)}$$

$$= 1.0 + 1 \qquad (4)$$

and

$$\alpha = \ln \mu - \frac{\beta^2}{2}$$

$$= \ln 2.08 - \frac{1.041^2}{2}$$

$$= 0.191 \qquad (5)$$

Truncating the Scale

The designer of the slotting scale must first decide upon the maximum task time to be considered by the slotting scale, i.e., what is the value of t_n? The maximum task time in the data base was 24.30 hours. However, for the sake of ease of administering the system, assume that is considered desirable that $t_n = 8$ hours, what will be the effect of this truncation upon the system? What proportion of work tasks in the work area are likely to be excluded from the system? The same questions may, incidentally, be asked when $t_o \geq o$. Since the method of calculation is substantially the same consider the case where $t_o = o$ and $t_n = 8$ hours, using the equations from Table 1, the data base and the Normal Probability Distribution Tables

$$P(t_o \leq t \leq t_n) = F\left[\frac{\ln t - \alpha}{\beta}\right] - F\left[\frac{\ln t - \alpha}{\beta}\right]$$

$$= F\left[\frac{\ln 8 - 0.191}{1.041}\right] - F\left[\frac{\ln 0 - 0.191}{1.041}\right]$$

$$= F[1.814]$$

$$= 0.9652 \qquad (6)$$

Thus, in the work area being studied we can expect to account for 96.52% of the tasks which may be encountered. The remaining tasks would have to be handled by some means outside of the time-slotting system. The designer has the option of changing t_o or t_n and quickly determining its effects.

Bases for Calculating the Disposition of the Time-Slots

The disposition of the time-slot i is defined by the two times t_{i-1} and t_i, where $i = 1, 2, ..., n$. There are two ways in which t_{i-1} and t_i can be determined using either of the following bases:

1. The width of all of the time-slots can be equal.

2. The probability of work tasks falling into each time slot is equal. Thus $a_1 = a_2 = ... = a_i = ... = a_n$. Where a_i is the area under the probability density curve between t_{i-1} and t_i.

Time Slots of Equal Width

Having determined the acceptable level of truncation and specified the value of t_o and t_n as shown earlier the designer selects the number of time-slots which will be used in the system. Take the case where the designer selects n to be equal to 10, $t_o = o$ and $t_{10} = 8$ hours, the disposition of time-slots are calculated based upon

$$t_i = t_{i-1} + \left(\frac{t_n - t_o}{N}\right) \qquad (7)$$

For time-slot 1 $t_{i-1} = 0$ thus

$$t_1 = 0 + \left(\frac{8 - 0}{10}\right) = 0.8 \text{ hour} \qquad (8)$$

For time-slot 2 $t_{i-1} = 0.8$, thus

$$t_2 = 0.8 + \left(\frac{8 - 0}{10}\right) = 1.6 \text{ hours} \qquad (9)$$

Clearly, this procedure is repeated through $i = 1$ to 10.

Time-Slots with Equal Probability

Again, an interactive process is used in these calculations. The parameters are as follows $n = 10$, $t_o = 0$, $t_{10} = 8$ hours, $\alpha = 0.191$ and $\beta = 1.041$. Remember that the area under the curve due to truncation, calculated in Equation (6) was 0.9652. Therefore, with $n = 10$,

$$a_i = \frac{0.9652}{10} = 0.09652 \qquad (10)$$

It follows from Table 2 that:

$$a_1 = F\left[\frac{\ln t_1 - \alpha}{\beta}\right] - F\left[\frac{\ln t_o - \alpha}{\beta}\right] \qquad (11)$$

Thus,

$$0.09652 = F\left[\frac{\ln t_1 - 0.191}{1.041}\right] \qquad (12)$$

$$t_1 = 0.312 \text{ hours} \qquad (13)$$

In the case of slot number 2, equation (14) would be solved for t_2, where

$$0.09652 = F\left[\frac{\ln t_2 - 0.191}{1.041}\right] - F\left[\frac{\ln 0.312 - 0.191}{1.041}\right] \quad (14)$$

From which

$t_2 = 0.503$ hours.

Calculating the Slot-Time

The equation for calculating the slot-time is a little more intimidating than those which have been used so far. This equation also appears in Table 1. To illustrate this, consider the case of slot 1, for a time-slotting scale based upon equal probabilities in each time-slot. The values quoted above will be used again.

$$T_i = \frac{e^{(\alpha + \frac{\beta^2}{2})}\left\{F\left[\frac{\ln t_i - (\alpha + \beta^2)}{\beta}\right] - F\left[\frac{\ln t_{i-1} - (\alpha + \beta^2)}{\beta}\right]\right\}}{F\left[\frac{\ln t_i - \alpha}{\beta}\right] - F\left[\frac{\ln t_{i-1} - \alpha}{\beta}\right]} \quad (15)$$

Thus,

$$T_1 = \frac{e^{(0.191 + \frac{1.041^2}{2})}\left\{F\left[\frac{\ln 0.312 - (0.191 + 1.041^2)}{1.041}\right]\right\}}{F\left[\frac{\ln 0.312 - 0.191}{1.041}\right]}$$

$= 0.206$ hours \quad (16)

Calculating the System Precision

The equation used to calculate the system precision, p', of a slotting scale is shown in Table 2 and is repeated in Equation (17).

$$p' = \left[\sum_{i=1}^{n} e^{2(\alpha+\beta^2)} \cdot \left\{F\left[\frac{\ln t_i - (\alpha+2\beta^2)}{\beta}\right] - F\left[\frac{\ln t_{i-1} - (\alpha+2\beta^2)}{\beta}\right]\right\}\right.$$
$$\cdot \left\{F\left[\frac{\ln t_i - \alpha}{\beta}\right] - F\left[\frac{\ln t_{i-1} - \alpha}{\beta}\right]\right\}$$
$$- e^{2(\alpha+\beta^2)}\left\{F\left[\frac{\ln t_i - (\alpha+\beta^2)}{\beta}\right] - F\left[\frac{\ln t_{i-1} - (\alpha+\beta^2)}{\beta}\right]\right\}^2\Big]^{1/2}$$
$$\cdot \left[\frac{F\left[\frac{\ln t_n - \alpha}{\beta}\right] - F\left[\frac{\ln t_o - \alpha}{\beta}\right]}{e^{(\alpha+\beta^2)/2} F\left[\frac{\ln t_n - (\alpha+\beta^2)}{\beta}\right] - F\left[\frac{\ln t_o - (\alpha+\beta^2)}{\beta}\right]}\right] \cdot 100 \quad (17)$$

The system precision for several designs of slotting systems, depending upon the number of time slots, whether the time slots had equal widths or probabilities and the level of truncations are shown in Figure 6. The significance of the level of truncation on the system precision is made very clear by this figure. In fact it is reasonable to argue that this is of far greater importance than the number of slots used in the system, since, by truncating to 8 hours less than 4% of the possible tasks are excluded.

Calculating the System Accuracy

The equation used to calculate the system accuracy, e', is even more intimidating than that in Equation 19, for this reason it has been reduced to four simple forms. Nevertheless, the computer program given in the Appendix is still used to make the calculations. Furthermore, it will be realized that if T_i, as calculated earlier is used, then

$$e' = 0 \quad (18)$$

For the sake of completeness the equations used to calculate e' are given as Equations (19) through (22).

$$e' = 100 \cdot \frac{D}{H} \sum_{i=1}^{n} G_i \quad (19)$$

Where $D = F\left[\frac{\ln t_n - \alpha}{\beta}\right] - F\left[\frac{\ln t_o - \alpha}{\beta}\right] \quad (20)$

$$H = e^{(\alpha+\frac{\beta^2}{2})}\left\{F\left[\frac{\ln t_n - (\alpha+\beta^2)}{\beta}\right] - F\left[\frac{\ln t_o - (\alpha+\beta^2)}{\beta}\right]\right\} \quad (21)$$

and

$$\sum_{i=1}^{n} G_i = e^{(\alpha+\frac{\beta^2}{2})}\left\{F\left[\frac{\ln t_i - (\alpha+\beta^2)}{\beta}\right] - F\left[\frac{\ln t_{i-1} - (\alpha+\beta^2)}{\beta}\right]\right\}$$
$$- \hat{T}_i \left\{F\left[\frac{\ln t_i - \alpha}{\beta}\right] - F\left[\frac{\ln t_{i-1} - \alpha}{\beta}\right]\right\} \quad (22)$$

CONCLUSIONS

This paper has demonstrated a method of designing time-slots, with a priori requirements, which can then be tested. The method recommended to determine the system precision and system accuracy of a slotting system is based upon analysis, whereas previous approaches have used simulation. In this way, it is possible for direct comparisons to be made between different designs of slotting scales than with other approaches.

This paper also brings to the readers attention the importance of truncation in time-slotting. Earlier authors emphasized the influence of the number of time-slots upon the behavior of the system. Here, the truncation is emphasized.

Figure 6: The Relationship Between System Precision, Number of Time Slots and Truncation.

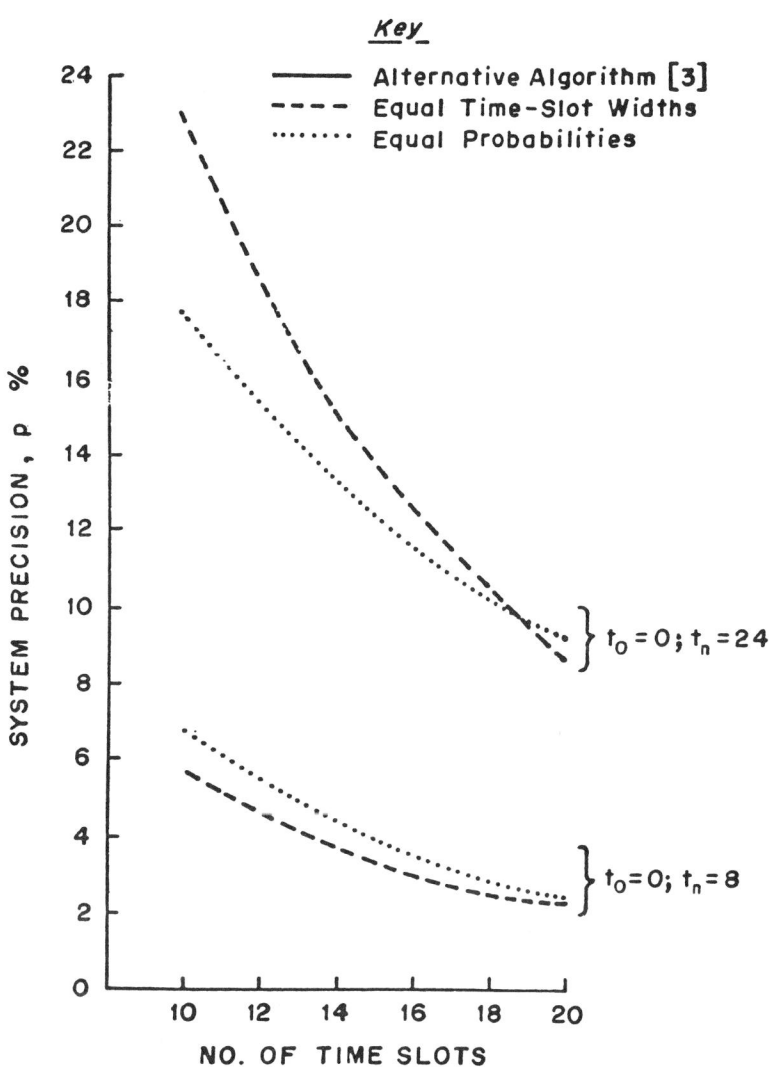

REFERENCES

[1] Brinkloe, W. D. and Coughlin, M. T. "Precision Analysis of MTM-1 and MOST," University Research Institute, Pittsburgh, Pennsylvania, 1975.

[2] Enscore, E. E. and Niebel, B. W. "Reliable Indirect Labor Standards Achieved Through Computerized Computation of Mean Slot Values of the Gamma Distribution," Third National Conference on Computers and Industrial Engineering, Orlando, Florida, October 22-24, 1980.

[3] Enscore, E. E., Knott, K. and Niebel, B. W. "A Comparison of Alternative Time Slotting Methods for Indirect Time Standards," Proceedings, Institute of Industrial Engineers Conference, New Orleans, 1982.

[4] Ferri, F. and Niebel, B. W. "Universal Maintenance Standard System, Industrial Engineering, pp. 26-28, February 1976.

[5] Hancock, W. W., "The System Precision of MTM-1," The Journal of Methods-Time Measurement, Vol. XV., No. 3, pp. 4-10, 1980.

[6] Knott, K. "A Simplified Approach to Comparative Estimating," <u>IMSE Working Paper 82-124</u>, Department of Industrial and Management Systems Engineering, The Pennsylvania State University, 1982.

[7] Knott, K. "An Investigation of Methods-Time Measurement Systems for Work Measurement Application," unpublished Ph.D. Thesis, <u>Department of Engineering Production</u>, The University of Technology, Loughborough, 1983.

[8] Knott, K. "An Examination of the Theory of Time-Slotting" <u>IMSE Working Paper 84-112</u>, Department of Industrial and Management Systems Engineering, The Pennsylvania State University, 1984.

[9] Knott, K., Chandra, J. and Enscore, E. E. "Time Slotting Based Upon an Assumed Log Normal Distribution of Sample Work Task Times - Part I: Theoretical Analysis" <u>IMSE Working Paper 84-114</u>, Department of Industrial and Management Systems Engineering, The Pennsylvania State University, 1984.

[10] Knott, K., Chandra, J. and Enscore, E. E. "Time Slotting Based Upon an Assumed Log Normal Distribution of Sample Work Task Times - Part II: Case Study." <u>IMSE Working Paper 84-117</u>. Department of Industrial and Management Systems Engineering, The Pennsylvania State University, 1984.

[11] Lewis, B. T. "<u>Developing Maintenance Time Standards</u>," Industrial Education Institute, Boston 1978.

[12] Maynard, H. B. and Stegmenten, G. J. "Universal Maintenance Time Standards," <u>Factory Management and Maintenance</u>, November 1955.

[13] McKenzie, H. G., "The Selection and Design of Slotting Scales," <u>Management Services</u>, pp. 10-15, December 1977.

[14] Pena, M. F. "Using the Properties of the Beta Distribution to Estimate the Standard Times of Indirected Jobs," unpublished Thesis, <u>Department of Industrial and Management Systems Engineering</u>, The Pennsylvania State University.

APPENDIX - COMPUTER PROGRAM FOR ANALYSING TIME-SLOTTING SCALES

```
10    REM THIS PROGRAM DESIGNS TIME-SLOTTING SYSTEMS.
20    DEFINT I-N
30    DIM A(300),T(50),TL(50),B(300),T20(50),F1(50),I50(50),E(50),N20(50)
40    DIM Y(6),C4(6),F10(50),F20(50),F30(50),F40(50),T1(50),T2(50)
50    DEF FN D5(YT)=B0+B9*YT : REM FUNCTION IN THE GAUSS-LEGENDRE FORMULA
60    DEF FN F(W)=1/(EXP((W^2)/2)*SQR(2*3.1415927#)) : REM STD NORMAL FN.
70    REM INPUT OF WEIGHING FACTORS FOR GAUSS-LEGENDRE FORMULA.
80    CLS
90    FOR I=1 TO 3
100   READ C4(I)
110   C4(7-I)=C4(I)
120   NEXT I
130   DATA .171324492,.360761573,.467913935
140   REM INPUT OF FUNCTION ARGUMENTS.
150   FOR I=1 TO 3
160   READ Y(7-I)
170   Y(I)=-Y(7-I)
180   NEXT I
190   DATA .932469514,.661209386,.238619186
200   OPEN "MTIME.DAT" FOR INPUT AS #2
210   OPEN "TIME.DAT" FOR APPEND AS #1
220   KIL=0
230   REM DATA IS READ FROM THE FILE MTIME.DAT.
240   J1=0
250   FOR I=1 TO 54
260   IF EOF(2) THEN 310
270   INPUT #2,A(J1+1),A(J1+2),A(J1+3),A(J1+4),A(J1+5)
280   J1=J1+5
290   NEXT I
300   CLS
310   LOCATE 8,10: PRINT " THE NO. OF TASK TIMES IN THE FILE=" J1
320   REM ALPHA AND BETA ARE CALCULATED FROM THE DATA.
330   SUM1=0
340   SUM2=0
350   FOR I=1 TO J1
360   SUM1=SUM1+A(I)
370   SUM2=SUM2+A(I)^2
380   NEXT I
390   VAR=((J1)*SUM2-(SUM1^2))/((J1)*(J1-1))
400   AMEAN=SUM1/(J1)
410   BL=EXP(2*(LOG(AMEAN)))
420   BL=LOG(1+(VAR/BL))
430   BL=SQR(BL)
440   AL=LOG(AMEAN)-(BL^2)/2
450   LOCATE 12,10: PRINT "ALPHA=" AL;
460   LOCATE 16,10: PRINT "BETA=" BL;
470   PRINT #1, "ALPHA=" AL;
480   PRINT #1, "BETA=" BL
490   LOCATE 20,10: INPUT "HIT RETURN KEY" ;M1$
500   CLOSE #2
510   CLS
520   PRINT "*************************************************"
530   LOCATE 6,10: PRINT " THE CALCULATIONS ASSUME THAT THE TASK TIMES ARE"
540   LOCATE 8,10: PRINT "LOGNORMALLY DISTRIBUTED AND THAT THE TASK TIMES ARE"
550   LOCATE 10,10: PRINT "IN THE DATA FILE MTIME.DAT."
560   LOCATE 12,10: PRINT " THE OUTPUT IS IN  THE FILE TIME.DAT."
570   PRINT "*************************************************"
580   LOCATE 14,10 : INPUT "HIT RETURN KEY TO CONTINUE.";L1$
590   CLS
600   LOCATE 6,10: PRINT "YOU HAVE THE FOLLOWING OPTIONS."
610   LOCATE 8,10: PRINT "CHOOSE ONE OF THEM."
620   LOCATE 10,10: PRINT " 1. USING ENSCORE METHOD."
630   LOCATE 12,10: PRINT " 2. USING KNOTT METHOD."
640   LOCATE 14,10: PRINT " 3. WANT TO QUIT."
650   LOCATE 16,10: INPUT "ENTER THE # OF THE OPTION YOU WANT.";I1
660   CLS
670   IF I1=3 THEN 2270
680   IF I1=1 THEN 720
690   FOR I=1 TO 300
700   B(I)=0
710   NEXT I
720   LOCATE 8,10:INPUT "ENTER THE MIN. VALUE THAT SHOULD BE CONSIDERED";TL(1)
730   LOCATE 10,10: INPUT "ENTER THE MAX. VALUE THAT SHOULD BE CONSIDERED.";G
740   IF I1=2 THEN 1440
750   L1=1
760   LI=1
770   FOR I=1 TO J1
780   IF A(I) => TL(1) AND A(I) =< G THEN J100=J100+1
790   B(I)=A(I)
800   NEXT I
810   LOCATE 12,10: PRINT " TOTAL # OF TASK TIMES IN THE INTERVAL=" J100
820   PRINT #1,"TOTAL # OF TASK TIMES IN THE INTERVAL=";J100
830   LOCATE 10,10:  INPUT " ENTER THE # OF SLOTS DESIRED.";N
840   CLS
850   GOTO 1490
860   JF=J100+1
870   IF I=M-3 THEN L1=1
880   FOR I=J1 TO 300
890   B(I)=0
900   NEXT I
910   REM CALCULATION OF THE SLOT TIMES AS PER ENSCORE'S METHOD.
920   N1=N+1
930   AN=N
940   R=(G-TL(1))/AN
950   X=(LOG(G)-AL)/BL
960   GOSUB 3390
970   F1=F
980   IF TL(1)>0 THEN 1020
990   X=(LOG(TL(1))-AL)/BL
1000  GOSUB 3390
1010  F1(0)=F
1020  F1=F1-F1(0)
1030  M2=0
1040  N5=N
1050  FSUM=0
1060  FOR I=1 TO N
1070  I2=I+1
1080  I1=I-1
1090  AI=I
1100  I50(I)=TL(1)+R*AI
1110  X=(LOG(I50(I))-AL)/BL
1120  GOSUB 3390
```

```
1130      F1(I)=P
1140      Z=J100*(F1(I)-F1(I1))/P1
1150      M=INT(Z+.5)
1160      T(I)=(F1(I)-F1(I1))
1170      IF M2=J100 THEN 1330
1180      IF 1 =N THEN 1240
1190      M1=M2+1
1200      M2=J100
1210      M=M2-M1+1
1220      AM=M
1230      GOTO 1360
1240      IF M=1 THEN M=1
1250      AM=M
1260      M1=M2+1
1270      M2=M1+M-1
1280      IF M2 <= J100 THEN 1360
1290      M2=J100
1300      M=M2-M1+1
1310      AM=M
1320      GOTO 1360
1330      M=0
1340      T20(I)=0
1350      GOTO 1410
1360      SUM12=0
1370      FOR J=M1 TO M2
1380      SUM12=SUM12+B(J)
1390      NEXT J
1400      T20(I)=SUM12/AM
1410      N20(I)=M
1420      NEXT I
1430      GOTO 1640
1440      CLS
1450      LOCATE 10,10: INPUT "ENTER THE # OF SLOTS DESIRED.";N
1460      L1=2
1470      CLS
1480      LOCATE 6,10: PRINT "THIS IS KNOTT'S METHOD."
1490      LOCATE 8,10: PRINT " YOU HAVE THE FOLLOWING OPTIONS."
1500      LOCATE 10,10: PRINT " 1. EQUAL INTERVALS IN KNOTT'S METHOD."
1510      LOCATE 12,10: PRINT " 2. EQUAL PROBABILITIES IN KNOTT'S METHOD."
1520      LOCATE 14,10: PRINT " 3.USING ENSCORE'S METHOD WITH THE SAME INPUT."
1530      LOCATE 16,10: PRINT " 4. IF YOU WANT TO QUIT."
1540      LOCATE 18,10: INPUT " ENTER THE # OF YOUR OPTION.";M
1550      CLS
1560      LOCATE 16,10: PRINT " PLEASE STAND BY.............."
1570      IF KM=1 OR KM=2 THEN L1=2
1580      IF KM=3 THEN L1=1
1590      IF KM=3 THEN 860
1600      IF KM=4 THEN 2370
1610      GOSUB 3120
1620      GOSUB 2400
1630      GOTO 1650
1640      GOSUB 2400
1650      CLS
1660      IF L1=1 AND KM=1 1770
1670      IF L1=2 AND KM=2 THEN 1860
1680      PRINT "*********************************************************"
1690      PRINT #1,"*********************************************************"
1700      PRINT "THIS IS AS PER ENSCORE'S METHOD."
1710      PRINT #1,"THIS IS AS PER ENSCORE'S METHOD.";
1720      PRINT "TRUNCATION IS AT    "  G;
1730      PRINT "THE # OF SLOTS =    "  N
1740      PRINT #1,"TRUNCATION IS AT    "  G;
1750      PRINT #1,"THE # OF SLOTS=    "  N
1760      GOTO 1960
1770      PRINT "*********************************************************"
1780      PRINT " THIS IS AS PER KNOTT'S METHOD WITH EQUAL INTERVALS.";
1790      PRINT "TRUNCATION IS AT    "  G;
1800      PRINT "THE # OF SLOTS=    "
1810      PRINT #1,"*********************************************************"
1820      PRINT #1,"THIS IS AS PER KNOTT'S METHOD WITH EQ. INTERVALS.";
1830      PRINT #1,"TRUNCATION IS AT    "  G;
1840      PRINT #1,"THE # OF SLOTS=    "  N
1850      GOTO 1970
1860      PRINT "*********************************************************"
1870      PRINT "THIS IS AS PER KNOTT'S METHOD WITH EQ. PROBABILITIES. ";
1880      PRINT "TRUNCATION IS AT    "  G;
1890      PRINT "THE # OF SLOTS=    "  N
1900      PRINT #1,"THIS IS AS PER KNOTT'S METHOD WITH EQ. PROBABILITIES. ";
1910      PRINT #1,"TRUNCATION IS AT    "  G;
1920      PRINT #1,"THE # OF SLOTS=    "  N
1930      PRINT "  #  SLT MAX TIME         SLT TIME "
1940      PRINT #1," # SLT MAX TIME          SLT TIME "
1950      GOTO 1980
1960      PRINT "  #  SLT MAX TIME         SLT TIME    AREA    FREQ.    SYS ERROR"
1970      PRINT #1," # SLT MAX TIME         SLT TIME    AREA    FREQ.    SYS ERROR"
1980      L1=1
1990      IF L1=1 THEN  2100
2000      FOR I=1 TO N
2010      PRINT USING "### ";I;
2020      PRINT USING "#####.###  ";T50(I);
2030      PRINT USING "#####.###  ";T20(I)
2040      PRINT #1,USING "### ";I;
2050      PRINT #1,USING "#####.###  ";T50(I);
2060      PRINT #1,USING "#####.###  ";T20(I)
2070      INPUT "HIT RETURN KEY.";N5$
2080      NEXT I
2090      GOTO 2250
2100      FOR I=1 TO N
2110      PRINT USING "### ";I;
2120      PRINT USING "#####.###  ";T50(I);
2130      PRINT USING "#####.###  ";T20(I);
2140      PRINT USING "#### ";N20(I);
2150      PRINT USING "####.###  ";E(I)
2160      PRINT #1,USING "### ";I;
2170      PRINT #1,USING "#####.###  ";T50(I);
2180      PRINT #1,USING "#####.###  ";T20(I);
2190      PRINT #1,USING "#### ";N20(I);
2200      PRINT #1,USING "####.###";E(I)
2210      INPUT " HIT RETURN KEY.";A5$
2220      PRINT " PRECISION=" PR;
2230      PRINT " ACCURACY=" AC
2240      PRINT #1," PRECISION=" PR;
2250      PRINT #1," ACCURACY=" AC
2260      INPUT "DO YOU WANT TO TRY WITH A DIFFERENT # OF SLOTS. (Y/N)";O$
2270      CLS
2280      IF O$="Y" AND L1=1 THEN 830
2290      IF O$="y" AND L1=1 THEN 830
2300      IF O$="Y" AND L1=2 THEN 1440
2310      IF O$="y" AND L1=2 THEN 1440
2320      IF O$="N" OR O$="n" THEN 590
2330      GOTO 2290
2340      STOP
2350      END
2360      REM CALCULATION OF PRECISION AND ACCURACY FACTORS.
2370      FOR I=1 TO N
2380      I1=I+1
2390      TL(I1)=T50(I)
2400      NEXT I
2410      A10=AL+(BL^2)
2420      A20=AL
2430      A30=A10+(BL^2)
2440      F10(1)=0
2450      F20(1)=0
2460      F30(1)=0
2470      D2=AL+(BL^2)/2
2480      A2=EXP(D2)
2490      B2=EXP(2*A10)
2500      C2=A2^2
2510      IF TL(1)=0 THEN  2640
2520      X=(LOG(TL(1))-A10)/BL
2530      GOSUB 3390
2540      F10(1)=P
2550      X=(LOG(TL(1))-A20)/BL
2560      GOSUB 3390
2570      F20(1)=P
2580      X=(LOG(TL(1))-A30)/BL
2590      GOSUB 3390
2600      F30(1)=P
2610      E10=0
2620      FOR I=1 TO N
2630      E(I)=0
2640      NEXT I
2650      E10=0
2660      FOR I=1 TO N
2670      I1=I+1
2680      T10=TL(I1)
2690      X=(LOG(T10)-A10)/BL
2700      GOSUB 3390
2710      F10(I1)=P
2720      X=(LOG(T10)-A20)/BL
2730      GOSUB 3390
2740      F20(I1)=P
2750      X=(LOG(T10)-A30)/BL
2760      GOSUB 3390
2770      F30(I1)=P
2780      GOTO 3390
2790      F30(I1)=P
2800      IF L1=1 THEN 2890
2810      T20(I)=A2*(F10(I1)-F10(I))/(F20(I1)-F20(I))
2820      H=(F30(I1)-F30(I))*(F20(I1)-F20(I))*B2
2830      F1=C2*(F10(I1)-F10(I))
2840      F1=F1*(F10(I1)-F10(I))
2850      F1=H-F1
2860      GOTO 2920
2870      E(I)=A2*(F10(I1)-F10(I))-T20(I)*(F20(I1)-F20(I))
2880      SUM10=SUM10+E(I)
2890      B10=F20(I)-F20(I)
2900      H=A2*(F10(I1)-F10(I))
2910      B20=A2*(F10(I1)-F10(I))
2920      B30=B2*(F30(I1)-F30(I))
2930      Z45=T20(I)^2
2940      Z46=T20(I)
2950      F1=H-(Z45*B10-2*Z46*B20+B30)
2960      SUMB=SUMB+F1
2970      NEXT I
2980      N=N+1
2990      D2=F20(N)-F20(1)
3000      H=A2*(F10(N)-F10(1))
3010      F1=SQR(SUMB)
3020      P1=(F20(N)-F20(1))*F1*100
3030      P2=A2*(F10(N)-F10(1))
3040      SUM=P1/P2
3050      PR=SUM
3060      AC=E10
3070      N=N-1
3080      RETURN
3090      REM CALCULATION OF SLOT TIMES AS PER KNOTT'S METHOD.
3100      TI=TL(N)^6
3110      AN=N
3120      A2=T1(N)/N
3130      X=-LOG(T1(N))
3140      X=-(X-AL)/BL
3150      GOSUB 3390
3160      P1=P/N
3170      AJ= .4
3180      FOR I=1 TO N
3190      I1=I-1
3200      AI=I
3210      A1=AI*P1
3220      GOSUB 3570
3230      AL=AI
3240      AI=AI*N+AL
3250      T2(I)=EXP(AI)
3260      IF I=1 THEN 3310
3270      T1(I)=T1(I1)+A2
3280      GOTO 3320
3290      T1(I)=A2
3300      IF KM=1 THEN 3350
3310      T50(I)=T2(I)
3320      GOTO 3360
3330      T50(I)=T1(I)
3340      NEXT I
3350      RETURN
3360      REM THIS FUNCTION CALCULATES THE AREA UNDER STD. NORMAL CURVE BETWEEN
3370              NEGATIVE INFINITY AND THE VALUE X USING GAUSS-LEGENDRE FORMULA.
3380      A5=0
3390      SUM=0
3400      IJ=0
3410      X1=X
3420      IF X1>0 THEN 3470
3430      IF X1=0 THEN 3540
3440      X1=-X1
3450      IJ=1
3460      B8=(A5+X1)/2
3470      B9=(X1-A5)/2
3480      FOR IM=1 TO 6
3490      SUM=SUM+C4(IM)*FN F(FN D5(Y(IM)))
3500      NEXT IM
3510      SUM=SUM*B9
3520      IF IJ=1 THEN SUM=-SUM
3530      P=.5+SUM
3540      RETURN
3550      REM THIS FUNCTION CALCULATES THE INVERSE STD. NORMAL FUNCTION.
3560      X=AJ
3570      SMALL=1000
3580      GOSUB 3390
3590      DIFF=ABS(AI-P)
3600      IF DIFF > SMALL THEN 3650
3610      SMALL=DIFF
3620      X=X+.01
3630      GOTO 3590
3640      AJ=X-.01
3650      RETURN
```

VII. Standards Auditing

Standards auditing may be defined as a procedure to ensure that current time standards meet an acceptable level of accuracy for a specific work location. Work standards rarely match the exact (correct) amount of time required for an average worker to perform a defined task using a prescribed method. Both the method used for determining the standard and the actual work environment typically cause some discrepancy between the standard and the correct actual time required. Standards with error above or below acceptable limits usually have a negative effect on the business. Production standards that are too low may result in overpricing the product and thus in losing market share, while too high standards may result in loss of potential revenue due to underpricing the product. Invalid standards also can damage employee motivation and morale. Workers who are unable to meet excessively tight standards may become discouraged, while those working with loose standards will usually underproduce. By utilizing the results of a correctly performed audit, standards may be revised, resulting in higher morale, productivity, and profits.

Management may request an audit of standards for a variety of reasons—for example, a company may want to check the standards after a change in operating procedures, or justify certain standards in response to a formal grievance from the employees' bargaining unit. In addition, random, ongoing auditing should be part of any work measurement program. Sampling time standards in an area or family of similar operations will indicate the probable reliability of that total population. If a periodic random sample indicates that the reviewed standards meet accuracy requirements, it is usually unnecessary to check further; conversely, a random sampling filled with discrepancies indicates the need for a more extensive review.

THE RATIONALE FOR AUDITING STANDARDS

Deterioration in standards, wage incentive programs, and measured daywork control plans are major reasons for conducting periodic audits. In addition, the inevitable tendency of work methods to change is itself a compelling reason for audits. Operators cannot realistically be expected to give notice of methods changes which they originate, and these changes may be too subtle for supervisors or technical support personnel to detect easily. Even in highly automated environments, people can still change work operations in ways that alter production outputs and introduce error into the standard times. Therefore, it is critical to check existing methods against those on which the standards were based. The value of a wage incentive program that is aimed at promoting extra manual effort can be deleteriously affected by increased mechanization unless proper auditing and standards reevaluation occurs. As automated equipment or paced lines go into effect, revisions in the existing incentive plan should be considered. Because measured daywork standards tend to establish performance ceilings, their validity needs to be checked. The level of performance on which standards are based is also an important factor.

THE AUDIT FREQUENCY

A complete standards review within a three-year period is a common management practice. For example, a company with 6,000 time standards can review them all in a three-year period if the organization audits about forty standards a week. Another practice for assuring effective periodic review of time standards is an audit schedule based on the total hours the standard is utilized on an annual basis. For example:

Hours per Year	Frequency of Audit
less than 50	3 years
50-250	2 years
251-600	1 year
over 600	6 months

The hours-per-year range for each frequency of audit should be established in relationship to the total distribution of time standards. As a general design guideline, the standards audited every six months should have an hours-per-year value that represents approximately 20 percent of the standards and should account for about 80 percent of the work volume.

The methods for checking specific standards will vary from situation to situation. If the values have been derived from time study, the original study can be reviewed for completeness, and the work methods currently in use can be compared with those of the time study. If the values have been set with a predetermined time system, a brief observation of the job cycle will usually identify any methods variations.

THE AUDIT RESULTS

The audit report should include a statement of variations from standard practice, an evaluation of these variations, and recommendations for changes and corrections. A wage incentive program should be based on carefully established policies and procedures, and unless these procedures are followed precisely, the plan may soon become unfair to management or to the employees, depending on the nature of the error. Historically, the majority of inaccurate standards favors the employee, and management consequently stands to gain most by comprehensive auditing procedures. The added cost of auditing standards may be regarded as insurance against eventual deterioration of planning and management control or of the wage incentive system. The cost of audit programs typically range between 5 and 10 percent of the total methods and standards budget.

SUMMARY OF ARTICLES

In "Auditing Standards by Sample," Gavriel Salvendy and George P. McCabe, Jr., propose a comprehensive method. The process begins with identifying the major job categories, or functions, in a plant and then randomly selecting departments within which these functions occur. The analyst then samples two jobs—one high volume, one low volume—in each selected department, using an analysis of variance model; the analysis is weighted by dollar volume to give more significance to high-volume jobs. From the resulting statistics, the analyst can calculate the percentage of improvement possible for each major function. The percentage can then be converted to dollars by multiplying by the product of yearly hours and average hourly wage rates. The authors state that confidence interval analysis will assure the reliability of the analysis and identify excessively loose or tight standards and associated cost figures.

"An Algorithm: The Use Of Sequential Sampling in Work Measurement Audits" by R.A. Bascom, G.S. Kalemkarian, and J.L. Miller III presents a statistical technique for sample size selection in work measurement audits when costs must be kept to a minimum without compromising statistical accuracy. The authors' sequential sampling technique is statistically rigorous, requires fewer samples than fixed-size sampling, takes half the time of conventional auditing, and is accepted by the Department of Defense for the manufacture of military hardware. The method also provides an early indication when the samples are inclined toward accept or reject. The article includes the necessary statistical equations and graphing methods to perform sequential sampling, and identifies the considerations involved in correctly selecting parameters.

"Work Measurement Standards Accuracy and Audit" by William F. Fielder, Jr., provides a thorough overview of the topic. The article provides examples of methods for statistically evaluating the results of a standards audit. It also offers guidelines for establishing accuracy requirements in a variety of circumstances.

An industrial engineer picks the five jobs nearest his office and examines the standards for accuracy. He discovers they average 20 percent loose. What conclusions can he draw regarding the overall accuracy of standards in his plant? Most likely, "none."

G. SALVENDY and G. P. MCCABE, Purdue University, West Lafayette, Indiana

Auditing standards by sample

Inaccuracies in work standards can have profound negative effects upon the operation of a plant, from the points of view of both management and workers. Since standards are used for deriving costs of products and services, errors can result in financial loss through underpricing or lack of competitiveness in the market through overpricing. Employees who are unable to meet excessively tight standards can become poorly motivated while those with excessively loose standards may consistently underproduce.

An audit of work standards can provide useful information regarding the accuracy of the standards at a plant. Of course there are inaccuracies with work standards which an audit cannot identify (box). By

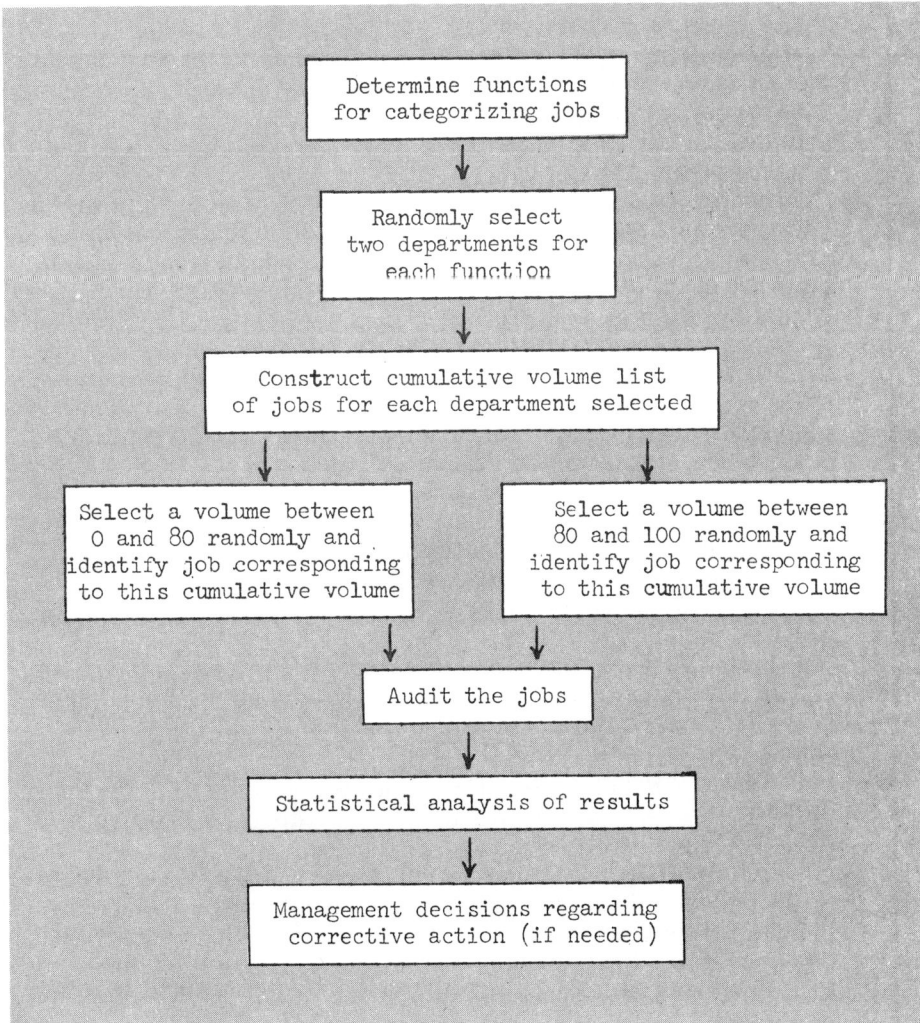

Flow chart for an audit.

using the methods decribed in this article, a measure of the overall accuracy of standards can be obtained and, in addition, specific problem areas may be identified. Corrective action can then be taken to adjust the standards, thereby stimulating worker motivation and productivity, providing more accurate pricing criteria, and increasing profits.

Statistical methodology

Since the number of work standards in many industrial situations is often very large, a complete audit of standards is usually impractical and too costly to be performed on a regular basis. In such situations, however, statistical techniques can effectively be used to assess the overall accuracy of standards and to pinpoint particular problem areas where corrective action is needed.

To design and analyze the audit, the information described in Table I must first be collected. Basically, an analysis of variance methodology is employed. For clarity of presentation, a particular case is discussed. Minor modifications would be required for other audits.

Selection of jobs for audit

Suppose that all jobs can be classified into one of six categories on the basis of industrial engineering considerations. These categories will be referred to as functions. Determination of the number and type of functions should be considered carefully. Similar jobs should be classified in the same function.

Within each function, it is assumed that there are a number of departments and within each department, a number of jobs. One can obtain adequate information by sampling two jobs within each department. A major difficulty which arises in the applicability of the results of an audit which uses randomly sampled jobs is related to the fact that all jobs do not have equal importance in terms of their individual contribution to the overall accuracy of standards. In particular, high-volume jobs should be given more importance (or weight) than low-volume jobs. This factor is incorporated into the statistical design and analysis by randomly sampling one high-volume and one low-volume job for each department selected.

In a number of cases, approximately 80 percent of the volume is accounted for by 20 percent of the jobs, while the remaining 20 percent of the volume is accounted for by the remaining 80 percent of the jobs. Thus, a high-volume job is one in the 80 percent of volume category while a low-volume job is one that falls in the 20 percent of volume category.

Statistical analysis of data

The analysis of variance model, Reference [2], for analyzing the

---

Dependability of work standards

Dependability encompasses reliability, homogeneity, constant and random errors, and change effects associated with the measure being investigated rather than individual performance studies.

Work standards are never 100 percent accurate. Both the method used for determining the standard and the actual production circumstances limit the dependability of work standards. Some sources of error are inherent in the entire process while others are avoidable and such errors can be corrected if properly diagnosed.

The following factors can limit the dependability of work standards:
1. Production variables.

Organizational variability in quality and supply of material; layout, mechanization and speed of machinery; effectiveness of supervision; working conditions; and recording of quality and quantity.

Individual variability in worker competence, worker morale, and worker's personal circumstances.

Of course, there is a certain amount of variation in production output which depends upon other factors. Correlations of individual production outputs between different time periods, e.g. odd and even weeks, are measures of the reliability of production output. Typical values vary from 0.7 to 0.9, Reference [1]. Indeed, it is not unusual to find worker performance varying as much as ten percent within a given working day.

2. Dependability of allowances for fatigue and unavoidable delays.
3. Grouping effect for work standards based on predetermined motion time system (PMTS); if a small number of elements constitute a work standard, errors due to "round-off" can occur.
4. Additivity of elemental times for standards based on PMTS. For certain tasks the times may be sub- or supra-additive.
5. Transferability. Work standards are not uniformly applicable for all types of jobs with the same degree of accuracy. Generally, manipulative skills can be predicted more accurately than decision making activities.
6. Personnel determining the work standard. Not only do individuals vary in skills but also an individual varies in accuracy from job to job.
7. Operator performance during time period when work standard is determined.
8. Analyst reliability. Even if the same analyst would study the same operator twice (with an elapse of time), small differences in determined standards would likely occur.
9. Variability in work place layout, tools, equipment, and design modifications. Of course, if substantial changes occur in this area, standards should be recalculated. However, often such changes take place so rapidly that a small industrial engineering staff is unable to keep modifying old standards while new jobs are continually being introduced.

above design is:

$$Y_{ijk} = \mu + \alpha_i + \beta_j + \gamma_{k(j)} + (\alpha\beta)_{ij} + (\alpha\gamma)_{ik(j)}$$

where
- $i = 1, 2$
- $j = 1, \ldots, 6$
- $k = 1, 2$
- μ is the overall mean
- α_i is the effect of the low (20, $i=1$) volume items versus the high (80, $i=2$) volume items
- β_j is the effect of the jth function
- $\gamma_{k(j)}$ is the effect of the kth department nested within the jth function
- $(\alpha\beta)_{ij}$ is the interaction effect of the ith level of volume with the jth function
- $(\alpha\gamma)_{ik(j)}$ is the interaction of the ith volume level with the kth department (within function j)
- Y_{ijk} is the value of the measured variable for the job corresponding to indices (i, j, k).

The α and β (and hence $\alpha\beta$) terms are considered to be fixed factors, while the γ term (and hence $\alpha\gamma$) is considered to be random.

The outline of the analysis of variance model is illustrated in Table II. Thus, for constructing F-tests, departments are used to test the mean and functions while the 80-20 by departments interaction is used to test the 80-20 effect and the 80-20 by functions interaction.

Interpretation

The measured variable for the analysis is the percentage improvement, which is defined to be $100(x_1 - x_2)/x_1$ where x_1 is the current standard and x_2 is the audited standard. Of primary interest, of course, is the estimation of the mean percentage improvement for each function and the translation of this quantity into dollars. This is accomplished by the following procedure. For each department, calculate a weighted average of the percentage improvements for the two jobs. Weights are determined by the dollar volume for the jobs. Thus, if the jobs correspond to $148 and $886, then the weights would be

1. Yearly wages and hours worked for each function.
2. Yearly wages and hours worked for each department – ordered from largest to smallest volume within each function.
3. Yearly wages and hours for each job that has a work standard – ordered from largest to smallest volume within each department selected for audit.
4. Average wage rate for each function.

Table I. Preparatory information required to audit work standards.

Source	Degrees of Freedom	Expected Mean Square
Mean	1	$24\sigma_\mu^2 + 2\sigma_\gamma^2$
80-20	1	$12\sigma_\alpha^2 + \sigma_{\alpha\gamma}^2$
Functions	5	$4\sigma_\beta^2 + 2\sigma_\gamma^2$
Departments	6	$2\sigma_\gamma^2$
80-20 by functions	5	$2\sigma_{\alpha\beta}^2 + \sigma_{\alpha\gamma}^2$
80-20 by departments	6	$\sigma_{\alpha\gamma}^2$
Total	24	

Table II. Analysis of variance model for auditing work standards.

0.14 and 0.86 respectively. For each function, the two department values are weighted similarly to determine the function average.

Calculation of the standard deviations for function averages is somewhat complicated. After pooling, if possible, estimates of the variance components σ_γ^2 and $\sigma_{\alpha\gamma}^2$ are determined. The estimated standard deviation for a function average is given by

$$\sqrt{(\hat{\sigma}_\gamma^2 + \hat{\sigma}_{\alpha\gamma}^2) \sum_{jk} W_{ij}^2}$$

where the w_{jk} are the weights assigned to the four jobs corresponding to the function.

The conversion from percentage improvement to yearly dollars is accomplished by multiplying the percentage improvement for each function times the product of yearly hours and average hourly wages. The standard deviation is, of course, also multiplied by this factor.

With the estimated means and standard deviations, confidence bounds for any given confidence coefficient can be calculated. A device for concisely presenting this information, which has been found useful by the authors, is to construct a table with means and standard deviations for each function and confidence limits for the 1, 5, 10, 20,

Function	Mean	Standard Deviation	1%	5%	10%	20%	50%	80%	90%	95%	99%
Sub-assembly	-3.93	11.42	-39.79	-26.08	-20.37	-14.27	-3.93	6.42	12.52	18.23	31.93
Power Press Operation	11.57	9.09	-16.97	-6.06	-1.52	3.33	11.57	19.8	24.65	29.19	40.09
Model assembly	-0.39	11.59	-36.77	-22.87	-17.07	-10.89	-0.39	10.11	16.30	22.09	35.99
Painting and finishing	-7.47	11.19	-42.62	-29.19	-23.59	-17.61	-7.47	2.67	8.65	14.24	27.67
Inspection	-8.74	9.71	-39.24	-27.59	-22.73	-17.54	-8.74	0.06	5.25	10.11	21.76
Testing	33.96	10.81	0.03	12.99	18.40	24.17	33.96	43.75	49.52	54.93	67.89
	6.11	4.59	-8.31	-2.81	-0.51	1.95	6.11	10.27	12.72	15.02	20.53

*This table can be used to calculate one or two-sided confidence intervals. For example, a two-sided 90% confidence interval for the sub-assembly percentage improvement is (-26.08%, 18.23%). See Sec. 2.3 for definition of percentage improvement.

Table III. Confidence limits for percentage improvements. This table can be used to calculate one- or two-sided confidence intervals. For example, a two-sided, 90 percent confidence interval for the subassembly percentage improvement is (-26.08 percent, +18.23 percent).

Function	Mean	Standard Deviation	1%	5%	10%	20%	50%	80%	90%	95%	99%
Sub-assembly	-11,723	34,088	-118,761	-77,855	-60,811	-42,608	-11,723	19,161	37,364	54,408	95,315
Power Press Operation	156,072	119,074	-217,820	-74,931	-15,394	48,191	156,072	263,952	327,538	387,075	529,963
Model assembly	-1,017	30,488	-96,751	-60,165	-44,921	-28,640	-1,017	26,605	42,886	58,130	94,716
Painting and Finishing	-37,073	15,009	-214,218	-146,729	-118,596	-88,550	-37,573	13,403	43,450	71,582	139,101
Inspection	-72,159	80,176	-323,914	-227,702	-187,614	-148,513	-72,160	480	43,294	83,382	179,594
Testing	198,180	63,003	351	75,955	107,501	141,099	198,180	255,260	288,903	320,404	396,008
	231,777	172,724	-310,576	-103,308	-16,946	75,289	231,777	388,265	480,499	566,861	774,130

*This table can be used to calculate two or one-sided confidence intervals. For example, a two-sided 90% confidence interval for the sub-assembly savings dollars is (-$77,855, +$54,408). Alternatively, there is 90% confidence that the savings dollars for sub-assembly is less than or equal to $37,364 (one-sided).

Table IV. Confidence limits for dollar savings. This table can be used to calculate two- or one-sided confidence intervals. For example, a two-sided, 90 percent confidence interval for subassembly savings is (-$77,855, +$54,408). Alternatively, there is 90 percent confidence that the savings for subassembly is ≤ $37,364 (one-sided).

50, 80, 90, 95, and 99 percentage points.

The validity of a statistically based audit as described is highly dependent upon a variety of factors. Of course, the competence of the individual performing the audit is crucial.

Precautions

To prevent bias, the auditor should not have access to present information regarding the work standard. A completely new determination of the work standard is required.

Since jobs are generally audited in a sequence, the order for auditing should be randomized.

Benefits of auditing

The results of a statistically based audit provide estimates of the accuracy of work standards for each function along with confidence limits. Thus, the procedure itself provides information regarding the reliability of the results.

Functions with excessively loose or tight standards can be identified and corrective action taken. As a result of this corrective action, the following benefits can be realized:
- More accurate costing for products and services.
- Improved production through increased worker motivation and reduction of excessive lateness, absenteeism, labor turnover and grievances.
- More equitable financial incentives
- More realistic production planning and control.

Cost of audit

We have found that, typically, 10 to 15 man days are required to design and analyze an audit.

The methodologies for auditing work standards described were applied to a number of work situations. The hypothetical case study of company XYZ is illustrative of our experience in this area.

Example

Company XYZ employs 3,000 hourly workers of which 2,500 are working on measured work standards utilizing a predetermined motion time system. The company has a total of 7,000 active work standards. Twenty percent of these work standards account for 80 percent of the work volume. The major functions of company XYZ may be classified as: subassembly, model assembly, power press operation, painting and finishing, inspection, and testing.

Following the process outlined above, a total of 24 work standards (four from each function, two from each department) were audited. The results of this audit are presented in Tables III and IV.

Table III illustrates the confidence probability limits for average looseness or tightness of work standards by function. For example, the work standards in the power press and sheet metal function are, on average, 11.57 percent loose. There is a five percent probability that the standards in this function average at least 6.06 percent tight and there is also a five percent probability that the standards in the same function are more than 29.19 percent loose. In other words, the probability that the average is between 6.06 percent tight and 29.19 percent loose is 90 percent. In some situations it may be the case that these confidence intervals are too large for making effective management decisions; this large variance of the data may be markedly reduced by increasing the sample size of the audit. The data portrayed in this table clearly indicate that the greatest inaccuracy in work standards exists at the testing function, where, on the average, the standards are 33.96 percent loose.

The extent to which money could be saved by the company by having more accurate work standards is portrayed in Table IV. This indicates that $231,777 may be saved annually by correcting all the work standards. Of this savings, $198,180 can be saved by correcting only the standards of the testing function.

A full discussion of all the results of an audit such as this would, of course, be lengthy and is not presented here. For example, a much more detailed analysis of loose and tight standards should be routinely performed.

References

[1] Salvendy, G., and W.D. Seymour, *Prediction and Development of Industrial Work Performance,* John Wiley & Sons, Inc., New York, New York, 1973.
[2] Anderson, Virgil L., and Robert A. McLean, *Design of Experiments,* Marcel Dekker, New York, New York, 1974.

Gavriel Salvendy is an Associate Professor of industrial engineering at Purdue University. Previously, he was an Assistant Professor of industrial engineering at State University of New York-Buffalo, a research associate in the department of engineering production at the University of Birmingham, England, and head of management services at Gunson Sortex Ltd., England. He has been a consultant to several corporations on industrial productivity and quality of working life.

Dr. Salvendy holds an MS in work design and ergonomics and a PhD in engineering production, both from the University of Birmingham, England. A Senior Member of AIIE, Dr. Salvendy is the 1973 recipient of the Phil Carroll Award for Outstanding Achievement in Work Measurement and Methods Engineering by the Work Measurement and Methods Engineering Division of AIIE.

George P. McCabe, Jr. is Associate Professor of statistics and Head of Statistical Consulting at Purdue University. He holds a PhD in mathematical statistics from Columbia University. He has published numerous articles on applied and theoretical statistics. As a consultant, Dr. McCabe has worked for various corporations as well as the Federal Energy Administration.

AN ALGORITHM: THE USE OF SEQUENTIAL SAMPLING IN WORK MEASUREMENT AUDITS

R. A. Bascom
General Dynamics' Fort Worth Division

G. S. Kalemkarian
Deere and Company

J. L. Miller III
General Dynamics' Fort Worth Division

INTRODUCTION

A basic problem in any comprehensive work measurement system is the cost-effectiveness of its auditing procedure for verifying the accuracy of (or determining) the data base for the work standards.

Although several very good approaches to determining accuracy have been proposed in the last few years, they have all assumed the use of fixed-size sampling. The trade literature has emphasized auditing criteria (size, frequency, costs) for fixed-size sampling plans.

Since the resources expended in inspection sampling are generally minimal, fixed-size sampling is appropriate in most inspection situations. Fixed-size samples have long been relied on by proponents of quality control systems. Indeed, most quality control textbooks teach fixed-size sampling techniques only.

However, when resources are scarce, fixed-size sampling is not cost-effective. For such situations, a less cost-consuming technique was needed to evaluate the population under consideration without compromising statistical rigor and reliability. Such a method has been found in the application of sequential sampling for work measurement audits.

Through the use of a sequential sampling algorithm, an auditing procedure has been developed and implemented that

o Is statistically and mathematically rigorous

o Requires significantly fewer samples than fixed-size sampling for the same accuracy and confidence levels

o Has been accepted by the USAF, after considerable scrutiny and testing by Department of Defense experts in the field of statistics and reliability.

SEQUENTIAL SAMPLING

Most hypothesis testing methods are based on the assumptions that a sample of fixed size is to be taken and that a choice is to be made in favor of one of two possible decisions after all of the samples have been taken. The number of observations required by the procedure is determined in advance of the experiment or test.

Sequential sampling, on the other hand, operates on the basis that samples are to be taken one at a time. The information thus generated is accumulated and evaluated as each additional sample is taken. As one might expect, the sequential accumulation-of-information procedure requires considerably less sampling, on the average, than the equally reliable fixed-size sampling methods. In fact, the sequential sampling procedure
  "frequently results in a savings of about 50 percent in the number of observations over the most efficient test procedure based on a fixed number of observations" [1].

The advantage of the sequential approach over the fixed-sample-size approach lies in the capacity of the sequential method to provide an <u>early</u> decision when the samples are extremely inclined toward either the "accept condition" or the "reject condition."
  "This ability to arrive at an early decision can be very useful in such fields as sampling inspection, where it is not uncommon for lots to be very bad when they are bad or very good when they are good" [2].

When samples are difficult to obtain, the savings in sampling cost will outweigh the added computations required by the sequential sampling method. Furthermore, it is not necessary to derive the density function of a statistic such as t or F in order to use sequential sampling. The sample size needed, n, becomes a random variable.

APPLICATION OF SEQUENTIAL SAMPLING TO AUDITING

Sequential sampling is a means to test the true percentage of standards meeting some predetermined accuracy requirement. The requirement for the percentage of standards meeting the accuracy criteria is set to equal the confidence level.

Conceptually, the use of sequential sampling as an audit/validation procedure in work measurement is rooted in the literal statement of the null hypothesis, as follows:

A time value (whether engineered or estimated) will, when compared to a standard value for the same element, operation, or task component, be within $\pm x\%$ of that standard value at least $y\%$ of the time.

The $\pm y\%$ is the inaccuracy permitted by the system. Time values either meet or fail the accuracy desired which makes the audit a series of Bernoulli trials. The $y\%$ is the percentage of samples which must meet the audit criteria, that is, fall within the $\pm x\%$ limits. That is, we want to ensure that at least $y\%$ of the time, values fall within the bounds permitted if we accept the null hypothesis (subject, however, to sampling risks).

Another way of stating the basic hypothesis is:

$y\%$ of the time our individual time values are within $\pm x\%$ of the standard value.

The $\pm x\%$ allowable inaccuracy is calculated each time values are compared. It is unique to each standard, or compared value, e.g.,

Given: Accuracy = $\pm 25\%$
Original value = 2.00 minutes
Audit value = 1.95 minutes

(1) $\pm x\%$ accuracy = $\pm 25\%$ x 1.95 minutes
Lower limit = 1.4625
Upper limit = 2.4375

Therefore, the value must be between 1.4625 minutes and 2.4375 minutes to satisfy the accuracy criteria. The value of 2.00 obviously is between the two values and therefore satisfies the criteria.

If the audit value had been 3.00 minutes with an original value of 2.00 minutes, then

(2) $\pm x\%$ accuracy = $\pm 25\%$ x 3.00 minutes
Lower limit = 2.25 minutes
Upper limit = 3.75 minutes

In this case, the original value does not lie between the limits and must be "rejected" as not meeting the criteria.

A basic, underlying assumption of this system is that in reality, events, parts, tools, data, etc., tend to be either bad or good and not in between. Well-trained, experienced engineers will tend to make good studies and produce good standards and estimates. In the same vein, bad tools produce bad parts, and poorly-trained, inexperienced employees tend to produce poor work. Therefore, a major advantage of sequential sampling is the ability to give an early indication of the condition of the sample (acceptable/unacceptable).

GENERAL PROCEDURE

In the application of sequential sampling to auditing work measurement data, the null hypothesis would be H_0: $p = p_0$, where p is the true percentage of data meeting the criteria and p_0 corresponds to the required confidence level. The alternate hypothesis would then be H_1: $p = p_1$, where p_1 is some value less than p_0.

With the null and alternate hypotheses stated, the next step is to set the levels of risk to be assumed. In sequential sampling, the risks are associated with two types of errors, which can be stated explicitly and can be at different levels (see Figure 1). A Type I error occurs when a reject decision is made when, in fact, the given criteria have been met. A Type II error occurs when an accept decision is made when, in fact, the given criteria have not been met. Type I errors are associated with the producer's risk while Type II errors are associated with the consumer's risk. The levels of risk assumed, generally, range between 5% and 20%.

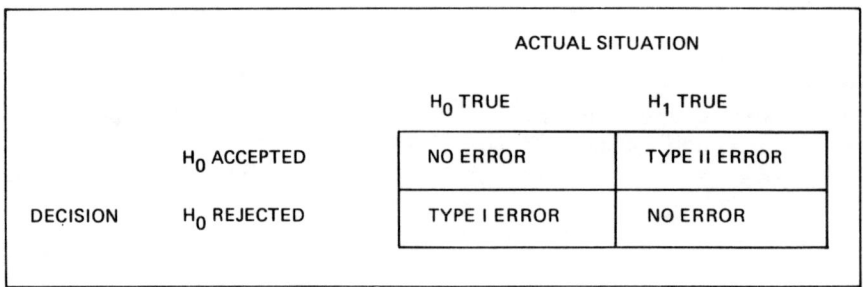

Figure 1 SAMPLING ERROR

With the four parameters described above: p_0, p_1, producer's risk level, and consumer's risk level, the acceptance and rejection lines used in sequential sampling can be plotted. With the acceptance and rejection lines plotted, sampling can begin. Each sample meeting the accuracy criteria is given a value of one, and each sample failing to meet the criteria is given a value of zero.

The sequential sampling process is illustrated in Figures 2 and 3. As each sample is taken, the sum of the samples meeting the criteria is plotted versus the total number of samples taken up to that point. Sampling is continued until this plot intersects either the acceptance or rejection line. When the plot intersects either line, the corresponding decision is made and sampling is terminated. If a large number of samples have been taken and no decision has been reached, sampling may be terminated by relaxing the producer's and consumer's risks and forcing the plot to intersect one of the decision lines.

ACTUAL APPLICATION

The purpose of the sequential probability ratio test is to test the hypothesis H_0: $p = p_0$ against the alternative H_1: $p = p_1$. To do this, we must first define the variables involved as follows.

Definitions:

n = Number of observations taken
p = True proportion of the population
p_0 = Proportion to be tested for
p_1 = Alternate proportion to be tested against
α = Probability of a Type I error, which is rejection of H_0 when H_0 is actually true

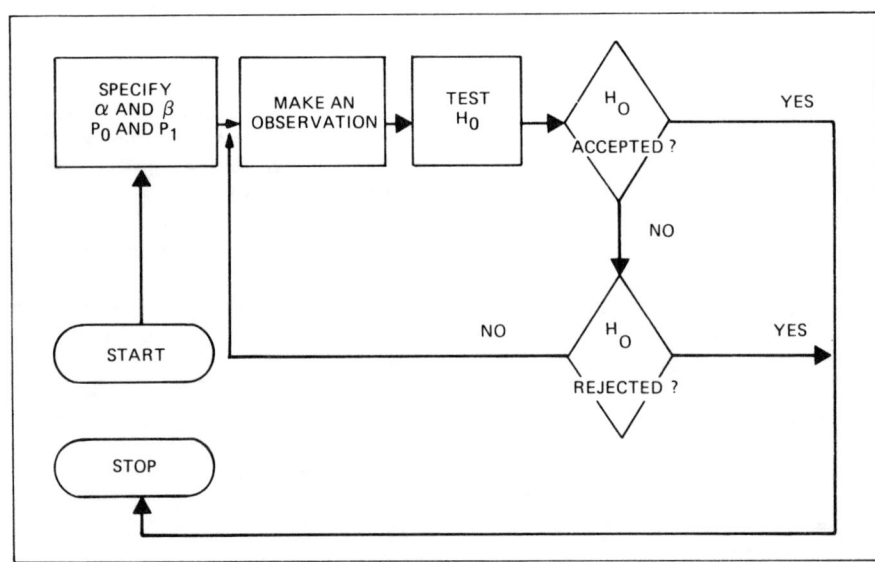

Figure 2 THE SEQUENTIAL SAMPLING PROCEDURE

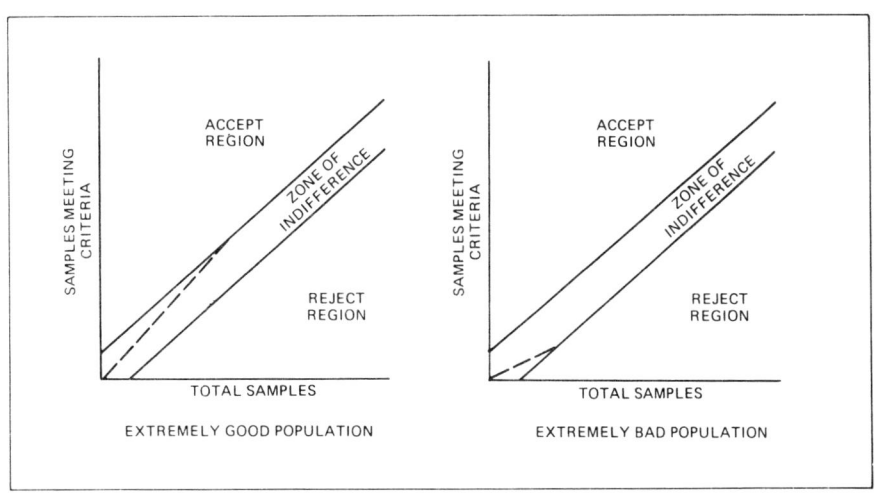

Figure 3 EXAMPLES OF SEQUENTIAL SAMPLES

β = Probability of a Type II error, which is acceptance of H_0 when H_1 is actually true

$T = \sum_{i=1}^{n} X_i$, where X_i = number of successes (failures).*

$h_o = -\ln \left| (1-\alpha)/\beta \right| / \ln\left[\frac{p_1}{p_0} \left(\frac{1-p_0}{1-p_1} \right) \right]$ = y intercept of the acceptance line

$h_1 = \ln \left| (1-\beta)/\alpha \right| / \ln\left[\frac{p_1}{p_0} \left(\frac{1-p_0}{1-p_1} \right) \right]$ = y intercept of the rejection line

$S = \ln \left| (1-p_0)/(1-p_1) \right| / \ln\left[\frac{p_1}{p_0} \left(\frac{1-p_0}{1-p_1} \right) \right]$
= slope of acceptance and rejection lines

Procedure:

A. Select an appropriate p_1. If H_0 is $p \geq p_0$, choose $p_1 < p_0$. If H_0 is $p \leq p_0$, choose $p_1 > p_0$.

B. Select appropriate α and β values.

C. Calculate h_0, h_1, and s and plot the accept and reject lines.

Accept line = $\dfrac{-\ln \left| (1-\alpha)/\beta \right|}{\ln\left[\frac{p_1}{p_0} \left(\frac{1-p_0}{1-p_1} \right) \right]}$ +

$\dfrac{(n)\ln \left| (1-p_0)/(1-p_1) \right|}{\ln\left[\frac{p_1}{p_0} \left(\frac{1-p_0}{1-p_1} \right) \right]}$ =

$h_0 + sn$

Reject line = $\dfrac{\ln \left| (1-\beta)/\alpha \right|}{\ln\left[\frac{p_1}{p_0} \left(\frac{1-p_0}{1-p_1} \right) \right]}$ +

$\dfrac{(n)\ln \left| (1-p_0)/(1-p_1) \right|}{\ln\left[\frac{p_1}{p_0} \left(\frac{1-p_0}{1-p_1} \right) \right]}$ =

$h_1 + sn$

The area between these lines represents the zone of indifference. The areas outside the acceptance and rejection lines represent the acceptance and rejection areas, respectively.

D. Begin sampling. At each sample, plot T versus n. Continue sampling until a decision to either accept or reject the population is made or until sampling is truncated. A decision to accept is made as soon as the sampling plot intersects the acceptance line. A decision to reject is made as soon as the sampling plot intersects the rejection line. If a large number of samples have been taken and the sample plot is still in the zone of indifference, sampling can be truncated and a decision reached by relaxing α and β.

E. When sampling is terminated at sample n_0, the decision rules are modified as follows:

---

*If p_0 is chosen as the desired proportion, X_i = the number of successes. If p_0 is chosen as one minus the desired proportion, X_i = the number of failures.

Accept H_0 if $h_0 + sn_0 > T \geq sn_0$
Reject H_0 if $sn_0 > T > h_1 + sn_0$.

F. Under the above decision rules, the upper limits for the revised α and β, $\alpha(n_0)$ and $\beta(n_0)$, respectively, are defined by the following equations:

$\alpha(n_0) \leq \alpha + G(\omega_2) - G(\omega_1)$

$\beta(n_0) \leq \beta + G(\omega_4) - G(\omega_3)$

where:

$E_0[Z] = p_0 \ln\left[\dfrac{p_1(1-p_0)}{p_0(1-p_1)}\right] + \ln\left[\dfrac{1-p_1}{1-p_0}\right]$

$E_1[Z] = p_1 \ln\left[\dfrac{p_1(1-p_0)}{p_0(1-p_1)}\right] + \ln\left[\dfrac{1-p_1}{1-p_0}\right]$

$\sigma_0(Z) = \left\{\left[\ln\left(\dfrac{p_1(1-p_0)}{p_0(1-p_1)}\right)\right]^2 (p_0 - p_0^2)\right\}^{1/2}$

$\sigma_1(Z) = \left\{\left[\ln\left(\dfrac{p_1(1-p_0)}{p_0(1-p_1)}\right)\right]^2 (p_1 - p_1^2)\right\}^{1/2}$

$A = \dfrac{1-\beta}{\alpha} \qquad B = \dfrac{1-\alpha}{\beta}$

$\omega_1 = \dfrac{-n_0 E_0[Z]}{(n_0)^{1/2} |\sigma_0(Z)|}$

$\omega_2 = \dfrac{\ln A - n_0 E_0[Z]}{(n_0)^{1/2} |\sigma_0(Z)|}$

$\omega_3 = \dfrac{\ln B - n_0 E_1[Z]}{(n_0)^{1/2} |\sigma_1(Z)|}$

$\omega_4 = \dfrac{-n_0 E_1[Z]}{(n_0)^{1/2} |\sigma_1(Z)|}$

$G(\omega)$ represents the cum area under a standard normal curve from $-\infty$ to ω.

PARAMETER SELECTION

Parameter selection for sequential sampling is done subjectively, considering the costs involved and the levels of risk allowable. The value of p_0 is generally fixed, leaving p_1 and the levels of risk to be selected.

The closer the value of p_1 is to p_0, the greater the expected number of samples becomes. The expected number of samples increases due to the increased level of discrimination required.

Lower levels of risk also require a greater expected number of samples. The values of p_0, p_1, and the levels of risk are generally selected in conjunction with one another. Higher levels of risk are usually assumed when p_1 is very close to p_0 since the magnitude of the possible errors is small. The opposite is true when p_1 is quite different from p_0, say 60% versus 80%.

The relationships between p_0, p_1, and the expected number of samples to be taken is demonstrated in Figure 4. The ability of the method to differentiate between acceptable and unacceptable samples is plotted in

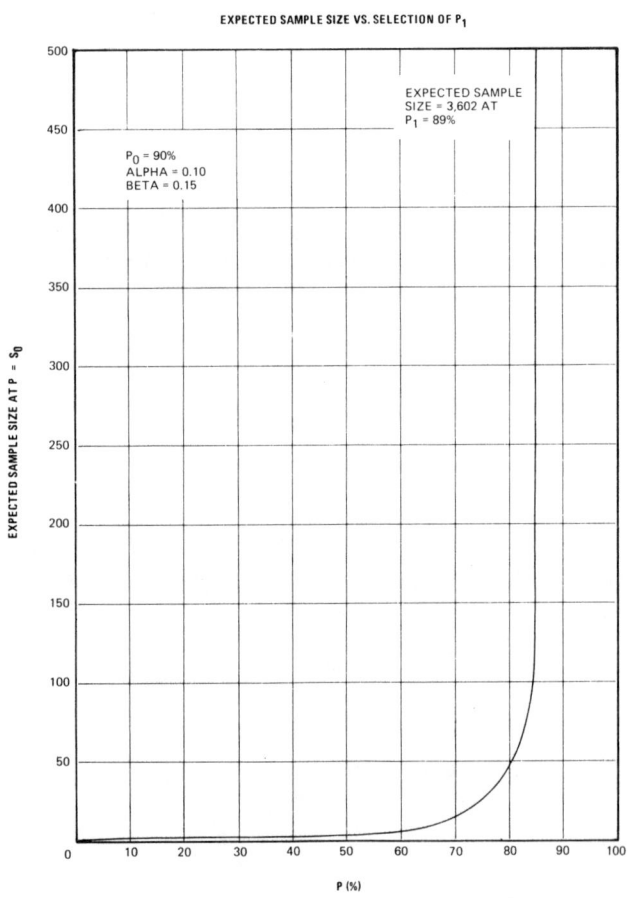

Figure 4 EXPECTED SAMPLE SIZE VS SELECTION OF P_1

Figure 5 for several values of p_1. Note from Figure 5 that the point of diminishing returns is soon reached. Sequential sampling variables for several parameter combinations are listed in Table 1 to show the effects of the levels of risk assumed on the expected number of samples to be taken.

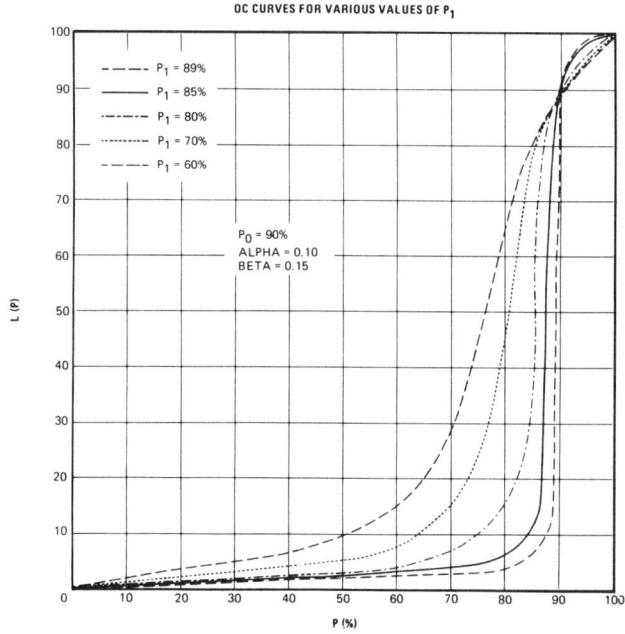

Figure 5 OC CURVES FOR VARIOUS VALUES OF P_1

In the application of sequential sampling to auditing work measurement, the percentage of samples required to meet the given criteria is generally fixed. The selection of the remaining three sequential sampling parameters is determined by subjectively considering the intended application.

Note from Figures 4 and 5 and Table 1 that the selection of parameters has a great effect on the number of samples required and also on the level of discrimination of the audit procedure.

CONCLUSIONS

The development and application of this algorithm has demonstrated an invaluable use for sequential sampling. It represents a substantial step forward in the area of work measurement audits. The procedure, which is straightforward and statistically rigorous, will undoubtedly have many additional applications.

However, the ultimate cost-effectiveness (hence, the usefulness) of the technique is critically dependent upon the willingness of the producer and the consumer to take a reasonable risk of making a wrong decision. (Any sampling procedure carries this innate risk.) Selection of most of the parameters is purely subjective and must be negotiated. If a recalcitrant attitude by either party exists toward the selection of the

TABLE I SEQUENTIAL SAMPLING PARAMETER COMPARISONS

P_0	P_1	α	β	H_0	H_1	S	ASN at S	MINIMUM ACCEPT	MINIMUM REJECT
0.95	0.85	0.15	0.15	1.4337	-1.4337	0.9081	24.62	15.60	1.58
0.95	0.85	0.10	0.10	1.8161	-1.8161	0.9081	39.51	19.76	2.00
0.95	0.85	0.05	0.05	2.4337	-2.4337	0.9081	70.95	26.48	2.68
0.92	0.87	0.15	0.15	3.2040	-3.2040	0.8968	110.90	31.05	3.57
0.92	0.87	0.10	0.10	4.0585	-4.0585	0.8968	177.95	39.33	4.53
0.92	0.87	0.05	0.05	5.4387	-5.4387	0.8968	319.56	52.70	6.06
0.90	0.85	0.15	0.15	3.7495	-3.7495	0.8764	129.83	30.34	4.28
0.90	0.85	0.10	0.10	4.7495	-4.7495	0.8764	208.31	38.43	5.42
0.90	0.85	0.05	0.05	6.3647	-6.3647	0.8764	374.09	51.49	7.26
0.90	0.80	0.15	0.15	2.1390	-2.1390	0.8548	36.85	14.73	2.50
0.90	0.80	0.10	0.15	2.2095	-2.6390	0.8548	46.97	15.22	3.09
0.90	0.80	0.10	0.10	2.7095	-2.7095	0.8548	59.13	18.66	3.17
0.90	0.80	0.50	0.0001	10.5030	-0.8546	0.8548	72.30	72.33	1.00
0.90	0.80	0.05	0.05	3.6309	-3.6309	0.8548	106.19	25.01	4.25
0.90	0.80	0.05	0.10	2.7762	-3.5643	0.8548	79.70	19.12	4.17
0.90	0.80	0.05	0.15	2.2762	-3.4938	0.8548	64.06	15.68	4.09
0.90	0.80	0.01	0.01	5.6665	-5.6665	0.8548	258.63	39.03	6.63
0.90	0.80	0.01	0.10	2.8270	-5.5489	0.8548	126.36	19.47	6.49
0.90	0.80	0.01	0.15	2.3270	-5.4785	0.8548	102.69	16.03	6.41
0.90	0.70	0.10	0.10	1.6277	-1.6277	0.8138	17.49	8.74	2.00

parameters, a situation will develop in which the effectiveness and use (if any) of the algorithm is doubtful.

Regardless of the parameter values chosen, sequential sampling generally results in significantly fewer samples being taken than with an equally rigorous fixed-size sampling plan. When the cost of resources required to carry out the auditing procedure is high, the savings potential of sequential sampling far outweighs the added computational requirements.

REFERENCES

1. Wald, Abraham, Sequential Analysis, John Wiley & Sons, New York, 1947.

2. Hoel, Paul G., Introduction to Mathematical Statistics, Fourth Edition, John Wiley & Sons, New York, 1971.

BIBLIOGRAPHY

Duncan, A. J., Quality Control and Industrial Statistics, Third Edition. Richard D. Irwin, 1965.

Mann, N. R., Schafer, R. E., and Singpurwalla, N. D. Methods for Statistical Analysis of Reliability and Life Data. John Wiley and Sons, New York, 1974.

Statistical Inference. U.S. Army Management Engineering Training Activity, Rock Island, Illinois.

BIOGRAPHICAL SKETCHES

R. A. Bascom, Industrial Engineer in the Industrial Engineering Department, General Dynamics' Fort Worth Division. BS, MS Kansas State University. He is currently working on a Ph.D. in Administration at the University of Texas at Arlington and is a senior member of AIIE.

G. S. Kalemkarian, Reliability Engineer, Deere and Company, Moline, Illinois. (He was with AMETA, Rock Island Arsenal, Illinois, at the time this algorithm was developed.) BA in Mathematics, Temple University. MA in Statistics, University of Iowa. He is a member of the American Society for Quality Control.

J. L. Miller, Production Management Specialist in the Industrial Engineering Department, General Dynamics' Fort Worth Division. BA, University of Texas at El Paso; MS, Industrial Engineering, Kansas State University; MBA, University of Texas at Arlington; ABD, Ph.D. in Administration at University of Texas at Arlington. He is a senior member of AIIE and Vice President of Fort Worth Chapter of AIIE.

WORK MEASUREMENT STANDARDS ACCURACY AND AUDIT

William F. Fielder, Jr., P.E.
Chairman - Industrial Engineering Corporate Advisory Group
Hughes Aircraft Company
El Segundo, California

ABSTRACT

Accuracy of work measurement standards has been a critical issue among industrial engineers since Professor Robert Franklin Hoxie made his report to a U.S. Commission on Human Relations in 1914. With the advent of predetermined time systems, computerization of work measurement, and the MIL-STD 1567A-Work Measurement, the issue has become more complex -- but no less controversial [1].

In the decade of the 1950s, a transition away from the classical application of work measurement systems to commercial, mass-production, high-volume, high-rate, short-cycle manual work began. At the present time, most of this type of work has either been automated out of existence or exported to third-world countries.

Current work measurement applications in most cases must be adapted to an entirely different environment. Today's applications typically must accommodate customized-production, low-volume, low-rate, long-cycle, man/machine work in a highly dynamic environment.

Fortunately, work-measurement technology has kept pace with the changing environment of its application. This paper highlights some of the new industrial engineering concepts and techniques which empower our profession to meet the current and evolving challenges cost-effectively and cost-beneficially.

THE MEANING OF "ACCURACY"

Prior to the 1950s, "standards accuracy" was a general term which meant that work measurement standards could be relied upon for such demanding purposes as wage-incentive payments -- a typical application of work measurement. The classical expression for "good standards accuracy" was ±5%.

With the advent of operations research and the application of statistical techniques to industrial engineering methodologies, the expression changed to "±5% at a 95% level of confidence".

Whatever the specific meanings such statistical terms may have had for practicing industrial engineers, our current understandings have become much more complex and analytical. No longer is there any single number, such as ±5%, which describes "good standards accuracy".

Today -- in light of the principles and techniques of statistics -- statements about standards accuracy (e.g., ±10% at a 90% confidence level) are statements of probability, based upon statistical inference, about a specific population of data based upon a sample randomly taken from that population.

MTM ACCURACY

Different MTM systems have different levels of accuracy, as shown in Figure 1. below. In addition, the accuracy of every MTM system -- actually every work measurement system -- varies inversely with the nonrepetitive cycle time (NRT) for the standard produced by such a system.

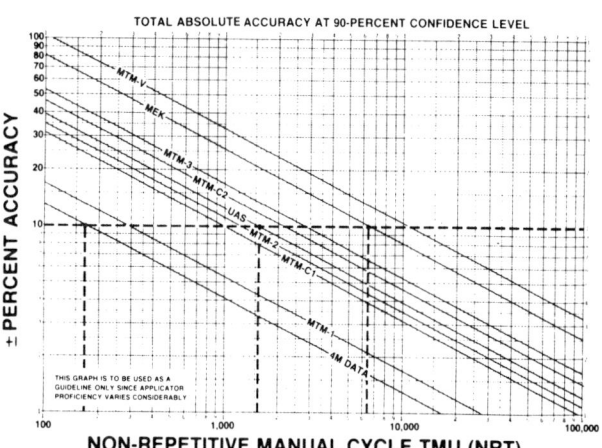

Figure 1. Accuracy of MTM Systems

This relationship of accuracy to non-repetitive cycle time derives from the operation of the law of large numbers. The larger the number of different manual motions combined within a standard, the less the percentage variation that is inherent in the resulting data.

MTM systems, concepts and techniques are being presented in this paper because they are the basis of work measurement systems within the Hughes Aircraft Company. Additionally, our active involvement in various MTM Association research projects has contributed to the understandings being presented. The basic principles involved, however, apply universally to any work measurement system.

DATA SIMPLIFICATION

Referring to Figure 1., two of the nine MTM systems represented on the chart demonstrate an important principle for the most economical application of work measurement. As shown, 4M Data is the most accurate MTM system available. It also reflects a relatively high level of detail required in its application.

On the other hand, MEK is a highly simplified predetermined time system which requires much less detailed information, and it requires only one-half to one-fourth the amount of time to apply. At the same time, as Figure 1. reflects, MEK has an absolute accuracy of ±25% at the same nonrepetitive cycle time at which 4M Data has an accuracy of ±4%.

This raises the obvious question: What level of accuracy is actually required in any particular application?

The answer to this question can permit substantial opportunities to simplify work measurement systems and reduce the cost for their development and maintenance. The principles involved not only guide the selection of specific MTM systems to be employed, but also apply to the level of simplification that is most economical in the construction of a company's own standard data system.

LEVEL OF ACCURACY NEEDED

The level of accuracy required in any particular application is determined by the use which will be made of the standards produced by the system. Two hypothetical examples will help demonstrate how accuracy requirements may be established.

Example 1. - Repetitive Operation

Generally speaking, standards which are used repetitiously require a higher level of accuracy than those which are used infrequently. An example of a repetitious usage would be an operation standard which covers all of the work of an employee for a complete week. At the end of the week the standard hours earned for the work completed will have identically the same ±% accuracy as that of the particular standard used. For example, if a particular standard had an accuracy of ±15%, the total standard hours earned for the week would also have an accuracy of ±15%.

Example 2. - Suboperation

On the other hand, a suboperation which represents only 10% of the total time for a complete operation may have an accuracy of ±15% while the level of accuracy of the complete operation will be substantially better. This is because the plus-and-minus errors of a number of different suboperation standards tend to offset each other. That is, one particular suboperation standard with a positive deviation will tend to offset the error of another suboperation standard which has a negative deviation. The interaction of such offsetting deviations can be predicted with statistical techniques.

Nomogram Solution

A series of statistical equations have been developed [2] which provide guidance for MTM system selection and the degree of standard data simplification which is permissible to meet specified standards-accuracy requirements. These equations in turn have been incorporated within the construction of the nomogram shown in Figures 2. and 3. below. This nomogram is included in the training manual for the MTM Association's course, "Principles of Standard Data Systems", and is utilized here to demonstrate the concepts involved.

The bold lines in Figure 2. reflect the determination of MTM system selection in Example 1. The nomogram is entered at 100% OF CYCLE TIME, on the left side of the chart. This reflects the fact that the operation standard accounts for 100% of the total cycle time for the work being measured.

The bold line extends diagonally to the vertical line for 10% TOTAL REQUIRED DEVIATION (R)%. This represents the accuracy level specified for the total cycle time. The bold line extends horizontally to the right through the scale on the vertical line in the middle of the chart. The 10% shown at this point represents the accuracy required for the specific standard involved. It is the same as the total required deviation for the reason explained for Example 1. above.

The right-hand half of the nomogram is a reproduction of Figure 1., ACCURACY OF MTM SYSTEMS. The vertical bold line at 1,000 TMU nonrepetitive cycle time represents the NRT for the standard involved in Example 1.

The circle shown at the intersection of the vertical and horizontal bold lines reflects the upper limit for the accuracy of the MTM system which may be utilized, achieving the level of standards accuracy required. In Example 1., 4M Data is required to be used. Neither the UAS nor MEK systems provide sufficient accuracy to satisfy the requirements of this particular application.

On the other hand, in the case of Example 2., the suboperation involved represents only 10% of the total cycle time, and for a suboperation time of 2,000 TMU nonrepetitive cycle time, any one of the three MTM systems may be used -- MEK, UAS or 4M Data.

Of special significance is the application time required to develop the standard. The numbers superimposed above each system-accuracy line represent the approximate number of hours required to apply the system for the nonrepetitive cycle time shown. In the case of Example 2., MEK is 0.3 hours, UAS is 0.6 hours, and 4M Data is approximately 0.9 hours*. Clearly, MEK would require the least time, assuming other requirements for its applicability are satisfied (i.e., methods - level determination -- a consideration outside the scope of this paper).

*Note: 4M Data application time varies between 0.3 and 1.2 depending upon the degree of multiple usage of existing elements on file. The 0.9 hours represents the specific situation in Example 2.

GENERALIZED GUIDELINES

The principles, concepts and techniques involved in cost-effectively designing and maintaining work measurement systems are complex and extensive. The subject of standards accuracy is fundamentally a central aspect of the work-measurement technology involved. This paper can reflect in only a cursory manner the principles, concepts and techniques involved.

However, a few general guidelines may be suggested here for consideration. It is hoped that they might lead the reader toward a more comprehensive consideration of the benefits possible from their application.

1. The larger the percentage of total time measured which a particular standard represents, the more accurate the standard must be to assure the overall level of accuracy of measurement required.

2. The more different standards which are included in the measurement of a particular operation, the greater the degree of data simplification which is permissible to achieve overall accuracy requirements.

3. The more dynamic the environment of a particular operation, the wider the range of variability of actual work methods required, and the wider the range of variation in time requirements for performing the work.

4. Where inherent variability of work methods is extensive, greater stability of the accuracy of overall measurement may be achieved through the use of moving averages to provide more reliable indications of trends and overall levels of performance.

5. Under any set of conditions, the more precise the level of accuracy of measurement, the greater the cost of measuring. Careful consideration must be given to the trade-off of benefits achievable from investments in industrial engineering effort for increased accuracy compared to the benefits achievable from increased effort in performance analysis and methods improvement.

6. For any level of standards accuracy required, investments in industrial engineering effort can be minimized by utilizing data-simplification techniques which are consistent with analytically determined standards accuracy requirements (e.g., as demonstrated in Example 2. above).

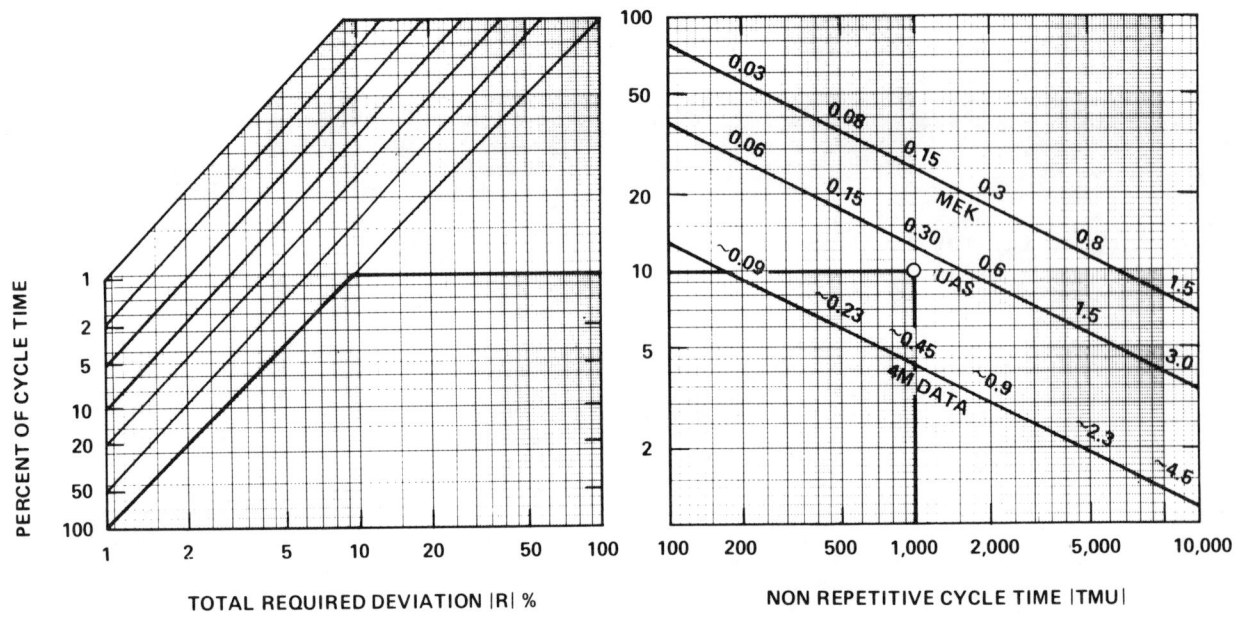

Figure 2. Example 1. (A Repetitive Operation)*

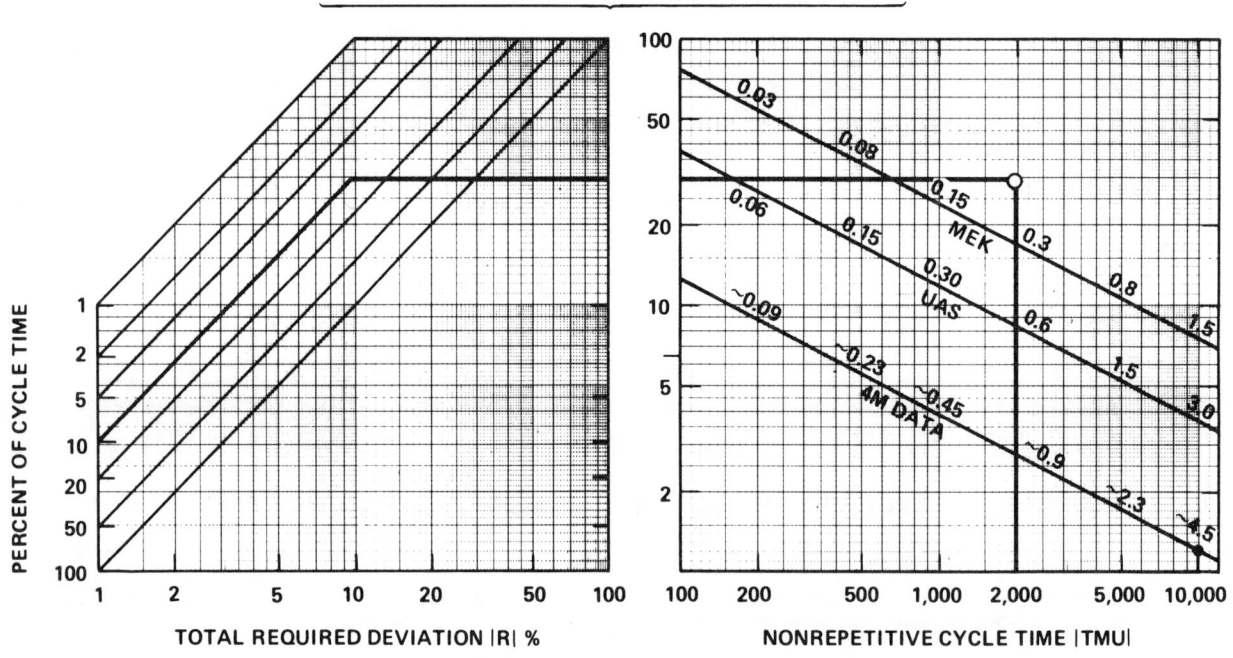

Figure 3. Example 2. (A Suboperation)*

*Nomogram by courtesy of the MTM Association for Standards and Research

STANDARDS-ACCURACY AUDITS

The discussion up to this point in this paper has dealt with establishing standards cost-effectively which meet specified statistical requirements for their accuracy.

Once standards have been established, maintaining required levels of accuracy is another matter. Maintaining standards accuracy is primarily a function of monitoring changes to operations and methods and making appropriate standards changes accordingly.

However, "creeping changes" are typical of every system, and without an objective audit of standards accuracies, "drift" in accuracy is inevitable.

Standards Accuracy Audit (Example A.)

Based upon random techniques, the audit sample shown in Figure 4. below is representative of the total population of standards in a given organization. It reflects the differences existing between established ("official") standards and the standard time which should be allowed based upon an audit of each operation included in the random sample.

A disciplined system is required to assure the objectivity and authenticity of the audit findings. This paper, however, will discuss only the technique for statistically evaluating the results of such an audit relative to standards accuracy requirements.

NO.	OFFICIAL* STANDARD	AUDIT* STANDARD	DELTA*	% ERROR
E1	20.08	21.23	−1.15	−5.42%
E2	26.92	24.13	2.79	11.56
E3	58.29	57.07	1.22	2.14
E4	7.19	6.53	0.66	10.11
E5	0.28	0.37	−0.09	−24.32
E6	0.30	0.29	0.01	3.45
E7	2.54	2.64	−0.15	−3.79
E8	2.54	2.64	−0.10	−3.79
E9	0.87	0.93	−0.06	−6.45
E10	6.29	6.22	0.07	1.13
E11	56.01	54.10	1.91	3.53
E12	16.20	13.38	2.82	21.08
E13	4.47	3.67	0.80	21.08
E14	31.43	24.80	6.63	26.73
E15	1.82	1.57	0.25	15.92

* TIME IN STANDARD HOURS PER 100 PIECES.

Figure 4. Example A (Accuracy Audit Sample)

From the sample data shown in Figure 4., the mean error ($\bar{X}$) is +4.91%. This indicates that, on the average, the standards of the parent population are indicated as allowing approximately 5% more standard time than perfectly accurate standards would allow.

Beyond mean error, there is obviously a substantial amount of plus-and-minus percent variation. Seven of the fifteen standards have errors greater than ±10%, and four of the fifteen are greater than ±20%. The standard deviation (s) is 13.32%.

The question is, does this population of standards meet the level of accuracy required? The answer to this question depends upon two considerations:

1. The specific accuracy requirements of the specific application of the standards, and

2. The statistical technique employed to evaluate the sample data.

Evaluation of Audit Results

The Hughes Aircraft Company has developed a statistical methodology for evaluating the results of standards accuracy audits. The approach was developed by Dr. Elmer M. Hsu. It is based upon the statistical techniques which were originally developed for evaluating the accuracy of world-wide crop surveys, utilizing satellite photography [3]. The basic objective of that program was almost identical to that required for evaluating standards accuracy audits in a work measurement system.

The criteria normally used at the Hughes Aircraft Company for audit evaluation is that a given population of standards reflects an accuracy of ±10%, compared to the audit standards, with a probability of at least 90%. The same methodology can also be used to determine the probability of accuracy relative to other limits (e.g., ±15%, ±20%, etc.), where such criteria are more cost-effective.

Figure 5. shows a graphic representation of the standards accuracy level reflected by the audit sample contained in Figure 4. As can be seen, the mean error (+4.91%) is well within the ±10% range. However, the statistical probability that individual standards fall within ±10% accuracy is only 51.7%, as shown.

One might ask, "Do these standards "PASS" or "FAIL" the audit? The answer is, it depends upon the accuracy requirements specified for their specific application.

To demonstrate the meaning of this answer, we will provide two hypothetical examples which represent the two extremes within a continuum of possible accuracy requirements (i.e., rather than a single arbitrary number, such as 90% or 95%).

Example 3. - Conveyorized Assembly Line

This hypothetical application of standards is for a mass-production, powered conveyor line having twenty work stations for progressive assembly of one product, assembled continuously over a six-week period.

The critical requirement for standards accuracy is that of line balancing. A "bottleneck" operation would force balancing delays upon the nineteen other work stations.

In this type of line-balancing application, standards accuracies of ±5% at a 95% or better level of confidence are required by many companies having this condition of standards application.

Clearly, the population of standards represented in Figure 4. would "FAIL" this accuracy requirement.

Example 4. - Receiving Inspection and Test

This hypothetical application of standards is for a department of inspectors and test technicians who process tens of thousands of different parts and assemblies for more than fifty different commodity groupings. Literally speaking, every individual package received is unique compared to every other package received.

Thus, specific work methods vary substantially from package to package.

The work of each inspector and test technician is measured on a weekly basis. If an individual person inspected, or tested, only twenty different items in one week, there is a 95.5% probability that his weekly measurement is within ±10% accuracy (Figure 5.). If the number of different items were 100, the probability of ±10% would be 99.9% (Figure 6.).

In an operation as varied as this hypothetical example, four-week-moving averages would be shown together with current-week performance reporting.

The population of standards represented in Figure 4. would "PASS" this accuracy requirement, given the environment being measured.

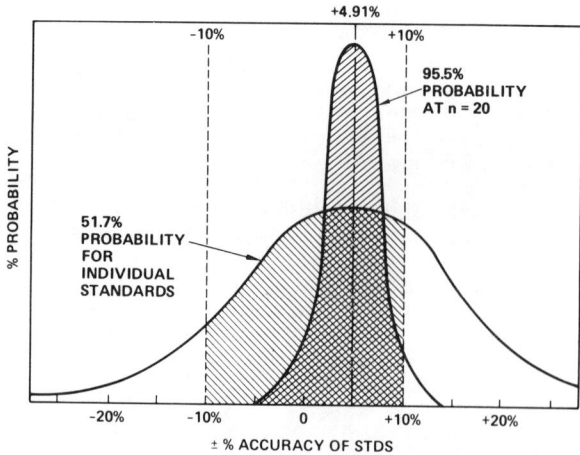

Figure 5. Probability of ±10% (20 Stds.)

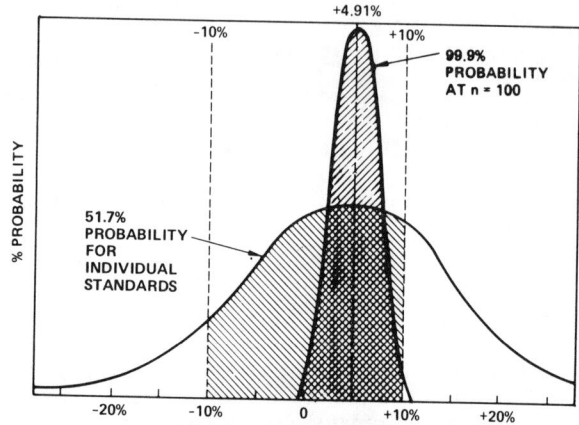

Figure 6. Probability of ±10% (100 Stds.)

GENERALIZED GUIDELINES (CONTINUED)

The same general principles and concepts discussed previously apply identically relative to the methodologies employed in conducting standards-accuracy audits. There are, however, several additional guidelines which are important to recognize in administering a cost-effective, cost-beneficial work measurement system.

7. Standards accuracy requirements are different for different types of operations, such as for the following examples:

    a. Balanced, progressive assembly line
    b. Continuous production of a single product
    c. Independent activities of individual workers
    d. Intermittent production of a variety of products

8. Standards accuracy requirements vary for different modes of work management, and different reporting systems, as in the following environments:

   a. Wage incentive payment system
   b. Measured day work system
   c. Measurement of performance of individual workers
   d. Group measurement systems

BASIC CONCLUSIONS

Several basic conclusions can be drawn from this analysis of work-measurement standards accuracy and audit techniques:

1. Specific principles, concepts and techniques must be considered in light of the nature of the specific environment for work measurement application. Different environments have different requirements.

2. Standards accuracy is a complex, statistical subject requiring technical understanding of the cost impact of alternative accuracy levels. Simplistic "rules of thumb" can be costly to administer.

3. Standards development, standards maintenance and standards audit are interdependent industrial engineering functions which must be considered collectively in designing work measurement systems.

4. Significant cost savings are potentially achievable by the judicious application of currently available work measurement technology (e.g., alternative predetermined time systems, data simplification, statistical methodologies, etc.).

5. Professionally determined standards accuracy requirements and disciplined work measurement systems assure,

   a. The reliability of work measurement data, and
   b. The cost-effectiveness of administration of work measurement systems.

References

[1] C.L. Brisley and W.F. Fielder, Jr., "Balancing Cost and Accuracy in Setting Up Standards for Work Measurement," *Industrial Engineering*, May 1982, p. 82.

[2] Aulanko, V.; Hotanen, Jr.; and Salonen, A. (English version by Karl Eady), *Standard Data Systems and Their Construction*, MTM Association, 1977.

[3] A.G. Houston, A.H. Feiveson, R.S. Chhikara, and E.M. Hsu, "Accuracy Assessment: The Statistical Approach to Performance Evaluation in LACIE," *Proceedings of Technical Sessions, Volume I*, The LACIE Symposium, October 1978.

BIOGRAPHICAL SKETCH

Mr. William F. Fielder, Jr., P.E., is Chairman of the Industrial Engineering Corporate Advisory Group of the Hughes Aircraft Company and is responsible for the Work Measurement Technology function within Hughes' Radar Systems Group, Product Operations Division. He is a Fellow of the MTM Association, is Secretary of the Association, and has been an active member of the Association's Research and Project Development Committee for the last ten years. His previous positions include Chief of Industrial Engineering, Head of Standards Administration, and Senior Industrial Engineer at the Hughes Aircraft Company. He was an industrial engineer in work measurement at U.S. Steel Corporation and McCulloch Corporation. Mr. Fielder has an MBA degree from U.C.L.A. and is a Senior Member of I.I.E.

VIII. The Future

Work measurement will continue to be an important aspect of productivity and profitability enhancement for most businesses well into the future. Both practice and theory will undoubtedly evolve, however, as the work environments undergo continued transformation. With greater automation will come a decrease in the number of low-skill jobs, coupled with a substantial increase in positions requiring high-tech training. A growing group of support, professional, and managerial personnel will perform irregular, complex, long-cycle work in the factory and the office. Tasks requiring production standards will include automated system maintenance, equipment setup and changeover, and a variety of office operations; measurement of the activities of knowledge workers will continue to grow in importance. Work measurement will also be affected by physical alterations in the work place, where design improvements will contribute to worker effectiveness and minimize health and safety problems. (See "The Work Station of the Future for Knowledge Workers," by Richard L. Shell, O. Geoffrey Okogbaa, and Thomas R. Huston, in the August 1985 *Industrial Engineering* magazine.)

THE IMPACT OF COMPUTERIZATION

Computers will continue to spread rapidly in the work measurement field and be pervasive by the end of this decade because of their inherent advantages. In general, computers are faster, cost less, are more accurate, and are smaller than their manual system counterpart. Three main factors will help the computer to become the primary work measurement tool during the next several years: decreasing cost of hardware and commercial software; specialized software development by end-users, and increasing use of artificial intelligence and expert systems in work measurement. A further trend which will advance the utilization of computers is the increasing amount of indirect labor in most organizations. Overall productivity measures and consequently indirect labor measurement will establish the need for knowledge worker standard data systems.

For selected work measurement tasks, computers will become more appropriate as they become smaller. Hand-held super microcomputers will allow the analyst to record and process data quickly and easily at the work-place location. Data could also be transmitted to a larger computer for final processing and testing. This mode of data collection and processing will free the analyst from mundane clerical duties and will tend to improve the accuracy of computations. The trend toward decentralized data collection and manipulation will in turn create a need for standardization in data handling, as data will have to be transferred quickly and easily to a central processor.

Larger computers will have their biggest impact through computer integrated systems. Computer-established time standards and manufacturing costs for an item just designed on a CAD/CAM system would be immensely valuable, allowing the designer to evaluate various designs based on manufacturing costs. Greater interaction with the shop floor or office would allow increased information transfer by the computer. The shop or office could send time data to the computer for comparison with standard times, thus alerting the analyst to inaccurate standards and to bottlenecks, assisting in record keeping for quality assurance, and improving individual accountability.

Programs for every work measurement technique will be available. Time study compilation with complete statistical analysis of data will be feasible. A future possibility is a computer-based system which could continuously monitor all workers. This system could comprise sensors to monitor brain waves and other physiological conditions. The values would be transmitted to the central computer, where the actual levels of mental processing and physical work would be compared with the appropriate production standard. The system could not only monitor the individual's performance, but also send suggestions that could improve productive output. For example, if brain wave sensors recorded a high alpha rhythm (indicating that the individual has a reduced ability to respond to mental stimuli), the computer could recommend a rest break or possibly a change of work task.

SUMMARY OF ARTICLES

"The Impact of Automation on Work Measurement" by Richard L. Shell provides an overview of automation and computer-aided design in manufacturing. In characterizing the automated factory of the future, Shell predicts several important changes in the theory and application of work measurement and methods engineering, including less direct observation timing and more work sampling. At the same time, he notes several traditional practices that should remain largely unchanged. The article concludes with recommendations for improving industrial engineering practices in the manufacturing environment of tomorrow.

In "Better Use of Better Tools Should Make Work Measurement Increasingly Valuable in Future," Clifford Sellie reviews work measurement as it is practiced today and discusses its future role. Using a Standards International survey of approximately 1,000 operations, Sellie reports the following trends in work measurement practice:

	YEAR	
	1980	*2000*
• No work measurement	30%	20%
• The use of historical data	3%	7%
• Judgment-based standards	5%	5%
• Stopwatch time studies	50%	40%
• Predetermined motion time systems for standard data	12%	28%

THE IMPACT OF AUTOMATION ON WORK MEASUREMENT

Richard L. Shell, P.E.
Department of Mechanical and Industrial Engineering
University of Cincinnati
Cincinnati, OH 45221

ABSTRACT

This paper provides a brief overview of automation and computer aided design and manufacturing developments with estimates of future growth. Common benefits and problems associated with automation are discussed along with future developments predicted for the automated factory. Work measurement and methods engineering activities and techniques that will likely remain unchanged and those that will likely change in the future are discussed. In addition, some of the major industrial engineering considerations likely to be associated with the automated factory of the future are identified.

AUTOMATION OVERVIEW

Computer aided design and manufacturing (CAD/CAM) is perhaps the heart and brain of automation and may be defined as the integration of the complete factory system from order entry through design engineering, manufacturing/industrial engineering, production, assembly, inspection and testing, to final storage/distribution of the completed product. The exact boundaries of CAD/CAM are not well defined for either hardware or software because of the widely diversed number of industries that produce segments of the total system, e.g., data processing/computers, machine tools, and material handling. Also, some CAD systems run on general purpose computers, and with CAM it is sometimes not clear where the computer controlled system begins or ends. Consequently, it is difficult to estimate the present market size of CAD/CAM and even more difficult to precisely forecast future growth and development.

The principal components of CAD include systems and applications software and computer hardware. All sizes of computers (micro, mini, and main frame) have been utilized for CAD activities. Estimates of the 1981 CAD market varied from $275 to $590 million sales. Forecasts for 1985 range from $1.5 to 2.0 billion sales.

The principal components of CAM include systems and applications software, computer hardware, machine tools, NC, DNC, CNC systems, robots, sensors, and automated material handling equipment. Estimates of the 1981 CAM market varies from $6.5 to $7.0 billion sales. Most forecasts for 1985 exceed $10 billion sales not including automated material handling equipment. The most conservative estimates predict that CAM sales will increase in excess of 10 percent annually.

A recent market research report by International Data Corporation surveying only CAD/CAM systems and applications software illustrates the rapid growth of the CAD/CAM industry. Sales in 1981 were reported to be $35.8 million. Sales for 1982 are expected to more than double and sales for 1986 are projected to be $543.5 million. It is interesting to note that low cost software configurations (below $100,000) accounted for about 5 percent of 1981 revenues. It is estimated that by 1986 low cost configurations will account for almost 20 percent of CAD/CAM industry revenues [6].

COMMON BENEFITS AND PROBLEMS OF CAD/CAM AND AUTOMATION

It is generally understood that computer based technologies are expanding rapidly and improving manufacturing productivity through more cost effective utilization of materials, labor, energy, improved inventory control, product quality and ultimately better customer service and marketplace performance [3,4,13,16,17,18]. While not true in all cases, when one considers the manufacturing cost equation, increased levels of automation usually improves overall productivity [9].

Benefits of CAD/CAM and automation include the following:

- Reduced through-put time for the design and manufacturing functions.

- Increased use of "standard" parts and tooling.

- Reduction of errors and changes in design and processing.

- Greater design and production flexibility permitting more product variation at affordable cost and response to market demand.

- Improved management control and understanding of design engineering and manufacturing.

- Greater utilization of equipment and physical plant.

Common problems and concerns associated with CAD/CAM and automation include the following:

- High capital investment.

- Rapidly changing technology.

- Failure to develop a master plan (system architecture) for all CAD/CAM elements.

- Hardware and software selection (purchase or lease) versus internal development versus complete turnkey system.

- Personnel problems, e.g., resistance to change, training older technical employees and production personnel, and properly developing management awareness and understanding.

While most of the problems cited above can be resolved, a few suggestions are offered:

- Work toward developing a combined CAD/CAM system by selecting elements that can later be interfaced with minimum cost and difficulty. Obviously, the combined system reduces duplicate technical activities, further lowers total through-put time and improves response to marketing, and in general increases productivity.

- Emphasize training for all employees, technical, manufacturing, and managerial.

- Interface CAD/CAM with the remaining management information system within the firm.

- Review and develop as needed any remaining support functions, e.g., improved basis for cost/benefit evaluation for future automation, material requirements planning, group technology, and specialized maintenance.

- Continue research and development for CAD/CAM to permit ongoing improvement in the system "intelligence" through human creativity.

FUTURE DIRECTION OF AUTOMATION

Clearly the future holds higher levels of automation for both the factory and the office. In addition, we have become a service oriented economy. The shift of employment from industrial manufacturing to service industries has steadily continued during the past three decades. It is projected that by 1985 less than one-fourth of the work force will be employed in the industrial manufacturing sector [16]. Figure 1 shows the projected distribution of an estimated United States total work force of 114.1 million in 1990.

Figure 1. Projected Distribution of Employment by Occupation in 1990 (Millions of Workers)

Before discussing the factory of the future, clarification recently offered by John White is useful. "The automated factory is not the same thing as the automatic factory. The automatic factory is a peopleless factory. In the automated factory, automation and mechanization dominate, but people are still needed to perform a limited number of direct tasks and a greater number of indirect tasks" [19]. Humans will also be required in the automatic factory for indirect support and managerial functions. The automated manufacturing factory is comprised of the following major areas: machining, metal forming, assembly and testing, quality assurance, material handling, robotics, computer control, and support functions.

The trend in automated manufacturing is the ability to process different components, models, or families of parts on the same machining system as contrasted to high production single purpose transfer machines. As Thomas C. Kennicott recently put it, "The dedicated machine of the past is rapidly becoming an anachronism. Increasingly, hard automation is becoming flexible"[7].

For most American companies, assembly accounts for over one-half of all the direct labor costs in manufacturing. Consequently, for the cost motivation alone, assembly automation will probably increase dramatically during the 1980's. With the exception of noncontacting sensors, no major breakthroughs in automated assembly equipment are forecast. However, it is predicted that productivity gains will be made through improved workplace designs. As Frank J. Riley recently stated, "Using ergonomic principles and simple automated processes, these work stations will integrate the operator with the simple machines"[15].

Material handling will become more important in the automated factory [19]. It has been predicted by Richard Polacek that the use of robots and pallet shuttles that operate around the clock will become commonplace. Even the job shop with small lot sizes and great varieties of parts will have unmanned operations [14].

The future development of factories has been very well summarized by Russel M. Loomis. "Business is preparing for a fundamental change in production techniques and management that, by the end of the 20th century, will rival the Industrial Revolution of the 18th century. This transition will herald the manufacturing facility of the future - the fully automated factory. These factories will operate at optimum levels of efficiency and productivity with virtually every operation, from ordering and accepting raw materials to manufacturing, to assembly, to quality control inspections, to shipping, and customer billing performed, monitored, and controlled by intelligent machines" [8].

FUTURE DIRECTIONS OF WORK MEASUREMENT AND METHODS ENGINEERING

Introduction

During the next decade, CAD/CAM and automation will significantly change the nature of the work force. The decrease of jobs for minimal skilled direct factory workers will be accompanied with a rapid increase in demand for highly trained (and perhaps better educated) specialists. There will be a large number of clerical/office support, technical, professional, and managerial personnel performing irregular, complex, long cycle work activities. Most of there time will be devoted to mental processes with much less time expended doing physical work.

Recently reported data on sources of productivity loss in high production manufacturing operations illustrates the importance of work measurement in the highly automated factory. For dedicated automatic equipment where set-up and tear down time losses are minimal, studies show that actual output is only 59 percent of the optimum capacity or losses of 41 percent. The losses are categorized as follows [2]:

Source	Percentage
Equipment Failure	42
Machine in Wait Mode	34
Work-force Control	16
Other	8

Based on the above percentages, effective work measurement to improve maintenance, scheduling, and support personnel utilization could greatly reduce productivity losses. For job shop operations where equipment changeover becomes a major activity and consumer of human resources, work measurement of the set-up procedure would produce additional benefits.

The major objective of work measurement and methods engineering will remain fundamentally unchanged as levels of automation increase and we move toward the automatic factory. Work measurement and methods engineers will still function to increase productivity and lower unit cost, thus allowing more goods to be produced for more people [11]. How industrial engineers realize this objective will indeed change. The following sections summarize what will likely remain unchanged, what will likely change, and what will be some of the major considerations of work measurement and methods engineering in the future.

What will Remain Essentially the Same

While numerous changes will occur in the theory and application of work measurement and methods engineering, it is believed that several traditional practices will remain. In the future CAD/CAM environment, the following practices will be conducted as in the past and present time.

- Perform the production engineering function to review product designs, and select raw materials.

- Perform the manufacturing engineering function to determine processes, manufacturing sequence, tool/fixture and test equipment design, quality assurance planning, and cost estimating.

- Review the product design and total manufacturing planning for possible safety/health problems and/or product liability exposure.

- Interact with other functional areas of the business, e.g., sales/marketing, finance/accounting, and general management.

The equipment utilized to conduct the above work activities as well as the work time expended will change as the automation level increases. For example, there will be a greater availability and use of computers and interactive graphics [5, 8, 20].

Examples of specific work measurement and methods engineering functions that will remain essentially unchanged include the following:

- Design of work place and human-machine interface considerations.

- Evaluation and measurement of indirect and support personnel, e.g., data processing, professional/technical staff, clerical/office, and maintenance.

- Evaluation and measurement of equipment change over.

- Determination of personnel allowances.

- Planning for manufacturing capacity and utilization.

Most of the existing work measurement and methods engineering techniques will be utilized in performing the above functions.

Probable Changes

The CAD/CAM automated factory of the future will cause several changes in the practice of work measurement and methods engineering including the following:

- The requirement to conduct direct observation timing and the associated rating or leveling of performance will decrease because of the increased machine and computer controlled processing cycles. Many cycle times will be automatically determined from internal clocks and calendars of the digital control device.

- While the determination of personal, fatigue, and delay allowances will be done as in the past, there will be changes. In general, physical fatigue will decrease while mental fatigue of workers will increase. Unavoidable delays in the highly automated factory will be different from traditional manufacturing plants.

- Work sampling will be simplified because of the ability to electronically monitor and record the activity state of most elements of the automated factory.

- While predetermined time systems will still be used in the traditional form for certain indirect and support personnel measurement, there is a need for further development in computerized synthetic time systems to facilitate application to the activities of knowledge workers. Robot Time and Motion (RTM) is an example of a special predetermined time system that will be useful in the highly automated CAD/CAM factory [12]. The trend to computerize predetermined time systems is likely to continue [10].

- Standard data will probably be developed for major segments of the automated manufacturing system as contrasted to small elements of an individual processing cycle. With major segment manufacturing times built into the CAD/CAM data base, one could easily simulate cost comparisons of alternate methods of manufacture and rapidly complete product cost estimates before production.

Other Considerations

The advent of the highly automated CAD/CAM factory will present a number of questions or difficult challenges for the work measurement and methods engineer. Among these are:

- The ability to easily compare the total cost effectiveness between various levels of automation to produce a given product. It is possible that certain direct production activities in the automated factory could be best (lowest cost) performed with a more labor intensive process.

- Should the automated system be provided with a "manual" backup during extended periods of breakdown? One must consider both the CAD/CAM system as well as the automated factory equipment.

- Inaccuracies in work measurement systems occur at every level of the total system, starting with the basic technique(s) [1]. Consequently, accuracy requirements must be determined for those techniques employed in the automated factory.

CONCLUSIONS AND RECOMMENDATIONS

The following conclusions and recommendations are offered for improving the practices of work measurement and methods engineering in the highly automated CAD/CAM factory of the future:

- The development of manufacturing automation is proceeding at an increasing rate, and the automatic factory is predicted to arrive by the year 2000.

- The ongoing development of CAD/CAM will greatly influence the practice of work measurement and methods engineering. The work activities of industrial engineers concerned with work measurement and methods in the automated factory will be significantly different, e.g., less time devoted to direct observation time study, more concern with indirect and support functions. Because of this, work sampling may become the most widely utilized of the presently known work measurement techniques.

- The overall work assignment of the industrial engineer in the automated or automatic factory will be fairly comprehensive and broad based. Consequently, the future engineer should be well educated and trained to understand and properly interact with the total complex CAD/CAM factory.

- There is a need to upgrade and expand existing work measurement techniques. Examples cited include specialized predetermined time systems, and standard data for automated manufacturing segments.

- Because of the increase of knowledge workers, there is a need to accurately assess mental output performance and the resulting mental fatigue.

REFERENCES

1. Brisley, Chester L., and Fielder, William F., "Balancing Cost and Accuracy in Settling Up Standards for Work Measurement", Industrial Engineering, Vol. 14, No. 5, May 1982, pp. 82-91.

2. Carter, Charles F., Jr, "Toward Flexible Automation", Manufacturing Engineering, Vol. 89, No. 2, August 1982, pp. 75-78.

3. Classen, Ronald J., and Malstrom, Eric M., "Effective Capacity Planning for Automated Factories Requires Workable Simulation Tools and Responsive Shop Floor Controls", Industrial Engineering, Vol. 14, No. 4, April 1982, pp. 73-78.

4. Engelberger, Joseph E., "Turning America Around with Robots", Manufacturing Engineering, Vol. 89, No. 2, August 1982, pp. 112-114.

5. Hosni, Yasser A., "Time Standards Using Micro Computers", Proceedings, Annual Industrial Engineering Conference, Institute of Industrial Engineers, 1982, pp 694-699.

6. "IDC Forecasts Low Cost CAD/CAM Systems Market", Typeworld, August 20, 1982, p. 15.

7. Kennicott, Thomas C., "Transfer Machines in the '80s", Manufacturing Engineering, Vol. 89, No. 2, August 1982, pp 111-112.

8. Loomis, Russell M., "The Programmable Controller Today and Tomorrow", Manufacturing Engineering, Vol. 89, No. 2, August 1982, pp. 117-119.

9. Malstrom, Eric M., and Shell, Richard L., "Projections for Future Manufacturing Work Force Productivity", Proceedings, 25th Annual Joint Engineering Management Conference, The American Institute of Industrial Engineers, 1977, pp. 41-47.

10. Martin, John C., "Program, Portable Microprocessor Allow Direct On-site Input of MTM-1 Data", Industrial Engineering, Vol. 14, No. 8, August 1982, pp. 50-53.

11. Niebel, Benjamin W., Motion and Time Study, 7th edition, Irwin, 1982, p.7.

12. Nof, Shimon Y., and Lechtman, Hannan, "Robot Time and Motion System Provides Means of Evaluating Alternate Robot Work Methods," *Industrial Engineering*, Vol. 14, No. 4, April 1982, pp. 38-48.

13. Ottinger, Lester V., "Questions Potential Robot Users Commonly Ask", *Industrial Engineering*, August 1982, pp 28-31.

14. Polacek, Richard, "Machining Centers: Today and Tomorrow", *Manufacturing Engineering*, Vol. 89, No. 2, August 1982, pp. 97-99.

15. Riley, Frank J., "The Look of Automatic Assembly", *Manufacturing Engineering*, Vol. 89, No. 2, August 1982, pp. 79-80.

16. Shell, Richard L., and Malstrom, Eric M., "Measurement and Enhancement of Work Force Productivity in Service Organizations", *Proceedings, 25th Annual Joint Engineering Management Conference*, The American Institute of Industrial Engineers, 1977, pp. 29-35.

17. Van Singel, Gary, "Tips on Selecting the Most Suitable Equipment Type", *Industrial Engineering*, Vol. 14, No. 5, May 1982, pp. 32-36.

18. Vasilash, Gary S., "General Electric: Bring the Factory of Tomorrow to Life", Manufacturing Engineering, October 1981, pp. 72-75.

19. White, John A., "Factory of Future Will Need Bridges Between its Islands of Automation", *Industrial Engineering*, Vol. 14, No. 4, April 1982, pp 61-68.

20. Wright, Marc B., "Work Measurement System Monitors The Output of A Word Processing Operation", *Industrial Engineering*, Vol. 14, No. 7, July 1982, pp. 70-72.

Better Use Of Better Tools Should Make Work Measurement Increasingly Valuable In Future

By Clifford Sellie
Standards, International Inc.

Productivity is essential for a prosperous economy, a prosperous company, a prosperous employee. The productivity of all three is influenced by work methods, work standards, and worker motivation.

The work measurement tools used for setting work standards can help determine work methods and can have an effect on worker motivation. Accordingly, let us take a close look at work measurement as it is practiced today and its role in the future.

Trends in work measurement

A company president told an industrial engineer: "I have good news and bad news for you! First, let me tell you that you have won a one-week all-expenses-paid trip to New York. What do you think of that?" The IE replied: "First, let me know; is that the good news or the bad news?"

Whether current and future trends in work measurement are good news or bad news will depend on the company you work for, our national economy and your particular talents and training.

Today's difficult times for many companies, and the national economy, have resulted in the current increased emphasis on work measurement. Table 1, which shows productivity under varying operating conditions, gives ample reason for this increased emphasis.

During tough and competitive times, increased attention is given to productivity improvement and work measurement. Work measurement, as one of the basic tools for productivity improvement, quite logically varies in popularity according to need. Our economic cycles accelerate the trends in work measurement. But technical advances determine the trends.

Based on Standards, International's study of approximately 1,000 operations, let me evaluate the current trend and predict the year 2000 trend as they are shown in Table 2. What will cause the shift in percentages? I doubt that it will be better understanding of techniques, or greater understanding of the importance of good work measurement. I believe the shift will result from minor but significant improvements in information gathering tools (see Table 3).

To put it bluntly, the quality of most work measurement, as currently practiced, is not equal to the tools available. We are like the farmer who declared: "I don't need education on better farming; I'm not farming now as good as I know how."

What we need to improve work measurement techniques is motivation, along with information on how to do a better job. Let us look at the various reasons there will and will not be improvements in work measurement trends.

No work measurement—Why?

There will continue to be instances in which work measurement is not practical. Many operations will be too small, or too complex, or so well automated and paced that work measurement will not be of benefit. There also will continue to be instances in which work measurement is practical, but management is uninformed or unwilling to change long established habits.

There will be sociologists, psychologists and consultants who advise that work measurement is unnecessary, un-American and anti-social. About every 10 years, "new discoveries in human nature" have been found that reportedly mitigate or substitute for the need for work measurement.

Accordingly, some time in the late 1980s or early 90s, it would be logical to expect an "exciting new discovery" that a proper combination of diet and environment, for instance, will result in an improved work ethic and dynamic productivity. Promoters of this exciting new discovery will advise that all other programs of productivity improvement or worker motivation including quality circles, work simplification, work measurement, wage incentives, etc.—are unnecessary.

This theory will run its course, greatly improving the sale of more nutritious diets and wall paint. After this solution-to-all-problems theory expires, it should be replaced within five to 10 years by another "new, exciting and unique discovery" about human nature and motivation.

In addition, skilled but bored work measurement practitioners will identify certain areas of work as being over measured, and will talk executives out of measuring large chunks of direct labor on the theory that they have already had all the attention necessary.

The above factors will contribute to the continuation of a large percentage of unmeasured work.

Stopwatch time study

Stopwatch time study will continue to be the most widely used and widely abused work measurement technique. There will be improvements in stopwatch tools, primarily in terms of convenience and ease of use. Electronic digital stopwatches will make reading and obtaining the times much simpler than using the old mechanical sweephand stopwatches.

New performance leveling films, probably in the area of office work, should come along to supplement those now available. However, the performance leveling bugaboo will continue to be as big a problem as ever. Problems associated with human frailties, such as difficulties with reading of time observations, can be corrected.

Problems associated with human judgment, such as those related to performance rating, will just have to be endured. Performance rating techniques in use today have shown no improvement over those used 30 years ago; there is no reason to suspect that they will in the next 30 years.

In fact, the more basic and simple the performance rating, the more effective it seems to be. More sophisticated performance leveling films and equations only seem to hinder the results.

Predetermined times

Again, the experience of the past 30 years predicts the future. There will be continued announcements of "new and better" systems based on condensed and abbreviated predetermined times. Great advantages will be proclaimed, supported by not-so-great arithmetic. One of the most outrageous examples of this, from a public relations viewpoint, was the predetermined time system that came on stream in the 1960s. This was a system touted mainly by public accounting firms. One of its proponents, in a major CPA firm, advocated its use "because it was so simple and easy to understand, even the accountants could use it." The system itself was as clever as that public relations effort and has sunk into the sea.

Many other systems will come and go, some deserving, others purely to meet an insatiable need for new tools or new names for old tools. MTM should continue as the foremost system, due mainly to the excellent support and organization of the MTM Association.

One of the most accurate systems, MTA—Segur's Motion Time Analysis system—will continue to dwindle in usage for two reasons. One will be lack of knowledge and promotion of the MTA system. The second and stronger reason will be decreased need for an extremely accurate system because most long-run, short-cycle work will be automated.

More and more of the work to be measured will be job shop or short-run work. Accordingly, there will be increased use of two effective work measurement shortcuts:

☐ Standard data programs—catalogs of motion time values for the work centers.
☐ Parts technology programs—setting standards by common parts characteristics or "same as" work methods.

Since these two shortcuts are most easily developed using predeter-

Table 1: Productivity under Varying Operating Conditions (Based on over 1,000 productivity audits and work measurement installations)

Performance measurement	Poor	Supervision: Average	Good
Incentives	80% to 125%	100% to 125%	115% to 125%
Measured	60% to 80%	70% to 90%	80% to 95%
Unmeasured	30% to 70%	50% to 75%	60% to 85%

Table 2: Trends in Work Measurement

Work Measurement Practices	1980	2000
☐ No work measurement.	30%	20%
☐ Historical records, many of them varying so widely they justifiably earn the often expressed term "hysterical records."	3%	7%
☐ Personal evaluations and memories by foremen and other informed people, commonly referred to as "reasonable expectancies."	5%	5%
☐ Stopwatch studies, with and without performance leveling.	50%	40%
☐ Predetermined motion times (MTM, etc.) and/or standard data.	12%	28%

Table 3: How are work standards set, and how well?

Work Measurement Techniques:	When Set	Customary Trends
Historical records	±30%	20% tight to 60% loose
Reasonable expectancies	±20%	10% tight to 45% loose
Stopwatch studies	±10%	5% tight to 35% loose
Predetermined times	± 5%	5% tight to 20% loose

Influences affecting standards reliability:
  Standard data formats   Competitive needs
  Methods specifications   Management experience
  Work sampling   Management controls
  Union knowledge   Line and staff knowledge

The three modifying forces that have the strongest influence on the quality of a firm's work standards usually are:
☐ Competitive needs.
☐ Methods specifications.
☐ Management experience.

---

mined times, this will increase PDT usage.

Table 4 outlines the benchmarks we have found helpful in guiding our clients as to the work measurement techniques they should use.

Computerized measurement

The explosive development of low-cost computers will have the most significant effect on work measurement trends. It will contribute greatly to successful work measurement applications and benefits. It will also result in amusing disasters.

A few of you may remember "Autorate." This was a computerized work measurement technique brought forth in the 50s as the answer to the industrial engineer's prayers. A highly sophisticated and expensive system, it did everything its proponents said it would except provide practical and understandable time standards.

Focused on the computer, this system did not force the industrial engineer to go to the work center to analyze methods and discuss results with the line supervisor and operator. Accordingly, it produced reams of "funny papers." The time values might or might not have been correct; nobody knew, and few cared.

There will be more over-promising and under-performing software systems. Most of the current computerized systems developed in the last 10 to 15 years are already sinking into the sea from the same flaws that sank Autorate. New ones will continue to come to the fore, have their glorious moments of promotion and high hopes, then fade away after the rush of enthusiasm and results.

It is axiomatic that any new system will succeed for a while from a combination of the "Hawthorne effect" and self-fulfilling expectations. However, this success will not continue if the basic system is illogical or cumbersome.

We already have the black box capacity to develop a variety of new and astounding systems. We might enjoy speculating on a logical development, a computerized time study system attached directly to the operator. This system—let's identify it as the "Bio-Science Time Study" system—would measure actual time taken by an operator in performing a task. It also would measure energy consumption, providing automatic performance rating of the time study.

This Bio-Science Time Study system would be promoted as the greatest scientific breakthrough ever, combining the best of Taylor, Gilbreth and Segur. The standards would be "pure," untouched by industrial engineering hands.

This system would enjoy a great flurry of popularity. Then it would be discovered that unions (yes, they will still be with us in the year 2000) were encouraging operators when attached to the computer to think vicious thoughts about the IE. This would raise energy consumption and make the performance rating worse (when possible) than with regular stopwatch studies. The Bio-Science Time Study system would then be condensed down to its initials and tossed in the garbage.

Let us hope this scenario does not occur in the future. However, it compares so well with other trends in computer systems that it probably will; therefore, let's hope its life will be short.

Does this mean computerized standards setting is a dream and will not affect the future of work measurement? No! Computers will have two definitive effects. Good systems will be developed that will simplify, speed up and improve the setting of standards. Computer improvements also will increase the acceptance and effectiveness of work measurement.

Simple standards-setting

Along with overcomplicated, oversophisticated computer systems for standards setting will come simple, direct and understandable computerized standards-setting programs. There are some already on the market. They are systems focused on standards setting from the user's viewpoint, not the software programmer's viewpoint. They are short; they are direct; they are modular.

These systems will continue to expand and improve. They will not eliminate the need for the industrial engineer to go to the work center. They will increase the speed with which IEs can do clerical work, recall past work, alternate work methods, etc. And when the work measurement engineer has done a proper job of analyzing methods and cataloging the studies, these computer systems will greatly reduce IE time wasted on "re-inventing the wheel." "Re-inventing the wheel" is, sad to say, one of the most common trademarks of current work measurement practices.

In addition to improving the technical competence of the work mea-

Table 4: Benchmarks for Work Measurement Techniques

Criteria		Appropriate work measurement techniques				
		P.D.T.	P.D.T. Standard Data	Pre-Rates®	Component Std. Data	Reasonable Expectancies
Documents 1	Instruction documents	Motion	Motion sequence purpose	Operation	Job	Complete order
	Drawings for part	Detailed drawings for each part		Drawing for group of parts	Overview of drawing	Sketch/notes/ simple part verbal description
Work flow structure 2	Shop/ organization/ work place descript	Optimized single purpose work place	Designated standard work place	Standard work place	Universal or general work place	Work place (not defined)
	Aids/tool organization	Optimized single purpose tools	Single purpose tools	Universal and simple tools		Without aids/tool organization
	Material organization	Material at work place		Materials obtained from storage or other locations		Total search for materials
Person 3	Personal situation	Trained for one operation		Trained for all operations or jobs at a given work place		All around trained

surement engineer, the computer can also increase the acceptance of work measurement.

Acceptance will increase

One of the common complaints of the industrial engineer is the hostility and opposition of line supervisors. When line supervision and staff are both measured by the same cost yardstick, there will be greater cooperation and interest in mutual contributions to cost reduction. This will not come automatically. The computer will simplify the mechanism.

Management, aided by accounting and industrial engineering, must coordinate the standard time system and the standard cost system. Computer improvement in information gathering and massaging will permit appropriate cost information to be furnished quickly to management, line supervisors and engineers.

Alert management will use bottom line labor costs as a measurement of the effectiveness of both supervisor and engineer. When this is done, the work measurement engineer's job can be more enjoyable and more effective.

There is hope

The results available from good work measurement provide a strong need for work measurement. The availability of better techniques, whether in stopwatch tools or in predetermined time tables—when combined with appropriate computer programs—can triple and quadruple the contributions of the industrial engineer involved in work measurement.

Accordingly, we hopefully visualize work measurement by the year 2000 contributing even more to the competitive position of progressive companies and our country. And since rewards go with results, we visualize these trends benefiting the position (and the pocketbook) of the IE involved in work measurement. IE

Clifford N. Sellie, P.E., is chairman and founder of Standards, International Inc., a Chicago-based management consulting and engineering research firm. Sellie has published numerous articles on work measurement, work simplification, wage incentives, supervisor bonus programs, union relations, cost reduction and management controls. He holds a BA from St. Olaf College and an MA from the University of Minnesota. He is a past president of the Industrial Management Society and is a senior member of the Institute of Industrial Engineers, the Society for Advancement of Management and the MTM Association for Standards and Research. He is listed in Who's Who in Commerce and Industry and in World Who's Who in Industry.

References

For the past several years, the Institute of Industrial Engineers has indexed all articles published in *Industrial Engineering, IIE Transactions,* and the *Proceedings* of the annual and fall industrial engineering conferences. Indices for all of the publications except *IIE Transactions* are included in a special section each year in the December issue of *Industrial Engineering.* The *IIE Transactions* index appears in that journal's December issue each year.

The subjects and their index codes that relate to work measurement topical areas are shown below:

Work Measurement Subjects	Index Code
Work measurement	1600
Allowances	1601
Computer-aided work measurement	1602
Measured day work	1603
Pace rating	1604
Predetermined time systems	1605
Standard data	1606
Standards auditing	1607
Standards development	1608
Time measurement	1609
Work Sampling	1610
General	1611

The following articles have appeared in the indices during the past ten years. Because of their diverse content, some articles that have been classified for one index contain substantial information about one or more other work measurement subjects. Since all *Proceedings* references are to those of IIE's biannual conferences, the titles of these volumes have been shortened for convenience, e.g., *1984 Annual Proceedings, 1985 Fall Proceedings.*

Work Measurement: Concepts and General Background (1600, 1611)

Otis, I., and K. K. Mead. "Industrial Work Standards and Productivity," *1975 Annual Proceedings,* p. 225.

Boyer, C. H. "Work Measurement: The Flap Over MIL-STD-1567 (USAF)," *Industrial Engineering* (November 1976), p. 14.

Williams, C. D., Jr. "Saving Money with a Computerized 4M Data System," *Industrial Engineering* (May 1977), p. 24.

Rice, R. S. "Survey of Work Measurement and Wage Incentives," *Industrial Engineering* (July 1977), p. 18.

Eady, K. "Today's International MTM Systems—Decision Criteria for Their Use," *1977 Annual Proceedings,* p. 483.

Mazzolla, D. P., and J. D. Kauffman. "Activity Measurement Program Promotes Productivity," *Industrial Engineering* (June 1978), p. 26.

Flowers, D. A., and L. F. Mulvehill. "An Application of Industrial Engineering to Literary Utilization," *AIIE Transactions* (September 1978), p. 315.

Varisco, K. M., and J. M. Pruett. "An Interactive Computer System and Work Measurement—An Innovative Approach," *1978 Fall Proceedings,* p. 342.

Otis, I. "Industrial Work Standards and Productivity," *1978 Fall Proceedings,* p. 291.

Nance, H. W., and R. L. Kreighbaum. "Using MCD-MOD-1 to Measure and Control Office Cash," *1979 Fall Proceedings,* p. 361.

Adams, K. S., and T. J. McGrath. "A Procedure for an Economic Comparison of Work Measurement Techniques, Part I—The Model; Part II—An Application," *AIIE Transactions* (September 1979), p. 229.

Young, R. T. "Forecasting, Monitoring and Controlling Productivity," *Industrial Engineering* (July 1980), p. 46.

Lee, J. A. "The Hawthorne Relay Studies: A Faith Impervious to Iconoclasm," *1980 Annual Proceedings,* p. 179.

Clare, R. A. "Banks Management Reporting System Sets Workloads and the Utilization of People," *Industrial Engineering* (February 1981), p. 78.

Fein, M. "How 'Reliability,' 'Precision' and 'Accuracy' Refer to Use of Work Measurement Data," *Industrial Engineering* (July 1981), p. 26.

Schantz, J. J. "Video Tape Recording Saves Time for IEs," *Industrial Engineering* (July 1981), p. 48.

Fein, M., and U. Halen. "Macro Work Measurement Using Micro Measurement and Techniques," *1981 Annual Proceedings,* p. 591.

Byrd, J., Jr., and L. T. Moore. "The Productivity of Industrial Engineering," *1981 Annual Proceedings,* p. 301.

Koshula, M., and R. J. Allshouse. "Time Slotting as a Measurement of Assigned Maintenance Incentives," *1981 Annual Proceedings,* p. 502.

Gage, H. "Unorthodox Methods for Assessing Efficiency in Physical Work," *1981 Annual Proceedings,* p. 600.

Byars, W. S. "Work Measurement in Inventory and Warehouse Operations," *1981 Annual Proceedings,* p. 577.

Russell, J. R. "Work Measurement and Performance Goals in Major Power Plant Construction," *1981 Fall Proceedings,* p. 518.

Ferrell, M. D. "A Plan of Action for Rehabilitating an Ailing Wage Incentive Program," *Industrial Engineering* (November 1982), p. 52.

Smith, P. J., T. J. Armstrong, and D. G. Lizza, "IEs Can Play Crucial Role in Enabling Handicapped Employees to Work Safely, Productively," Part Three of a Series, *Industrial Engineering* (April 1982), p. 98.

Young, R. T. "New Training, Work Analysis Methods Needed to Manage Office of Future," *Industrial Engineering* (July 1981), p. 66.

Wright, M. B. "Work Measurement System Monitors the Output of a Word Processing Operation," *Industrial Engineering* (July 1981), p. 70.

O'Brien, J. P. "Attitude of Employee and Employer—Key to Productivity Improvement," *1982 Annual Proceedings,* p. 289.

Malzahu, D. "Job Modification & Placement Strategies for Persons with Physical Disabilities Using the Available Motions Inventory," *1982 Annual Proceedings,* p. 179.

Schultz, C. L., and W. W. Graham. "Work Measurement on the Santa Fe Railway," *1982 Annual Proceedings,* p. 625.

Globerson, S., and E. Darom. "Comparison of Three Techniques for Weighting Importance of Performance Indicators," *1982 Fall Proceedings,* p. 348.

Shell, R. L. "The Impact of Automation on Work Measurement," *1982 Fall Proceedings,* p. 348.

Landon, J. H. "Measuring Electric Utility Efficiency," *1982 Fall Proceedings,* p. 3.

Engwall, R. L. "There Is More to Industrial Engineering than Traditional Work Measurement," *1982 Fall Proceedings,* p. 840.

Snyder, F. "IEs Must Convince Arbitrators that Work Measurement Data Are Fair and Accurate," *Industrial Engineering* (July 1983), p. 28.

Brisley, C. L., and W. F. Fielder. " 'Unmeasurable' Output of Knowledge/Office Workers Can and Must Be Measured," *Industrial Engineering* (July 1983), p. 15.

Redmond, G. R. "The Evaluation of a Flexible Manufacturing System—A Case Study," *1983 Annual Proceedings,* p. 749.

Bonello, F. J., J. G. Beverly, and W. I. Davisson. "Impact of Government and Defense Sales on Labor Utilization by Firms in the Aerospace Industry," *1983 Annual Proceedings,* p. 13.

Brisley, C. L. "Measuring the 'Unmeasurable'," *1983 Fall Proceedings,* p. 598.

Shell, R. L., and O. G. Okogbaa. "The Effect of Mental Fatigue on Knowledge Worker Performance," *1983 Fall Proceedings,* p. 631.

Daugherty, J. C. "The Peace Keeper MIL-STD-1567 Work Measurement Program," *1983 Fall Proceedings,* p. 16.

Johnson, S. L. "Work Measurement: Present and Future," *1983 Fall Proceedings,* p. 191.

Sellie, C. "Better Use of Better Tools Should Make Work Measurement Increasingly Valuable in Future," *Industrial Engineering* (July 1984), p. 82.

Anthony, G. M. "IE's Measure Work, Write Standards for White Collar Workers at Financial Institution," *Industrial Engineering* (January 1984), p. 77.

Conn, H. P. "Improving Use of Discretionary Time Raises Productivity of Knowledge Workers in Offices," *Industrial Engineering* (July 1984), p. 70.

Jackson, L. "Office Automation Provides Opportunity to Examine What Workers Actually Do," *Industrial Engineering* (January 1984), p. 91.

Magliola-Zoch, D., and R. G. Weiner. "Plan Applies IE Concepts to Improve Productivity and Measure Creative Process of Professionals," *Industrial Engineering* (September 1984), p. 46.

Laker, J. K., and W. M. Hancock. "Work Environment Survey Generates Ideas on Increasing White Collar Productivity," *Industrial Engineering* (July 1984), p. 60.

Malmborg, C. J., and S. J. Deutsch. "Performance Measurement in the Presence of Interdependent Organizational Activities," *1984 Annual Proceedings,* p. 434.

Mital, A., and R. L. Shell. "Determination of Rest Allowances for Repetitive Physical Activities That Continue for Extended Hours," *1984 Annual Proceedings,* p. 637.

Souder, H. R., W. Leigh, and N. Damachi. "The Utilization of a Decision Support System for Work Measurement," *1984 Fall Proceedings,* p. 437.

Overby, W. M. "A Case Study of a Bank Teller Staffing Model," *1984 Fall Proceedings,* p. 457.

Easton, J. "Output Related Manpower Performance Measures in Power Situations," *1984 Fall Proceedings,* p. 529.

Newmiller, C. E. "Performance Evaluation Processes and Perceptions of Productivity: What You See Is What You Get," *1984 Fall Proceedings,* p. 284.

Shoen, J. L. "A Participative Approach to Work Measurement," *1985 Annual Proceedings,* p. 18.

Oberle, L. R., and P. M. Scott. "Manpower Requirements Forecasting: A Case Example," *1985 Fall Proceedings,* p. 526.

Downes, W. A., and D. E. Lutz. "Work Measurement for the Transportation Supervisor," *1985 Fall Proceedings,* p. 454.

Time Measurement, Pace Rating, Allowances, and Measured Day Work (1609, 1604, 1601, 1603)

Akiyuki, S. "New Insight to Pace Rating," *Industrial Engineering* (July 1975), p. 32.

Moore, C. G. "Manpower Planning, Scheduling and Control System for Airline Service Commissaries," *1975 Annual Proceedings,* p. 337.

Corningmore, E. G., Jr. "Platform Management System—A Case Study in Productivity Improvement," *1975 Fall Proceedings,* p. 180.

Nedridek, C. A., Jr. "Timing Those Group Operations," *1976 Annual Proceedings,* p. 16.

Reuter, V. G. "New Way to Learn Pace Rating," *Industrial Engineering* (July 1977), p. 16.

Schutz, R. K., and T. H. Campbell. "The Worker's Perception of a Change in Work Rate," *1978 Annual Proceedings,* p. 95.

Nof, S. Y., H. Gershoni, S. D. Ansell, and A. Cohen. "Industrial Worker Pace Variability—A Study with Real Time and Posterior Analyses," *AIIE Transactions* (September 1978), p. 321.

Fein, M. "Let's Return to MDW for Incentives," *Industrial Engineering* (January 1979), p. 34.

Phillips, D. T., and D. R. Smith. "Determination of Probabilistic Time Standards for Tasks Performed Under Uncertainty," *1979 Annual Proceedings,* p. 451.

Poage, S. C. "Man Machine Charts: New Applications and Analysis," *1979 Annual Proceedings,* p. 445.

Douglas, W. J., and R. A. Pimentel. "Operations Analysis and Modeling Through Underground Time Study Techniques," *1979 Fall Proceedings,* p. 227.

Graham, R. B. "Performance Measurement of Customer Service Personnel," *1979 Fall Proceedings,* p. 297.

Pape, E. S. "Work/Activity Sampling—Contemporary Design Analysis Methodology and Applications: Part II—Work Sampling Calculations Revisited," *1979 Fall Proceedings,* p. 371.

Salvendy, G. "Effects of Job Pacing on Job Satisfaction, Psychophysiological Stress and Industrial Productivity," *1980 Annual Proceedings,* p. 433.

Richards, J. "The Major Unit of Activity—A Simplified Approach to Measuring and Controlling Indirect Labor," *1980 Annual Proceedings,* p. 455.

Christian, P. H. "Setting Rates at Crayola," *Industrial Engineering* (September 1981), p. 36.

Coe, D. D., and R. Nailwajek. "The Use of Electronic Desk Top Computers and Time Study Equipment in Industrial Engineering," *1981 Annual Proceedings,* p. 522.

Das, B. "Performance Rating Training in Industry: Using Das's Programmed Learning Method," *1982 Annual Proceedings,* p. 687.

Corser, M. "Utility Program Summarizes Stopwatch/Pace Data," *Industrial Engineering* (July 1983), p. 18.

Johnson, E. N. "T&D Crew Time Reporting in the American Power System," *1983 Annual Proceedings,* p. 735.

McDermott, K. J. "Microcomputer and Spreadsheet Software Make Time Studies Less Tedious, More Accurate," *Industrial Engineering* (July 1984), p. 78.

Wygant, R. M. "An Analysis of Performance Rating," *1984 Annual Proceedings*, p. 632.

Predetermined Time Systems (1605)

Zandin, K. B., and R. M. Weiss. "MOST Systems for Work Measurement," *Industrial Engineering* (June 1977), p. 43.

Zandin, K. B. "MOST Systems for Improved Industrial Engineering Productivity," *1978 Annual Proceedings*, p. 486.

Brisley, C. L. "Comparison of Predetermined Time Systems (PTS)," *1978 Fall Proceedings*, p. 13.

Weaver, R. F., and E. A. Boepple, Jr. "WOCOM and Quick Work-Factor: The State-of-the-Art in Predetermined Time Systems," *1978 Fall Proceedings*, p. 31.

Garg, A. "Methods for Estimating Physical Fatigue," *1979 Annual Proceedings*, p. 68.

Nof, S. Y., and H. Lechtman. "Robot Time and Motion System Provides Means of Evaluating Alternate Robot Work Methods," *Industrial Engineering* (April 1982), p. 30.

Nof, S. Y. "Decision Aids for Planning Industrial Robot Operations," *1982 Annual Proceedings*, p. 46.

Singleton, S. K., A. Masud, and D. Malzahn. "High Level Predetermined Time Standard System and Short-Cycle Tasks," *1985 Annual Proceedings*, p. 656.

Work Sampling (1610)

Kinack, R. J. "Work Sampling Tables," *Industrial Engineering* (March 1975), p. 43.

Brisley, C. L. "Past-Present-Future of Work Sampling," *1975 Annual Proceedings*, p. 440.

Allen, D. S. "Multiple Activity Work Study Needs X Samples," *Industrial Engineering* (December 1979), p. 20.

Bascom, R. A., G. S. Kalemkarian, and J. L. Miller III. "An Algorithm: The Use of Sequential Sampling in Work Measurement Audits," *1979 Annual Proceedings*, p. 3.

Brown, R. A., and W. F. Sowder. "Performance Rating for Service Jobs," *AIIE Transactions* (June 1979), p. 121.

Robertsen, J. A. "Multi-Dimensional Work Sampling at the Executive Level," *Industrial Engineering* (August 1980), p. 70.

Mundel, M. E. "Sampling, Using Self-Reporting Methods, Established Standards for Federal Bureau and Saves Millions," *Industrial Engineering* (August 1980), p. 44.

Hanna, S., and S. Konz. "Determination and Analysis of the Productivity Gap in a Measured Daywork Plant," *1980 Annual Proceedings*, p. 444.

Moder, J. J. "Selection of Work Sampling Observation Times: Part I—Stratified Sampling," *AIIE Transactions* (March 1980), p. 23.

Moder, J. J., and H. D. Kahn. "Selection of Work Sampling Observation Times: Part II—Restricted Random Sampling," *AIIE Transactions* (March 1980), p. 32.

Kulonda, D. J. "Method Measures Productivity in Non-Traditional Work Situations," *Industrial Engineering* (July 1981), p. 34.

Adlfinger, A. C. "Sequential Sampling Technique Reduces the Fixed-Size Sampling Number and Corresponding Labor," *Industrial Engineering* (May 1981), p. 74.

Whitehouse, G. E., and D. A. Washburn. "Work Sample Size Analyzer," *Industrial Engineering* (April 1981), p. 22.

Whitehouse, G. E., and D. A. Washburn. "Work Sampling Observation Generator," *Industrial Engineering* (March 1981), p. 16.

Brown, T. G. "Work Sampling Inside General Motors," *1981 Fall Proceedings*, p. 229.

Reisman, A., S. K. Kotha, L. Gonanza, T. Murray, and W. T. Kidd. "IEs Find Ways to Improve Throughput, Worker Morale on a Firm's Warp Line," *Industrial Engineering* (September 1982), p. 70.

Parsons, H. R. "Work Sampling During Power Plant Scheduled Outages," *1982 Annual Proceedings*, p. 16.

Overby, W. M. "Technique for Group Time Measurement Simplifies Indirect Labor Observations," *Industrial Engineering* (July 1983), p. 34.

McNall, G. R. "Program Calculates Results of Work Sampling," *Industrial Engineering* (July 1984), p. 25.

Seemer, R. H., L. G. DeLuca, and G. A. McBean. "Work Sampling—An IIE Utility Industry Survey," *1985 Fall Proceedings*, p. 514.

Standard Data (1606)

O'Neil, T. J., A. G. Shah, A. N. Heyman, and C. C. Holmes. "System 18—A Programmable Computer System for Standard Data Applications," *1977 Annual Proceedings*, p. 34.

Williams, D. C., Jr. "Introduction to 4M Data System and Its Application at Baxter Travenol Laboratories Inc.," *1977 Annual Proceedings*, p. 414.

Gambrel, D. W. "Department of Defense Engineered Performance Standards—A Contemporary Work Measurement Tool for Evaluating and Establishing Efficiency in Maintenance Work," *1979 Fall Proceedings*, p. 426.

Hosni, Y., and D. Linton. "Performance Standards in Development of Computer Projects," *1979 Fall Proceedings*, p. 433.

Nada, R. "Using Learning Curves in Integration of Production Resources," *1979 Fall Proceedings*, p. 376.

Clark, D. O. "Standard Data and Its Maintenance Today," *1981 Annual Proceedings*, p. 594.

Filter, E. H. "Time and System Helps Firm Compare Department Efficiency on Weekly Basis," *Industrial Engineering* (August 1982), p. 54.

Necastio, N. P. "A 3 Step Approach to Weld Standard Data," *1982 Annual Proceedings*, p. 681.

Bostion, W. H. "Standard Data Coding Techniques," *1983 Annual Proceedings*, p. 29.

Zielinski, E. L. "Computerized Vehicle Routing and Driver Work Standards in a Delivery Fleet Operation," *1984 Annual Proceedings*, p. 561.

Goodgame, D. T. "The Use of Performance Standards in Appraising Employee Performance in Service Occupations," *1984 Fall Proceedings*, p. 277.

Caldeira, E. "Integration of CAPP and Computerized Standard Data," *1985 Fall Proceedings*, p. 106.

Computer Aided Work Measurement (1602)

Balakrishnan, N. T., R. Fischer, and R. Kapin. "How Much Manual Handling Per Shipment?" *Industrial Engineering* (August 1979), p. 37.

Moon, W. W., and J. Jenney. "Work Sampling Meets COBOL," *Industrial Engineering* (November 1979), p. 26.

Brisley, C. L., and R. J. Dosselt. "Computer Use and Non-Direct Labor Measurement Will Transform Profession in the Next Decade," *Industrial Engineering* (August 1980), p. 34.

O'Neil, M. N., and C. Moore. "Multi-Plant Computer System for Standards Provides Tool for Overall Manufacturing Control," *Industrial Engineering* (August 1980), p. 54.

Stafford, R. C., and M. J. Riley. "Videotapes Expand IE Ability," *Industrial Engineering* (January 1980), p. 32.

Towne, D. M. "Automated Analyses of Clerical Productivity," *1980 Fall Proceedings*, p. 342.

Fiske, T. S. "Program Utilizes Learning and the Length of Run to Determine Tightness of Line Balance," *Industrial Engineering* (August 1981), p. 58.

Anderson, J., and Y. A. Hosni. "Time Standards by Microcomputers," *Industrial Engineering* (September 1981), p. 18.

Montroy, D. D. "A Statistical Approach to Solving Paper Making Process Problems," *1981 Annual Proceedings*, p. 213.

Klewin, E. M. "Guiding the Installation of an Incentive Wage System," *1981 Annual Proceedings*, p. 622.

Ice, L. R. "Rehabilitating an Incentive System: A Blend of Human Relations Skills, Modern Measurement Techniques and Computerization," *1981 Annual Proceedings*, p. 608.

Towne, D. M. "ADAM: A System for Computer-Based Work Measurement," *1982 Fall Proceedings*, p. 225.

Martin, J. C. "Program, Portable Microprocessor Allow Direct On-Site Input of MTM-1 Data," *Industrial Engineering* (August 1982), p. 50.

Galonek, J. "Use of Computerized Indirect Labor Reporting System Boosts Productivity," *Industrial Engineering* (August 1982), p. 46.

Lindenmeyer, C. R., and M. M. Sykes. "Computer Aided Work Sampling: Description and Application," *1982 Annual Proceedings*, p. 265.

Hosni, Y. A. "Time Standards Using Microcomputers," *1982 Annual Proceedings*, p. 694.

Heward, J. H. "Achieving Low Cost Accurate Work Standards Using Micro-Computers and Statistical Concepts," *1982 Fall Proceedings*, p. 331.

Murray, J. "A Computer Assisted Video Stopwatch and Participative Approaches to Job Situations," *1983 Annual Proceedings*, p. 761.

Zandin, K. B. "Computerized Work Measurement and Process Planning—Vital Ingredients in a Computer-Integrated Manufacturing System," *1983 Annual Proceedings*, p. 481.

Antoniewicz, J. R. "Influence of Computers and Automation on Work Measurement," *1983 Fall Proceedings*, p. 612.

Mercer, J. J. "New Developments in Computer Applied Work Standards and Routings," *1983 Fall Proceedings*, p. 589.

Savage, D. F., and R. J. Keevan. "Automating Your Current Work Measurement Data on Spreadsheet Software," *1984 Annual Proceedings*, p. 626.

Wilerson, D. L. "Computer Aided Standard Development for Work Measurement," *1984 Fall Proceedings*, p. 410.

Uzi, M., and J. Fischer. "Computer Program Development Estimation Model," *1984 Fall Proceedings*, p. 418.

Lindner, C., and W. M. Hancock. "Computerized Work Measurement Can Help Hospitals Identify Cost Reduction Possibilities," *Industrial Engineering* (March 1985), p. 70.

MacMillan, J. S., and A. P. Walker. "Portable Microcomputers Function As Data Controllers," *Industrial Engineering* (January 1985), p. 18.

Gibson, D. C., and N. F. Schmeidler. "Development of Computer-Aided Time Standards," *1985 Annual Proceedings,* p. 646.

Gerber, D. L., and G. C. Heyde. "Modapts Plus,® A Superior System for Work Analysis," *1985 Annual Proceedings,* p. 212.

Dao, T. M. "Cinematic Motion Modeling and Simulation on Micro-computer: An Economic Approach for Robot Time Performance Analysis," *1985 Fall Proceedings,* p. 463.

Banta, W. E. "Computer-Aided Production Standards Calculations," *1985 Fall Proceedings,* p. 102.

Standards Development and Applications (1608)

Lane, W. G. "Measuring Crew Operations," *Industrial Engineering* (May 1975), p. 20.

Vincent, C. A. "Measures Productivity of Fork Truck Operators," *Industrial Engineering* (June 1975), p. 36.

Wallis, R. A. "Employees Participate in Indirect Work Measurement," *Industrial Engineering* (August 1975), p. 10.

Smerilson, H. H. "Standards for Engineers," *Industrial Engineering* (Ocober 1975), p. 12.

Shurtz, S. "Maintenance Standards in a Public Utility," *Industrial Engineering* (November 1975), p. 14.

Sookasian, H. G. "The Development and Use of Labor and Material Standards in a Clinical Laboratory," *1975 Annual Proceedings,* p. 173.

Ferri, F., and B. W. Niebel. "Universal Maintenance Standard System," *Industrial Engineering* (February 1976), p. 26.

Benton, L. B. "Application of Labor Standards to Typing Tasks, " *Industrial Engineering* (October 1977), p. 32.

Kelleher, D. P. "A Different Approach to Measuring Maintenance Work," *1977 Annual Proceedings,* p. 453.

Schlender, B. H., and H. H. Smerilson. "Low Cost Measurement of Indirect Labor Productivity," *1977 Annual Proceedings,* p. 472.

Kiddney, R. P. "A Small Company's Guide to Computerized Shop Standards," *Industrial Engineering* (September 1978), p. 54.

Weaver, R. F., J. J. Kollman, and E. A. Boepple, Jr. "Developing Standards by Computer," *Industrial Engineering* (January 1978), p. 26.

Fein, M. "Establishing Time Standards by Parameters," *1978 Annual Proceedings,* p. 470.

Stevens, M. D. "An Easier Way to Measure Variable Standards," *Industrial Engineering* (December 1979), p. 38.

Gardner, T. M., and J. Blok. "Generate Standards Information Faster," *Industrial Engineering* (July 1979), p. 32.

Eliassen, R. A. "Measuring the Nth Job Economically," *1979 Annual Proceedings,* p. 460.

Kyser, R. C., Jr. "MTM-C Clerical Standard Data," *1979 Annual Proceedings,* p. 423.

Williams, D. D., Jr. "Optimizing Product Transfers and Start-Up Utilizing the 4M Data System, Work Place Layout and Video Taping," *1979 Annual Proceedings,* p. 200.

Philbin, L., and B. Renegan. "Federal Express Corp. Develops Performance Measurement Index for Indirect Labor Sector," *Industrial Engineering* (August 1980), p. 48.

McGuire, T. R. "Standards for Radiology Are Made Easier," *Industrial Engineering* (April 1980), p. 37.

Stafford, R. E., and M. J. Riley. "Videotapes Expand IE Ability," *Industrial Engineering* (January 1980), p. 32.

Das, B. "Effects of Production Feedback and Standards in a Repetitive Industrial Production Task," *1980 Annual Proceedings*, p. 72.

Vasiliou, G. "Improving the Productivity of Assembly Operations Through Basic Industrial Engineering Techniques," *1980 Annual Proceedings*, p. 450.

Tang, K., and J. Alvaraz. "Manpower Planning and Control System Installation," *1980 Annual Proceedings*, p. 631.

Ricks, L. "MTM and Hospital Nursing," *1980 Annual Proceedings*, p. 166.

Huth, S. "Putting a Handle on Clerical Productivity in the 1980's," *1980 Annual Proceedings*, p. 231.

Banglore, N. R. "Regression Analysis Aids Setting Standards for Data Entry Operations," *1980 Annual Proceedings*, p. 415.

Bostion, W. H. "Development of Personal, Fatigue & Delay (PF&D) Allowances," *1980 Fall Proceedings*, p. 457.

Williams, C. D. "How You Can Optimize the QC Function in Your Firm," *Industrial Engineering* (March 1981), p. 48.

Ferderber, C. J. "Measuring Quality and Productivity in a Service Environment," *Industrial Engineering* (July 1981), p. 38.

Bihr, R. A. "Navy Adopts Some Industrial Techniques to Plan Man-hours for Maintenance," *Industrial Engineering* (July 1981), p. 58.

Rai, P. "A Computerized Productivity Improvement and Measurement System for Food, Beverage and Chemical Manufacturing Plants," *1981 Annual Proceedings*, p. 333.

Dickel, F. W. "GMP/RPM for Delay Work Improvement," *1981 Annual Proceedings*, p. 408.

Green, R. R. "Productivity Improvement Strategies for the J. C. Penney Catalog Division," *1981 Annual Proceedings*, p. 489.

Geyser, G. J. "Successful Approaches in Industrial Engineering When Industrial Engineers Are in Short Supply," *1981 Annual Proceedings*, p. 698.

Davis, W., and F. Young. "Work Standards for Highly Variable Operations: A Statistical Approach," *1981 Annual Proceedings*, p. 8.

Lelak, F. M. "The Development and Implementation of Meter Reading Time Standards," *1981 Fall Proceedings*, p. 496.

Foncel, E. F., Jr. "Development and Use of Maintenance Time Standards in Generating Stations," *1981 Fall Proceedings*, p. 505.

Bostion, W. H. "Standard Data Development and Application," *1981 Fall Proceedings*, p. 117.

Globerson, S. "Developing a Multiple Factor Incentive Plan Involves Selection, Weighting, Standard Setting, Calculation," *Industrial Engineering* (November 1982), p. 74.

Brisley, C. L., and W. F. Fielder, Jr. "Balancing Cost and Accuracy for Setting Up Standards for Work Measurement," *Industrial Engineering* (May 1982), p. 82.

Graham, W. W. "Applications of Time Standards on the Santa Fe," *1982 Annual Proceedings*, p. 637.

Enscore, E. E., K. Knott, and B. W. Niebel. "A Comparison of Alternative Time Slotting Methods for Indirect Time Standards," *1982 Annual Proceedings*, p. 711.

Betke, R. L. "Improving Inspection Productivity with Engineered Labor Standards," *1982 Annual Proceedings*, p. 22.

Bobillo, T. E., and L. R. Ice. "Microcomputer Applications for Manufacturing and Industrial Engineering," *1982 Annual Proceedings*, p. 33.

Lichtenberg, N. "A Staffing Standards System for a Geographically Dispersed Maintenance Activity," *1982 Fall Proceedings*, p. 336.

Lauro, M. V., and R. L. Pruitt. "Business Office Clerical Standards," *1983 Annual Proceedings*, p. 537.

Niebel, B. "Establishing Standards in the Service Area," *1983 Fall Proceedings*, p. 607.

Smith, C. "Awareness, Analysis, and Improvement Are Keys to White Collar Productivity," *Industrial Engineering* (January 1984), p. 84.

Tremblay, P. F. "Improving Employee Productivity Through the Detailed Definition of a Work Standard," *1984 Annual Proceedings*, p. 617.

Merritt, T., and L. S. Aft. "Meeting the Requirements of MIL-STD-1567 and Lockheed-Georgia Computerized Standard Development System," *1984 Annual Proceedings*, p. 18.

Schwinn, D. R. "Work Standards—Aid or Impediment to Productivity?" *1984 Annual Proceedings*, p. 259.

Uzi, M., and J. Fischer. "A Self-defining Slotting Technique for Estimating Standard Times," *1984 Fall Proceedings*, p. 418.

Kugemann, L. E. "The IE's First Commandment—Thou Shall Not Use Work Standards to Set Production Quotas," *1984 Fall Proceedings*, p. 329.

Scarfuto, J. B. "Engineered Meter Reading Route Standards," *1985 Annual Proceedings*, p. 626.

Brady, T. F. "Minimizing Work Standards and Job Evaluation Plan Conflicts and Disputes Through Training," *1985 Annual Proceedings*, p. 333.

Johnson, L. R., and G. A. Fort. "Estimating Drafting Time," *1985 Fall Proceedings*, p. 109.

Knott, K., and M. F. Pena. "An Improved Method for Setting Time Standards for Maintenance Work," *IIE Transactions* (March 1985), p. 17.

Standards Auditing (1607)

Salvendy, G., and G. P. McCabe. "Auditing Standards by Sample," *Industrial Engineering* (September 1976), p. 25.

Arnwine, W. C. "Auditing Time Standards in Aerospace," *1977 Annual Proceedings*, p. 13.

Eley, N. G., and H. Villalobos. "The Application of Work Sampling to Short Cycling Assembly Operations," *1979 Annual Proceedings*, p. 429.

Laurence, G. G. "Answering Audit Recommendations to Engineer Work Standards," *1982 Annual Proceedings*, p. 671.

Organ, D. "Arbitration: The Final Test of Labor Standards," *1982 Annual Proceedings*, p. 707.

Kelley, R. E. "Misinterpretation of Labor Contract Can Lead to Inflated Piece-Work Rates," *Industrial Engineering* (July 1983), p. 48.

Brayton, G. N. "Simplified Method of Measuring Productivity Identifies Opportunities for Increasing It," *Industrial Engineering* (February 1983), p. 48.

Wade, L. F. "Update on MIL-STD-1567 and Proposed Applications Guide Describes Contractor Concerns," *Industrial Engineering* (February 1984), p. 74.

Skala, G. D. "Preparing for a Mandated Management Audit," *1984 Fall Proceedings*, p. 541.

About the Editor

Richard L. Shell, Ph.D., P.E., is professor and director of industrial engineering in the Department of Mechanical and Industrial Engineering at the University of Cincinnati. He is also Director University Liaison for the Institute of Advanced Manufacturing Sciences, a nonprofit corporation for manufacturing research designed as a joint enterprise among industry, local and state government, and the University of Cincinnati. Dr. Shell's past business experience has included engineering and management positions with Bourns, Ampex, and IBM. During the past several years, he has served as a consultant for government and private industry. He has authored or co-authored over eighty publications, including several books and monographs.

Dr. Shell has been active in the Institute of Industrial Engineers for over twenty-five years. He has held offices at the chapter and regional levels and is a past director of the Work Measurement and Methods Engineering Division. He has received numerous awards, including the Phil Carroll Achievement Award in 1978 from the Work Measurement and Methods Engineering Division, and the Distinguished Engineer Award for 1986 from the Engineers and Scientists of Cincinnati. Dr. Shell holds engineering degrees from the University of Iowa, the University of Kentucky, and the University of Illinois.